The Greek
Hoplite Phalanx

The Greek Hoplite Phalanx

Richard Taylor

Pen & Sword
MILITARY

First published in Great Britain in 2021 by
Pen & Sword History
An imprint of
Pen & Sword Books Ltd
Yorkshire – Philadelphia

Copyright © Richard Taylor 2021

ISBN 978 1 52678 856 6

The right of Richard Taylor to be identified as Author of this work has been asserted by him in accordance with the Copyright, Designs and Patents Act 1988.

A CIP catalogue record for this book is
available from the British Library.

All rights reserved. No part of this book may be reproduced or transmitted in any form or by any means, electronic or mechanical including photocopying, recording or by any information storage and retrieval system, without permission from the Publisher in writing.

Typeset by Mac Style
Printed and bound in the UK by CPI Group (UK) Ltd,
Croydon, CR0 4YY

Pen & Sword Books Limited incorporates the imprints of Atlas, Archaeology, Aviation, Discovery, Family History, Fiction, History, Maritime, Military, Military Classics, Politics, Select, Transport, True Crime, Air World, Frontline Publishing, Leo Cooper, Remember When, Seaforth Publishing, The Praetorian Press, Wharncliffe Local History, Wharncliffe Transport, Wharncliffe True Crime and White Owl.

For a complete list of Pen & Sword titles please contact

PEN & SWORD BOOKS LIMITED
47 Church Street, Barnsley, South Yorkshire, S70 2AS, England
E-mail: enquiries@pen-and-sword.co.uk
Website: www.pen-and-sword.co.uk

Or

PEN AND SWORD BOOKS
1950 Lawrence Rd, Havertown, PA 19083, USA
E-mail: Uspen-and-sword@casematepublishers.com
Website: www.penandswordbooks.com

Contents

Introduction		viii
Chapter 1	Origins	1
Chapter 2	Arms and Armour	54
Chapter 3	Organization and Drill	98
Chapter 4	Hoplites and Citizens	156
Chapter 5	Command and Control	215
Chapter 6	Hoplites at War	254
Chapter 7	The Hoplite Battle	309
Chapter 8	Fighting in the Phalanx	391
Chapter 9	Decline of the Hoplite	463
List of Battles		482
Glossary		487
Notes		490
Abbreviations and Translations		520
Bibliography		521
Index		528

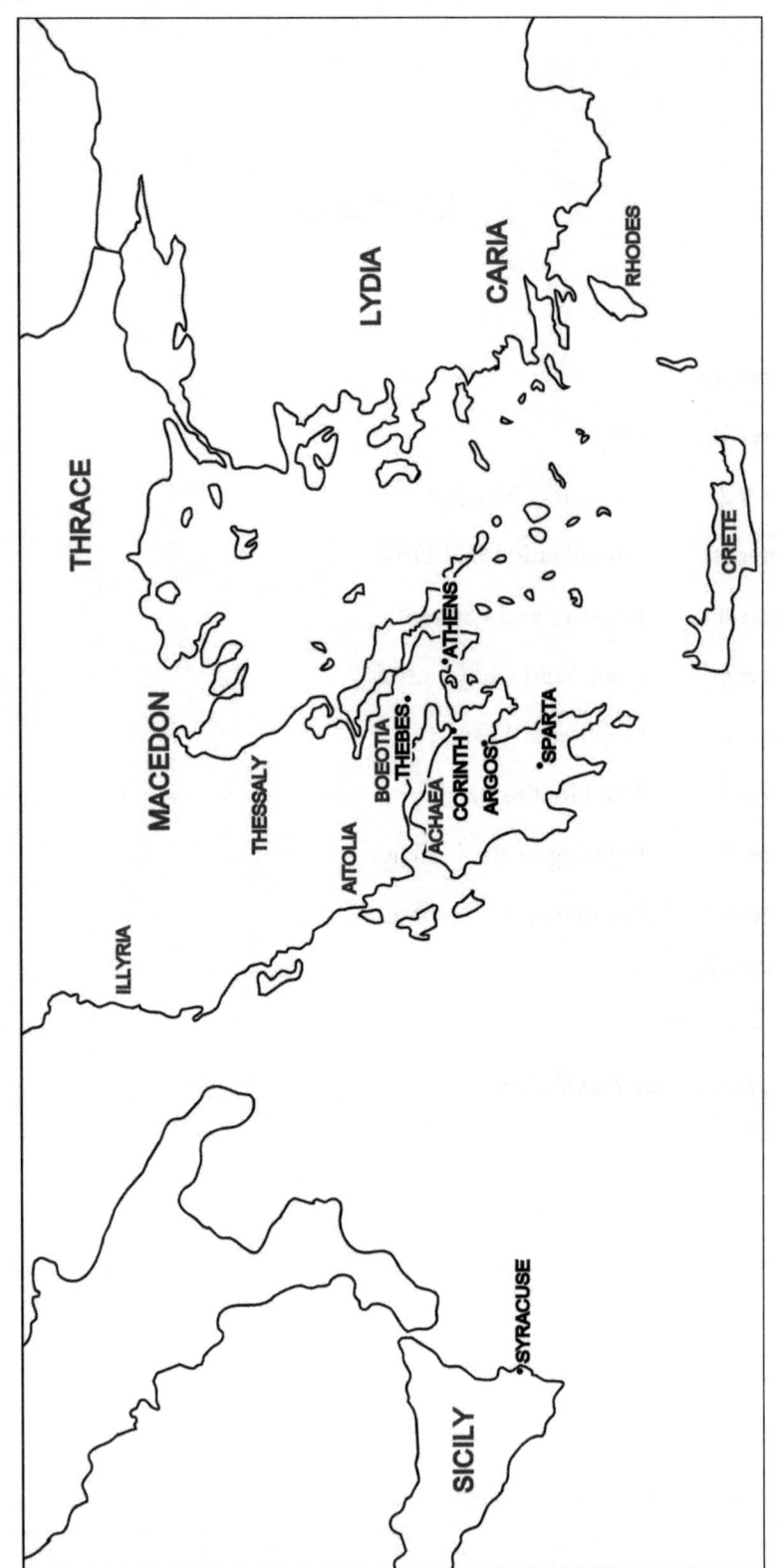
General map.

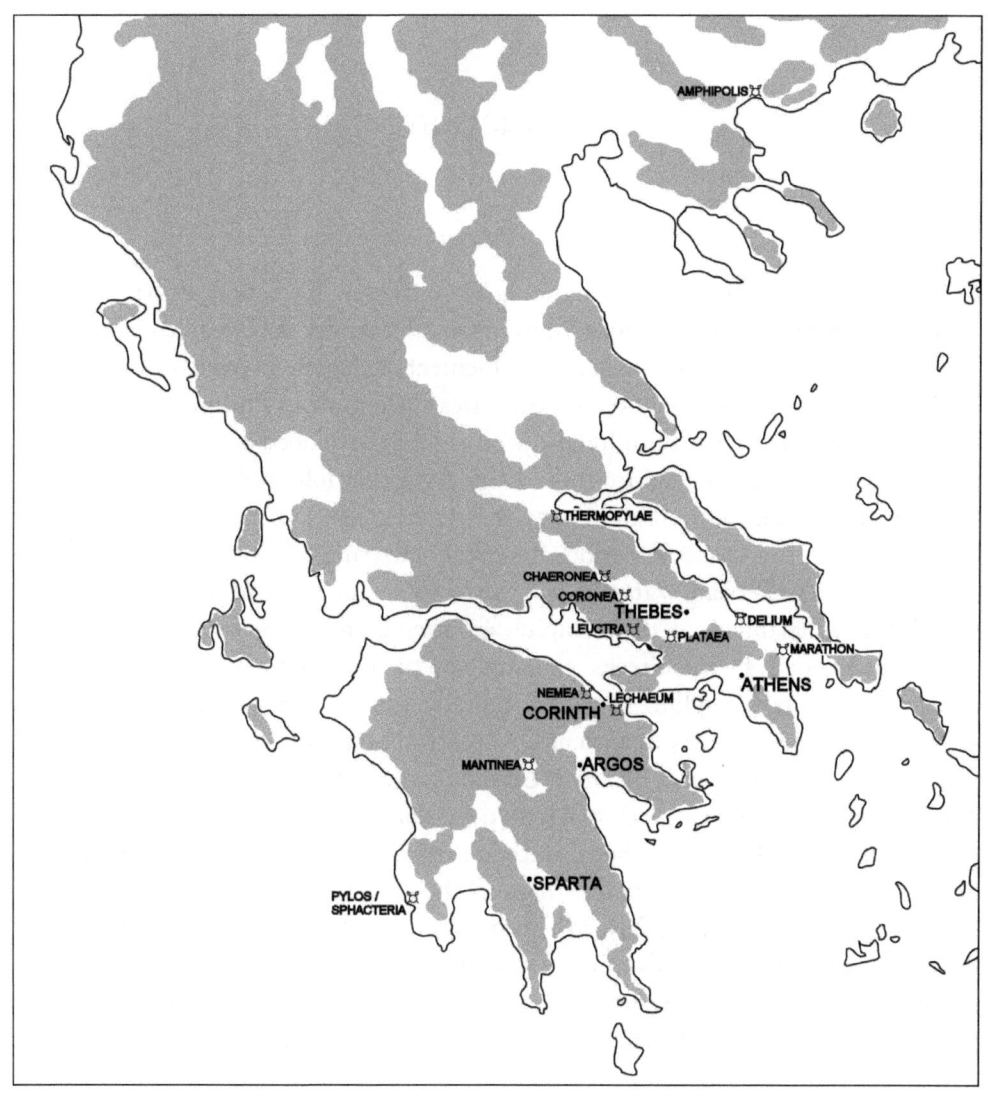

Major battles. Shaded areas indicate high or mountainous land.

Introduction

The Macedonian phalanx, the heavy infantry formation devised by Philip II of Macedon and used by his son Alexander the Great to conquer the Persian Empire, was itself a refinement of an earlier Greek formation, with roots already dating back several centuries at the time of Philip's invention. The Greek phalanx (it has traditionally been argued) developed after the obscure period when Greece emerged from the Dark Ages that followed the end of the Mycenaean era, was active throughout the Archaic age and reached its most refined form in the Classical period, the great flowering of Greek culture that occurred after the Persian Wars (the failed attempt by the Persians to incorporate the Greek homelands into their empire, an attempt foiled in large part, it is usually understood, by the Greeks' phalanx tactics). The men who fought in the phalanx were called 'hoplites' (*hoplitai*), literally 'armed men' or 'men-at-arms', and were the heavy infantry of ancient Greece, the backbone of all Greek armies.

The Archaic era of Greek history is devoid of any written historical accounts. There are poems – chiefly those of Homer, the *Iliad* and *Odyssey*, written down probably in the Archaic period but with their origins, as an oral tradition, perhaps dating back centuries before then, perhaps all the way back to whatever historical conflict, near the end of the Mycenaean period around the twelfth century BC, may have inspired the legends of the Trojan War. The extent to which Homer's poems reflect actual military practice at any period, whether of the Mycenaean past or the Archaic time of their written composition, remains controversial, and will form part of the subject of the first chapter of this book. It is also unknown whether Homer was a single historical figure or a representative of a wider oral tradition, but for convenience I will here assume he was a real man and composer of the *Iliad* and *Odyssey*. Other written evidence is limited to a few fragments of poetry, chiefly the works of Tyrtaeus, a Spartan poet (or one who worked in Sparta) of the seventh century BC (all dates hereafter are BC unless otherwise stated), who composed poems extolling the martial virtues of the Spartans and the joys of dying in battle for the fatherland. Among these fragments are traces of the type of fighting prevalent at the time – the extent to which these provide evidence for phalanx warfare will also be considered in the first chapter.

Aside from these sources, we are entirely dependent for Archaic evidence on art and archaeology. For the latter, the Greek practice of making shields and armour from bronze (although this was well into the Iron Age, iron was generally reserved for weapons, with bronze continuing in use for defensive equipment), and of either burying such arms in high-status graves or, later, dedicating captured spoils in religious sanctuaries (especially Delphi and Olympia), means that we have a fair sampling of such equipment in a reasonable state of repair (iron corrodes more rapidly than bronze, and of course organic materials – wood, leather, linen – are usually lost completely). For artistic evidence, there is a tremendously rich source in the Greek practice of producing painted vases; at first these were adorned with abstract geometric forms, but increasingly it became the fashion to depict human figures in various scenes of everyday life, often military, or in depictions of legendary scenes, often involving combat. The latter provide plentiful visual evidence for arms and armour, as well as some intriguing hints about formations and tactics, as we will see.

The Archaic period transitioned to the Classical with the Persian Wars, at the start of the fifth century, and here too we have a transformation in the quality and availability of literary evidence. The Persian Wars, and the Persian Empire against which they were fought, were the subject of the first ever writer of continuous literary history (at least, the first whose works have survived), the *Histories* of Herodotus, often dubbed 'the father of history'. Herodotus wrote in the mid-to-late fifth century, at a time when eyewitnesses to the events he was describing were still alive and whose accounts no doubt form the basis of his work, as Herodotus was a believer in 'autopsy', of seeing for oneself, and travelled and interviewed widely through the Eastern Mediterranean world of his day. He also liked a colourful tale, and declared that his duty was to set down what he heard but not necessarily to believe it, and he was not without some political bias (inevitable in the politically divided fifth-century Greek world), resulting in some calling him the 'father of lies'. Despite the criticisms, and despite the fact that Herodotus did not take much interest in technical military matters and took for granted (as all Greek historians did) a certain level of familiarity with the subject among his readers, there are many valuable clues in his writings as to the nature of infantry combat at the time.

More valuable still was the work of another late fifth-century author, Thucydides. An Athenian general exiled for a botched operation early in the Peloponnesian War (the great war, or series of wars, fought in the late fifth century for dominance in Greece between the two major powers, Athens and Sparta, and their respective allies), Thucydides set himself to write a complete, sober and detailed account of the war, and naturally his *History* contains a vast amount of valuable material on military matters, along with tantalizing glimpses of the formations and tactics that he assumed were familiar to his readers. Thucydides had his limitations and biases, as do all historians, yet his work is one of the finest pieces of history writing from the whole of antiquity, for its breadth, careful use and evaluation of evidence, and thoughtful judgement. Thucydides died leaving his work (and the war it described) uncompleted, but his history of the war, and of the events that followed, was continued by another soldier-turned-historian, Xenophon.

Like Thucydides, Xenophon was an Athenian who saw active service in the Peloponnesian War, though his sympathies lay to a large extent with Sparta. Following Spartan victory in the war, at the end of the fifth century, Xenophon signed up with a large Greek mercenary army in support of a Persian pretender's attempt to seize the imperial throne. When defeat in battle left this Greek army stranded deep within the Persian Empire, it marched back through hundreds of miles of hostile territory to the Greek cities on the coast, a feat immortalized in Xenophon's *Anabasis* ('Expedition'), which contains a wealth of information on military matters, especially small irregular operations against the various native peoples the Greeks encountered. Returning to Greece and serving now alongside a Spartan king, Xenophon proved a prolific author. His *Hellenica* ('Greek Affairs') continued Thucydides' work from the point he left off to the defeat of the Spartans in the early fourth century by the Thebans. He also wrote (or is traditionally considered to be the author of) a number of minor works, most importantly the *Cyropaedia* ('Education of Cyrus'), the world's first historical novel, a fictionalized account of the early career of Cyrus, the founder of the Persian Empire, in which Xenophon set out many of his ideas on the ideal ruler and on military theory. There were also two military handbooks, *The Cavalry Commander* and *Horsemanship* – though sadly nothing (that has survived) on infantry – along with other works such as those on hunting, household economics and Spartan institutions, and several accounts of the teachings of Socrates (of whom he was a friend and student).

These three historians and their works form the backbone of our knowledge of Greek military matters in the Classical period, and their coverage of the fourth century is particularly valuable, as the fashion for figurative vase painting seems to have died out around this time, resulting in a much reduced body of artistic evidence for this later period. There is little other contemporary literary evidence – the most intriguing being the traces of fourth-century military handbooks, exemplified now by Xenophon's cavalry works and by one describing defence against siege, written by Aeneas 'the Tactician', who is also known to have written other military handbooks, though none have survived. Any that covered drill and military organization may be reflected in the later Hellenistic tradition of writing works of 'tactics' describing the organization of the Macedonian phalanx. While these later Hellenistic works may trace their roots back to a fourth-century tradition, they can be used to provide details of Greek, rather than Macedonian, military practice only with caution. There were also a number of later historians who wrote histories of the Classical period, notably the *Library of History* of Diodorus Siculus ('the Sicilian'), written in the first century, and the *Parallel Lives* of Plutarch, biographical sketches of notable figures in Greek and Roman history, composed in the first to second centuries AD. Diodorus was not a particularly careful historian, but he did have access to some good sources, now lost, while Plutarch was more interested in morally uplifting stories than rigorous history, but is sometimes all we have. Also of occasional value is the *Description of Greece* of Pausanias, the first travel guidebook, composed in the second century AD but containing many scraps of information on otherwise unknown Archaic and Classical events.

The other major sources of information on Classical history in particular are the numerous inscriptions uncovered throughout the Greek world. It was normal practice for official documents and decrees to be literally set in stone, and these form a vast body of information on many official and institutional topics, though they do not tell us very much about military matters.

The Greek hoplite phalanx continued in use into the Hellenistic period (that is, the period following Alexander the Great's conquest of the Persian Empire), and so Hellenistic histories and other sources of information provide a few more scraps, but we must be extremely cautious when using them not to take information about the Macedonian phalanx and apply it to the Greek; the two are related but are also different in important ways, as will be explored in the coming chapters.

In this book, I aim to give an overview of many aspects of the Greek hoplite and of the phalanx in which he fought. This is of course a potentially vast topic and there are many areas where I can do little better than provide a brief outline – the notes and bibliography will, I hope, lead readers to more in-depth studies. I will concentrate on the evidence provided by the primary literary sources, and I rely heavily on direct quotation, since I find it more valuable to read the original words (in translation) than a precis or a statement of fact with accompanying reference. The modern secondary literature I have not attempted to review in any depth within the text as it is often vast and highly specialist, but I will so far as possible reference the most recent or accessible general books on the subject, to which there are further references in the bibliography. This will I hope be enough to direct the interested reader toward further information.

I also perhaps need to justify the writing of yet another book on the Greek hoplite phalanx, a topic that has already produced a shelf-worth of books in English alone during the past few decades. I believe that the time is right for a reappraisal of the hoplite, who has been elevated into a position of uniqueness and exceptionalism that is not perhaps fully justified, or not for the right reasons.

I believe that hoplite studies in future will need to set the hoplite much more securely in the context of his world and his times, unlike the specialist Greek-centred works (in English) that have tended to dominate this field. This present work is not such a study, but I hope that it may help, alongside other recent studies by the next generation of researchers in this field, and if only by its errors and omissions, to indicate the path down which hoplite studies should travel.

As noted, dates in what follows are BC unless otherwise stated. Names are usually (but with imperfect consistency) given in their Latinized or Anglicized form where such a form exists (for example, Thucydides for Thoukudidēs). Technical terms and important words are generally transliterated directly from the Greek using the Latin alphabet. Where possible, I quote directly from the ancient sources using the Loeb edition translations (or as otherwise noted), sometimes modified slightly to make them more literal or to retain or better reflect technical military terms.

Chapter 1

Origins

The Greek hoplite phalanx has its origins sometime in the Archaic period, perhaps as early as the eighth century, though this is very much disputed, as we will see. It was fully developed by the Classical period (fifth century), and was defeated by its Macedonian successor in the later fourth century, though it continued in use at least into the late third century. Before considering the origins and development of the phalanx in more detail, it may be helpful to sketch the historical background of this long period and to examine the meanings of the terms used.

Historical background

The Greeks – meaning in this context the inhabitants of ancient Greece – were the peoples inhabiting the southern Balkan peninsula and surrounding regions from around the second millennium (perhaps earlier). 'Greeks' is the English name, borrowed from the Latin *'Graeci'* (probably the subset of Greeks that the Romans first encountered); the Greeks called themselves 'Hellenes' (after the legendary Hellen, son of Deucalion, the Greek equivalent of Noah), from which the term Hellenic (for anything Greek) and Hellenistic ('Greek-ish', for the period after Alexander the Great) are derived. The migrations that brought the Greeks into what is now Greece took place over several phases in the second and first millennia, resulting in a number of different ethnic groups speaking differing dialects of a common language. In the late second millennium, Greece was dominated by the Mycenaean culture, named after its major centre, the city of Mycenae in the Peloponnese (the land mass forming the southern section of the Greek peninsula as a whole, connected to the rest by the narrow Isthmus of Corinth). Mycenaean civilization was based around numerous kingdoms, each centred on a single city or palace complex, and the people were also termed the Achaeans (in later periods, this name was applied specifically to a people of the northern Peloponnese). The Achaeans were traditionally thought to have fought a war, perhaps some time in the twelfth century, with the city of Troy (or Ilium) in Asia Minor (modern north-west Turkey) and its allies; this Trojan War formed the subject of the Greek foundational epic poems, the *Iliad* and *Odyssey*, composed probably sometime in the early Archaic period by Homer.

2   The Greek Hoplite Phalanx

Sometime around the twelfth to eleventh centuries, Mycenaean civilization collapsed, perhaps as a result of a new wave of migration of Greek-speaking peoples, and Greece entered a Dark Age that lasted until about 750. From around this time, the Archaic civilization begin to emerge. The defining feature of Archaic Greece was its division into numerous politically separate city states (*poleis*, singular *polis*) each with its own lands, system of government, and feuds and grievances with its neighbours, resulting in regular warfare between neighbouring cities (the details of which are mercifully obscure to us now). Greece is a mountainous country, with good agricultural land split into plains surrounded by mountains, and Greek city fought Greek city over these plains (figuratively and literally) for several centuries. The Greeks of this period were demographically dynamic, with (it seems) expanding populations that exhausted the agricultural possibilities of the constrained plains, and they were also great seafarers, sailing their mainly oared ships throughout the Mediterranean. Demand for land, ready access to the sea and an adventurous spirit led to the Greek colonization of many of the coastlines of the seas around Greece (always the coastlines – the Greeks were reluctant to venture too far inland). Greek cities sent out colonists to parcel out land in colonies along the Black Sea coast, the Ionian coast of Asia Minor, Sicily and southern Italy, and with scattered settlements as far afield as Egypt and the south of (what is now) France. Greeks also found their services in demand as mercenaries, with Greek soldiers of fortune serving in a number of foreign kingdoms' armies throughout the period.

During this period, the Greek cities also underwent a process of constitutional change, at different rates and with differing start and end points from city to city. The original monarchies were replaced by oligarchies (rule by a council of rich and elderly men, the form of government we are familiar with today), that in turn sometimes fell to 'tyrants' (populist authoritarian leaders, but without, necessarily, the negative overtones of the word in English) and finally perhaps to some form of democracy, with rule by the people (or at least by a larger subset of the people). The latter was characterized by the holding of political office by a wider group than in the oligarchies, by political power lodged in assemblies with wide (though far from universal) suffrage, and by a powerful concept of rule by consent and the rule of law that the Greeks themselves contrasted with what they considered slavish obedience to monarchs in the Eastern (particularly Persian) style. The development of these new forms of government was believed in ancient times to be intimately connected with the development of hoplite equipment and tactics, and this is still thought to be the case today, though the details remain obscure and controversial and will be the subject of the later portions of this chapter.

The establishment of the Ionian colonies led the Greeks into conflict with the Persians, the latest in a succession of empires to dominate the Near East.

First the Persians conquered Ionia, and then, when the city of Athens supported an Ionian revolt, the Persian kings Darius and Xerxes attempted to subjugate (or at least punish) Greece in the early fifth century with invasions that were defeated first at the Battle of Marathon (490) and then (after the rearguard action at Thermopylae in 480) in naval battle at Salamis (480) and land battle at Plataea (479). Success in these Persian Wars briefly united the Greeks, or at least some of them, chiefly Athens and Sparta (or Lacedaemon, the ancient name for the territory of Sparta, hence the frequent use of 'Lacedaemonians' to mean Spartans). Other cities – particularly the Thebans in Boeotia – bowing perhaps to necessity, joined the Persian invaders, 'Medizing' to use the Greek term (the Greeks used the names 'Mede' and 'Persian' largely interchangeably). The Persian Wars were considered in many ways the high point of Greek power and culture, a glorious episode to which later Greeks looked back with pride, a special status which continues to this day, extended now to encompass 'the West' rather than just the Greeks, and forming the basis of many modern 'East versus West' historical narratives, often fantastical in nature. The Persian Wars also mark the traditional point of transition from the Archaic period to the Classical.

Following Persian withdrawal, the Greeks were once more free to fight amongst themselves. For most of the fifth century, two cities in particular grew to relative superpower status, two cities which provide striking contrasts. Athens was the most democratic of the city states, relying chiefly on naval power which it used to establish a league (or empire) of allied (or subject) cities throughout the Aegean, and powered a dynamic, flourishing culture of philosophy, drama and architecture (responsible for the great monuments still standing, just about, on the Acropolis in Athens). Sparta was almost the exact opposite; a constitutional monarchy (with two simultaneous kings and a council of elders), a land power, with a deeply conservative and fanatically military culture based on discipline, duty and the brutal exploitation of permanently enslaved neighbouring Greek peoples (the 'helots'). Sparta dominated the Peloponnese, with an alliance of city states, of which the most important was Corinth. Rivalry and tension between Sparta and Athens led to the outbreak of the Peloponnesian War in the later fifth century. The Spartans hoped to use their formidable army to settle the war in the traditional way, by marching into Athenian territory, defeating the Athenians in pitched battle and then dictating terms. But the Athenians, able to supply themselves by sea and linked to their harbour by long walls (the Spartans were noted for their lack of ability in siegecraft), and following the advice of their leader, Pericles, refused to fight on Sparta's terms, instead just sitting out the annual Spartan invasion, while launching seaborne raids of their own around the Peloponnese and into the northern Aegean. The result was a lengthy stalemate, punctuated by a few battles such as Delium (424, Athens v. Thebes) and Mantinea

(418, Sparta v. Athens and allies) and minor engagements around the fringes of the main theatres (Pylos in the western Peloponnese in 425, Amphipolis in northern Greece in 422) that, while locally important, did not alter the overall balance of the war. Yet eventually the Athenians overreached themselves in their desire to establish a naval empire, attempting to subdue Syracuse, a Corinthian colony in Sicily. The Syracusans proved a tougher nut to crack than expected; the Athenians were beset by poor command and mismanagement, and ended up losing both their fleet and their army. This was not the end of the war, as the Athenians in due course built a new fleet and raised a new army, but the Spartans were now able to take the war into the Aegean, backed by Persian money which allowed them to raise a fleet of their own (the Spartans being happy to Medize to this extent), and reduce Athens' allies and cut off its food supplies. Finally, the Athenians lost another fleet at Aegospotami in 405 and were at last forced to surrender. The Spartans did not destroy the city, as others might have done, but did impose an oligarchic constitution more to their liking.

The end of the Peloponnesian War might have been expected to settle the balance of power in favour of the Spartans for good, but Spartan hegemony quickly proved unpopular among its old allies. In the fourth century, further wars broke out, between Athens and Sparta, Athens and Thebes (which now made its own bid for hegemony), and Thebes and Sparta, with major battles at Nemea (394, Sparta v. Athens, Corinth and Thebes) and Coronea (394, Sparta v. Thebes). The Spartans were also active in Asia Minor, with campaigns against their erstwhile friends the Persians, following the unsuccessful mercenary expedition in which Xenophon took part, which had ended in defeat in battle at Cunaxa (401). There were also some minor engagements, as there had been during the Peloponnesian War, in which light infantry armed with javelins ('peltasts', named after their light shield, the *pelte* or *pelta*) defeated Spartan heavy infantry, events considered remarkable at the time. Eventually the Theban general Epaminondas won two major victories over Spartan armies, at Leuctra (371) and Mantinea (362), which set back Spartan power considerably, especially when the Thebans founded a new home city for the Messenians, who formed most of the helot population of Sparta.

The end of Greek independence was now looming, as the formerly marginal kingdom of Macedon in northern Greece, under the dynamic leadership of its king Philip II, finally overcame its natural weaknesses (a rural population, hostile aggressive neighbours and chronic dynastic instability) and made use of its strengths (demographic growth and rich natural resources) along with a newly organized force of heavy infantry – a modification and refinement of the Greek phalanx – to extend its power over the whole of Greece, defeating the Thebans and Athenians at Chaeronea (338) and then, under Alexander the Great, conquering the Persian Empire. Military developments in Greece after this period are somewhat obscure, but so far as we can tell, cities – which were now sometimes under more or less direct Macedonian control, and which increasingly formed into leagues – continued to use the hoplite phalanx, sometimes supplemented or perhaps replaced by different lighter troop types, especially *thureophoroi*, infantry similar to peltasts but armed with larger, Celtic-style shields, until the late third century, when several Greek armies re-equipped themselves in the Macedonian style. The second century saw the Macedonian kingdom succumb to the Romans, with Greece as the primary battleground, and the end both of Greek independence and of the distinctive Greek method of arming and fighting, as Greece became a Roman province.

Definition of terms

I have referred already in passing to the 'hoplite phalanx' which is the subject of this book, and it is time to define what is meant by these two words. Firstly, 'hoplite' is the Anglicized form of the Greek word *hoplitēs* (ὁπλίτης), plural *hoplitai* (ὁπλῖται), which means something like 'armed' or 'equipped' ('man' being understood). In the ancient world, this meant something more than just any man fighting on foot; it meant in particular a heavy infantryman, one carrying a large shield (*aspis*) and wearing armour of some sort, equipped primarily with weapons for hand-to-hand combat. Light infantry, who might carry a lighter shield (*pelta*) or no shield at all, who fought without armour and were equipped with bow, sling or javelins for fighting at a distance, were variously called peltasts (*peltastai*), *psiloi*, *gymnetes* ('naked'), or depending on their armament, archers, slingers or javelinmen.

Some caution is necessary, however, before declaring (as so many do) that *hoplitēs* in Greek means the heavily armed infantryman with round shield and spear that we will meet in the pages of this book. As we will find frequently in the coming pages, Greeks (or at least, the surviving Greek authors) did not tend to use technical terminology with the exactitude we might like (in this of course they are no different from many modern English authors, who will use military terms such as 'lance', 'pike' or 'rank' with abandon, and without regard for their technical meanings). So in Greek usage, rather than a 'hoplite' being a specific type of infantryman with a particular set of equipment (the equipment that will be examined in Chapter 2), it was a much more general term for a heavily equipped, hand-to-hand specialist, as opposed to the various forms of (lightly equipped, missile-using) light infantry or cavalry. As such, the word could be applied by Greek authors to all sorts of other, non-Greek infantry, including Persians, Romans and various other 'barbarians' (non-Greek peoples).

6 The Greek Hoplite Phalanx

In our main three authors (Herodotus, Thucydides and Xenophon), the word is most often applied to Greeks and to those we consider typical hoplites (with the equipment we will soon examine), but that is because they were primarily describing wars of Greek against Greek, and in the Greek world the armament of heavy infantry was remarkably consistent and (apparently) unchanging. In these authors, 'hoplites' are usually Greek heavy infantry with spear and heavy shield, but that is not because this is the meaning of the word 'hoplite'; this strict technical meaning is something the word has acquired only in its modern, English usage. To an ancient Greek, a hoplite was just a heavy infantryman, and his fighting style, equipment and ethnicity are not defined by the term. Modern authors have taken this general Greek word and given it a more narrow technical meaning (something that we will encounter again in a number of cases). Note also that the word only appears even with this sense in the fifth century (although the scarcity of earlier literary sources in which the word might appear makes it hard to draw firm conclusions from this).[1]

Another common misconception is that the hoplite was named after the shield he carried, supposedly called a *hoplon*. This derives in large part from a statement of Diodorus, in the context of the fourth century 'Iphicratean reform' (which I will examine further in Chapter 9). The historical context is that the Athenian general Iphicrates introduced a new light shield (*pelta*) for the infantry under his command, and Diodorus notes:

> 'After a trial of the new shield its easy manipulation secured its adoption, and the infantry who had formerly been called "hoplites" [*hoplitai*] because of their shield [*aspis*], then had their name changed to "peltasts" [*peltastai*] from the light *pelta* they carried.' (Diodorus 15.44.3)

This implies that Diodorus at least thought that the hoplite's shield was called a *hoplon* (if not, the infantry should surely have been called *aspistai*), but that he rather clumsily uses the general term *aspis* rather than being specific and using *hoplon* in this passage. Note that *aspis* is a general word for many types of shield, and as usual is not used with any specific technical meaning, although in practice most Greek hoplites did carry shields of the same type, as we will see, and *aspis* is often used to specify a heavy shield as opposed to the light *pelta* or various other types of shield such as the Persian *gerra*; but *aspis* remains a general term – *peltai* and *gerrai* are both types of *aspis*, in this general sense. However, there are good reasons to doubt Diodorus' statement in this case. For one thing, the historicity of the Iphicratean reform is doubtful, and Diodorus' apparent statement that all hoplites were renamed peltasts does not stand up to scrutiny (hoplites continued to be called hoplites after the reform, and peltasts had existed for a long time before it). Furthermore, there are very few other cases (if any) of the word *hoplon* being used for a shield, and the word *aspis* is used vastly more often. There is reason to doubt that *hoplon* means shield at all, even though the word could, on occasion, be used to indicate a shield. *Hopla*, in the plural, is by far the most common usage, and here the word means something like 'tools' or 'equipment', depending on context. The *hopla* of a soldier are his weapons, shield and armour, collectively, his equipment or kit. In the singular, the word could indicate a particular piece of kit (such as the shield), but *hoplon* means 'a tool', 'a piece of kit', and there is little reason to suppose that it ever meant 'a specific type of heavy shield'. When Greek authors did want to specify what type of *aspis* they meant, they would use the adjective 'Argive' (or 'Argolic'), indicating that the heavy, round shield was supposed to have been first used in Argos. I will return to this matter in the next chapter.[2]

So rather than a 'hoplite' being 'a carrier of a *hoplon*', it means something much more general, an 'equipped' or 'kitted' man, and often with the understanding of 'a fully equipped man', as opposed to the partial equipment of the *peltasts*, *psiloi* and *gymnetes*, who variously lacked items of armour, heavy hand-to-hand weapons and the heavy shield. It is in this sense that the word first occurs in Greek authors; in Homer we do not find reference to hoplites, and infantry are referred to by a variety of terms, most commonly simply *aner*, 'man', which from the context can best be translated as 'warrior'. In Archaic sources (such as they are), *aner* is still often present (rather than being understood, as later) and '*hoplit-*' stemmed words are used adjectively, so we see 'equipped men' explicitly. Only with the Classical authors Herodotus and, particularly, Thucydides and Xenophon does the word *hoplitēs* on its own, as a noun, to indicate heavy infantry, become common. Given the paucity of literary sources pre-Herodotus, not too much can be read into this in terms of the historical origin of the hoplite (as we understand the term), though I will return to that question below.[3]

At any rate, I will be using 'hoplite' and 'hoplites' (unquoted and unitalicized) throughout this book to refer to something quite specific, the heavily armed, hand-to-hand-fighting Greek infantryman, but this is according to the modern English usage, and does not imply that *hoplitēs* (the Greek word) had such a restricted meaning. Where it is important to indicate that the Greek word is being used (such as when it is applied to non-Greeks), I will point it out specifically.

Next, the 'phalanx' (in Greek, *phalangx*, φάλαγξ). The same caveats apply as for *hoplitēs* and *aspis* above. 'Phalanx' is a Greek word which has been adopted in English to mean something quite specific, but which in Greek had a much more general meaning and usage. The word means something like 'block' or 'log', and could be applied to many longish, thinnish objects, but in a military context it means a block or line of soldiers. The word occurs quite often in Homer in this sense, all but one time in the plural, to mean a body of men, presumably formed up in a wide line of some (unknown) depth. As we will see below, the nature of infantry fighting that Homer had in mind in his battle descriptions is very unclear, but at any rate we can be fairly certain that Homer would not have envisaged the phalanx precisely as it is usually understood in the modern English usage.[4]

Curiously enough, after Homer the word 'phalanx' (or *phalanges*, the plural) occurs a couple of times in Archaic poets (in the plural, and once to refer to cavalry), and then not again until the writings of Xenophon in the early fourth century. For the Archaic period this is not too surprising, since there are very few literary sources anyway, but that Herodotus and Thucydides never used the word seems to me rather remarkable, given that the usual assumption is that the phalanx (as we understand it) was devised or developed before the end of the sixth century, and possibly as early as the seventh century (or even earlier). Thucydides' entire description of the Peloponnesian War never uses the word 'phalanx' once, but Xenophon's continuation of his history, and his other works – especially the *Anabasis* – contain numerous uses of the word. Whether this is because the phalanx was in fact only invented around this time (late fifth to early fourth century), or whether the word was only then adopted (by Xenophon, or by Greeks more generally) to describe a phenomenon that already existed, we can only speculate. At any rate, in Xenophon the word occurs frequently, and as with *hoplitēs* it is used most often to describe a Greek battle formation, but is also used for other formations and nationalities (it is used frequently for Cyrus' Persians, for example), and also in its more general sense of a battle line of unspecified troop type and formation, including cavalry. Later authors also use the word in the same way, to describe both a specific Greek formation and any battle formation or line more generally. With the invention of the Macedonian phalanx (mid-fourth century), the word is used specifically to refer to this new formation, but also continues in its more general sense of a battle line.[5]

So these caveats must be kept in mind – the word 'phalanx' was used by Greek writers in a non-technical and non-specific sense, and though it is generally assumed (with some justice) that in the context of Greek heavy infantry it means something more specific, it is not certain that this is so, and the English usage is more specific than is truly warranted before the time of the Macedonian phalanx. So what is this more specific English usage? In English, 'phalanx' indicates a linear formation, wider than it is deep but still of some considerable depth (eight men deep is the normally quoted figure), and dozens or hundreds of men wide, with formal ranks (lines of men side by side) and files (lines of men one behind the other), in close order (the men standing close together, perhaps shoulder-to-shoulder or at any rate shield-to-shield), and intended exclusively for fighting hand-to-hand, face-to-face with the enemy (who are assumed to be in a similar

The parts of an infantry formation.

formation), and continuing to fight in this way until one side is defeated and runs away, a final and decisive state of victory and defeat (unlike light infantry or cavalry, who might run from each other, then rally and return repeatedly). As we will see in the coming chapters, there are also various subsidiary beliefs about the nature of the phalanx, such as that it had a particular method of fighting by literally shoving, but for now this definition – a close order, organized, moderately deep linear formation of heavy infantry, fighting hand-to-hand – will be sufficient.

The hoplite debate

So we now have definitions of 'Greek', 'hoplite' and 'phalanx', but one more area of the historical background needs to be examined before delving further into the nature of the Greek hoplite phalanx. I do not intend, in the course of this book, to refer constantly to the views of modern historians (such discussion will be left to the footnotes), but in the case of the Greek hoplite phalanx, there is a long-running and wide-ranging debate among historians as to the origin and nature of the phalanx, its method of fighting and the political, social and economic causes and consequences of the adoption of hoplite equipment and fighting styles among the Greeks; an understanding of this background is helpful in the discussion that follows. This debate goes under various names, frequently 'the hoplite debate'.[6]

In the mid-nineteenth century (AD), Grote, in his *History of Greece*, set out what was to become the standard view of the hoplite phalanx, in terms of tactics and equipment:

> 'The Hoplites, or heavy-armed infantry of historical Greece, maintained a close order and well-dressed line, charging the enemy with their spears protended at even distance, and coming thus to close conflict without breaking their rank.' (Grote, 1846, II, p.106)

('Well-dressed' means, of course, well-drilled and regular, rather than anything to do with clothing.) Grote contrasted this with the type of fighting described by Homer, which was dominated by chariot-borne heroes, who fought sometimes with thrown spears and sometimes hand-to-hand, sometimes on foot and sometimes from their chariots. While there is mass infantry combat in Homer, as Grote saw it,

> 'The mass of the Greeks and Trojans, coming forward to the charge, without any regular step or evenly-maintained line, make their attack in the same way [as the heroes] by hurling their spears.'

So the difference between Homeric and Archaic-Classical warfare lay in the adoption of close order, of drilled lines that maintained their formation in combat, and of the exclusive use of hand-to-hand rather than missile weapons, and this forms the definition of the hoplite phalanx. But this was not simply a technical, military change, as Grote related it, as the military revolution marked by the adoption of the hoplite phalanx coincided with political changes (the 'hoplite revolution') – with monarchies giving way, not yet to democracies, but to oligarchies in which political power was more widely shared among a still elite group. The formation during the eighth to seventh centuries of *poleis* – city states with a concept of citizenship, the rule of law and the accountability of magistrates – contrasted with the arbitrary kingships that had gone before. A second revolution was to follow when populist tyrants, as we now know them, took over control from the oligarchic elites and gave greater power to a wider body of citizens, the small farmers and wealthier artisans, and was marked by the final replacement of (aristocratic) cavalry by heavy armed infantry as the main military force of each state. The rule of tyrants was only a temporary stage as the politically empowered and now militarily dominant heavy infantry acquired real political power, leading to the establishment of broad-based oligarchies and democracies by the late sixth century. Grote saw a causal link between the rise of the hoplite phalanx, made up of politically empowered citizens of moderate economic means, and the replacement of monarchies, narrow oligarchies and tyrannies by broader oligarchies and democracies rooted in the concepts of citizenship, shared political power and the rule of law.[7]

The military aspects of this protracted revolution were elucidated by Grundy in his *Thucydides and the History of his Age* (1911). In Grundy's view, heavy armed infantry, heavily burdened by their equipment and fighting in close, regular formations, were wholly unsuited to the terrain of Greece, which is mostly mountainous. But the small farmers of each *polis* were required – both by honour and necessity, since their livelihoods and indeed lives depended on it – to defend their agricultural plains, and so would of necessity, faced by any threat to their crops and farms, march out to meet an invader in battle, rather than sheltering behind their city walls or harassing the enemy with light troops on the mountain passes. This led to a particular style of warfare, of decisive pitched battle between similar and equally matched forces of heavy infantry (the hoplite phalanx of each *polis*), conducted according to, if not strict rules, then definite conventions, and fighting in a way that was new and unique:

> 'Under ordinary circumstances the hoplite force advanced into battle in a compact mass, probably at the slow step, breaking, it may be, into a run in the last few yards of advance. When it came into contact with the enemy, it relied in the first instance on shock tactics, that is to say, on the weight put into the first onset and developed in the subsequent thrust. The principle was very much the same as that followed by the forwards in a scrimmage at the Rugby game of football.'[8]

This picture of how hoplites fought, and in particular of its similarity to a rugby scrum (as it is now called), has been extremely influential. It forms – along with the relationship outlined above between the rise of the hoplite and the political changes seen in Greece from the eighth to the sixth centuries, and the formation of the phalanx from farmer-soldiers who could neither be away from their land for long, nor afford the time for training in complex military tactics – the 'hoplite orthodoxy'. Hoplite armies were made up of farmer-citizen-soldiers, who engaged in a form of warfare that made few demands on their time or technical skills, relying instead on mass effect and brute force against opponents with similar equipment and a similar set of priorities, economically and politically. This would produce a short, sharp conflict with decisive results, that would settle the current dispute (which was usually about ownership of land) and allow the surviving hoplites to return to their farms.

The precise timing of the transition from pre-hoplite (Homeric or heroic) warfare to true hoplite warfare remained unclear. Nilsson in 1929 used the evidence of archaeology and vase depictions, particularly the Chigi vase (which I will examine further below), to date the emergence of the hoplite phalanx to 'the seventh century at the latest', although this represents perhaps the end of a longer, slower adoption and perfection of hoplite tactics. Greater precision was offered in 1947 by Lorimer, who made a firm connection between the adoption of the hoplite shield (the round, two-handled Argive *aspis* we encountered above) and hoplite tactics:

> 'The momentous change from the essentially long-range fighting of the eighth century involved a single structural alteration in the round shield slung on a *telamon* [strap] which was in vogue ... The change consisted in the substitution for the single central hand-grip previously in use of a central arm-band of metal (*porpax*), through which the bearer thrust his arm up to the elbow, and a hand-grip (*antilabe*) ... which he grasped with his left hand.'

This shield – or rather this way of holding the shield using two handles, as opposed to a single central hand grip – Lorimer proposed, was wholly unsuitable for any sort of open-order, individual fighting, and was only possible in a close-order mass formation in which individual weapons play was not required. The adoption of this shield and related armour could be dated to the early seventh century, and the Chigi vase, dated around 650, gives 'the earliest reliable evidence for the new armature [armament]'.[9]

With this refinement stressing the importance of the shield (and its handles) and the date of its adoption, the 'hoplite orthodoxy' has continued as the dominant view of Archaic and Classical Greek warfare probably to this day, though over the past forty years it has faced increasing challenges. To Adcock, writing in 1957, 'the character and use of [the hoplites'] shields were of the essence of their fighting in battle'. Hoplite battles were decided by the physical shock and shove between heavy armed infantry (though Adcock did see a role for spear fighting), followed certain formal (though unwritten) 'rules', were fought by farmers over farmland and decided the outcome of the war.[10]

This orthodox view was forcibly restated by Hanson in his *The Western Way of War* in 1989. Hanson reassessed the agricultural nature of hoplite warfare, arguing that since the amount of damage an invading army could do to farmland was probably limited, the reason for the defenders marching out to meet an invader was more to do with status and honour than agricultural necessity. But the essential nature of hoplite armies was that they were made up of free, independent property owners, who adopted a style of fighting which gave a decisive result with minimum expenditure of time or money on complex military manoeuvres. The Greek style of fighting, Hanson argues, was unique:

> 'Heavily armed militiamen crashing together on flat plains ... each side after the initial collision seeking quite literally to push the other off the battlefield through a combination of spear thrusting and the shove of bodies.'

Like earlier writers, Hanson stresses that the hoplite shield and the Corinthian helmet (on which see below) were heavy, hot, awkward and restrictive, wholly unsuited to individual combat or any sort of weapons play, and adopted only because of their suitability for mass shoving in the phalanx.[11]

Hanson differs in some respects from the orthodoxy, however, especially in the further arguments set out in his *The Other Greeks*. Hanson places greater stress than others on hoplites as landowners and farmers (rather than the wealthy, or artisans and traders), and sees an agricultural revolution driven by independent farmers operating in a free-market environment (Hanson's views are more overtly motivated by modern political ideas than is the norm in this field) as driving population growth and economic changes from the eighth century onwards. He also envisages hoplite tactics (the close-order phalanx) arising before the adoption of full hoplite equipment (shield and armour) – the question of which came first, hoplite equipment or phalanx tactics, has long formed a rather unsatisfactory, 'chicken or egg' element of the hoplite debate.[12]

This 'hoplite orthodoxy' has come under attack from a number of quarters and over an extended period, and it is this conflict between the orthodoxy and the revisionist or 'heretical' views that has shaped the hoplite debate. One aspect of this attack focuses on the adoption of hoplite equipment, which is seen (for example by Snodgrass) as taking place in a long, drawn-out process rather than a military revolution, a process that was not complete before the middle or even the end of the seventh century. The earliest 'hoplite' armour (the Argos panoply) dates from the end of the eighth century, before the usual time proposed for the adoption of phalanx tactics, and is seen by Snodgrass as continuing an armament tradition dating back into the Bronze Age, breaking the link between 'hoplite' arms and phalanx tactics. Snodgrass also doubted that the hoplite shield was as restrictive in use as proponents of the orthodoxy claimed, and he sees the adoption of phalanx tactics also not as marking a sudden social revolution, but as a process in which political power was acquired gradually and not directly as a result of the hoplites' military role. Heavy armour and shields were adopted by a slowly expanding circle of soldiers, and 'possibly not much before the fifth century, the Greeks coined a word to define the status that the heavy infantryman had reached – *hoplitēs*'.[13]

This gradualist position – that hoplite equipment and tactics were adopted gradually over the course of a long period, and that wider political powers were likewise acquired in a piecemeal fashion and varying greatly from one *polis* to the next – provides the main alternative to the orthodox view. A more heretical view has also gained ground in recent decades. Krentz, in a number of articles, argues that the close order, the rigid nature and the exclusivity of hand-to-hand fighting that we seem to see in the later Classical phalanx were late-developing features, and that there was therefore no phalanx-driven impetus for political change in the Archaic period. Rather than the phalanx being a brute force, mass-shove formation, Krentz sees it as being made up of individual fighters in a more open formation than is usually assumed, and he does not see the double grip shield as being any impediment to individual duelling. The phalanx was already present in Homeric warfare, though in a different form than the orthodoxy insists, and just became more standardized later. Similar arguments about the nature of fighting – more open order, more dependent on individual duelling – were made by Cawkwell in the 1980s.[14]

Van Wees has developed the most complete version of this heretical view by revisiting the nature of fighting in Homer. Mass infantry formations are present in Homer, though they are not rigid, close-order formations. They are open enough for individual heroes to make their way to the front and pick out individual opponents, and to duel with them (and strip them of their armour when fallen), but behind them stands a mass of other fighters (who might make their own way to the front in turn). This style of fighting would have applied also to the early phalanxes, as reflected in the poems of Tyrtaeus, for example. The Classical phalanx we meet in Xenophon (and in Thucydides, though not by name) was not centuries old by the fifth century, but was the latest stage in a long, slow process of development as infantry equipment became more standardized, formations closer and fighting more exclusively hand-to-hand. Hoplite-led political revolutions could therefore not have taken place in the earlier period.[15]

At the time of writing, there is as yet no clear winner (if it is appropriate to think in such terms) in this hoplite debate. In the past decade or so, the heretical or gradualist position has probably become the stronger in academia, not so much through anyone changing their mind, as by an influx of new researchers unconvinced by the orthodox view. This has also given rise to a new outlook, a comparative position, according to which the Greek phalanx should not be seen in isolation, as it so often has, but as a part of a larger world of Greeks and non-Greeks, whose military and political institutions informed and shaped each other. Among the general public, however (so far as they are aware of or care about such debates at all), the wide popularity of Hanson's *The Western Way of War* has tended to dominate understanding of the hoplite, and the idea of the close-order, rigid phalanx, made up of 'middling farmers', competing with a mass shove rather than by fighting with weapons, is probably still the most common view.[16]

I do not believe, given the nature of the evidence, that it is possible to make much progress on many of the fundamental, underlying questions of the hoplite debate, especially in its wider social and political aspects. Because continuous literary histories only begin in the fifth century, artistic depictions are limited to vase paintings which provide limited evidence before the sixth century, archaeology can tell us a lot about equipment but little about how it was used, while the use of Homer to understand tactics is fraught with difficulty (see below), the chances of ever gaining a clear understanding of the development of hoplite tactics in the period from the eighth to the fifth century are slim, and it is difficult enough understanding the hoplite phalanx in the vastly better-documented fifth and fourth centuries. I will therefore attempt no more than to sketch in this chapter a possible picture of phalanx development before the fifth century, stressing the unknowns (known and unknown) and the limitations of the evidence, and the bulk of this book will consider the Classical hoplite phalanx of the fifth century and later, when evidence is more plentiful. To lay out my cards at the start, I am generally unconvinced by the orthodox view of hoplite tactics and fighting methods (this will be the subject particularly of Chapter 8). I also suspect that the political and economic details of the 'hoplite revolution' (if any) are lost to us, but that the link between infantry tactics and the political role of citizen soldiers is highly important. I also agree with the comparative view, that the Greek phalanx should be seen in the context of earlier and contemporary formations, Greek and non-Greek. Again, the accessibility of evidence is a stumbling block here, as ancient historians naturally tend to specialize, and those who write about ancient Greece tend, naturally, to specialize in Greek history, and in Greek history as viewed through the lens of Greek literary sources in particular (following the long tradition of Classical studies). A full treatment of the hoplite phalanx would require in-depth knowledge not just of the ancient Greek literary sources, but also of the art and archaeology, and of those of the neighbouring contemporary peoples as well. I do not have the expertise in these areas to offer new evidence or draw firm conclusions, so all I can do is to be aware, and hopefully make any readers of this book aware, of the limitations of the traditional approach and of the opportunities offered by a more holistic approach to the subject.

Homeric warfare

The descriptions of fighting in Homer, particularly the *Iliad*, are of great importance in understanding early Greek combat, since they are the first written accounts of Greek warfare, and indeed the earliest detailed and complete accounts of warfare of any sort. Earlier descriptions of combat among the Egyptians, Hittites or Assyrians, for example, are composed almost exclusively from the point of view of kings and rulers, in which individual combatants are merely faceless cyphers performing (or failing to perform) the monarch's will. The main problem with using Homer to reconstruct early Greek warfare is the difficulty of establishing which elements within Homer reflect the long oral tradition of which the written works we now have represent only the final form – and which might therefore date back into the Bronze Age – and which elements are drawn from Homer's own time and the contemporary experience of his audience, or indeed are later interpolations. We do not even know with any certainty when Homer's works were composed (or when the oral tradition they represent took a more solid form), nor when they were first written down, which could be as late as the sixth century, nor when the text took its final form, which could be as late as the second century. Even if we could establish which elements in Homer were contemporary with the composition rather than the origins of his work, we would not know if this reflected warfare as it was in the eighth, the seventh or the sixth centuries or even later. Use of Homer is therefore fraught with difficulty, and has led some to conclude that nothing useful can be extracted from Homer's accounts.[17]

I think this is overly pessimistic, however. It is certainly not possible to separate out and date with any certainty particular elements of combat in Homer, but what Homer describes, unless his work was one of pure fantasy, must reflect combat at some period; if Homer did describe phalanx warfare, then we can tentatively conclude that phalanx warfare existed by the sixth century at least. What is more, in antiquity it was widely assumed that Homer described combat accurately, and that his works could be used to form the basis of a contemporary analysis of tactics. The formation of the Macedonian phalanx in the fourth century was specifically said to have been inspired by Homer, and Homer is always cited by the Hellenistic tacticians (who set out the formal organization and drill of the Macedonian phalanx) as the first writer on tactics. Homer also provided a more general inspiration for all those who sought excellence in military matters – Alexander the Great always kept a copy of the *Iliad* to hand, and the Greeks generally saw themselves as fighting in a Homeric style, even when we today might be more inclined to note the differences between Homeric warfare and the Classical phalanx.[18]

There are some elements in Homer, particularly references to particular items of equipment – such as the 'tower shields' which are carried by a strap on the back and tap against the neck and ankles of their bearer, or helmets decorated with boars' tusks – which are known from archaeological finds to be accurate descriptions of Mycenaean equipment, so these elements at least may have been transmitted across the intervening centuries from the Bronze Age to Homer's written work. A central element of combat in Homer is the chariot, which was certainly central to Mycenaean warfare (there exist palace records and inventories detailing the number of chariots available), but which had fallen out of use (at least in mainland Greece) by Homer's time. Chariots were also central to Bronze Age warfare throughout the Near East, though the use of chariots in Homer is rather different from the normal Bronze Age use. In the armies of the Egyptians and Hittites, the Mycenaeans' contemporaries, chariots are used in mass formations as archery platforms and perform (so far as we can tell) a similar role to light (or sometimes heavy) cavalry in later armies; in Homer, the chariot chiefly serves as a transport for a single warrior, who usually dismounts in order to engage in hand-to-hand fighting with a similar opponent who has been individually selected for a duel, while the chariot waits nearby ready to take the warrior to safety if the fight goes against him. We can only guess whether this form of chariot warfare accurately represents Mycenaean practice – perhaps a variant form more suited to the more mountainous terrain and smaller number (and possibly size) of horses available in Greece, and the accordingly greater value attached to such horses and unwillingness to risk them in combat – or whether this is closer to a Dark Age form of warfare where mass chariot forces were not available, or possibly is Homer's (or the oral tradition's) invention simply to provide a dramatic and familiar but still exotic form of fighting to an Archaic audience.[19]

But it is infantry combat in Homer that is of most relevance to the origins of the hoplite, and despite the central role played by heroic champions in single combat and by chariots as their transports, massed forces of infantry certainly do appear in Homer, going by the name of 'phalanges' as we have seen, and play an important part in the combats described. Homer's combat descriptions contain an enigmatic mixture of massed infantry and heroic duels which have long perplexed scholars, since the close-order massed infantry apparently described do not sit easily with the ability of individual warriors to come and go at will in combat, moving to the front ranks to select and kill an opponent, stripping the dead of their armour or rescuing a fallen comrade, then falling back to rest, boast or argue with another hero, or to collect fresh weapons or armour. This fluidity of fighting is the main reason why Homeric infantry are not thought to have fought in a phalanx in the later Classical sense of the word – a massed, deep, close-order formation would not, it is assumed, allow a warrior to pick out an individual opponent, nor to advance to and retire from the front ranks the way they do. Yet there are occasions in Homer where such massed formations are apparently described, and indeed one such well-known passage was believed (in antiquity) to have formed the inspiration for the Macedonian phalanx:

> 'Thereon round the two Ajaxes there gathered strong bands [*phalanges*] of men, of whom not even Ares nor Athena, marshaller of hosts could make light if they went among them, for they were the picked men of all those who were now awaiting the onset of Hector and the Trojans. They made a living fence, spear to spear, shield to shield, buckler to buckler, helmet to helmet, and man to man. The horse-hair crests on their gleaming helmets

touched one another as they nodded forward, so closely locked in battle were they; the spears they brandished in their strong hands were interlaced, and their hearts were set on battle.' (Homer, *Iliad* 13.126–135)

This passage is quoted by Polybius (writing in the second century) in his description of the Macedonian phalanx (Pol. 18.28.6), and Diodorus (16.3.2) records that Philip II of Macedon 'devised the compact order and the equipment of the [Macedonian] phalanx, imitating the "locked shields" [*synaspismos*] of the warriors at Troy' (though Homer does not himself use the word *synaspismos*). This passage from Homer is also reflected in the poems of Tyrtaeus, writing perhaps in the late seventh century:

> 'Let a man learn how to fight by first daring to perform mighty deeds,
> Not where the missiles won't reach, if he is armed with a shield,
> But getting in close where fighting is hand-to-hand, inflicting a wound
> With his long spear or his sword, taking the enemy's life,
> With his foot planted alongside a foot and his shield pressed against shield,
> And his crest up against crest and his helm up against helm
> And breast against breast, embroiled in the action – let him fight man to man,
> Holding secure in his grasp haft of his sword or his spear!'
> (Tyrtaeus 11.27–34)

Neither the vocabulary nor the situation are identical (Homer is describing a formation pressed close to their comrades, Tyrtaeus one apparently pressed close to their enemy), but the similarities in language are apparent; whether Tyrtaeus is inspired by Homer, or both are part of a common tradition of combat description, is open to debate.[20]

But the Classical phalanx also could be directly compared with Homeric infantry fighting, since Xenophon (quite apart from his adoption of the word 'phalanx') is clearly echoing Homer in some of his battle descriptions, the clearest example being his account of the fighting at the Battle of Coronea in 394:

> 'Thrusting shield against shield, they shoved and fought and killed and fell. There was no shouting, nor was there silence, but the strange noise that wrath and battle together will produce … Now that the fighting was at an end, a weird spectacle met the eye, as one surveyed the scene of the conflict – the earth stained with blood, friend and foe lying dead side by side, shields smashed to pieces, spears snapped in two, daggers bared of their sheaths, some on the ground, some embedded in the bodies, some yet gripped by the hand.' (Xenophon, *Agesilaus* 2.12–14)

This passage – with obvious embellishments – seems inspired by Homer:

'When they were got together in one place shield clashed with shield and spear with spear in the rage of battle. The bossed shields beat one upon another, and there was a tramp as of a great multitude – death-cry and shout of triumph of slain and slayers, and the earth ran red with blood.' (Homer, *Iliad* 4.446–51; see also 8.60–67)[21]

So Classical and Hellenistic writers had no problem envisaging the *phalanges* of Homer as equivalent to, and even the inspiration for, the phalanx of their own day. Yet even in the following passage from Homer, the type of fighting is not what we would expect from a close-packed phalanx. After describing the close order of the Achaeans, Homer recounts how Hector was halted by them:

'The Trojans advanced in a dense body, with Hector at their head pressing right on as a rock that comes thundering down the side of some mountain from whose brow the winter torrents have torn it ... even so easily did Hector for a while seem as though he would career through the tents and ships of the Achaeans till he had reached the sea in his murderous course; but the closely serried battalions [*phalanges*] stayed him when he reached them, for the sons of the Achaeans thrust at him with swords and spears pointed at both ends, and drove him from them so that he staggered and gave ground.' (Homer, *Iliad* 13.137–149)

(Note that 'drove him from them' translates the verb *otheo*, 'thrust', 'push', the significance of which will become apparent in Chapter 8.)[22]

Yet after a short speech from Hector, the fighting that follows is a series of single combats of which the first example will suffice to give an impression:

'Deiphobos son of Priam went about among them intent on deeds of daring with his round shield before him, under cover of which he strode quickly forward. Meriones took aim at him with a spear, nor did he fail to hit the broad orb of ox-hide; but he was far from piercing it for the spear broke in two pieces long ere he could do so; moreover Deiphobos had seen it coming and had held his shield well away from him. Meriones drew back under cover of his comrades, angry alike at having failed to vanquish Deiphobos, and having broken his spear. He turned therefore towards the ships and tents to fetch a spear which he had left behind in his tent.' (Homer, *Iliad* 13.156–168)

There follow many (very many) similar descriptions of individual fights between heroes who are able to select an opponent, engage them in a brief fight and either strip their body or fall back under cover of the ranks. Note also that not every named hero is an important figure (like Hector himself, or Deiphobos, both Trojan princes). In some cases, men appear for the first and last time in the poem, killing or being killed, and it is apparent that, in this case at least, rather than a dense mass of infantry in front of which chariot-borne heroes seek out single combats with each other – as seems to be the case on some other occasions – the named figures duelling are the same men as those who make up the dense formation, the ordinary infantry, or at least the leaders among them. This has led to the conclusion that what we see here in Homer, despite the stress laid on the density of the formation, and despite the way this formed the inspiration for the later and very dense Macedonian phalanx, is really a description of a more mobile, flexible and open-order type of fighting, in which the ranks and files are, or become, open enough to allow men to move to and fro, advancing to the front rank, fighting, then falling back if need be, and extracting the bodies of the fallen or stripping off their armour.

This latter point should give us pause before we conclude that the Classical phalanx was something fundamentally different from what Homer describes. I will examine this question in more detail later on, but for now would just observe that the extraction from the front ranks of the bodies of dead or wounded kings and generals is also very much a feature of Classical battle descriptions (we do not hear that the same was done for ordinary hoplites, which is not too surprising, whether because they were not important enough to rescue in this way or because any such acts of recovery were not important enough to record). To give a single example from Xenophon:

> '[T]he fact that Cleombrotus and his men were at first victorious in the battle [of Leuctra] may be known from this clear indication: they would not have been able to take him up and carry him off still living, had not those who were fighting in front of him been holding the advantage at that time.' (Xenophon, *Hellenica* 6.4.13)

This does not suggest a dense, rigid formation, still less one that is pressed hard up against a similar formation of the enemy in a shoving match.

Be that as it may, the nature of infantry fighting in Homer has been likened to that of 'primitive' societies in more recent years, particularly the relatively well-documented clashes of tribesmen in Papua New Guinea. Rather than delving too deeply into the pros and cons of these arguments, I will assume that this comparison has some merit and attempt a sketch of infantry combat as it is depicted by Homer and illuminated by the more recent examples. Infantry were formed irregularly, that is they did not have formal ranks and files, but they could, when occasion demanded, form very densely, crowding closely together. There has been a tendency among scholars to get too hung up on precise file intervals (for the Archaic and Classical phalanx in particular), extrapolating from the fact that we know, from the Hellenistic tacticians, that the Macedonian phalanx had such formal intervals. I will examine this question in greater detail in Chapter 3, but for now it is sufficient to observe that it is unlikely that there were any such formal intervals for the earlier phalanx, and especially so for the *phalanges* of Homer, which were probably undrilled and irregular, and would form up at whatever density seemed appropriate in the circumstances.[23]

In the passage we have been considering, faced by Hector's onslaught, the Achaeans would huddle together into a dense formation, a phenomenon we will encounter in many later accounts when men are faced by a dangerous enemy, usually cavalry. But they would not be following a formal drill in doing this, and most of the time – and indeed after repelling Hector in this passage – they were probably more widely spaced. This does not necessarily mean that they were as widely spaced as the New Guinea tribesmen, who in photographs are in very open order indeed. It was probably sufficient to open out to around 2 metres per man or so, but the important point is that these intervals were not formally defined or centrally controlled, but rather arose naturally as men crowded together or spread apart according to the exigencies of the fighting. Infantry would form up in such *phalanges*, with an unknown (to us) and also probably unspecified depth, certainly several men deep, but without formal ranks there was no way to define and enforce a particular depth. These formations could be crowded densely together, as in the passage considered above, or on a similar later occasion when Achilles' Myrmidons under Patroclus engage the Trojans:

> 'With these words [Achilles] put heart and soul into them all, and they serried their companies [*stiches*] yet more closely when they heard their king. As the stones which a builder sets in the wall of some high house which is to give shelter from the winds – even so closely were the helmets and bossed shields set against one another. Shield pressed on shield, helm on helm, and man on man; so close were they that the horse-hair plumes on the gleaming ridges of their helmets touched each other as they bent their heads.
>
> ...
>
> 'Meanwhile the armed band that was about Patroclus marched on till they sprang high in hope upon the Trojans ... [Patroclus urges them on] ... With these words he put heart and soul into them all, and they fell in a body upon the Trojans. The ships rang again with the cry which the Achaeans raised, and when the Trojans saw the brave son of Menoitios and his squire all gleaming in their armour, they were daunted and their battalions [*phalanges*] were thrown into confusion ... Patroclus first aimed a spear into the middle of the press where men were packed most closely,

by the stern of the ship of Protesilaos. He hit Pyraikhmes who had led his Paeonian horsemen from the Amydon and the broad waters of the river Axios; the spear struck him on the right shoulder, and with a groan he fell backwards in the dust; on this his men were thrown into confusion, for by killing their leader, who was the finest warrior among them, Patroclus struck panic into them all.' (Homer, *Iliad* 16.211–219 and 16.257–274)

Yet the dense formation was a temporary state:

'The fight then became more scattered, and the chieftains [*hegemones*] killed one another when and how they could. The valiant son of Menoitios first drove his spear into the thigh of Areilykos just as he was turning round; the point went clean through, and broke the bone so that he fell forward.' (Homer, *Iliad* 16.306–310)

There follows a long, detailed list of the various ways in which named Myrmidons and Trojans killed each other. These are not chariot-borne heroes outside the *phalanges*, but just particularly noteworthy members of the *phalanges*. The formations become denser or looser as the fight progresses, loose enough for individual warriors to pick out and fight with individual opponents, but these are the front-rankers (*hegemones*, leaders, or *promachoi*, front-fighters) at the head of a large, more or less loosely ordered body of infantry (the *plethos*, mass or multitude). This body of infantry provides replacements for fallen *promachoi*, a safe haven for *promachoi* who need a rest or replacement weapons, and a body of helpers to drag away wounded men or to help despoil fallen enemies.

Fighting is also conducted with a mixture of hand-held or thrown spears, which means of course that two or more spears must be carried, or replacements sought. Archers are also sometimes present among the *phalanges*, picking out targets of opportunity in the enemy formation. Of necessity then, the two formations are not in constant close contact with each other – they must maintain some separation, an area of no man's land into which *promachoi* can step to attack an opponent hand-to-hand or across which spears and arrows can be thrown or shot. Again, this separation is not necessarily a permanent state, since a determined attack could be mounted by a body of men behind a particular leader – as in the case of Hector and Patroclus above – and the opposing force might tighten up its formation to resist such an attack. But with the attackers driven off, both sides could allow their formation to loosen up again and re-establish the neutral zone between them, with a return to fights between individuals who choose to step forward to engage or throw their spears.[24]

This is, I believe, the nature of the infantry fighting that we see in the *Iliad*. Its key features (and differences from later forms of infantry fighting, as we will see) are the absence of defined ranks and files and of defined intervals between ranks and files; the ability of men to move forward to engage the enemy as a matter of individual choice; the lack of any shame attached to men who choose to fall back from the front line, whether to rest or to fetch fresh weapons; the existence of a neutral zone, an area of separation between the main masses of men on each side, in which fighting between individuals could take place; the use of both hand-held and missile weapons in the same fight; and the ability of each side to loosen up or cluster together into more or less dense formations as occasion, and the degree of aggression of the men and their leaders, demanded (and probably according to their own choice or the dynamics of combat, rather than in response to drill commands from officers).

The question remains whether this type of fighting is a feature of the Bronze Age setting of the *Iliad*, of the Dark Age during which the oral tradition took shape or of the time – perhaps the seventh century or later – when the poems were first written down. This question remains unanswered and unanswerable with any degree of certainty, but I am generally swayed by the argument that this represents the nature of infantry fighting in Greece as it was around the

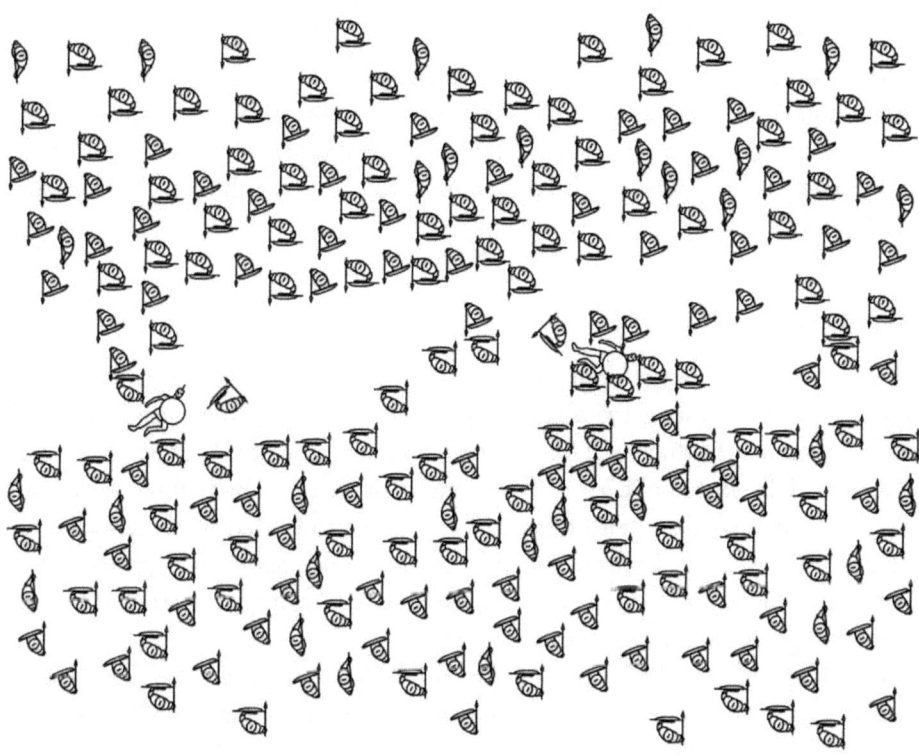

A representation of Homeric or Archaic warfare.

mid-seventh century. This makes it broadly contemporary with the poems of Tyrtaeus, and the features of fighting reflected in Tyrtaeus – as seen in the passage quoted above – fit well with the model of combat I have outlined. The references to missiles, the presence of missile-armed men within the formation, the ability of individual soldiers to choose whether or not to fight among the *promachoi*, and at the same time the closeness of the fighting when it does take place ('With his foot planted alongside a foot and his shield pressed against shield') all reflect what we also see in Homer, where close-packed masses of men could still fight at a distance and where fighting involved clash of shield as well as thrust of spear. Doubtless Homer also layered in features of warfare borrowed from other periods – or simply imagined – such as, perhaps, the use of chariots for transport, but it is probably the case that infantry combat as described by Homer had not changed greatly for many centuries anyway, and that the type of fighting I have described above is in fact typical of infantry combat throughout the ancient world, with numerous variations by time and location. I will have more to say on this topic later, but I would take the type of infantry combat I have outlined – fairly fluid, switching between open or close (but never strictly defined) formations according to circumstances, with an element of choice for individuals as to how they fight, with no shame attached to fighting for a while then retreating to rest, and with a gap or no man's land between the two sides in which fighting takes place – to be the default throughout antiquity (and perhaps beyond). Except where we have definite evidence of fighting taking some other form, this is probably what we should imagine when two bodies of infantry come into contact with each other on the battlefield.[25]

Hoplite equipment

There is another way of approaching the question of the origin of phalanx fighting (as we see it in the Classical period), and that is the introduction of hoplite equipment (which will form the main subject matter of Chapter 2). As we have seen, Classical hoplites are characterized by three main features: exclusive use of a thrusting spear, with a sword as backup, rather than thrown spears or other missile weapons; bronze armour, at least a bronze helmet and greaves, and frequently bronze body armour (later often replaced by linen or leather armour); and the large, round, heavy wooden shield.[26]

   This equipment is known from art and archaeology from the late eighth century onwards. In terms of archaeology and actual surviving examples, bronze body armour and helmets are known from earlier, Mycenaean tombs, such as the famous Dendra panoply, a very complete and heavy set of bronze plate armour for a (presumably) wealthy, perhaps chariot-borne, Bronze Age warrior. A late eighth-century equivalent, forming the earliest known example of what was to become typical hoplite equipment, is the Argos panoply, consisting of a bronze cuirass (body armour) and helmet. The Bronze Age practice of burying the dead with their weapons and armour was not extended with any frequency into the Archaic period, but instead captured equipment was dedicated at one of the major religious sanctuaries after a successful battle. Into the seventh century, we begin to see the characteristic hoplite equipment – bronze helmets and armour, and round bronze shield coverings (the wooden core of the shield has invariably perished). What it is not possible to tell from these items is how typical they were, and how many of the defeated army would have been carrying such equipment. Lighter armour made from organic materials, and lighter leather shields such as those described by Homer, might not have been considered splendid enough to dedicate at a sanctuary, and anyway would not have survived to be discovered by archaeologists, so there is a natural tendency to discover only the best and most expensive bronze equipment. It may be that such equipment was indeed carried by the mass of combatants, but it is possible too that only the wealthiest warriors (*hegemones*, *promachoi*) had the best equipment, with the mass (*plethos*) being padded out by men with lesser, lighter and organic equipment. It is indeed likely that, with social and economic changes through the centuries following the first appearance of hoplite equipment, the circle of those with full bronze equipment grew wider, so that the proportion of those with a full bronze panoply increased (at some point encompassing the whole of the phalanx), but that to begin with a lighter-armed mass was fronted by bronze-equipped aristocrats, as may have been the case in Homer's time.[27]

In terms of artistic depictions, there are similar uncertainties. When realistically depicted warriors appear in vase paintings, during the seventh century, they tend to have full bronze panoplies, though we cannot be certain that this was not an artistic convention to depict the best and brightest rather than the inferior kit of the masses. Nevertheless, mid-seventh-century vase paintings depict bronze-clad men in full hoplite panoply, and the hoplite shield appears even earlier during the seventh century.

The 'Argos panoply', late eighth century (*Argos Museum*)

A couple of other points about the shield are worth noting. One is that ancient tradition, at least as recorded by Herodotus, attributes much of what we recognize as Greek equipment to other peoples, such as the Carians, a people of south-western Asia Minor with whom the Greeks came into contact when they colonized the Ionian coast. According to Herodotus:

> '[The Carians] invented three things in which they were followed by the Greeks: it was the Carians who originated wearing crests on their helmets and devices on their shields, and who first made handles [*ochanes*] for their shields [*aspisi*]; until then all who used shields carried them without these handles, and guided them with leather straps [*telamosi*] which they slung round the neck and over the left shoulder.' (Herodotus 1.171)

How true this statement is, is open to doubt – helmet crests at least seem to be of greater antiquity. Furthermore, it is uncertain exactly what Herodotus means by 'handles'; whether by this he means the *porpax* and *antilabe* combination of the Argive shield, or some other carrying system. It is also not clear whether the Carians also invented a different shield, or that the Greeks adopted the new

Hoplites with Boeotian (left) and Argive shields, late sixth century (*Louvre, Paris*)

method of carrying a shield and applied it to their own type of shield, the Argive *aspis*. It is relevant to note in this context that a second type of Greek shield appears quite commonly in art into the fifth century – the 'Boeotian' or *dipylon* shield. This is a more elongated type of shield than the round Argive shield, with indentations or cutouts on each side (the function of which, if any, is not clear), and possibly a descendant of the Bronze Age 'figure of eight' shield. It is not known how this shield was constructed – possibly from leather or wicker, like earlier shields, which would explain why no physical examples of such shields have survived. This shield is often depicted being carried by a two-handle system similar to that of the Argive shield.[28]

This early appearance of hoplite equipment in art and archaeology has been a driving force behind the theory that the Greek hoplite phalanx, as we see it in Classical times, had an early origin in the Archaic period. The theory goes that heavy equipment, and in particular the heavy shield, required fighting in a dense, close-order phalanx; in particular, it is often asserted that the double-grip method of holding the hoplite shield (with two handles, *porpax* and *antilabe*) was totally unsuitable for single combat, and that therefore men holding their shield in this way must have been fighting in a close-order phalanx in which there was no scope or requirement for individual fighting and which was entirely dependent on the mass effect of close-packed ranks and files and weight in a literal shove. I am profoundly unconvinced by this line of argument (see further in Chapters 6 and 8), and it does appear that the absolute unsuitability of hoplite equipment for fighting other than in a densely packed close-order phalanx has been assumed and asserted without ever being demonstrated either from ancient evidence or practical experience.[29]

So far as ancient evidence goes, Greek art, particularly vase painting, is replete with examples of men in hoplite equipment engaging in single combat. Very often these scenes have a mythological or Homeric setting, but the equipment depicted is contemporary (with the creation of the depiction), and it seems to me unlikely that Greek art would have produced so many depictions of a type of fighting (single combat in hoplite equipment) which a Greek audience would have known was impossible. The combat scenes, even if not real, must have been satisfying to an audience that knew very well what it was like to fight in hoplite equipment. There are also descriptions of single combats among hoplites – the well-known duel of Eteocles and Polyneices in Euripides (*Phoenissae* 1405) being the most often quoted example – and while it is true that Classical Greeks did not think highly of training for single combat, and felt that this was unnecessary in the confines of the phalanx (see Chapter 4), this does not mean that equipment alone rendered such combat impossible.

In terms of modern practical experience, it should be possible for experimental archaeologists and re-enactors to shed light on this question, since there are a number of groups who have reconstructed hoplite equipment and experimented with its handling. Sadly, as so often, the findings of such groups tend to be inconclusive or contradictory. Nevertheless, there are enough examples of those successfully recreating single combat in hoplite equipment for it to be very hard to maintain that such a thing was impossible, or even unusually difficult. Attempts have also been made to draw lessons from, for example, riot police equipment, with the conclusion that riot shields are too unwieldy to be used other than in a dense, tight formation. But again, the lessons learned tend to vary from one police force to another, and it is not sufficiently clear that the circumstances of riot control (in which there is, hopefully, an absence of lethal force, and great stress is laid on the safety of the police and, to a lesser extent, the rioters) really provide a good model for infantry combat, in which killing the enemy was the main means to achieving a successful outcome.[30]

I am therefore not convinced that the existence of hoplite equipment – which was a feature of Greek warfare from at least the seventh century onwards – is evidence of the fully formed Classical phalanx at this early date. It is, I believe, perfectly possible to envisage combat of the type described by Homer – with its mass of fighters in varying density and informal formation, fronted by *promachoi* who may use missiles as well as spears – being conducted by men in hoplite equipment, and this I believe is how we should also envisage seventh- and perhaps sixth-century infantry combat. Bronze armour and heavy shields were increasingly adopted by men who still fought in this more fluid default style, and only over a longer period was there a move toward the more formal organization and homogenous fighting of the later phalanx.

We might wonder what the reasoning behind the adoption of this new equipment would have been. It seems obvious that the main incentive would have been the desire for greater personal protection, since a heavy wooden shield would provide greater defence against penetration by enemy weapons, whether thrust or thrown. In Homer, the shield was (we can assume) usually made of leather stretched over a wicker frame, and penetrating it or attempting to penetrate it is a common feature of combat:

> 'Harpalion son of King Pylaimenes then sprang upon [Menelaus] ... He struck the middle of Menelaus' shield with his spear but could not pierce it, and to save his life drew back under cover of his men, looking round him on every side lest he should be wounded.' (Homer, *Iliad* 13.642–649)

Compare this with Deiphobus' experience in *Iliad* 13.168 quoted above, where 'Deiphobos had seen [the spear] coming and had held his shield well away from him' so that any penetration would not reach through to his body. A solid wooden shield, while it could be penetrated by a spear, was much less likely to be so, and one held strapped to the forearm cannot be (and does not need to be) held so far away from the body as a shield held in the hand. The adoption of the Argive shield – which we must remember was not universal, with the presumably lighter Boeotian shield continuing to be preferred in some areas into the fifth century – represents a preference for protection over lightness and individual mobility. This does not mean that a heavier shield could not be used for lighter, more mobile fighting, since as argued above, single combat in hoplite equipment remained very much a feature of warfare, and (as we will see in Chapter 6) hoplites still needed to fight in all sorts of circumstances and situations outside of the phalanx – including aboard ship, where there was no question of forming a close-order phalanx. It does mean that the surrender of a certain amount of mobility was considered a reasonable price to pay in return for greater protection. The eventual replacement of the Argive shield might shed some light on this process in reverse. In the fourth century, the Athenian general Iphicrates is said to have altered hoplite equipment, in a passage we have already encountered above:

> 'He devised many improvements in the tools of war, devoting himself especially to the matter of arms. For instance, the Greeks were using shields [*aspisi*] which were large and consequently difficult to handle; these he discarded and made suitably sized light shields [*peltas summetrous*], thus successfully achieving both objects, to furnish the body with adequate cover and to enable the user of the *pelte*, on account of its lightness, to be completely free in his movements. After a trial of the new shield its easy manipulation secured its adoption.' (Diodorus 15.44.1–4)

The principle that a lighter shield of suitable size (Diodorus does not make clear what size this was) could provide adequate cover while also being less burdensome to the carrier, is clear enough, and in the third century there was widespread adoption of other lighter shields, the Celtic-inspired *thureos* and the Macedonian shield, which at least in some of its forms was lighter than the Argive shield and was sometimes also called a *pelta*. So we can see changes in fashion, in practice and in priorities, with heavier or lighter shields being preferred at various times; yet there was also considerable conservatism concerning military equipment in antiquity, not surprisingly when there was little chance to experiment, and trial and error was an imperfect method in a matter of life and death. Custom and tradition are likely to be as important as practical considerations, particularly when it is a matter of weighing up different strengths and weaknesses (protection versus mobility), rather than more straightforwardly advantageous technological improvements (such as iron versus bronze). For several centuries, the emphasis was on protection, but we cannot conclude from this that combat itself took a totally different form in that period. Nevertheless, I think it is reasonable to conclude that the desire for greater protection from the shield does represent a change in emphasis in infantry combat, if not a change in its nature, a change in which better protection in hand-to-hand fighting (represented by the adoption of more bronze armour as well as a heavy shield) began to assume greater importance than mobility in moving into and retreating from close combat. Consequently, this represents not a revolution in, but a steady development of the nature of infantry combat, as looser infantry *phalanges* became a more rigidly ordered and structured phalanx.[31]

Two further strands of evidence shed further light on this question. One is the changing role and use of missile weapons, as depicted in art, and the other is the depiction in art of what look very much like formal phalanx formations from the mid-seventh century.

Missiles and phalanxes

In Homer, the spear was frequently thrown by warriors, and archers were also present among the *phalanges*, the bow being carried by some as a matter, so far as we can tell, of personal choice. The passage from Homer quoted above as inspiring Xenophon's description of Coronea – 'shield clashed with shield and spear with spear ... the bossed shields beat one upon another' – is followed by a long series of encounters mixing thrust and thrown weapons with other unnamed missiles:

> 'First Antilokhos slew an armed warrior of the Trojans, Ekhepolos, son of Thalysios, fighting in the foremost ranks. He struck at the projecting part of his helmet and drove the spear into his brow ... King Elephenor, son of Khalkodon and leader of the proud Abantes began dragging him out of reach of the darts [*beloi*, missiles in general] that were falling around him, in haste to strip him of his armour. But his purpose was not for long; Agenor saw him hauling the body away, and smote him in the side with his bronze-shod spear ... Forthwith Ajax, son of Telamon, slew the fair youth Simoeisios, son of Anthemion ... he was cut off untimely by the spear of mighty Ajax, who struck him in the breast by the right nipple as he was coming on among the foremost fighters ... Thereon Antiphos of the gleaming corselet, son of Priam, hurled a spear at Ajax from amid the crowd and missed him, but he hit Leukos, the brave comrade of Odysseus, in the groin, as he was dragging the body of Simoeisios over to the other side ... Odysseus was furious when he saw Leukos slain, and strode in

full armour through the front ranks till he was quite close ... His dart was not sped in vain, for it struck Demokoön, the bastard son of Priam ... Odysseus, infuriated by the death of his comrade, hit him with his spear on one temple, and the bronze point came through on the other side of his forehead ... Then fate fell upon Diores, son of Amarynkeus, for he was struck by a jagged stone near the ankle of his right leg. He that hurled it was Peirous, son of Imbrasos, leader of the Thracians, who had come from Ainos ... But Peirous, who had wounded him, sprang on him and thrust a spear into his belly, so that his bowels came gushing out upon the ground, and darkness veiled his eyes. As he was leaving the body, Thoas of Aetolia struck him in the chest near the nipple, and the point fixed itself in his lungs. Thoas came close up to him, pulled the spear out of his chest, and then drawing his sword, smote him in the middle of the belly so that he died.' (Homer, *Iliad* 4.446ff.)

The scene is typical of combat scenes in Homer, with *belos*, the generic word for a missile of any type, being used frequently, though from context it is usually apparent that a thrown spear is meant (arrows and rocks are usually identified as such when they are used).

When men carrying Argive shields begin to appear in art from the seventh century, they are frequently depicted either carrying two spears, or spears that have a throwing loop or *ankyle*, a thong wrapped around a javelin to impart spin and to give it more force when thrown. Clearly, these early hoplites still expected to throw one or both of their spears, but as we see from Homer, this is not at all incompatible with fighting in a close formation, or with coming to close quarters with the enemy. Indeed, the Romans, centuries later, were to develop a system of heavy infantry based around the use of a preliminary barrage of javelins, followed by closing to close quarters, in their case with swords. The Greeks appear to have thrown one spear and retained the other for thrusting, which does not mean that the second spear could not also be thrown if occasion demanded, nor that the second spear might not also have a throwing loop for that eventuality. Infantry with throwing spears of this kind do not just appear in scenes of single combat; one of the earliest and best-known (and most analyzed and discussed) scenes of early hoplite combat is the Chigi vase, dating to the mid-seventh century (properly speaking this object is an *olpe* or jug). I will have more to say below on the nature of the combat depicted in this scene, but note here that lines of similarly equipped hoplites in close order are depicted carrying throwing spears (with loops), which provides evidence that increasing close order and formalization of infantry formations did not mean the end of the use of throwing spears by the hoplites themselves. Throwing spears and second spears disappear from art by

The Chigi Vase, mid-seventh century (reconstruction after Connolly (2012)).

the fifth century, however, and do not appear at all in literary descriptions from Herodotus onward, so what we seem to be seeing is a gradual process where the throwing and thrusting of the seventh-century phalanx (reflected in Homer) gradually gave way to the pure thrusting of the fifth-century phalanx we see in Herodotus, Thucydides and Xenophon. As with shields and armour, this is a process of gradual change and increasing specialization, rather than revolution.

In addition to the use of missiles by the hoplites themselves, there is the question of missile-users who are not themselves hoplites, but who fight as part of the (early) phalanx. In the Homeric *phalanges*, archers and javelineers appear to mingle among the other fighters and there is no apparent distinction between them (socially or organizationally). In Tyrtaeus, we see missile-users still among the phalanx, but distinguished from the heavy-armed fighters:

'You, light armed, squatting under a shield here and there, must throw great rocks and hurl smooth javelins while you stand close by the heavy armed.' (Tyrtaeus Fr. 11.35–38)

The shield in question is that of the heavy armed, so we see light infantry still intermingled among the phalanx, and the phalanx in open enough order, at least some of the time, to allow space for these light-armed men. Depictions of archers in Archaic art show them mostly in scenes of fairly chaotic combat, and it is not certain that actual formations can be reconstructed from such images; but they are consistent with the idea that light-armed men still fought among the phalanx, rather than being separated out into different units and given a subordinate role.

Herodotus records a tradition of the origins of the separation of heavy- and light-armed combatants:

Origins 33

'At his death [Phraortes] was succeeded by his son Cyaxares. He is said to have been a much greater soldier than his ancestors: it was he who first organized the men of Asia in companies [*telea*] and posted each arm apart, the spearmen and archers and cavalry: before this they were all mingled together in confusion.' (Herodotus 1.103.1)

Herodotus is talking about the Medes, and Cyaxares was king in the late seventh to early sixth centuries. It is not certain if Herodotus means that Cyaxares was first to organize the Medes this way, or first to organize any infantry, though early infantry such as Egyptians and Assyrians were surely similarly arranged into separate units. At any rate, the practice may have been new to the Greeks at around this time, and adopted by the Greeks following the Median or Persian example (Greek colonists and mercenaries in Ionia and Lydia bringing the innovation back to Greece). We might therefore tentatively date the separation of light-armed fighters from heavy-armed to the early sixth century, while noting that the evidence from art is that the change was neither immediate nor complete. In Herodotus' accounts of the Persian Wars, it is possible that the

Archers in Archaic infantry combat (after Snodgrass (1964) plate 15b).

Spartan hoplites at least were still mixed, with large numbers of light-armed, since he records the following at the decisive Battle of Plataea (479):

> 'On the right wing were ten thousand Lacedaemonians; five thousand of these, who were Spartans, had a guard of thirty-five thousand light-armed [*psiloi*] helots, seven appointed for each man ... as regards the number of the *psiloi*, there were in the Spartan array seven for each hoplite, that is, thirty-five thousand, and every one of these was equipped for war.' (Herodotus 9.28.2; 9.29.1)

It is often understood that these *psiloi* formed a separate unit or units (and one which played little part in the actual fighting), but Herodotus certainly suggests (for example at 9.61.2) that the *psiloi* and the other Spartans remained together. An alternative interpretation is therefore that the Spartan formation was a mixed one of hoplites and *psiloi*, with a preponderance of *psiloi*, although Herodotus also notes that the Spartans were more heavily armed than the Persians (9.63.2) – I will have more to say on these contradictions later on.[33]

By the time of Thucydides there is no sign of light infantry being incorporated in any phalanx, and all forms of light-armed troops, including peltasts (the javelin-throwers who rose to considerable prominence in the fourth century), are clearly drawn up separately from and fighting independently from the hoplites of the phalanx, as is especially clear in the many minor engagements described in Xenophon's *Anabasis*. We seem then to again be seeing a gradual transition in which light-armed combatants at first were an integral part of the infantry, then become more distinguished from them (in terms of equipment and, we can surmise, socially) but still fought amongst them (on occasion), before finally being completely excluded from the now specialist hoplite phalanx. The dates of these transitions are impossible to determine with any accuracy, but it is possible that the final transition to a pure hoplite phalanx had not yet fully occurred by the time of the Persian Wars.

Inclusive and exclusive infantry

We have seen that the most important distinction between types of infantry was between the 'heavy-armed' (hoplites, as they came to be known) and the 'light-armed' (*gymnetes*, *psiloi*, *peltastai*), but this distinction was not just a technical and military one. After all, hoplites continued to fight, like the light-armed men, with missile weapons, the only items that were universally carried by hoplites were the Argive shield and helmet, and heavy- and light-armed infantry probably fought in mixed units at least into the sixth century. The distinction between the two was also a social, economic and political one. Economically, because infantry provided their own equipment, the hoplites were those able to afford the full panoply of shield, body armour and helmet (and, perhaps at various times, greaves), while those unable to provide this kit would have to fight 'naked' as light infantry. Politically, full citizenship in Greek cities depended on a wealth qualification, which in practice usually meant a property qualification, since land and the rent and agricultural income from it was the main source of wealth in Greece, as in all ancient societies. The political rights that this citizenship entailed varied from city to city, according to its constitution, and also varied through time as various more or less monarchical, oligarchic or democratic groups within each city had the upper hand, in a process of political turmoil (*stasis* in Greek) that typifies the Greek world. Some cities, such as Sparta, had well-established constitutions that did not change significantly over time. Others, like Athens, developed more democratic institutions over time, extending the franchise to a wider class, where participation as rowers in the navy was almost as important as fighting as hoplites in the phalanx. Other cities, at least in the late fifth and fourth centuries when we know most about them, might see-saw between the oligarchic and democratic parties, with internal conflicts between the two that would also often pull in outside support from other cities, based on their own oligarchic or democratic leanings. It is also not the case that all hoplites had the exact same political rights as citizens, nor that only citizens could serve as hoplites. In the first case, the particular constitution in each city would determine the particular rights of the citizens, and in the second, it was common – even in such strictly delineated societies as Sparta – to employ non-citizens as heavy infantry, and therefore as hoplites in a military sense, whenever it was necessary to make up numbers or to provide forces for overseas expeditions, for example. Nevertheless, the fact that hoplites were of the same elevated economic class and that they had common political rights (within a city) meant that hoplites considered themselves equals to each other (literally in the case of Spartans, with the citizen hoplite class of *homoioi*, 'peers' or 'equals') and superior to the light infantry of the lower classes. Combined with the fact that writers of history were themselves drawn from the hoplite class, this perhaps leads to a skewing of our image of the importance of hoplites in war. As we saw above, there were very many more Spartan non-hoplites than hoplites at Plataea, though only the exploits of the hoplites in defeating the Persians were thought worthy of record, while the defeat of the Persian navy by the Greek fleet, manned by rowers from the lower orders, was arguably more important in saving Greece than the hoplites' efforts at Thermopylae and Plataea.[34]

A central part of the 'hoplite debate' is this political dimension of the rise of the phalanx. This connection between military formation and political institutions dates back well before the nineteenth century AD, when the hoplite debate took shape. Aristotle, writing in the fourth century, provides a statement of the significance of the rise of heavy infantry as understood in his day:

> 'And although it is proper that the government should be drawn only from those who possess heavy arms [*hopla*], yet it is not possible to define the amount of the property-qualification absolutely and to say that they must possess so much, but only to consider what sort of amount is the highest that is compatible with making those who have a share in the constitution more numerous than those who have not, and to fix that limit … And indeed the earliest form of constitution among the Greeks after the kingships consisted of those who were actually soldiers, the original form consisting of the cavalry (for war had its strength and its pre-eminence in cavalry, since without orderly formation [*syntaxis*] heavy-armed infantry [*to hoplotikon*] is useless, and the sciences and systems dealing with tactics did not exist among the men of old times, so that their strength lay in their cavalry); but as the states grew and the bearers of heavy arms had become stronger, more persons came to have a part in the government. Hence what we now call constitutional governments the men of former times called democracies; but the constitutional governments of early days were naturally oligarchical and royal, for owing to the smallness of the populations their middle class was not numerous, so that because of their small numbers as well as in conformity with the structure of the state the middle class more readily endured being in a subject position.' (Aristotle, *Politics* 1297b)

The term 'middle class' is accurate, in that such men did form a middle level between the aristocrats (rich enough to provide horses and therefore to serve as cavalry, although in practice they often fought on foot and used their horses merely for transport to the battlefield) and the labourers (unable to provide full hoplite equipment and usually fighting as light infantry or as rowers in the navy, or providing logistical support), though the nineteenth-century AD and later usage of the term 'middle class' perhaps gives the wrong impression. The hoplite class were ideally landowners and men of leisure, wealthy enough in land not to have to work for a living. Yet in practice the hoplite qualification did often extend low enough to include what we might call smallholders, independent landowners who would have had to work their own land, for all the implications this has for long-term military campaigns. These two groups – landowning wealthy men of leisure and self-sufficient farmers – might both qualify to fight as hoplites, though they really represent two quite distinct social and economic classes, and would often have to fight alongside men of even more modest means mobilized for the occasion. Plato provides an amusing image of the political and social tensions that could arise from this situation:

'And when, thus conditioned, the rulers and the ruled are brought together on the march, in wayfaring, or in some other common undertaking, either a religious festival, or a campaign, or as shipmates or fellow-soldiers or, for that matter, in actual battle, and observe one another, then the poor are not in the least scorned by the rich, but on the contrary, do you not suppose it often happens that when a lean, sinewy, sunburnt pauper is stationed in battle beside a rich man bred in the shade, and burdened with superfluous flesh, and sees him panting and helpless – do you not suppose he will think that such fellows keep their wealth by the cowardice of the poor, and that when the latter are together in private, one will pass the word to another "our men are good for nothing"?' (Plato, *Republic* 9 556c–d)

The fear, as Plato expresses it, is that this may lead the inhabitants of the city to turn on each other, 'the one party bringing in allies from an oligarchical state, or the other from a democratic'.[35]

The picture of the hoplites as a middling class in each city is thus true, up to a point, but is not the whole story. The ideal was for the hoplites to be a distinct, wealthy and leisured class, but in practice, economic reality and military necessity meant that the situation varied from city to city, and the heavy infantry were in reality drawn from a much wider proportion of the population, even in rigidly hierarchical and traditionalist states like Sparta. The prevalence of mercenary service makes the picture even more complicated. In the fourth century in particular, hoplite mercenary service became common, and this is often linked to economic hardship forcing impoverished members of the hoplite class to try to supplement their income. It is undoubtedly true that some mercenaries were poor, but others, as Xenophon (who himself took up mercenary service with the Persians) was keen to stress, were still men of 'quality' (of good family and often of some wealth), and mercenary service dates back to the very beginning of the hoplite period and before, with Greek mercenaries appearing in Egypt and the Near East in the seventh century or earlier. So hoplites could come from a variety of backgrounds, and not necessarily only from the wealthy middle, and the idea of a 'hoplite revolution' in which the middle classes took over power from their aristocratic superiors, perhaps under the temporary leadership of a tyrant, seems oversimplified. There was undoubtedly a move toward a wider franchise and extended political rights, and a connection between this and service in the heavy infantry phalanx, but the details varied greatly from city to city and through time.[36]

Nevertheless, the principle of the decline of aristocratic power and the rise of more democratic forms of government, and the link between this and the rise in power of infantry as envisaged by Aristotle in the quote above, does seem basically convincing. It is a process that has been seen often throughout history, where there is a link between military dominance, particularly in pitched battle, and political power, the one both justifying and enabling the other. In Medieval European history we see a similar process, with the military power of Dark Age cavalry leading to the political and social dominance of the feudal knightly class, to be eventually overthrown by the rise (or resurgence, if we see this as reflecting the return of the infantry of antiquity) of infantry, driven both by technical (weaponry that cancelled the dominance of cavalry, particularly the pike, longbow and firearms) and social factors (infantry forces recruited from growing city populations able to stand against cavalry in a way that rural peasantry were not). However, the example of Medieval Europe warns us that this process was not simple, with many forward and backward steps and proceeding at a different pace in different regions, and we should expect the 'hoplite revolution', covering as it did a period of some three centuries and dozens or hundreds of polities, to be similarly complex.[37]

In some areas, particularly Thessaly and Macedon in northern Greece, where horse breeding was easier due to the presence of large areas of pasture, and where the population was predominantly rural rather than urban, a powerful infantry developed late or not at all, and aristocratic cavalry remained the dominant force, politically and on the battlefield. When Macedon did develop its own infantry, in the fourth century, it was as a result of a specific policy from its king, Philip II, and was tied to a deliberate, centrally organized process of urban development, bringing the rural population out of the hills and settling them in cities, as well as arming them in a particular way to create the Macedonian phalanx, a refined, perhaps perfected, form of the Greek phalanx. In southern Greece, pasture was in shorter supply and horse breeding more limited, with the aristocracy correspondingly fewer in number and less willing to risk their valuable horses in battle. With the population becoming concentrated in urban centres, this allowed infantry to dominate cavalry from an earlier date.

Aristotle's observations about the importance of 'orderly formation [*syntaxis*]' and 'tactics' is particularly interesting. From earliest times, Greek infantry could have been plentiful, but the open, fluid formations of the early phalanxes, as reflected in Homer, would have been unsuited to defeating cavalry consistently, something which, as we will see in Chapters 6 and 7, required a tight formation and the drill and discipline to maintain this formation in the face of onrushing horsemen (or chariots). The implication of this is that the military and political dominance of infantry was dependent on its adopting a disciplined phalanx formation – more so than its adopting heavy hoplite equipment – and this might cast into doubt the idea that the phalanx did not develop its full, close-order form until as late as the fifth century. Aristotle does not say when the 'old times' were in which cavalry were still dominant and infantry disorganized, but it seems unlikely that he meant the early fifth century.[38]

It is also worth noting that aristocratic cavalry were not the only mounted enemy faced at this time, particularly by the Ionians and Carians, where hoplite equipment (and possibly tactics) may well have originated. The poet Mimnermus, writing in Ionia in the late seventh century, extols the valour of a warrior:

> 'Not his were such feeble might and poor nobility of heart, say my elders who saw him rout the serried ranks of Lydian cavalry in the plain of Hermus, rout them with a spear; never at all would Pallas Athene have had cause to blame the sour might of the heart of such as him, when he sped forward in the van, defying the foeman's bitter missiles in the thick of bloody war.'
> (Mimnermus, fr.14 West)

Perhaps Aristotle's model of infantry achieving political power when they were able to overcome the aristocratic cavalry of their own city is oversimplified. It was more important that a steady infantry was able to defeat mounted forces in general (especially in the East), and therefore was able to establish itself as the dominant force on the battlefield, and that this is what gave political power to the infantry classes.

Formed and unformed infantry

This leads us back to the central question of the hoplite debate, the timing of when the phalanx took something closer to its final form and infantry became a disciplined, organized force fighting in a tight formation, rather than the more flexible formation reflected in Homer and apparently depicted in much Archaic art and literature. The honest answer to this may simply be that we cannot tell with any certainty, given the nature of the evidence. One ancient tradition may be worth quoting in this context, that recorded by Pausanias, writing in the second century AD, describing the events of the First Messenian War, a semi-legendary war between the Spartans (Lacedaemonians) and the Messenians in the late eighth century. Pausanias describes the course of the decisive battle of this conflict:

> 'When the leaders on either side gave the signal, the Messenians charged the Lacedaemonians recklessly like men eager for death in their wrath, each one of them eager to be the first to join battle. The Lacedaemonians also advanced to meet them eagerly, but were careful not to break their ranks. When they were about to come to close quarters, they threatened one another by brandishing their arms and with fierce looks, and fell to

recriminations ... And now with their taunts they come to deeds, mass thrusting against mass [*athrooi pros athroous othismoi*], especially on the Lacedaemonian side, and man attacking man. The Lacedaemonians were far superior both in tactics and training, and also in numbers ... The Messenians were inspired alike by desperation and readiness to face death, regarding all their sufferings as necessary rather than terrible to men who honoured their country, and exaggerating their achievements and the consequences to the Lacedaemonians. Some of them leapt forth from the ranks, displaying glorious deeds of valour ... The Lacedaemonians refrained from exhorting one another, and were less inclined than the Messenians to engage in striking deeds of valour. As they were versed in warfare from boyhood, they employed a deeper formation and hoped that the Messenians would not endure the contest for so long as they, or sustain the toil of battle or wounds. These were the differences in both sets of combatants in action and in feeling.' (Pausanias 4.8.1–7)

We need not have much faith in the historical reality of this episode, since it seems to describe phalanx fighting (in the Classical style) yet falls before the period for which there is good artistic or archaeological evidence of such fighting. But the story is interesting in what it reveals of Pausanias' understanding of the different tactical systems in use in the past (even if it was not, in reality, such a distant past as he supposed). This sounds very much like a description of an organized phalanx (the Lacedaemonians) fighting against a traditional 'Homeric' infantry formation of the type we saw earlier (the Messenians). The Messenians form a solid battle line, but they do not maintain it for the fighting, but run out individually to perform 'deeds of valour', while the Spartans rely on discipline and on the depth of their formation, fighting as a unified force rather than as a gaggle of individuals. Intriguingly, Pausanias also uses the word *othismos*, a very rare word among Classical writers (Herodotus, Thucydides and Xenophon), though one which has been taken very much to heart by modern scholars and theorists of hoplite warfare (see Chapter 8). It is difficult to see precisely how 'mass thrusting against mass' could happen 'mostly on the Spartan side' (they must have had a mass to thrust against, even assuming 'thrust' is a good translation), but at any rate we see again a distinction between two types of fighting: the individual duelling, man against man, of the Homeric phalanx and of the Messenians, and the more organized mass fighting in formation of the later phalanx and of the Spartans. The whole passage is reminiscent in many ways of Herodotus' account of the fighting of Spartans and Persians at Plataea:

'Now [the Spartans] too charged the Persians, and the Persians met them, throwing away their bows. First they fought by the fence of shields and when that was down, there was a fierce and long fight around the temple of Demeter itself, until they came to *othismos*. For the barbarians laid hold of the spears and broke them short. Now the Persians were neither less valorous nor weaker, but they were unarmoured [*anoploi*]; moreover, since they were unskilled and no match for their adversaries in craft, they would rush out singly and in tens or in groups great or small, hurling themselves on the Spartans and so perishing.' (Herodotus 9.62)

The detailed implications of this passage will be the subject of Chapters 7 and 8, but we can note here (besides the rare use of the word *othismos* by Herodotus) the apparent use by the Persians, once their fence of shields was down, of the traditional infantry tactics of rushing out of formation individually or in small groups (to perform 'deeds of valour', we may suppose), while the Spartans, with greater skill, maintain their formation to defeat them. If we were to take Pausanias' account at face value, then we would see the Spartans at least using a drilled and organized phalanx, equivalent to the one they used at Plataea, at a much earlier date than usually assigned to the introduction of the phalanx, and well before the late date espoused by the 'gradualists' in the hoplite debate and suggested by the evidence of missile use and of individual combat in art.

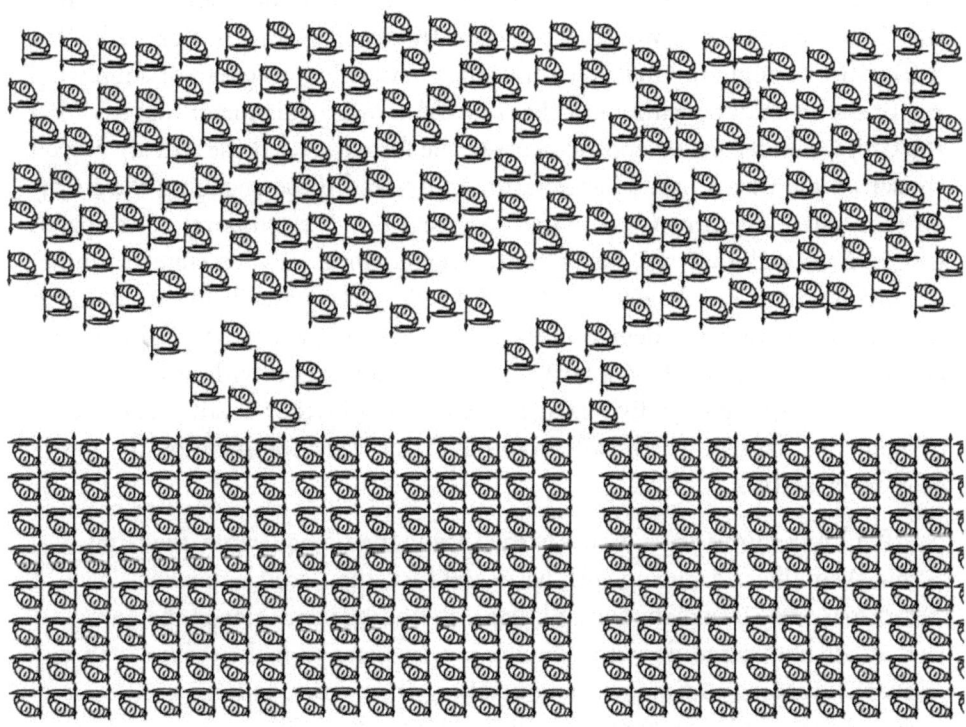

A representation of combat between a regular and irregular phalanx ('Spartans v. Messenians')

Nevertheless, Pausanias does seem to reflect a genuine tradition of two different but related forms of infantry fighting.[39]

To further date the development of the close-order phalanx we might turn again to vase paintings; we have already encountered the Chigi vase, which is among the earliest depictions of what appears to be a formed close-order phalanx and dates from the mid-seventh century. Although this vase depicts hoplites armed with throwing spears, it also shows them in what appears to be close order and formed ranks, and therefore (it is argued) in an ordered phalanx. Extracting tactical details from vase paintings is, it should be said, fraught with difficulty. Most painters were not interested in depicting (or did not feel able to depict) a mass formation of any sort, and the vast majority of vase paintings depict single combats, even if between large numbers of participants. Partly this is because they depict mythological or legendary scenes, often from the Trojan War, where single combat is appropriate, but there are also all sorts of technical problems with depicting a formation of any sort in the side view and profile depiction which all vase paintings use. The Chigi painter succeeds admirably in giving an impression of an infantry formation given the limitations of the format, but it is highly unlikely that details such as the number of ranks or the density of the formation can be extracted from this painting. In particular, I am not aware of any art that depicts a deep formation of several ranks, even from the fifth century when we can be certain that such formations existed – the difficulties of depicting multiple ranks may have been too great, or simply did not fit the artistic conventions of the time. Also, while it is sometimes said that the Chigi vase depicts several elements of the formation – those still arming on the left, some advancing to battle in the centre and those engaging the enemy on the right – it seems to me more likely that this is a temporal not a spatial progression, and that the forces are presented in successive states (arming, advancing, fighting). At any rate, I do not believe it is possible to extract much tactical detail from the depiction, but the fact remains that the painter does appear to be trying to show a fairly close-order formation with ordered ranks, rather than a group of individuals. There is also a further detail, the presence of a piper or flautist, which calls to mind Thucydides' comment on the Battle of Mantinea (418):

> 'After [the preliminary speeches to each army] they joined battle, the Argives and their allies advancing with haste and fury, the Lacedaemonians slowly and to the music of many flute-players – a standing institution in their army, that has nothing to do with religion, but is meant to make them advance evenly, stepping in time, without breaking their order, as large armies are apt to do in the moment of engaging.' (Thucydides 5.70)

Mantinea, incidentally, seems to be another example of the phenomenon of an ordered phalanx keeping order and staying together, fighting against a phalanx using the traditional default tactics of each man rushing out at will, like the Persians at Plataea or the Messenians. The presence of the flute (or pipe) player on the Chigi vase has been taken as evidence that what is depicted is an ordered phalanx, and would push the date of such order back to the mid-seventh century. It is, however, impossible to be certain that this is what the flautist implies – it could be that flute players had long accompanied armies, to provide encouragement or for religious reasons as Thucydides mentions, and that the Spartans at some later date adapted this practice to their purpose of keeping in step. It is also noteworthy that Thucydides points out that the Spartans were unusual in using flute players; presumably most armies did not, so we cannot assume that the presence of flutes implies the presence of close-order drill (or that their absence implies its absence).[40]

There are other vase depictions of apparently close-order formations dating from a similar period. Again, not much precise detail can be extracted from them, but at any rate they provide further indications that painters from the seventh century were interested in depicting a formation, not just single combats, which again provides a suggestion that a more ordered phalanx – not just an occasional ad hoc tightening of a loose formation, as we see in Homer – might have existed at this point. Another intriguing example is the Amathus bowl, which dates from around 700 and was produced somewhere in the Near East, in an 'Asian' (rather than Greek) style, and which depicts four infantrymen who may well be Greek mercenaries, who appear to be armed as hoplites, and who are formed in closer and more regular order than the other figures depicted, who

'Hoplites' on the Amathus bowl, seventh century. (*British Museum*)

all seem to be fighting individually. Again, this could push the dating of a more ordered phalanx back into the seventh century at least (assuming, that is, we are willing to extract such tactical details from a depiction of four men).[41]

So it does look as if – despite the artistic and literary (poetic) evidence for the continued existence of light infantry within the phalanx, and the suggestions of a more fluid formation in which individual soldiers could choose their own position in the formation, and change it at will – there is also artistic evidence and a literary tradition which suggests that an ordered phalanx, at least among the Spartans, is much older, and dates back closer to the first adoption of hoplite equipment. The existence of such conflicting evidence is of course the reason why the hoplite debate has dragged on inconclusively for so long, and seems unlikely to be finally resolved any time soon. In the absence of further evidence, we may simply have to accept that the exact date of the emergence of the phalanx in its Classical form is unknown (or, as is more likely, we may have to stick to our existing preconceptions without hope of changing the mind of anyone with different views). Perhaps a more important question to ask is in what ways, if any, the Greek phalanx differed from other contemporary or earlier infantry formations.

What was distinctive about the hoplite phalanx?

The origins of hoplite equipment – bronze armour and helmets, but most importantly the heavy, round Argive *aspis* – certainly date back to the seventh century or before. However, the often-repeated claims that hoplite equipment was useless outside the drilled close-order phalanx (in its Classical guise), and that therefore the presence of hoplite equipment requires the existence of such a phalanx, do not stand up to scrutiny. It appears that hoplite equipment was developed before the ordered phalanx was universally (or even widely) adopted, even if we cannot be sure that the ordered phalanx did not exist in some places and in some circumstances from an early date. Furthermore, men with hoplite equipment, or with variations on it such as the Boeotian shield, seem to have continued to fight, at least on occasion, in more fluid formations, ones which incorporated light-armed troops (rather than relegating them to separate formations), and in which the individual soldier had some freedom of movement, to move into contact with the enemy or to fall back again at will.

At the same time, there is some reason to suppose that the close-order, formed phalanx did have its origin before the fifth century, and did not come into existence only at or after the time of the Persian Wars. It may be that close-order, ranked formations are depicted in art from the seventh century, and Greek traditions transmitted by Aristotle (that the rise in importance of hoplites was instrumental in political change in Archaic Greece, and that an ordered phalanx was an essential element in this), though they are certainly not correct in every detail and contain broad generalizations, do provide evidence that the Greeks at least thought that the ordered phalanx had an early origin.

It is difficult to reconcile these two strands, but the way to do so seems to be to abandon the notion, which itself is of no great antiquity and has its origins in Edwardian Britain, that the hoplite phalanx was something unique, revolutionary, militarily decisive and quite unlike anything that had gone before it, and that once developed (in seventh-century Greece, for the sake of argument), all Greek cities had to 'go hoplite' and adopt the same armament, formation, drill and tactics themselves. Rather, as so often in military history, we should see a slow and piecemeal adoption of different elements of the hoplite phalanx, at different times, building on rather than overturning what had gone before, and proceeding at a different pace in different locations and circumstances.[42]

As we have seen, Homeric infantry warfare – whether it represents seventh-century (or later) practice, a memory of Dark Age or even Bronze Age combat or a confused amalgam of all three – may well represent the default form of infantry combat in antiquity. Formations are irregular and undrilled, but they still exist – the *phalanges* of Homer may be irregular, but they are still distinct bodies of men. The men mostly fought in a more open order, but they could close up into the tightest of formations, with 'helm up against helm', when occasion demanded. They also fight in some depth, without formal ranks perhaps, but there is no question of infantry of this period forming a single line – there is a mass behind the *promachoi*, giving the formation depth. Bronze armour and large shields were used throughout the Bronze and Dark Ages, and Archaic armour was just a more developed form of such protection, with head-enclosing helmets and especially strong shields, as protection came to be more important than mobility. It is a process that we then see being reversed already by the fifth century with the adoption of lighter armour and helmets and eventually, in the third century, of lighter shields, with, finally, the Macedonian phalanx privileging offensive capability over defensive (see Chapters 2 and 9 for these developments). In the course of the Archaic period, the cultural preference for close-order, hand-to-hand fighting, and an emphasis on defensive armour over individual mobility, caused other elements – throwing spears, a fluid formation, open order, integrated light infantry – to fall out of use, but the change need not have been abrupt or comprehensive.

Furthermore, the Greeks were by no means the first to see their infantry-fighting techniques move in this direction. Body armour and large shields were already a feature of the contemporary Assyrian (strictly speaking, Neo-Assyrian) armies, as were separate formations of missile and hand-to-hand specialist infantry (so far as we can tell). The Persians too used body armour, shield and spear, with deep, close-order infantry formations. For all that the playwright Aeschylus characterizes the Persian Wars as a battle of spear against bow (e.g. *Persians* 86; 148–149), this does not match the Persians' own view of how they fought, and Persian depictions of their own infantry emphasize large shields and spears. When they fought behind barricades of shields which gave cover to their archers, presumably they could not engage in the individual charges that characterize the default style of infantry fighting, but they are still said to have done so on specific occasions, such as Thermopylae and Plataea. Xenophon (*Cyrop.* 6.4.17–19; *Anab.* 1.8.9) presents Egyptian infantry (note that they are called 'hoplites') as fighting in very deep formations and with large shields suitable for pushing, in a way that seems identical to the common view of Greek hoplite formations and fighting techniques. Indeed, Egyptian and Persian armies are described forming up much deeper than was ever the case for Greeks, so depth of formation cannot be a distinctive feature of the Greek hoplite phalanx. We do not know what file intervals (if any formal file intervals at all) were used by Persians or Assyrians, or any infantry forces contemporary with Archaic or Classical Greek warfare. But then we also do not know what file intervals were used by Archaic or Classical Greeks. It may be true that Greek formations used a closer order than other contemporary or earlier infantry, but this is nowhere stated, so it remains an assumption derived from the supposed unique properties of the hoplite shield. Yet hoplite shields, supposedly diagnostic of a particular form of fighting, appear

The 'Stele of Vultures', from Lagash (Iraq), around 2500 BC. (*Louvre, Paris*)

Origins 47

Persian spearmen on the Apadana Palace, sixth century. (*Persepolis, Iran*)

Assyrian heavy infantry, late seventh/early sixth century. (*Nineveh, Iraq*)

to have originated in Egypt, Caria or elsewhere in the Near East, while bronze body armour may have a Central European origin (see also the following chapter for these points). The earliest depiction in art of a close-order infantry phalanx with locked shields and serried spears is the 'Stele of the Vultures', dating from mid-third millennium Mesopotamia, 2,000 years before the 'hoplite revolution'. It is not clear what, if anything, was so very different about the hoplite phalanx, aside perhaps from a modest increase in the weight and protective properties of shields and armour.[43]

The one feature of hoplite fighting that is supposed to be truly unique, the assumed 'shove', is controversial in the extreme, and Chapter 8 will be devoted to examining the evidence for the idea. Without shoving, but fighting in a deep, sometimes close-order formation (how close is open to debate) with spears and shields and wearing armour, as other infantry had long done, the hoplite phalanx loses its uniqueness. Even with an element of pushing, other infantry were thought, by the Greeks themselves, to be similar to their hoplites, as Xenophon's Egyptians illustrate. When Herodotus gives explanations for Greek victories in the Persian Wars, they tend to be related to armour and weapons or to the Persians' lack of 'skill' and 'craft' (as at Plataea, Hdt 9.62.3), not to fundamental differences in military systems or tactics. As I will discuss in Chapter 7, providing simple explanations of this sort for battlefield victories is anyway a difficult undertaking; after a victorious battle, there was no detailed debrief (so far as we know), and anecdotal evidence of what happened and why would tend to accrete around a few popular stories or features. Yet Herodotus' claim that the Persians' lack of armour, skill or craft were the decisive factor in their defeat by the Greeks, is a long way away from modern theories of close-order drill and mass shoving. Plutarch for one was outraged that Herodotus should so devalue the superior courage of the Greeks, the true cause of their victory, as he thought ('What then is there left great and memorable to the Grecians of those fights, if the Lacedaemonians fought with unarmed men', Plut. *De Herod.* 43). According to Herodotus, reasons for victory lay not in a different type of fighting, but in Greeks being better at, and better equipped for, the type of fighting both sides shared. The difference is in degree, not kind. The hoplite orthodoxy, with its insistence on an early revolutionary change to a unique style of fighting (the nature of which is somewhat under-specified in modern accounts, but depth, close order and shoving in combat seem to be the key features), adopts Aristotle's identification of the political importance of the development of infantry tactics, but ignores his reasons for why this was important (as a defence against cavalry) and his definition of what was different (the 'science of tactics'). At the same time it ignores Herodotus' reasons for victory over the Persians (better armour, more skill), and the traditional Greek explanation (better men), and replaces it with a special formation necessitated by a shield with two handles. In my view, those shield handles are doing a lot of work here.

So rather than a hoplite revolution, there was more likely a long series of hoplite innovations, running in parallel with and closely related to political developments in Greek cities. There seems no reason to doubt that Archaic Greek infantry could form up in deep formations, could mass together into close order and could fight hand-to-hand using pure melee weapons. This is a type of fighting we see in Homer, and it was also common to armies throughout the Eastern Mediterranean and Near East. It is also the type of fighting for which the heavy hoplite equipment of large shield, body armour and helmet were developed. Such deep, relatively close formations could be called *phalanges*, or individually a phalanx. But there was at first no rigid structure of file and rank, no set intervals enforced and imposed across the formation, no defined position in the file in which each man always stood, no formal drills for transitioning between different depths and densities or for changing facing or formation (all the things that Aristotle termed the 'sciences and systems dealing with tactics'). Men could fight with a variety of weapons, including throwing spears, and archers and light-armed fighters could station themselves inside such formations, sheltering behind the large shields of the heavy-armed men. These were heavy infantry, not groups of skirmishers, but there were opportunities to move into and out of direct contact with the enemy. Formations probably fought across a narrow safe zone, with brave or aggressive (or well-equipped) individuals and their followers stepping forward to fight in short flurries, able to fall back to the safety of the mass behind them.

These early *phalanges* meet the modern definition of a phalanx, if by that we mean a wide, close-order, deep formation of heavy-armed infantry. But they are not yet identical with the more strictly ordered phalanx of the fifth and fourth centuries. A number of other developments were required to meet this stricter definition; more defined intervals, formal files and ranks, fixed positions in the ranks, formal drill, homogenous weapons specialized for hand-to-hand fighting, with the light-armed formed into separate bodies, and a fighting style in which the whole formation would close to hand-to-hand contact all at once, with little scope for individual prowess or display. These extra features were adopted early on by some cities – notably Sparta – and only partially or not at all by others, with some continuing to fight in a more traditional style, or with elements of the traditional style, right into the Hellenistic period. These features were also not unique to the Greeks, so far as we are able to tell, but some Greek armies became experts in their use. It is a mistake to see 'a phalanx' as being a single strictly defined entity that is the same every time it appears across several centuries. The heavy infantry formations of the Archaic period can correctly be called a phalanx, as can the (in some cases) more rigidly ordered and drilled formations of the Classical period. When one transitioned into the other varies considerably from place to place, and it may be impossible (and pointless) to argue whether one formation is a phalanx and another is not – there were many features to a phalanx, some, all or none of which may be present on any given occasion. Even by the time of the Peloponnesian War in the late fifth century, elements of earlier practice – in particular the disordered charge to contact of individuals rather than the steady advance of the mass, and some use of missiles – continued among many hoplite armies, as we will see in Chapters 6, 7 and 8, while other than by the Spartans, strict drills and formations were probably not widely adopted by most city armies. Close order, drill and formed ranks and files were frequently honoured more in the breach than in the observance.[44]

Indeed, it may be that Plutarch had a point in his criticism of Herodotus. Ancient Greek historians sometimes adopted technical rather than human explanations for military success, and Herodotus' emphasis on spear length and armour, just as much as modern emphasis on drill and tactics, may both conceal what was really distinctive about the Greek phalanx. What that was will form the subject of much of the rest of this book, but to provide a brief preview, we should observe that what changed during the Archaic period was not just equipment and tactics, but also the mindset of those who made up the phalanx. Homeric warriors, in their more fluid formations, could advance, strike and retire at will, and no shame was attached to them for retreating, for declining battle or even for returning to their tents for a rest (or a sulk, in Achilles' case) in the heat of battle. In Tyrtaeus, we see exhortations to advance to the front and stand toe-to-toe with the enemy, which suggests that this was still optional, or at least avoidable. But by the fifth century, we see a fetishization, not so much of discipline (in the sense of obedience to orders, as is often the case in better-organized armies) but of bloody-minded stubbornness, of standing one's ground, as reflected in Herodotus' tale of the Spartans at Plataea. Pausanias, the Spartan regent and overall Greek commander, ordered the Greek army to withdraw to a more advantageous position:

> 'Thereupon, all the rest of the captains being ready to obey Pausanias, Amompharetus son of Poliades, the leader of the Pitanate battalion, refused to flee from the barbarians or (save by compulsion) bring shame on Sparta; the whole business seemed strange to him, for he had not been

A representation of combat between two ordered phalanxes.

present in the council recently held. Pausanias and Euryanax were outraged that Amompharetus disobeyed them. Still more, however, they disliked that his refusing would compel them to abandon the Pitanate battalion, for they feared that if they fulfilled their agreement with the rest of the Greeks and abandoned him, Amompharetus and his men would be left behind to perish. Bearing this in mind, they kept the Laconian army where it was and tried to persuade Amompharetus that he was in the wrong ... Though Euryanax and Pausanias reasoned with Amompharetus, that the Lacedaemonians should not be endangered by remaining there alone, they could in no way prevail upon him. At last, when the Athenian messenger came among them, angry words began to pass. In this wrangling Amompharetus took up a stone with both hands and threw it down before Pausanias' feet, crying that it was the pebble with which he voted against fleeing from the strangers (meaning thereby the barbarians). Pausanias called him a madman.' (Herodotus 9.53–55)

The historicity of this episode has been doubted, but even if it did not take place exactly as described, it does give an insight into the mentality of the most drilled and disciplined of all Greek armies of the time. Far from displaying unquestioning obedience to orders, Spartan officers were prepared to argue a point with their commanders; the driving motivation, if Amompharetus' attitude is to be taken as typical, was in all circumstances to give not an inch of ground (literally and figuratively). Greeks were certainly not the first nor the only infantry to be stubborn and courageous in battle, or to give their lives in pursuit of what they saw as their duty, but they do seem on occasion to have carried this to an unusual extreme. Phalanx fighting, in its ideal form, was not characterized by a particular set of weapons and armour, or of tactics and style of fighting, but rather by a particular attitude, a dogged determination to stand and fight alongside one's fellows, and not to give way at any cost. This is exemplified in the criticism of the hero Heracles that the playwright Euripides gives to one of his characters:

> 'Are these your weapons for the hard struggle? Is it for this then that Heracles' children should be spared? A man who has won a reputation for valour in his contests with beasts, in all else a weakling; who never buckled shield to arm nor faced the spear, but with a bow, that coward's weapon, was ever ready to run away. Archery is no test of manly bravery; no! he is a man who keeps his post in the ranks and steadily faces the swift wound the spear may plough.' (Euripides, *Heracles* 155–164)

Standing firmly alongside one's fellows and face-to-face with whatever death or wounding might ensue were the hoplite ideal. The hoplite phalanx calls to mind Tennyson's words:

> 'One equal temper of heroic hearts,
> Made weak by time and fate, but strong in will
> To strive, to seek, to find, and not to yield.' (Tennyson, *Ulysses*)

So what was distinctive about the Greek phalanx depended primarily not on weapons, drill or tactics, and Greeks excelled on the battlefield not due to their advantages in these areas. Rather, the Greeks' successes are due to cultural and behavioural differences, an attitude of stubborn determination to fight to the finish alongside one's fellows and (more or less) equals. This was hardly revolutionary, since many militarily successful peoples had a combination of high self-esteem and an enthusiasm for fighting in their preferred style, but it did make the Greeks particularly successful in the conditions that applied in the Eastern Mediterranean world between the seventh and fourth centuries. Greek hoplite equipment developed around the start of the seventh century and went through a slow process of adoption, perfection and eventual alteration, in which only the heavy shield remained constant. Greek hoplite tactics were a modification of standard infantry tactics common throughout antiquity, but refined in favour of an emphasis on static hand-to-hand fighting, close order and a degree of formal organization, at different rates in different cities, reaching a more or less final form (which still saw many regional variations) in the fifth century. As they developed their equipment and ways of fighting, the Greeks developed a military culture which proved particularly suitable and successful for the type of battles in which they engaged, and it was these human qualities rather than the technical ones which were chiefly responsible for their success.

Chapter 2

Arms and Armour

The equipment and tactics of the phalanx appear to have developed gradually or at differing rates over the eighth to fourth centuries (and beyond, since equipment continued to be modified into the Hellenistic period). Some items of equipment appeared early and remained largely unchanged throughout this period, most notably the shield, while armour and weapons underwent a greater process of change. The poet Alcaeus (seventh to sixth centuries) provides a vivid early image of a well-equipped household with its arms and armour hung up for display:

> 'The great house gleams with bronze; the whole ceiling is adorned for war with bright helmets, down from which horsehair crests nod, ornaments for the heads of men. Bright bronze greaves conceal the pegs they hang around, a fence against the strong missile. New linen corselets and hollow shields are cast down; beside them are Chalcidian swords; beside them are many belts and tunics.' (Alcaeus 140 V, fr.19)

First of all, we will look at arguably the hoplite's most distinctive piece of equipment, his shield.

The shield

As we saw in the previous chapter, the oft-repeated claim that the hoplite's shield was called a *hoplon* and gave the hoplite his name appears to be incorrect. The word *hoplon* was used on occasion to describe the shield, but it does not appear to have been – at least in the period in which the hoplite was active – a specific name for the hoplite shield, which was instead known by the generic term *aspis*, shield, sometimes qualified as an Argive *aspis*. One piece of evidence which is sometimes adduced for the shield being called a *hoplon* is from Xenophon:

> 'While [the Spartan King Cleombrotus] was on the homeward way, however, an extraordinary wind-storm beset him … it hurled down the precipice great numbers of packasses, baggage and all, while very many *hopla* were snatched away from the soldiers and fell into the sea. Finally many of the men, unable to proceed with all their *hopla*, left their *aspides*

behind here and there on the summit of the ridge, putting them down on their backs and filling them with stones ... on the following day they went back and recovered their *hopla*.' (Xenophon, *Hellenica* 5.4.17–18)

Certainly, *hopla* in this case is used to indicate shields (only shields would offer enough wind resistance to be snatched away, and specifically *aspides* are abandoned and are the *hopla* recovered next day). But it seems impossible to conclude from this that *hopla* means 'shields' – if it did, Xenophon would have no need to use the word *aspides* at all. Rather, this seems another case where Xenophon is using *hopla* to mean 'equipment' or 'kit', in general, and *aspides* when he wants to say 'shields' specifically. As a matter of style, Xenophon would vary the vocabulary used. Translating the above passage accordingly – 'a lot of kit was snatched away from the soldiers ... many of the men, unable to proceed with all their kit, left their shields behind ... on the following day they went back and recovered their kit' – expresses this just as well in English as in Greek. A distinction between shields and other *hopla* is made elsewhere by Xenophon:

'And now, while the enemy were advancing, Thrasybulus ordered his men to ground their shields [*aspidas*] and did the same himself, though still keeping the rest of his arms [*hopla*], and then took his stand in the midst of them and spoke as follows.' (Xenophon, *Hellenica* 2.4.12)

Despite this lack of a specific name for the hoplite shield, it was (as it developed up to the fifth century) of a common type, used throughout Greece and beyond, wherever Greek colonists settled on the Mediterranean coasts. It was adopted also by many of the peoples the Greek colonists encountered, including the Etruscans and Romans in Italy, becoming the common shield of the Eastern Mediterranean world. I will refer to this shield as 'the hoplite shield' for convenience, though it should be remembered that not all hoplites carried such Argive *aspides*, and that the Greeks themselves did not apply any such specific terminology.[1]

The first appearances of the hoplite shield are difficult to pinpoint precisely, but are generally dated to the years around 700. In the absence of surviving dated examples, we are dependent on depictions in art – which in practice means vase painting. The difficulty is that a hoplite shield is not obviously different from any other type of shield of the same shape unless the depiction is quite detailed, showing ideally the carrying arrangement (the two-handled grip) and the cross-sectional shape (bowl-shaped, with an offset rim). Otherwise, any sort of round shield, whatever the precise details of its shape, construction and carrying materials, will look much like any other. Greek art only really becomes developed enough to depict the sort of details that we require to identify the type of shield around the seventh century; the fact that the earliest depictions of hoplite shields appear around this time may then be due to this fact, rather than because this is when the shield reached this form. So it is usually assumed that the first vase depictions of what are clearly identifiable as Argive shields also mark the point of the first development of these shields, but this cannot be known with absolute certainty.[2]

Round shields are common in art throughout antiquity, naturally enough, since a round shape is an obvious shape for a shield. From the early seventh century, shields start to be depicted with blazons (painted or possibly relief pictures, probably intended to identify the bearer in the manner of Medieval European heraldic devices, and on which more below). Since these often need to be carried the right way up, it has been plausibly claimed that they indicate that the shield was carried with the double grip – which both requires the shield be a particular way up and keeps it that way, since the arm is held relatively rigidly. This is in truth a slightly weak argument, since the handle of any shield would determine which way it was held, but at any rate it is possible that such blazoned shields may represent the first appearance of hoplite shields as such.[3]

From the early seventh century, artistic depictions have developed to the point where hoplite shields can clearly and certainly be identified as such, from four particular features: their size and round shape, their rims, their convex cross-section and their two-handled carrying arrangements. A number of early to mid-seventh-century vase paintings depict hoplites in battle (of which the Chigi vase is one example – these are the source of much debate about the origins of the phalanx, as we saw in the previous chapter) and show what are unequivocally hoplite shields. Art of the sixth and fifth centuries contains very many more depictions, some showing aspects of construction in great detail, and allow the nature of the hoplite shield to be known with a great deal of accuracy.[4]

Some details cannot be recovered from such depictions, however, particularly the materials used and internal details of construction. For such details, a better source of information are the actual surviving examples discovered by archaeology. These consist mainly of shields dedicated at Olympia as spoils of war, and consist almost entirely of the bronze covers of the shields (and occasionally bronze internal fittings such as the armband), the wooden and leather parts having long-since decayed. To these examples from Olympia

Shield interior and armour, early fifth century. (*Royal Museums of Art and History, Brussels*)

can be added a few others, perhaps most interestingly, as it can be connected to a specific historical event, the Spartan shield captured by the Athenians on Pylos (424) and now in the Agora Museum in Athens. The very best-preserved example, however, is not Greek at all but Etruscan (Italian) and now in the Vatican Museum in Rome (the Bomarzo shield). This preserves extensive elements of the shield's wooden core, and comparison with other more fragmentary examples suggests that it was typical of how such shields were constructed. Some caution is necessary, however; as with all ancient military equipment, there was no enforced uniformity, not even within armies and certainly not between different cities and different geographical regions, nor across the four centuries such shields were in use. There must have been considerable variation of materials, construction, weight, size and (in detail) shape, much of which is now lost to us, reliant as we are on a handful of fragmentary examples and mostly not very detailed artistic depictions. It is common for books about the Greek hoplite to offer a description of the typical hoplite shield (and this one will be no exception), but it must be kept in mind that, while the overall shape and form did remain remarkably constant across the period, there would have been innumerable variations in detail, and any description of a typical hoplite shield cannot be taken as completely accurate for every shield that existed.[5]

With that disclaimer in mind, the typical hoplite shield was around 90cm in diameter – examples exist ranging from around 80cm up to an enormous 125cm. Bearing in mind that most diameters are estimated from artistic depictions (surviving examples mostly being crushed or deformed to some extent, making estimates of their original diameter difficult), and that shields presumably varied somewhat according to the size of their bearer (since the length of the forearm would ideally define the radius of the shield, and naturally varies from man to man), it is fair to say that shield diameter was usually around 90cm, plus or minus 10cm. It is generally convenient to assume a diameter of 1 metre, which is close enough to typical and simplifies calculations of file widths as we will see in Chapter 3. Note incidentally that no ancient literary source ever states the size of the shield, nor provides any details of its construction; in this the hoplite shield differs from its successor, the Macedonian shield, for which the Hellenistic tacticians do give dimensions and suggestions as to shape and construction (though not, it has to be said, uncontroversial or especially helpful ones). For the hoplite shield, despite its four-century period of existence and ubiquity in the Greek and Italian world, we are dependent entirely on passing hints in literary sources and on the artistic and archaeological record.[6]

So the shield was round and about 1 metre in diameter. Its next most distinctive feature was its cross-sectional profile: it had the shape of a shallow bowl, with a flattish central surface, turning back sharply to produce a pronounced perimeter wall. Around the outside circumference was a rim, parallel to the central surface and more or less perpendicular to the walls of the bowl. The purpose, if any, of these features will be considered below. The wall of the bowl was, so far as we can tell from the surviving examples, the thickest part of the shield; the wood of the central section of the bowl may have been around 0.5–1cm thick, with the walls 1–1.5cm. The rim was around 4cm wide, and the bowl of the shield had a depth of around 10cm. This basic structure was made from wood, though precisely of what construction (solid wooden planks or layered laths, again presumably with no great consistency) we cannot be certain, and there is no reason to suppose that it was made exactly the same way every time – no doubt there were experiments in construction and personal preferences, for all that the final shape turned out very similar. The shield was apparently turned on a lathe, at least on occasion, as the comic poet Aristophanes refers to the (presumably imaginary) profession of *torneutolruaspidopegoi*, 'turners-of-lyres-and-shields' (*Birds*, 491). It is, however, most unlikely that the shield was lathed out of a solid block, and more likely that the lathe was used to give the final finish to a shape built up from component planks or laths.[7]

The type of wood used is also not known (and, again, was unlikely to have been consistent throughout the range of the shield's construction). Wood from outstanding examples (the Bomarzo and Basel shields) has been identified as poplar and willow, respectively, and the Roman natural historian Pliny (*NH* 16.209) records that such woods were the best for making shields in his day (first century AD). On occasion, 'willow' seems to be used as a word for shield (much as we might use it as a word for a cricket bat). The advantage of these types of wood is that they are relatively light, resist splitting well and also tend to close up around any penetration, an obviously useful quality in a shield.[8]

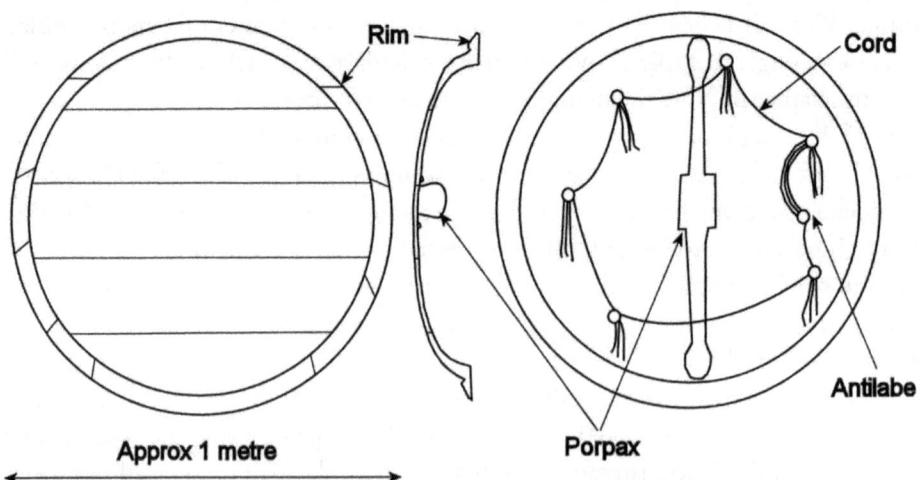

The parts of a shield (after Blyth (1982)).

The outside surface of the shield was usually covered with a thin sheet of bronze. Whether this was always the case cannot be judged with certainty – painted depictions of shields on vases might be plain wood, painted (possibly painted a solid colour with the blazon painted on top) or perhaps covered with linen (which would help protect the wood from splitting), which would also be whitened and painted. However, the fact that so many extant examples do have bronze covers (indeed often now consist solely of the bronze cover), and that there are many references to sunlight flashing off shields, or shields being used to send signals, makes it very likely that it was usual for the shield to be bronze-covered. The layer of bronze would have been very thin – perhaps 0.5–1mm – which would have kept the weight down, and would not in itself have provided very significant protection against penetration. It would, however, have protected against splitting, and of course would have looked impressive. The rim was also bronze-covered, as well as often reinforced with extra wooden material, which would help it to resist blows from edged weapons which might otherwise have separated the components of a plank- or lath-constructed shield. The inside of the shield might also be covered in leather, again not adding much to protection, but looking decorative and providing further protection against splitting.[9]

The overall weight of such a shield is another great unknown, and the subject of considerable debate. Figures (based on study of the Bomarzo shield) as high as 7–8kg have been quoted, or (based on modern reconstructions) as low as 5.5kg. Of course the important point is not the absolute weight but the degree of encumbrance experienced by the hoplite. It is very difficult to judge how difficult a piece of equipment would be to handle based only on such raw figures. I do not have any clear feel for how difficult a shield of 7kg (taking that as an average) would be to handle, since it depends on the carrying arrangements and also on the carrying position, and (perhaps more importantly) on the position when in combat, the length of time it needed to be wielded (rather than just carried) and the amount of motion required of the shield when it was in use (whether it was swung vigorously from side to side to ward off incoming blows, or was simply held statically forward as part of a continuous, and largely immobile, shield wall). The answers that scholars have come up with to these questions tend, unsurprisingly, to depend on their position in the hoplite debate and on their view of the nature of hoplite combat. Revisionists and gradualists with a belief in open-order phalanxes with individual duelling find shields to have been generally lighter and their encumbrance level not very high, while supporters of the orthodoxy find that the shield was exceptionally heavy and awkward, and that hoplites were restricted by their shields to a static, immobile form of combat based on mass pushing, not individual fighting. Indeed, it is a central tenet of the orthodox view that the hoplite shield was so awkward to handle that as soon as it appeared, hoplites must have adopted (or must have already adopted) a particularly close-order, static form of fighting which would have greatly reduced the physical demands of using such an awkward shield.[10]

The question of the nature of hoplite combat is one I will return to in Chapter 8, so I will not cover it in depth here. For now, just considering the weight and encumbrance of the shield, it has to be said that many modern pronouncements as to the extreme awkwardness of the shield seem to be based on very little evidence, and often simply on a statement of the assumed weight of the shield (in its upper range, 7–9kg). Yet out of context it is very hard to draw certain conclusions from such figures. The most telling comparison is perhaps the draw weight of the Medieval longbow, which was frequently well above 100lb. Such a weight is triple the typical draw weight of a modern sports archery bow, and might seem inconceivably high to a modern library-bound scholar. Yet we know such bows were used and drawn, and specialist modern archers and re-enactors are able to handle such bows, with suitable training and practice. So we might turn to modern hoplite re-enactors, of whom there are a good number throughout Europe and North America, and experimental archaeologists, who have constructed hoplite shields to high levels of accuracy, and experimented with wielding them. Unfortunately, however, the results tend to be contradictory (some re-enactors find the shield easy to handle, while others report it as very awkward), and are also heavily dependent on the physical fitness and training of the individual and on their expectations (it is likely that people in the ancient world generally were able to endure levels of physical hardship well beyond what would be considered reasonable to twenty-first-century Europeans or Americans).[11]

We might hope to look instead to ancient evidence for the difficulty of handling the shield. Certainly, there is evidence that it was considered heavy. Xenophon records the punishment of a Spartan officer, who:

> 'was compelled ... to stand sentry, carrying his shield – a thing which is regarded by Lacedaemonians of character as a disgrace; for it is a punishment for insubordination.' (Xenophon, *Hellenica* 3.1.9)

This seems to me weak evidence that the shield was exceptionally or unusually heavy. Standing on sentry duty holding any object would be tiring (and it is not recorded for how long or in what temperatures), and the usual preference would probably be to put the shield on the ground propped against the legs, an 'at ease' position which is recorded elsewhere. We can well imagine that holding a shield of any weight would be punishing after an hour or two, and the fact that Xenophon notes that the punishment was considered disgraceful also suggests that the corporal aspect of the punishment was not necessarily paramount (it is perhaps more akin to being forced to stand at attention in a public place than to being flogged).[12]

Another strand of evidence is the very frequent occurrence of *rhipsaspia*, the act of throwing away one's shield in order to run away from the field of battle more quickly. Certainly, this act was common (and severely frowned upon in many cities), but it need come as no surprise that a large wooden shield was an encumbrance to a man wanting to run for his life, and that it was possible to run faster without a shield than with one. The whole dynamic of combat between hoplites and their lighter-armed tormentors, particularly in the late fifth and fourth centuries (see Chapter 7), depends on the fact that hoplites were unable to chase and catch peltasts or *psiloi* who were not encumbered by shields or armour. This does not, however, tell us much about how difficult the shield was to wield in combat, where running at top speed was not a requirement and where both sides were similarly encumbered, and there is plentiful evidence that hoplites certainly could run, both from charges at the run in battle – which Herodotus tells us first happened at Marathon (490, Hdt. 6.112) – and from the fact that the 'race with arms' (*hoplitodromos*) was an event at the Olympic games from early times. Certainly, the shield would have slowed down the hoplite, compared to an unencumbered man, but if running was still a possibility, we need not doubt that vigorous fighting was too.[13]

Another line of argument is comparison with the shields of other periods, in antiquity and later. We know too little about earlier shields – which were mostly made of perishable materials – to be able to make meaningful comparisons, but the later Roman shield is well known and much studied. The weight of the best-surviving example, an Imperial Roman (third century AD) shield discovered at Dura-Europos in modern Syria, has been estimated at around 5.4kg without its central metal boss (which would increase the weight by maybe half a kilo or so). Republican Roman (third to first century) shields, of which a possible example was discovered at Kasr el-Harit in Egypt, are estimated to have weighed from 6–10 kg. These shields were made from wooden strips, as were at least some hoplite shields, and though they lacked the reinforced bowl and rim and the bronze cover of the hoplite shield, they were generally larger in size and had a central metal boss. They were also held in one hand and were specifically said by Polybius to have been used for active, mobile fighting ('as their method of fighting admits of individual motion for each man – because he defends his body with a shield, which he moves about to any point from which a blow is coming' (Pol. 18.30.7), in contrast to the Macedonian phalangite), so it is evident that this sort of weight was no obstacle to such fighting.[14]

Later peoples also used shields similar in size and construction to the hoplite shield. The Vikings and their contemporary enemies around the eighth to tenth centuries AD made use of round shields constructed from wooden planks, with leather coverings and a central metal boss. Modern reconstructions of such shields weigh around 5–7kg. They too were held in one hand by a central handgrip. Viking fighting techniques are not especially well understood, and did include the quintessential static defensive tactic, the 'shield wall', but modern re-enactors have no difficulty using such shields in much more mobile fighting, and it seems generally implausible that the Vikings, shipborne fast raiders as they were, were particularly encumbered by their shields.[15]

Taking more modern comparisons, attempts have been made to draw lessons from riot police shields – those of the Danish riot police being taken as an example, weighing in at a very modest 2.74kg. This was apparently considered too heavy and awkward for anything but static defensive use, but given the examples above, it seems that this must be more a question of differing uses and the expectations of their users rather than absolute physical necessity; the requirements of riot policing are probably not close enough to the realities of ancient combat to make a valid comparison. In terms of the raw weight of equipment, around 5kg is typical for the weight of a modern combat rifle; the SA80, standard rifle of the modern British Army, weighs 3.82kg empty, going up to 6.58kg loaded and with telescopic sight fitted. The Lee-Enfield, standard rifle of the British Army in the First and Second World Wars, weighed around 4.19kg. For total encumbrance, the modern infantryman often far exceeds anything known in the ancient world, with British infantry carrying weights in Operation *Herrick* (Afghanistan) of up to 56kg, a weight that often needed to be carried into combat, not just on the march. Modern armies recognize the disadvantages of such great weights and are working to reduce them, but even so, typical combat carried weights approach 20kg. Outside the military sphere, we might note for example that the 1953 expedition carried packs weighing some 20kg to the summit of Everest.[16]

So it seems unlikely that the hoplite shield – even at the upper end of its weight estimates at 9kg – was considered dramatically more heavy and awkward than other heavy infantry shields of antiquity. Certainly, it was much heavier than deliberately lightly constructed contemporary shields, the *peltai* of the light infantry, which were smaller and also made from lighter materials (probably wicker or leather over a wooden frame), and was also probably heavier than contemporary Persian shields, which though large in size were also made from wicker. A man could run faster without his shield than with it, and light infantry could run faster than heavy (as we would expect), but there is nothing about the construction or weight of the hoplite shield to suggest that it was unusually difficult to wield or that special fighting techniques (putatively, close-order, static and depending on mass shoving) would have been required to use it.

What was more unusual about the hoplite shield was its double-grip method of carrying and wielding in combat. Most shields, and certainly the Roman and Viking examples considered above, were held in one hand (the left hand) by a central handle or grip, which was often protected by a metal boss. The hoplite shield instead had a ring (*porpax*) in the centre, often made of bronze, through which the left arm was thrust up to the elbow, with the hand then grasping a handle (*antilabe*) near the edge of the shield, which was frequently extended by a length of rope running around the inner circumference of the shield bowl. Concerning nomenclature, this combination of *porpax* and *antilabe* is in fact known only from one, late source, the geographer Strabo (late first century BC to early first century AD), who when describing the Lusitanians of Spain noted that:

> 'The small shield they make use of is two feet in diameter, its outer surface concave, and suspended by leather thongs [*telamosi*]; it neither has rings [*porpakas*] nor handles [*antilabas*].' (Strabo, *Geography* 3.3.6)

This is a late reference – a century or two after the hoplite shield had fallen out of use – and one not directly referencing Greek shields at all. Nevertheless, these terms have been universally accepted as applicable to the hoplite shield (even though they do not appear in that context in the pages of Herodotus, Thucydides or Xenophon), and probably reasonably so. The *porpax* at least is known from a few references in drama, and it was evidently normal practice to decommission shields in storage, or those seized from enemies, by removing the *porpax* so they could not be used by rebels, slaves in revolt or upstart democrats:

> 'Sausage-Seller: Of the shields [*aspisi*]! Hold! I stop you there and I hold you fast. For if it be true that you love the people, you would not allow these to be hung up with their rings [*porpakas*]; but it's with an intent you have done this. Demos, take knowledge of his guilty purpose; in this way you no longer can punish him at your pleasure. Note the swarm of young tanners, who really surround him, and close to them the sellers of honey and cheese; all these are at one with him. Very well! you have but to frown, to speak of ostracism and they will rush at night to these shields, take them down and seize our granaries.
>
> 'Demos: Great gods! what! the shields retain their rings [*porpakas*]! Scoundrel! ah! too long have you had me for your dupe, cheated and played with me!'
>
> (Aristophanes, *Knights* 846–858)

It appears that the *porpax* and *antilabe* could also collectively be given the name 'handles' (*ochanon* or *ochane*). As we have seen, Herodotus notes that the Carians first invented the *ochanon* as a way of carrying the shield:

'[I]t was the Carians who originated wearing crests on their helmets and devices on their shields, and who first made handles [*ochanes*] for their shields [*aspisi*]; until then all who used shields carried them without these handles, and guided them with leather straps [*telamosi*] which they slung round the neck and over the left shoulder.' (Herodotus 1.171)

Much later, the Christian historian Julius Africanus, in the second to third centuries AD, says of Classical Greek equipment:

'For the Greeks are fond of heavy, full armour: they have a double [crested?] helmet, a breastplate covered with scales, a concave bronze shield held by two handles [*ochanois*] (of which the one surrounding the forearm avails for pushing [*othismon*], while the other is grasped by the end of the hand), two greaves, a hand-held javelin, and a spear for hand-to-hand combat.' (Julius Africanus, *Kestoi* 7.1.10)

All of this would suggest that *ochanon*, *porpax* and *antilabe* could be used largely interchangeably; again, we are struck by the absence of a precise technical terminology in Greek, at least among later authors. To set against this, Plutarch describes the conversion of the Spartan army to Macedonian equipment in the late third century:

'[Cleomenes] thus raised a body of four thousand hoplites, whom he taught to use a *sarisa*, held in both hands, instead of a spear, and to carry their shields [*aspides*] by an *ochane* instead of by a *porpax*.' (Plutarch, *Cleomenes* 11.2)

Here an *ochane* is specifically contrasted to a *porpax*. Clearly, certainty is impossible, but it seems likely that *porpax* was a quite specific word for the rigid arm ring in the centre of the shield, with *antilabe* a more general word for the handle at the edge, that could also be applied to other types of handle; both could also be referred to as an *ochanon*, in the general sense of a handle of unspecified type, unless some specific contrast is being made, as by Plutarch. The implication of *ochanon* seems to be a more general handle made of perishable materials (likely leather), as suggested by one of Herodotus' tales of an Assyrian attack on Egypt:

'[The Assyrians] came there, too [to Pelusium], and during the night were overrun by a horde of field mice that gnawed quivers and bows and the handles [*ochana*] of shields, with the result that many were killed fleeing unarmed the next day.' (Herodotus 2.141.5)

Aeneas 'the Tactician' records the clandestine arming of a force with equipment made from woven osiers (wicker):

'And by day they wove other kinds of basketry, and by night they wove armour [*hopla*], such as helmets and shields [*aspidas*], to which they attached leathern and wooden handles [*ochana*].' (Aeneas 29.12)

A *porpax* was probably more specifically of bronze, and it would likely offer less flexibility but greater solidity (hence its value for pushing, on which see Chapter 8).[17]

We should also note that it is unlikely that the *telamon*, shoulder strap, was ever the only means by which a shield was carried. Shields in the *Iliad* are notable for being carried across the back, which could be especially useful when running away (or making a tactical withdrawal to one's tent). A shoulder strap would have made this possible by simply pulling the shield round to the back so that the strap passed across the chest. This would also have left both hands free and protected the back. But it is highly improbable that anyone ever advanced into combat with a shield just dangling from a strap, as it would have offered no rigidity, and would also have been impossible to manoeuvre to deflect blows or missiles, so pre-hoplite shields must also have had a handle, presumably a central handle like those of most other shields, by which they were wielded in combat. Herodotus' comments on the Carian invention would only make sense if what he meant was that they invented the double grip, not simply that they were the first to use shield handles at all, and that the innovation was to replace the combination of shoulder strap and single central handle with a combination of arm ring and edge handle. Note also that the idea that the Macedonian phalanx reverted to this shoulder strap is, I firmly believe, a modern invention with no basis in the evidence at all. Such straps might still have been used for carrying the shield (Argive or Macedonian) on the march.[18]

Another common feature of hoplite shields – not described in any literary source, but often depicted in vase paintings – is a cord running around the inside of the rim of the shield. Various attempts have been made to provide a practical explanation for this cord: that it might have been a strengthening cord to provide tension to the shield rim (though it is always depicted hanging loosely); that it may have itself formed the *antilabe* (sometimes it appears to do so, but other depictions show a separate handle); or that neighbours in a phalanx would have held each other's shield cords to keep the shield wall together (perhaps they did, but it is odd that Thucydides' account of Mantinea, for example, does not mention this, where it could have been highly relevant to his point about each man seeking protection from his neighbour). I think it likely that this cord was used either to hang the shield (in camp, for example, where shields are sometimes shown hanging from trees, and in the home, where shields were probably hung on the wall) or to carry the shield (on the back, and often by an accompanying servant) on the march – serving in other words the same role as the earlier *telamon* (but not used in combat). It might also, of course, have been purely decorative, a fossil remnant of the functional *telamon*, along with the tassels often shown hanging from it at its attachment points.[19]

One clue for the reason behind the change in grip – from *telamon* to *porpax* and *antilabe* – comes from Julius Africanus' comment that the forearm handle 'avails for pushing'. The subject of pushing in combat, so central to the hoplite debate, is one to which I will return in detail in Chapter 8. For now, we might note one other sort of shield said to have been useful for pushing, in Xenophon's fictional account of a battle between Egyptians and Persians:

> 'The Egyptians, however, had the advantage both in numbers and in weapons; for the spears that they use even unto this day are long and powerful, and their shields cover their bodies much more effectually than corselets and targets [*gerrai*], and as they rest against the shoulder they are a help in pushing. So, locking their shields together they advanced and pushed.' (Xenophon, *Cyropaedia* 7.1.33 f.)

Concerning such Egyptian shields, we might note that Herodotus records a charming local tradition in Libya (North Africa) where the village girls were set to fight each other with sticks and stones:

> 'Before the girls are set fighting, the whole people choose the fairest maid, and arm her with a Corinthian helmet and Greek panoply, to be then mounted on a chariot and drawn all along the lake shore. With what armour they equipped their maidens before Greeks came to live near them, I cannot say; but I suppose their arms were Egyptian; for I maintain that the Greeks took their shield and helmet from Egypt.' (Herodotus 4.180.3–4)

How this squares with the Carian origin of shield handles, and if there is any truth in either story, is unknown. Herodotus describes Egyptians in Xerxes' army with shields like Greek shields:

> 'The Egyptians furnished two hundred ships. They wore woven helmets and carried hollow shields with broad rims, and spears for sea-warfare, and great battle-axes. Most of them wore cuirasses and carried long swords.' (Herodotus 7.89.3)

The shields Xenophon describes Egyptians carrying at Cunaxa and in the *Cyropaedia* are said to reach to their feet (Xen., *Anab.* 1.8.9; Xen., *Cyrop.* 6.2.10), which suggests a different shape ('tower shields', rather than Herodotus' round shields). However, assuming Herodotus' Egyptian shields and Xenophon's are related, it suggests that the Greek shield was held the way it was because this brought the whole arm into contact with the shield ('shoulder' is a slightly too specific translation in the Xenophon passage above, the sense being more of 'arm', particularly 'upper arm'). A shield held in the hand could be swung and punched with, as the Romans did with their large hand-held shields, which could not so effectively be done with a shield strapped to a bent arm, but such a shield would allow for a hard and effective push with arm and shoulder, although at shorter range (as a handheld shield can be extended to the full length of the arm, but a shield on the forearm only half as far). This then may have been the purpose of the double-grip shield – to provide a more powerful push and a more solid defence against such pushing. Early hoplites fighting in increasingly dense masses may have found this preferable to the more flexible, mobile shield use of the earlier handheld shield, particularly given that they were mostly fighting opponents similarly armed and organized.[20]

As we saw in the previous chapter, the Argive *aspis* did not immediately and universally supplant the other main heavy infantry shield in use in Archaic Greece, the Boeotian shield, a development of the earlier *dipylon* shield. This tended to have a more elongated shape than the Argive shield, rather than being round, and is distinguished by two 'cutouts' on each side, the purpose of which, if any, is not clear. That on the right side of the shield would have allowed the spear to be projected forward through the hole, but this would have been of doubtful efficacy (it would have rendered both shield and spear immobile), and does not explain the hole on the left of the shield. It may be that these shapes are simply decorative, a hangover from the earlier shield which in turn may have developed from the Bronze Age figure-of-eight shield, and arose from its construction around a crossing pair of wooden stretchers, which gave the shield a narrow 'waist'. The construction details of these shields are unknown, since none have survived and they are known only from depictions in art. They may have been built from wooden strips or planks, like the Argive shield, or might have been formed from hide stretched over a frame, like earlier shields. They do at any rate appear to have a flatter profile, without the characteristic bowl shape or rim of the Argive shield. The holding arrangements are also not clear. Some depictions which show the inside have a double grip just like the Argive shield, with the forearm running down the long axis of the shield, which, if the shield was longer than an Argive shield was wide, would have been awkward in a close formation unless the arm was held out straight. The shield might also have been held in the hand. Some scholars doubt the reality of this shield, seeing it as an indication that the subject is mythical or heroic and not representing real contemporary shields, but as all other equipment in such paintings does tend to be accurate (so far as we know), and hoplites are sometimes depicted with a mix of Boeotian and Argive shields, this seems unfounded. There is no obvious reason to doubt the reality of the Boeotian shield, and its frequent depiction on Boeotian coins makes most sense if it was a real and familiar type of shield. The appellation 'Boeotian', though, is a modern one, there being no written record of any special name for such a shield, which was presumably just called, like other shields in general, an *aspis*.[21]

The method of wielding the Argive shield has already been partly defined by the discussion of the handles: it was held on the left arm with the elbow typically bent at 90 degrees, resulting in the shield extending from about neck to upper thigh, and on the left of the body (resulting in the familiar concept of the 'unshielded side', as we will see in Chapter 7). Standing facing forward with the shield on the left arm, it would have faced perpendicular to the hoplite's front, so that the edge was presented to the enemy. This side-on view is common in vase paintings, though it is very likely that depictions in art preferred such a side view because it was easier to

Hoplite with Boeotian shield, late sixth century. (*Berlin Antiquities Collection*)

draw, presenting the shield as a circle, and allowed the shield blazon to be seen. Depictions of hoplites in single combat, which are extremely numerous, show a wide variety of dynamic poses, often with the arm extended forward so that the shield is in advance of the body, though still perpendicular to the direction of facing of its bearer. The most common way of holding the shield among modern re-enactors and in artistic depictions is to half rotate the body to the right and bring the left arm in front of the bearer, so that the shield is facing directly forwards and the bearer is more or less behind its centre. This is not a posture commonly seen in ancient depictions, but again this could be because it leaves the shield edge-on to the viewer, which is less appealing artistically. There are sixth-century depictions of side-on shields, though the way they are held is often not very clear and seemingly artists had not quite worked out how to depict them this way. Another very common way for the shield to be held is directly in front, but with the top edge leaning back against the left shoulder and the bottom edge protruding forwards – this is a common way of holding the shield in vase paintings and also in sculptural representations (where the sculptor is freed from the artist's difficulties of representing the shield as viewed from a single angle, so perhaps could depict it more realistically). This bottom-forward, top-back hold for the shield appears much more frequently than the

Arms and Armour 69

Hoplite fighting Amazon, demonstrating slanted shield hold, fifth century.

flat hold assumed to be typical by modern practitioners, and is relevant to the question (to be examined in Chapter 8) of shield clashing and pushing.[22]

The comment of Thucydides referred to above, concerning the Battle of Mantinea (418), is also relevant to the way the shield was carried (as well as to the formation of the phalanx, to which I will return in the next chapter):

> 'All armies are alike in this: on going into action they get forced out rather on their right wing, and one and the other overlap with this their adversary's left; because fear makes each man do his best to shelter his unarmed side with the shield of the man next him on the right, thinking that the closer the shields are locked together the better will he be protected. The man primarily responsible for this is the first upon the right wing, who is always striving to withdraw from the enemy his unarmed side; and the same apprehension makes the rest follow him.' (Thucydides 5.71.1)

I think that this has been taken rather over-literally by some scholars. It need not mean that each man's shield only covered half his body as seen from the front, with the other half unprotected unless he sheltered behind the protruding half of his neighbour's shield. Rather, the shield would be easier to use to block on the left side, and could be swung only to a limited extent to the right, and naturally it totally covered the left arm, while the right arm was unprotected. It is clear that a hoplite's 'unshielded side', frequently referred to in literary accounts, means his right side, not the right half of his front. In these circumstances, a hoplite would naturally feel more nervous about attacks from his right, and his desire to stick closer to his right-hand neighbour would be a natural attempt to prevent any enemy attacking from that direction, rather than an attempt to actually shelter behind his neighbour's shield. Based on this passage, there is no reason to suppose that each man's shield literally covered his left-hand neighbour, just that each man would not want his right-hand side to be open to the enemy. How close a hoplite would need to be to his neighbour to feel close enough is a subject to which I will return. Each man therefore used his own shield to defend himself. Another comment often quoted in this context is that of Plutarch, talking of the Spartans:

> 'When someone asked why they visited disgrace upon those among them who lost their shields, but did not do the same thing to those who lost their helmets or their breastplates, he said, "Because these they put on for their own sake, but the shield for the common good of the whole line."' (Plutarch, *Moralia* 220a)

Again this should not be taken over-literally. The meaning is clearly not that the shield protected someone other than oneself, but that a proper continuous frontage of shields was important to the safety of the formation, with any significant gaps or disorder in the line being open to exploitation by the enemy.[23]

An innovation of the late sixth to early fifth centuries was the introduction of a skirt or apron, presumably of fabric or leather, that was attached to the bottom of the front of the shield so as to hang down and cover the legs. It seems clear that the purpose of this was to provide extra protection against missiles (it would have been of limited value in close-quarters fighting), and is a reminder that the shield was difficult to manoeuvre to intercept missiles coming in low. Such an apron would have given some of the advantages of the earlier tower shield or later body-covering shields like the Gallic *thureos* or Roman *scutum*, without adding greatly to the weight of the shield. We can imagine the apron might have been restricted to front-rankers (it never appears to be used by all hoplites in any depicted group) and that its use

Shield apron, early to mid-fifth century. (*Altamura Painter, British Museum*)

was prompted by the greater volume of archery encountered by the Greeks when they started fighting Persian armies.[24]

I have mentioned shield devices or blazons already. The earliest depictions of shields in art tend to show them undecorated, covered with raw cowhide, or perhaps studded with small rivets or lines. From the seventh century, when Argive shields start to be depicted, they are always shown with highly detailed shield decorations in a variety of forms (as noted above, Herodotus credits the Carians with this idea, rightly or wrongly). These may have been painted onto the shield, especially if shields were not covered with bronze but with linen or leather (the application of paint to metallic surfaces is not something that has received much study, to my knowledge, and it is not known how well Greek paints would have adhered to a bronze surface). A number of examples have been discovered in Olympia of bronze relief or moulded blazons – it is possible that these were stuck to the surface of a wooden or bronze-covered shield, although as these were dedicated in a sanctuary they may not be typical examples. A vast array of subjects are depicted in these blazons in Archaic art – they may be grouped roughly into animal subjects (lions, horses, bulls, snakes, scorpions), mythological (Gorgon's head, Pegasus), symbolic (eyes, cups, tripods) and abstract (starbursts, lines, whirls). Where units are depicted (as on the Chigi vase), there is no commonality of device shown, so it is unlikely that they are city or unit identifiers, at this stage at least, and are presumably personal emblems of the bearer, which may represent his house or family, or may just be personal choices. A story is told of a Spartan who painted a life-size fly on his shield, explaining that given how close he intended to get to the enemy, they would be able to see it (Plut. *Mor.* 234c–d), while Aeschylus devotes a fair chunk of his *Seven Against Thebes* (ll. 385–648) to detailing the different, terrifying shield devices of the Seven. In some cases the significance of the device is obvious (fierce lions, dangerous scorpions), in others less so. In the course of the fifth century, and perhaps prompted by 'friendly fire' incidents where hoplites on the same side accidentally killed each other (unsurprising where both armies are of the same ethnicity and with the same equipment, and there is no uniformity of dress), it became common to decorate shields with a device or single letter representing the city, the best-known examples being the club of Thebes and the lambda (Λ) of Sparta (standing for Lacedaemon). Whether these devices were picked out in bright colours, like Greek sculpture, or with a more simple palette (as they appear on vases, which usually have at most three colours) is also unknown.[25]

So, Argive shields were common throughout Greece from the seventh to the third century, and were also used in Italy and Asia Minor. The Boeotian shield seems to have fallen out of use around the start of the fifth century. The

Typical shield blazons, sixth century.

Argive shield might possibly be based on a shield form, or at least a way of carrying the shield, from Egypt or Asia Minor. The construction was robust, producing a shield that was heavy in comparison with the smaller, lighter shields of light infantry, but was not at all remarkable in comparison with other heavy infantry shields in antiquity, and may have been lighter than the Roman *scutum*, a shield that was considered perfectly suitable for individual, mobile action. Modern accounts of the hoplite shield that emphasize its extreme awkwardness and unwieldiness appear to be in error – the shield could be used in the many and various activities in which a hoplite would be required to participate. It is important to remember that pitched battle formed only a small proportion of the military activities of any hoplite, and minor irregular actions on campaign (as detailed so vividly and extensively in Xenophon's *Anabasis*) along with assaults on cities and actions aboard ship for the hoplites enrolled as marines, which would also have involved amphibious operations, would have been much more common for them. The shield must have been suitable for all these types of action, even if it was optimized for fighting in the phalanx, and it is absurd to argue as some have that the shield was so uniquely unwieldy that the hoplite would have been helpless outside of the phalanx. Certainly, the shield emphasized protection over mobility, and had some features, particularly the means of wielding it, that may have been specially designed with close-quarters fighting in mind. At the same time, other solutions to the problem of balancing protection against ease of use were also possible, hence the move in the fourth and third centuries to adopt lighter shields and equipment generally, as discussed in Chapter 9.[26]

Body armour

Body armour was used throughout the Greek Bronze Age and Dark Ages. Bronze Age armour has been discovered in graves, and could be remarkably all-encompassing and sophisticated. Particularly notable is the 'Dendra panoply', discovered in a Mycenaean grave in Dendra (in the hills above the Argive plain near Mycenae), which consists of multiple articulated plates and hoops of bronze, covering the warrior from a large neck guard down to a series of lobster-like plates around the groin and upper thighs. Such armour may have been intended for a chariot-borne warrior rather than an infantryman, and may, at several removes, have inspired the recurring theme in Homer of a warrior falling among a 'clatter of bronze' (and also explain the desirability of stripping fallen warriors of their armour, another recurring Homeric theme).

Hoplite armour as such makes its first appearance in Argos in the late eighth century. The 'Argos panoply' consists of a bronze 'bell cuirass' (*thorax*, the Greek word for body armour generally) very similar in appearance to those depicted on the Chigi vase (and many other Archaic artistic depictions), along with a bronze helmet to which I will return below. A 'bell cuirass' is an item of body armour, made in two pieces – a chest plate and a back plate – that completely covers the torso from throat to waist, leaving the arms uncovered. It is shaped to roughly match the torso, with a flare outwards at the waist (giving rise to the 'bell' comparison, and hinting that it might have originated as a horseman's armour, shaped to sit down in), and is decorated with moulded or incised features representing, in a highly abstract form, the chest and stomach muscles of the wearer. The two parts would be held together with laces or pins, and there might have been a felt or leather lining. The bronze is fairly thin, but would give decent protection against cutting blows or missiles, if not necessarily against strong thrusts from a pointed weapon. Rather surprisingly perhaps, no protection is provided below the waist for the lower abdomen or groin, and depictions of hoplites equipped with such armour show them bare-thighed and with no more than a tunic to cover the groin (wounds to the lower abdomen with extrusion of the intestines are a common cause of death in the *Iliad*). In terms of vocabulary, Greek writers seem to have used the general word *thorax* for all body armour without regard to the material or details of construction – sometimes it is specified to be 'bronze armour' (e.g. Hdt. 1.215.2 – in this case horse armour; Xen., *Cyrop.* 7.1.2), sometimes 'linen armour' (see below), and most often just 'armour', and we are left to guess how it was made, most often presuming it to be of bronze. The same word is used for cavalry armour (e.g. Xen., *Hell.* 7.2.21; Xen., *Anab.* 3.4.47) and infantry armour, and Persian as well as Greek (Xen., *Anab.* 1.8.3; 6; 26; Xen., *Cyrop.* 1.2.13). Modern translations variously render this as 'cuirass', 'corslet', 'breastplate' or 'armour', and I will stick with the neutral 'armour' here.[27]

Armour of this type is quite common among the finds in Olympia from the seventh century, and has been taken to be typical of that worn by hoplites in this period. It is also possible that this bronze armour represents only the top-of-the-range armour of the richest hoplites, and that the bulk of the rank-and-file would have been equipped with cheaper armour or no armour at all. Vase depictions do show all hoplites in bronze armour, but such depictions might have preferred to show only the best-equipped, or might have deliberately depicted the front-rankers (*promachoi*) of the upper classes and suppressed the role of the less well-equipped lower orders. It is also possible that only the best armour was dedicated at Olympia, leading to a disproportionate amount of top-quality armour, and no representation at all of cheaper common armour. Two possibilities arise from this surmise: that the early phalanx consisted of a mass of less well-equipped, poorer hoplites (still richer and better-equipped than the missile-throwing light

Archaic bell cuirass, seventh to sixth century (National Archaeological Museum of Spain), with Chigi vase.

infantry), fronted by a cutting edge of bronze-equipped richer men; or that the early phalanx consisted solely of the better-equipped men and was correspondingly smaller in numbers than the massed phalanxes of the fifth century. Whether either of these possibilities were true – or indeed the simple assumption that vase paintings depict the equipment of the mass accurately, and that the early phalanx was already large and uniformly well equipped – seems impossible to determine given the state of the evidence, although it has been noted that body armour is less common in the archaeological record than shields and helmets, which suggests that hoplites with no armour or with non-metallic armour may have been common.[28]

Whatever the case, there was some development of bronze armour in the sixth and fifth centuries, with the decorative musculature on the armour becoming steadily less stylized and more lifelike, and with a (fairly subtle) change in the shape, fitting closer to the figure of the wearer, with an elongated front section that gave better protection to the lower abdomen, though not to the groin. Such armour is generally known (today) as a 'muscled cuirass'. To protect the groin, a system of leather or fabric strips called *pteruges* ('feathers' or 'wings') was developed, either attached to the bottom of the armour (they often appear integral to metal armour when depicted stored rather than worn) or perhaps as part of a garment worn under the bronze armour. Usually consisting of two overlapping

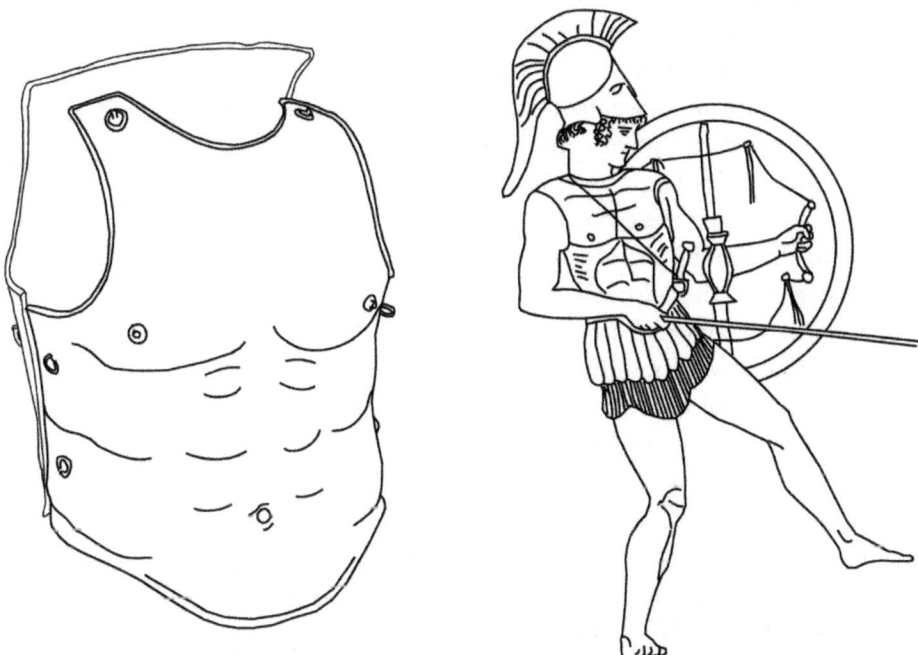

Greek muscle cuirass from Italy, fourth century, with mid-fifth-century kalyx crater.

Linen or leather cuirass, mid-fifth century. (*Achilles painter*)

layers of flaps (so that the upper layer covered the gaps in the lower layer), these provided flexible protection for the groin and upper legs without limiting movement. It may also be that some wore a skirt or apron of thicker cloth or perhaps leather, a *perizoma*, around the waist, protecting the groin area.

Perhaps developing from this, or perhaps independently, in the sixth to fifth centuries, bronze body armour was increasingly replaced by armour that was entirely made from leather or fabric. This would (presumably) have been cheaper to produce (so might perhaps always have been the armour of the masses), and also lighter, cooler and more flexible to wear; bearing in mind that pitched battle formed only a tiny proportion of a hoplite's time on campaign, it presumably provided sufficient protection in battle in return for the greater ease of wearing it the rest of the time. Such

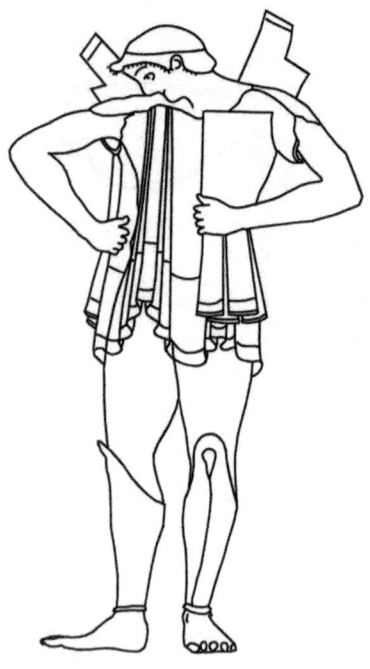

Linen cuirass, showing the method of fastening, end of sixth century. (*Antikensammlungen, Munich*)

armour is known only from numerous depictions in art, since, being organic, none has survived. It took the form of a tube of stiff material that was wrapped around the body and fastened on the left side (where it was protected by the shield), with two shoulder pieces (*epomides*) that in their natural state stuck up above the hoplite's head, and could be drawn down over the shoulders and tied to the chest. This construction leads to this type of armour being termed 'tube and yoke' by some modern authors. A set of *pteruges* would be integral to the tube section, extending over the groin and upper legs.[29]

The material used for this armour is never clearly identified, but it is known that linen was used for armour throughout the ancient Near East, and like other items of hoplite equipment, the Greeks could well have borrowed this armour from their experiences of mercenary service abroad. Homer mentions linen armour on a couple of occasions:

> 'Aias the less, in no wise as great as Telamonian Aias, but far less. Small of stature was he, with corslet of linen, but with the spear he far excelled the whole host of Hellenes and Achaeans.' (Homer, *Iliad* 2.529)

> 'Adrastus and Araphius, with corslet of linen.' (Homer, *Iliad* 2.830)

We also see linen armour in the poem of Alcaeus quoted at the head of this chapter. The word Homer uses, *linothorax* (or *linothorex*), means 'linen-corsletted' and refers to the wearer, not to the piece of armour itself. However, the word *linothorax* has been adopted in English as a noun for the piece of armour ('the linothorax'), and intense modern study, reconstruction and literature has been devoted to it. As is becoming familiar, there does not seem to have been a Greek word applied specifically to this type of armour – the general word *thorax* is used for all armour, and sometimes it is qualified as *lineos thorax* (or indeed *thorax lineos*), 'linen armour', 'armour of linen'. Where not so qualified, we cannot be certain whether armour of bronze, linen or some other material is meant. The frequency of references to linen armour, though, does make it more likely that linen was the material of choice for armour (rather than, say, leather), although there are questions about how widely available linen was in Greece.[30]

As with other items of hoplite equipment, linen armour was also used in other Near Eastern armies. Herodotus records a noteworthy Egyptian example:

> 'Moreover, Amasis [Pharaoh of Egypt 570–526] dedicated offerings in Hellas. He gave ... to Athena of Lindus, two stone images and a marvellous linen *thorax*.' (Herodotus 2.182.1)

A similar item was sent to the Spartans (Hdt. 3.47). The marvellousness of this armour lay in its gold decoration and the particularly fine and complex weave of the linen, which would of course not be typical of ordinary linen armour. Herodotus also records other peoples of the Persian Empire using linen armour:

> 'The Assyrians in the army wore on their heads helmets of twisted bronze made in an outlandish fashion not easy to describe. They carried shields and spears and daggers of Egyptian fashion, and also wooden clubs studded with iron, and they wore linen armour.' (Herodotus 7.63)

> '[T]he Phoenicians with the Syrians of Palestine [on the Persian ships] furnished three hundred; for their equipment, they had on their heads helmets very close to the Greek in style; they wore linen armour, and carried shields without rims, and javelins.' (Herodotus 7.89.1)

> 'But the Persians more than all men welcome foreign customs. They wear the Median dress, thinking it more beautiful than their own, and the Egyptian *thorax* in war.' (Herodotus 1.135)

We may guess that 'the Egyptian *thorax*' is of linen. Xenophon also suggests that linen armour was traditional Persian gear:

> 'And Abradatas's chariot with its four poles and eight horses was adorned most handsomely; and when he came to put on his linen *thorax*, such as they used in his country, Panthea [his wife] brought him one of gold.' (Xenophon, *Cyropaedia* 6.4.2)

Xenophon records (Xen., *Anab.* 1.8.9) that the Persian cavalry of Tissaphernes at the Battle of Cunaxa (401) were 'white-corsletted' (*leukothorakes*) – we may be justified in assuming that this means linen armour – and that the Chalybians of the Black Sea coast 'had corselets of linen [*thorakas linous*] reaching down to the groin, with a thick fringe of plaited cords instead of flaps [*pteruges*]' (Xen., *Anab.* 4.7.15). Alexander the Great (late fourth century) adopted a Persian linen *thorax*, according to Plutarch:

> 'After sending this message to Parmenio, he put on his helmet, but the rest of his armour he had on as he came from his tent, namely, a vest of Sicilian make girt about him, and over this a *thorax* of two-ply linen from the spoils taken at Issus [that is, captured from the Persians].' (Plutarch, *Alexander* 32.5)

So the Greeks may have adopted textile armour of this sort from their experiences in mercenary service abroad or their contacts in Asia Minor, or it may have been in continuous use in Greece from the time of Homer or before, with possibly new styles being introduced from abroad. At any rate, non-bronze armour came by the fifth century to almost completely replace bronze armour, at least among infantry, judging by depictions in art. We should note that the Roman author Cornelius Nepos (*Iphicrates* 1.4) attributed a move from 'bronze linked' armour to linen *lorica* (Latin for armour) to Iphicrates in the early fourth century, though our other account of Iphicrates' equipment reforms (Diod. 15.44) doesn't mention this, so it may be that Nepos is just attributing a general earlier move from bronze to linen armour to Iphicrates as a noted reformer. We cannot be certain that all such armour depicted in art is linen rather than some other material (see further on leather armour below), though linen armour is certainly favoured in most modern accounts of hoplite equipment. Cavalry may have continued to wear bronze armour, since before the fourth century they did not carry shields and so might have preferred a higher level of protection (plus, they did not have to carry the weight of the armour themselves).

The details of the construction of linen armour are unknown, though plausible hypotheses have been suggested. We have seen references to 'two-ply linen', and when the Roman Emperor Caracalla (second to third centuries AD) equipped a Macedonian phalanx, he used a three-ply linen *thorax* (Cassius Dio 78.7.2). This presumably refers to the number of layers of material used. The Alexander Mosaic, a depiction of a battle of Alexander the Great against the Persians, depicts the Persians in 'tube and yoke' armour that appears to be quilted (judging by the diagonal pattern on the material), though depictions in Greek art do not generally show evidence of such quilting; the main modern hypothesis for the construction of this armour is that it was made from glued layers of linen, which may well be so (though it remains only a theory). This produces a surprisingly stiff and hard armour – perhaps too stiff, since the whole needs to be flexible enough to wrap around the body and for the shoulder pieces to pull over the shoulders; modern reconstructions tend to be rather bulky and awkward-looking. This armour also performs well enough in tests that have been carried out by shooting or striking it with various reconstructed weapons. These tests have as outputs quantified results which I do not find very helpful. The tests are often not carried out rigorously or consistently enough to make the data very reliable, and we have no measure of what level of protection would have been required. Suffice it to say that evidently, from its widespread adoption, the protection was considered good enough, although ancient evidence is that the level of protection was (as we might expect) less than that of metal armour. Pausanius, speaking of the later Sarmatians, wrote:

> 'Linen *thorakes* are not so useful to fighters, for they let the iron pass through, if the blow be a violent one. They aid hunters, however, for the teeth of lions or leopards break off in them.' (Pausanias 1.21.7)

Pausanias goes on to note (1.21.6) that the Sarmatians reinforced their linen armour with horse-hoof scales, and we see examples in Greek art of armour reinforced with scales or plates of (presumably) metal armour. There is an enormous range of variations in the shape, quantity and positioning of these metal reinforcements, which presumably were up to the whim (and budget) of their owner and often took the form of panels protecting the side (especially the right-hand, shieldless side) or the chest.[31]

It is also possible that the 'linen' armour we see depicted in art was actually a form of composite armour, in which metal (bronze, or maybe iron) plates were sewn into the panels of a fabric or leather armour. This would in effect be an extended version of the insertion of the metal scales or plates, and may have been similar to the iron armour found in the

Composite (scale) cuirass from Baseggio Cup (after Sekunda (1992)).

'tomb of Philip II' (Macedonian, late fourth century). If so, there would be no way of telling visually when we are seeing pure linen, other fabric or leather armour, and when composite armour. In practice, there may have been little difference between them.[32]

This reminds us that armour would need to be made specifically for its wearer to ensure a good fit, both for comfort and protection. Xenophon comments on the cavalryman's *thorax*, with advice which must have applied also to that of the infantry:

> 'We say, then, that in the first place his *thorax* must be made to fit his body. For the wellfitting *thorax* is supported by the whole body, whereas one that is too loose is supported by the shoulders only, and one that is too tight is rather an encumbrance than a defence. As for the pattern of the *thorax*, it should be so shaped as not to prevent the wearer from sitting down or stooping. About the abdomen and middle and round that region let the *pteruges* be of such material and such a size that they will keep out missiles.' (Xenophon, *On Horsemanship* 12.1)

Xenophon also presents a dialogue of Socrates and an armourer which makes similar points, and is worth quoting for its military if not its philosophical interest:

> 'On visiting Pistias the armourer [*thorakopoios*], who showed him some well-made *thorakes*, Socrates exclaimed: "Upon my word, Pistias, it's a beautiful invention, for the *thorax* covers the parts that need protection without impeding the use of the hands. But tell me, Pistias," he added, "why do

you charge more for your *thorakes* than any other maker, though they are no stronger and cost no more to make?"

"Because the proportions of mine are better, Socrates."

"And how do you show their proportions when you ask a higher price – by weight or measure? For I presume you don't make them all of the same weight or the same size, that is, if you make them to fit."

"Fit? Why, of course! A *thorax* is of no use without that!"

"Then are not some human bodies well, others ill proportioned?"

"Certainly."

"Then if a *thorax* is to fit an ill-proportioned body, how do you make it well-proportioned?"

"By making it fit; for if it is a good fit it is well-proportioned."

"Apparently you mean well-proportioned not absolutely, but in relation to the wearer, as you might call a shield well-proportioned for the man whom it fits, or a military cloak – and this seems to apply to everything according to you. And perhaps there is another important advantage in a good fit."

"Tell it me, if you know, Socrates."

"The good fit is less heavy to wear than the misfit, though both are of the same weight. For the misfit, hanging entirely from the shoulders, or pressing on some other part of the body, proves uncomfortable and irksome; but the good fit, with its weight distributed over the collar-bone and shoulder-blades, the shoulders, chest, back and belly, may almost be called an accessory rather than an encumbrance."

"The advantage you speak of is the very one which I think makes my work worth a big price. Some, however, prefer to buy the ornamented and the gold-plated *thorakes*."

"Still, if the consequence is that they buy misfits, it seems to me they buy ornamented and gold-plated trash. However, as the body is not rigid, but now bent, now straight, how can tight *thorakes* fit?"

"They can't."

"You mean that the good fits are not the tight ones, but those that don't chafe the wearer?"

"That is your own meaning, Socrates, and you have hit the right nail on the head."

(Xenophon, *Memorabilia* 3.10.9–15)

Presumably (but not certainly) it is a bronze *thorax* being spoken of here. The general points are that ideally armour (and, note, the shield) was made to fit the individual, and that a well-fitting piece of armour was not an encumbrance to wear. On the other hand, not everyone could afford the best, and there must have been plenty of hoplites with second-hand, hand-me-down or cheap armour, for whom it would have been an encumbrance – one which they would just have to bear, unless they were to do without armour altogether.

For it is likely that some hoplites did go unarmoured: from the origins of the phalanx, there must have been variations in levels of wealth amongst hoplites, and we can be by no means certain that every man could have afforded the best (originally bronze) armour. For a man whose station was towards the back of the phalanx (once position in the phalanx became fixed, rather than up to each individual), there must have been little need for armour of any sort since he was unlikely to come into direct contact with the enemy (although outside of pitched battle, armour would still have been useful). Along with the general lightening of equipment into the fifth and fourth centuries, there is evidence for hoplites (that is, men armed with the hoplite shield and spear) who do not wear armour – a mix of economic necessity (with poorer men admitted to the hoplite class to make up numbers) and the physical convenience of being unhindered by armour may have meant that the shield was felt to be sufficient protection on its own. Identifying unarmoured hoplites in art – particularly, for example, funerary monuments – may be complicated by the fact that we cannot be sure whether the man is depicted in battle gear or in campaign or informal wear, in which the

Unarmoured hoplites on Attic tombstone, late fifth century. (*Berlin*)

armour might be dispensed with. Also, unlike in the later Macedonian army – where there were regulations stipulating what items of equipment were to be worn – there were no rules (that we know of) for Greek city contingents, with the exception of the shield, since the laws against throwing away the shield in flight imply also that a shield was required in order to fight. The protection provided by the shield was presumably considered a bare minimum, below which no hoplite could be allowed to fall (since that might imperil the whole formation), but which individuals were free to supplement according to their own needs and desires.[33]

Xenophon tells a story of a forced march to seize a hill by a picked group of men that may be indicative of such unarmoured hoplites:

> 'Then they set out with all possible speed. But no sooner had the enemy upon the hill observed their dash for the summit of the mountain than they also set off, to race with the Greeks for this summit. Then there was a deal of shouting from the Greek army as they urged on their friends, and just as much shouting from Tissaphernes' troops to urge on their men. And Xenophon, riding along the lines upon his horse, cheered his troops forward: "My good men," he said, "believe that now you are racing for Greece, racing this very hour back to your wives and children, a little toil for this one moment and no more fighting for the rest of our journey." But Soteridas the Sicyonian said: "We are not on an equality, Xenophon; you are riding on horseback, while I am desperately tired with carrying my shield." When Xenophon heard that, he leaped down from his horse and pushed Soteridas out of his place in the line, then took his shield away from him and marched on with it as fast as he could; he had on also, as it happened, his cavalry *thorax*, and the result was that he was heavily burdened. And he urged the men in front of him to keep going, while he told those who were behind to pass along by him, for he found it hard to keep up. The rest of the soldiers, however, struck and pelted and abused Soteridas until they forced him to take back his shield and march on.' (Xenophon, *Anabasis* 3.4.44–49)

Xenophon's force was a mix of peltasts and 'picked men' (hoplites), with Soteridas presumably one of the latter since he carried an *aspis* not a *pelta*. Clearly, Xenophon with his cavalry *thorax* (of bronze?) and shield was more encumbered than the other men with just their shields – these must have had lighter armour, or no armour at all. We can also recognize that bronze armour and a heavy shield would be a heavy burden for someone racing to the top of a hill, though this does not mean that either shield or armour were exceptionally heavy in general – Soteridas' colleagues managed well enough.

Xenophon on other occasions mentions an item of equipment called a *spolas*. The officers of the Ten Thousand proposed that they form a scratch force of slingers and cavalry:

'These proposals also were adopted, and in the course of that night a company of two hundred slingers was organized, while on the following day horses and horsemen to the number of fifty were examined and accepted, and *spolades* and *thorakes* were provided for them.' (Xenophon, *Anabasis* 3.3.20)

Presumably, rather than each man having a *spolas* and a *thorax*, a mix of *spolades* and *thorakes* were provided, one per man. On another occasion, an infantryman was killed:

'Then it was that a brave man was killed, Leonymus the Laconian, who was pierced in the side by an arrow that went through his shield and *spolas*.' (Xenophon, *Anabasis* 4.1.18)

The word is uncommon and it is hard to be certain of the meaning. It is sometimes translated as 'cuirass' or 'jerkin', and we will perhaps be not too far off in guessing that it means leather armour of some sort, perhaps in appearance similar to linen armour, since there are no candidates in art of non-metallic armour that are not usually currently identified as linen (the 'tube and yoke' form). Julius Pollux, a Greek dictionary compiler of the second century AD, records that 'the *spolas* is a *thorax* of leather, hanging from the shoulders, so that Xenophon says "and the *spolas* instead of *thorax*"' (*Onomast.* 7.70; it is not known where or if Xenophon says this). The word also occurs in the comic plays of Aristophanes, but its meaning is unclear there. For what it is worth, the second-century AD Roman author Aelian, writing in the tradition of the Hellenistic tacticians, notes (in one manuscript) that Hellenistic peltasts had lighter equipment than the hoplites, wearing 'a *stolas* in place of a *thorax*' (Ael. *Tact.* 2, and assuming a *stolas* is a variant on *spolas*); whether he means a *stolas* is lighter than a bronze *thorax* or than a linen one is not clear. Intriguingly, Aelian's words – '*stolas* instead of *thorax*' (*stolas anti thorakos*) – are the same as those attributed by Julius Pollux to Xenophon, except for the spelling of *spolas*.

The most revealing passage may be from Aeneas 'the Tactician' writing in the fourth century on the smuggling of arms into a city under siege:

'To the unarmed citizens who were to be accomplices there were brought in *thorakes lineoi*, *stolidia* and helmets, *hopla*, greaves, short swords, bows, arrows, stowed away in chests like those of merchants.' (Aeneas Tacticus 29.4)

Of the words I have not translated, *thorakes lineoi* are the linen armour we are familiar with, and *stolidia* (or *stolidas* in one reading) presumably are *spolades*. What the nature of the distinction was between *thorax lineos* and *spolas/stolas*, other than the material of construction, we do not know, and it may be that at least some – and conceivably most – of the armour now identified in art as linen (and a *thorax*) is really leather (and a *spolas*). We should also note here the use of *hopla*, which might mean shields here, since no other obvious item of equipment is missing, though shields are listed specifically a few sentences later 'in baskets of chaff and wool, *peltai* and small *aspidia*' (Aen. Tact. 29.6).

We have thus seen that body armour went through a process of change, from the simple bronze bell cuirass of the early Archaic, through more sophisticated bronze armour, to non-metallic armour of linen or leather, and perhaps in some cases to no armour at all. We must also remember that in all periods, as hoplites provided their own equipment (in most cases) and would only wear what they could afford, there must have been a good deal of variety. Consequently, it is possible that artistic representations, concentrating on presenting an appealing image and often depicting the combat of heroes, give a false impression of the uniformity and quality of armour worn. At the same time, Greek hoplites (and those equipped like them), at least in the early days (seventh to sixth centuries), undoubtedly had a reputation for being heavily armed and for being 'men of bronze', as related in a tale of Herodotus, where the Egyptian Pharaoh Psammetichus (Psamtik I, mid-seventh century), deposed from his throne, asked an oracle how he might regain it:

> 'The oracle answered that he would have vengeance when he saw men of bronze coming from the sea. Psammetichus did not in the least believe that men of bronze would come to aid him. But after a short time, Ionians and Carians, voyaging for plunder, were forced to put in on the coast of Egypt, where they disembarked in their armour of bronze; and an Egyptian came into the marsh country and brought news to Psammetichus (for he had never before seen [bronze] armoured men) that men of bronze had come from the sea and were foraging in the plain.' (Herodotus 2.152.3–4)

Psammetichus duly hired these adventurers as mercenaries (or volunteers with promise of reward). Greeks were by no means the first or only infantry in antiquity to wear bronze armour – note that the non-Greek Carians were apparently equipped the same as the Greek Ionians – but they may have worn more than most, and solid bronze body armour is uncommon in antiquity in general. This element at least is perhaps unique to the Greeks (and Carians) and those who copied their equipment. It should also be noted that while there was a steady increase (or according to the orthodox view, a sudden and sustained increase) in the degree of close-order and hand-to-hand fighting experienced by the phalanx, armour seems to have evolved in the opposite direction, becoming lighter. The theory that early adoption of very complete and heavy bronze armour is evidence for the early adoption of the fully developed phalanx seems undermined by the fact that armour then became progressively lighter, while still being used in the same close-order phalanx.

Helmets

Helmets are well represented in art and also in surviving examples, since they were invariably made from metal (bronze in the vast majority of cases) and were frequently dedicated in sanctuaries. This means that the styles of helmet available, and their development, are more clearly understood than is the case for body armour.[34]

A metal helmet had been an essential part of infantry (and often cavalry) equipment for centuries before the hoplite period throughout the Mediterranean and Near East (and beyond), and the earliest Greek helmets were similar in form to those of their neighbours, particularly in Asia Minor. The earliest helmet associated with hoplites, that of the eighth-century Argos panoply discussed above, was of a type known today as Kegel (or Kegelhelm), which developed into the Illyrian and was used throughout Greece and the Balkans (the Illyrians were a Balkan people in what is now the Albania–Croatia region). Note incidentally that most names for types of helmets are modern ones, sometimes based on where they are mostly found (as is the case for the 'Illyrian') or depicted, and in most cases we do not know how Greeks designated different helmet types, if at all. The Kegel helmet took the form of a conical cap on top of the head, with cheek pieces on each side to provide some protection to the sides of the face. Protection for the back of the neck was limited and the face was open, making the helmet fairly easy to wear but not providing great protection in a close frontal fight.

From the seventh century, the Kegel/Illyrian helmet was almost completely replaced by the quintessential Greek helmet, the Corinthian. This name does seem to have been applied by the Greeks themselves (as for the Argive shield), judging from, for example, Herodotus (4.180.3), and we may not go too far wrong in surmising that this helmet style was first developed in Corinth. The earliest Corinthian helmets were like a pot fitting over the entire head, with openings for the eyes and a narrow slit running down in front of the mouth. The helmet was (apparently) hammered from a single sheet of bronze, which must have been a highly skilled task given the complexity of the shape (especially in its later developments). It must have been excruciatingly hot and awkward to wear, although modern reconstructions show that visibility out of one was as good as or better than that of other helmet styles, since the relatively wide eye slits could allow better peripheral vision than the cheek pieces of Illyrian or later similar helmet styles. This helmet maximized, to an extent not seen in helmets of other infantry before (or since), forward protection to the face. It also completely covered the ears, making hearing difficult. It is unwise to attempt to draw strong conclusions about fighting techniques from helmet style, but the early Corinthian does suggest that its wearers (who, as with the best quality armour, may have been only a subset of the hoplites) were more concerned about close-range protection to the front than awareness of their surroundings or vigorous activity.[35]

The Corinthian helmet remained in use throughout the lifetime of the Greek phalanx, though its form developed considerably. The short nose guard of earlier helmets, that was little more than a tab between the eye holes in the earliest forms, was extended until it covered most or all of the nose, while the face pieces were extended downwards to provide some protection for the neck. From quite early on, a flare

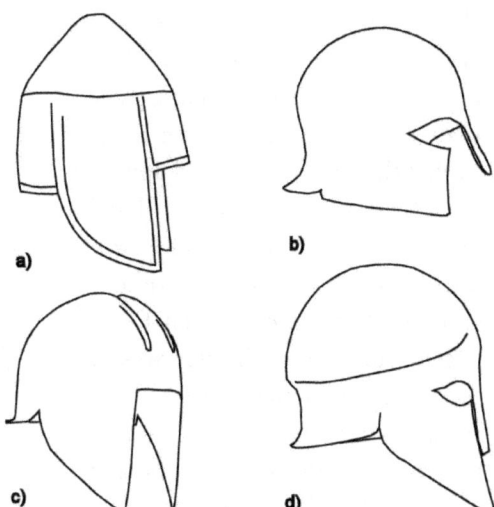

Early helmet types, a) Kegel, b) early Corinthian, c) Illyrian, d) later Corinthian.

was added to the back of the helmet, which would help to deflect downward blows away from the back of the neck. The whole helmet also acquired a more refined shape, often with decoration picked out in relief, and a more elegant form generally, making some of the later examples real works of art. From the fifth century, helmets also acquired ear holes rather than covering the ears in metal. This suggests that there must have been something worth listening to from this period, and we may surmise that this something was shouted (or trumpeted) orders, or the playing of the flute, which might suggest in turn that it was around this time that the phalanx became less a mass of individuals, however closely massed, and more a drilled, organized and controlled formation.

The Corinthian helmet must have been too hot and uncomfortable to wear over the face outside of battle. On the march, most helmets were presumably carried from a strap rather than worn, but the Corinthian helmet leant itself to a halfway position in which it was worn on the head but pushed up to reveal the face. This is a position familiar from art, and when this helmet was adopted in Italy along with other elements of the Greek panoply, particularly by the Etruscans, it gave rise to a variant form in which the helmet was permanently worn pushed back on the top of the head. The eye holes and mouth slits were now completely unnecessary (being on top of the head), but remained part of the design, shrunk down to be so small as to be merely decorative, thus forming the Etrusco-Corinthian helmet.

In Greece, meanwhile, a new helmet developed around the sixth to fifth centuries, the Chalcidian. This took the refined head form of the later Corinthian along with the extended nose guard, but replaced the face-covering lower portions with separate cheek pieces, that could in some cases be hinged so that they could be folded upwards out of the way when not in battle. Such helmets left the face more open but also less encumbered, and also left the ears open for the hearing of orders. Similar was the Attic helmet, which had no (or a much shorter) nose guard and was popular in Italy. Also in the fifth century, the Thracian helmet was devised, which became extremely popular and, in various forms, the quintessential infantry helmet of the Hellenistic period, from the fourth century onwards. This had a higher crown than the Chalcidian, and the cheek pieces swept forward and down further; the crown was also often extended forward into a bulb or crest (a variant also called Phrygian). An unrelated form was the Boeotian helmet, which appears to have developed as a representation in bronze of a fabric hat and was in common use in the Hellenistic period as a cavalry helmet. It may also have been used by Classical cavalry, being recommended by Xenophon (Xen., *Hipp.* 12.3), and perhaps for Boeotian hoplites too.

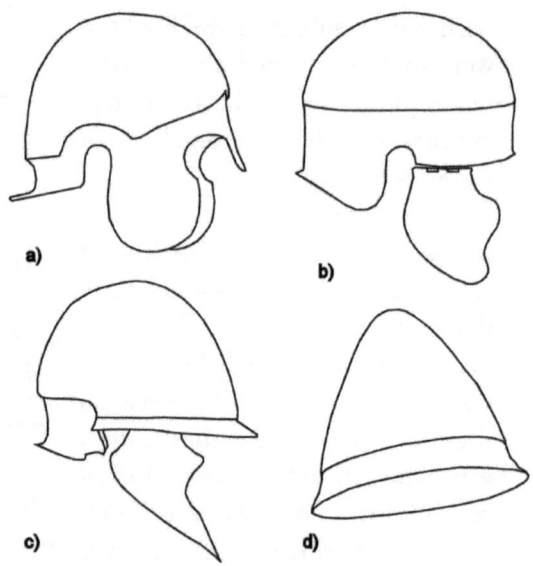

Later helmet types, a) Chalcidian, b) Attic, c) Thracian, d) *pilos*.

Also appearing in the fifth century was the last major form of helmet, the *pilos*. This had originally been (and probably continued to be) a conical felt cap (the literal meaning of *pilos*), and may have been worn by some hoplites in place of a metal helmet. The shape was then adopted in bronze to give a very simple, conical helmet without any protection for face or neck (though cheek pieces were sometimes added, especially in the Hellenistic period). This helmet, like the felt version, is particularly associated with the Spartans. Thucydides (4.34.3) records that the Spartan hoplites assailed by light infantry on Sphacteria (425) were particularly vulnerable as 'their *piloi* would not keep out the arrows', which could mean that they were wearing only felt caps, though it is unclear why hoplites would not have helmets, so Thucydides may instead be referring to the lack of overall protection for face and neck of the *pilos* helmet.

The existence of the *pilos* cap reminds us that a metal helmet could not be worn directly against the head, especially in a climate like that of Greece. It was probably normal to wear a cloth cap of some sort – either a skull cap or a *pilos* –

under the helmet. Helmets were probably also usually lined with felt or leather, both for comfort and to keep hot surfaces off the skin; many helmets retain a set of small holes around their edges where the lining would be attached. A felt cap would also provide some shock absorption for blows to the helmet, and ensure a snug fit, since (as with body armour) a loose-fitting helmet would be uncomfortable to wear and would also shift when struck, transferring force to the face or skull.

All types of helmet (less often the *pilos* or Boeotian) also often had a horsehair crest (which according to Herodotus was invented by the Carians). The earliest helmets – such as that in the Argos panoply – had metal crests attached to the crown, but it is easy to imagine that this would be heavy and unwieldy. Instead, a raised or flat bronze bracket would be attached to the top of the helmet and a crest attached to this – either short and stiff, standing vertically, a stylized horse's mane, or hanging loosely downwards, like a horse's tail. Often these would be combined into a stiff 'mane crest' with a 'tail' plume at the back. The hair would presumably be dyed – whites, reds and blacks are generally assumed, though colours are not well represented in vase paintings – and either with solid blocks of colour or stripes. Crests could also be used as a badge of office, as they were in Hellenistic times, when the number of crests and plumes would indicate rank. There is a statuette of a Spartan with a transverse crest, so this may also have indicated an officer (though of what rank we can only guess).[36]

Greaves

The final regular item of hoplite equipment was the greave, or shinguard. These are depicted in the earliest representations of hoplites, such as the Chigi vase. Archaic greaves were slightly shorter than Classical ones, ending at or below the knee. Classical greaves extended above the knee, being cut away at the back so that the leg could still flex. They were made from sheet bronze, and usually relied on the natural springiness of the bronze to be clipped to the leg, without any other means of attachment, though Hellenistic greaves were more often tied on, so this might have also been an earlier practice. Like other armour, they probably had a fabric lining. As with bronze body armour, a stylized outline of the muscles was moulded (or beaten, or etched) into the greaves, and like all armour, they would need to be made to measure to ensure a good fit and comfort. Greaves seem to have continued in use throughout the period (at least for some hoplites – as usual we cannot be certain that everyone could afford or needed them), and their value for men armed with a round shield that extended only to about mid-thigh, and was held on a crooked arm, is obvious. Unlike earlier or eastern tower shields, a hoplite's shield would not naturally protect his lower legs, and it would be difficult to lower the shield to intercept incoming missiles. In close combat, with spears held overarm, most blows would be directed against the head or body, but during the approach march or in the more open combat of the early phalanx or non-pitched battle encounters, the legs must have been quite vulnerable.[37]

Perhaps surprisingly, most hoplites did not have protection for their thighs or feet. There are some examples of thigh, ankle and foot guards, and armour for the arms – although these may have been intended for cavalry, as Xenophon tells us that cavalry would ideally wear armour on their left arm (Xen., *Horse*. 12.5–7). At any rate, such extra armour does not appear to have been widespread, probably being worn only by the wealthiest, and fell out of use entirely by the end of the sixth century. Evidently it was felt that the shield provided enough protection for the extremities, and that the advantage in reduction of weight, cost, complexity and encumbrance made up for the slightly reduced protection.

The spear

The hoplite, having provided for his own protection, now needed the means to harm his enemy, and for this the main tool was the spear. Hoplites were first and foremost spearmen, and while swords were carried and used (see below), they were much more a weapon of last resort, or perhaps of particular situations such as city and street fighting, or for all the day-to-day acts of violence that a campaign might require and where a spear might be less useful. The length of the hoplite spear is difficult to judge with certainty, partly because there are few extant examples (wooden spear shafts do not survive readily in archaeological contexts), partly because in artistic representations the requirements of the composition, in particular the space available, would be a stronger determinant of the length of spear depicted than its actual length, and partly because there was, of course, no standardization. In the centralized royal armies of the Hellenistic period, weapons were manufactured centrally and distributed, so could be made a uniform length. But hoplites armed themselves, and as such we must expect that each man chose a length of spear that suited his own fighting style, and that within any phalanx there must have been spears of a wide variety of lengths. So all we can do is give a range of lengths, and it appears that spears were around 1.5–3 metres long, averaging perhaps something over 2 metres. Some spear depictions on vases, such as that of the 'Achilles painter', are quite long, at least 3 metres, but it would be difficult to conclude that this was typical.

More important than the length of the spear is its reach (or 'measure', to use a fencing term), and here we are even more uncertain, since the reach of a spear depends of course on how it is held. Vase depictions show the spear being held by its rear half, as we would expect, and generally closer to the centre than to the rear end. The important thing, since the spear was held one-handed, would be to hold it at or near the point of balance (otherwise the spear would be difficult to handle), but where on the shaft the balance was would depend on the details of the construction of the shaft, in particular how much it was tapered toward the point, something about which we can again only guess. Modern re-enactors, who love to quantify such things, have done much experimentation, measurement and calculation on the reach of the spear. In combat, the absolute reach of a spear within a metre or so each way cannot have been of enormous importance (but on this topic see further in Chapters 7 and 8). Once an enemy was past the point of a spear, then greater length would be only a hindrance, and hoplites in a close formation did not always have the freedom of movement to step back to maintain a greater separation from an opponent, as a longer spear would require. Consequently, I do not believe that size mattered particularly in combat, and personal preference, practice and familiarity were probably the main factors in determining spear length.[38]

The way the spear was held is a question that has provoked much debate. Vase paintings depict three main holds: low or underarm, in which the spear was held at around waist height with the hand above the shaft, thumb forward; high or overarm, in which the spear was held at or above head height with the hand below the shaft, thumb backward; and couched or high underarm, in which the spear was held under or near the armpit in a raised version of the low position. The fact that spears are depicted held in all these ways in art suggests that all such holds were used, at one time or another, according to the exigencies of combat. Nevertheless, the most common hold, and in particular the most common when hoplites are depicted in formation, is the high or overarm hold. In a close formation, this would give the best chance of not accidentally poking the man behind with the butt end of the spear, and would also allow the greatest freedom of movement of the spear. It would also allow the spear to be held clear of the line of shields, and used to stab downwards at the face and neck of opponents. There is also some reason to suppose that the overarm hold produced a more powerful strike than the other options. I think, therefore, that the long-held consensus view that the overarm hold was usual in combat is the correct one.[39]

While standing to attention or marching, spears were probably held vertically, which could be done either point down – allowing an easy transition to the overarm hold – or point up (as they are depicted in the vast majority of cases). This produces one possible problem: holding a spear vertically, point up, requires holding it underarm, that is with the thumb towards the point, while holding it overarm to strike requires having the thumb towards the butt end. Some have been perplexed as to how the transition between the two was achieved, but this appears to be a non-problem. It seems easy to switch grip, either by briefly placing the end of the spear on the ground, by using the 'throw and catch' demonstrated by some re-enactors, by holding the spear with the left hand, or (most simply) by manipulating the fingers to reverse the grip. I am able to do the latter easily with minimal practice and almost without thinking, so I am sure that hoplites could too.[40]

The wood from which spears were made probably varied. Cornel wood (the wood of *cornus mas*, the Cornelian cherry) was favoured

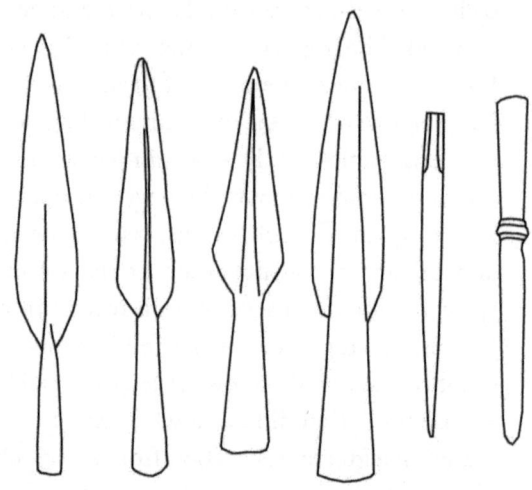

Typical spearheads (left four) and spear butts (right two).

for javelins because of its strength and hardness, but it was also quite a heavy wood and not easy to grow straight to any great length. It seems likely that ash, the traditional wood for spears across many centuries, was preferred as it is light, reasonably strong and grows straight. In practice there was probably some variation, especially when on campaign overseas. Spears were also fitted (at least some of the time) with handholds, presumably leather bindings, to provide a secure grip, and might also have been waxed or oiled to maintain the strength of the wood. Despite such precautions, spears could and did often break in combat; once broken, some use could be made of the butt spike, or else the hoplite would have to take to his sword.[41]

The butt spike and the blade were the two metal fittings, at each end of the spear. Spear blades were generally leaf-shaped blades of iron, and came with an enormous variety of shapes, sizes and patterns, with no particular uniformity, as we would expect. The butt spike, *styrax* or *sauroter* at the other end was usually of bronze and had a number of uses: it protected the end of the spear from the elements, it allowed the spear to be pushed upright into soft ground, it provided a counterbalance to the blade, allowing the point of balance to be moved further back along the spear, and it provided an alternative spike to use if the blade broke off, or if opportunity arose to jab downwards such as at an enemy on the ground. Both blade and sauroter had hollow fittings into which the spear shaft could be inserted, and presumably were fixed in place with pitch.[42]

The primacy of the thrusting spear as the hoplite weapon in the Classical period should not obscure the fact that the Archaic hoplite was probably frequently armed with two spears, one of which was intended for throwing (and hence is, properly speaking, a javelin). Throwing spears were fitted with a throwing loop or *ankyle*, which gave greater leverage to the throw and imparted spin to the shaft, and also, fortuitously, allows such spears to be identified in art. As we saw in the previous chapter, such spears continued in use for some time, though with only a single spear for throwing, hoplites could never have engaged in prolonged skirmishing the way specialist javelineers and skirmishers (such as peltasts) could. Rather, the throwing spear was probably used for picking out targets of opportunity in an open melee, in the fashion described by Homer, or, in closer formations, would presumably have been thrown before charging into contact in a manner similar to the later Roman *pilum*. With the increase in protection offered by shields and armour, and the greater emphasis on close formation and an orderly advance from the sixth century, the diminishing returns from the extra encumbrance of a throwing spear (which presumably required the other spear to be held in the left hand along with the shield) meant that they were abandoned, and hoplites became single thrusting spear users. Impromptu missile weapons, however, were still used, particularly stones when fighting in rocky areas or when defending a fortified position or city, so hoplites never completely abandoned their missile capability.[43]

Swords and sidearms

If the spear was broken, the hoplite would take to his sword (*xiphos*). Unlike some peoples, particularly in the Western Mediterranean, for whom the sword was the primary weapon, for the hoplite the sword was always a weapon of last resort (in the context of pitched battle). Greek swords were made of iron, and mostly had a gently tapered leaf-shaped blade, of a form that dates back to the Bronze Age. Such swords could be used both for cutting and thrusting, though art suggests that cutting was the usually preferred method, often with a fairly wide forehand or backhand swing. Fallen enemies, prisoners and unarmed civilians in sacked cities might, however, be dispatched with a thrust, as is frequently depicted.[44]

From the fifth century, a second type of sword, the *kopis* (or possibly *machaira*), came into fashion. This was more like a sabre, with a curved blade with a cutting edge on the inside of the curve, and weighted toward the point. Such a sword was a specialist cutting weapon and was often associated with cavalry. Xenophon's recommendation of it as a cavalry weapon provides a good example of the use (or absence) of technical vocabulary:

> 'For harming the enemy we recommend the *machaira* rather than the *xiphos*, because, owing to his lofty position, the rider will find the cut with the *kopis* more efficacious than the thrust with the *xiphos*.' (Xenophon, *Horsemanship* 12.11)

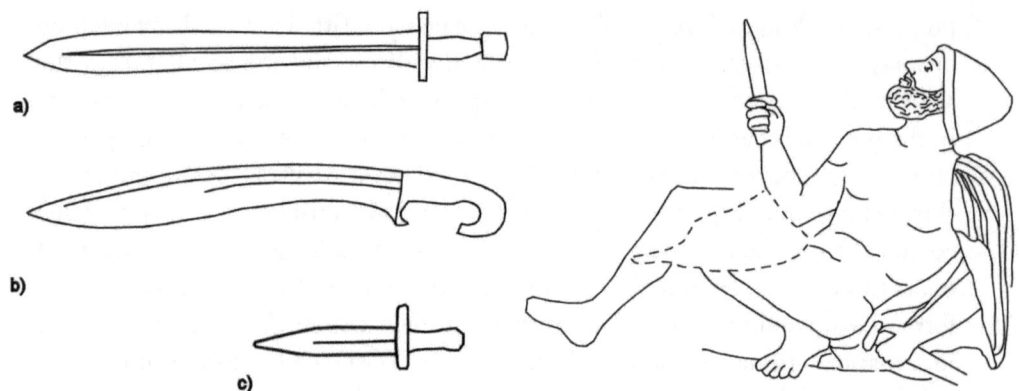

Sword types, a) *xiphos*, b) *kopis*, c) *encheiridion*.

Fallen Spartan with cloak, *pilos* helmet and *encheiridion*

It thus appears that *kopis* is another word for *machaira* (or perhaps that a *kopis* is a type of *machaira*). The design may well have been borrowed from the Persians (Xen., *Cyrop*. 1.2.13; 'in their right hands a *machaira* or *kopis*').

Finally, there was the *encheiridion*, which appears to be a knife, dagger or short stabbing blade, and which would have been useful in a close and confused melee. This came to be associated particularly with the Spartans, as Plutarch records:

'King Agis, accordingly, when a certain Athenian decried the Spartan swords [*machairas*] for being so short, and said that jugglers on the stage easily swallowed them, replied: "And yet we certainly reach our enemies with these daggers [*encheiridiois*]." And I observe that although the speech also of the Spartans seems short, yet it certainly reaches the point, and arrests the thought of the listener.' (Plutarch, *Lycurgus* 19.2)

Note the switch in terminology mid-paragraph from *machaira* (sword in general) to *encheiridion* (specifically a short sword). Herodotus records numerous peoples of the Persian army also being armed with *encheiridia* (e.g. Hdt. 7.61, 63, 64, 67, 72, 75, 85, 92, 93), so their use was probably quite general (possibly being adopted by Spartans from Persians, as with other equipment, but a short sword was surely pretty universal).

All such swords were carried in a scabbard, made of wood or leather, often with a decorated metal chape (fitting at the end of the scabbard), and worn slung over the shoulder so that it hung on the left side, quite high, under the shield arm.

Clothing

The traditional Greek male clothing for outdoor activities, such as soldiering, was the *chiton* or tunic (sometimes *chitoniskos*, 'little tunic'), in form like a short sleeveless (or sometimes short-sleeved) dress, that was worn with a belt through which loose folds of material would be pulled up to hang over the belt. From the fifth century, a variant on this, the *exomis*, became popular, which could be worn in such a way as to leave the right shoulder bare, with folds of cloth hanging down the right side. Over this might be worn a *chlamys* or cloak, a rectangular or semi-circular length of cloth (usually wool) which could be worn as a cloak over one shoulder or in cold weather wrapped around the body. This could also be used as a combination blanket and groundsheet when camping (such garments are variously also called a *tribon* or *himation*, the latter being more common in the Archaic period). Lengths of cloth can of course be worn around the body in a huge variety of ways, and there are depictions of cloth garments wrapped around the waist, draped over shoulder or arms, or tied to the arm or leg as an impromptu shield (or perhaps bandage). There was probably no uniformity in the way such garments were made or worn.[45]

There was probably also no uniformity in the colour of such clothing. As we have seen with shield blazons, these were originally up to the individual, with a huge variety of designs used, and only later were standardized designs for each city

Unarmoured hoplites in *exomis* and *pilos*; that of the left-hand figure apparently made of felt (left, Attic grave from Megara, late fifth century; right, tomb of Lisas the Tegean)

introduced (and then only in some cases). Clothing was presumably the same, at the whim of the individual. This meant of course that with identically equipped hoplites facing each other in a random assortment of clothing, misidentification was possible and 'blue on blue' or 'friendly fire' incidents could and did occur (e.g. Thuc. 4.96.3 at Delium, 424). The exception as usual is the Spartans, who were said to have all worn a red garment from earliest times, along with other uniform features:

> 'In the equipment that he [Lycurgus] devised for the troops in battle he included a red cloak [*stolē*], because he believed this garment to have least resemblance to women's clothing and to be most suitable for war, and a bronze shield, because it is very soon polished and tarnishes very slowly. He also permitted men who were past their first youth to wear long hair, believing that it would make them look taller, more dignified and more terrifying.' (Xenophon, *Constitution of the Lacedaemonians* 11.3)

The actual shade of red – variously translated as scarlet, crimson, purple or just red – indicated by the Greek word *phoinix*, is not certain, nor is the garment translated 'cloak' (*stolē* – see also comments above on *stolas* and *spolas*), perhaps in fact a tunic. At the Battle of Coronea (394), it was said that the Spartan king Agesilaus:

> 'brought into the field an army not a whit inferior to the enemy's; he so armed it that it looked one solid mass of bronze and scarlet.' (Xenophon, *Agesilaus* 2.7)

Hoplites arming, and (left) removing the cover from a shield.

Into the fourth century, uniform clothing may have become more common. Xenophon records the Ten Thousand mercenaries of Cyrus' expedition (401) on parade:

> 'Cyrus ... inspected the Greeks, driving past them in a chariot, the Cilician queen in a carriage. And the Greeks all had helmets of bronze, crimson tunics, and greaves, and carried their shields uncovered.' (Xenophon, *Anabasis* 1.2.16)

Note that armour is not mentioned. This was, however, the private mercenary army of a wealthy individual laying claim to an Imperial throne, so the uniformity may not be typical of city or mercenary forces. Note also the mention of shield covers – presumably, to protect the bronze or the paintwork, shields were carried in covers (of fabric or leather) which would be removed before combat.

So there was a high degree of commonality in hoplite equipment – the shield, helmet, spear and greaves at least being common among many hoplites covering a huge geographical area – but little uniformity of colours, clothing or designs, with some exceptions. But to be really effective, as Aristotle reminds us, an infantry army needed not just to be well equipped, but also to be well organized, and it is to that subject that we turn in the next chapter.

Chapter 3

Organization and Drill

The nature of the evidence

The hoplite phalanx is chiefly known to us from the detailed literary histories that begin in the early fifth century (in terms of dates of coverage) or late fifth century (in terms of dates of composition), and is most fully described in the works of Xenophon, in the early fourth century. Yet as we have seen, the phalanx had its origins at an earlier date – according to some arguments, much earlier – and by any reckoning it began to develop around the late seventh century, even if it did not reach its final form for two centuries. This poses an obvious problem: all the detailed knowledge we have of the phalanx is from the late fifth and early fourth centuries, and before this time we have only artistic depictions, references in poetry and archaeological remains. Art and archaeology are excellent for studying items of equipment, but for organization, tactics and drill they are obviously somewhat limited. This means that the picture we have of the phalanx is inevitably unbalanced – we see it in detail only when it had attained its most advanced form, and it is difficult to establish the validity of extrapolating backwards to earlier ages. Many modern accounts take as their starting point the descriptions of Thucydides and Xenophon and talk as if this applies to the Archaic phalanx, but there must at the very least be considerable doubt as to whether the Archaic phalanx was really similarly organized. There was (perhaps) an increase in the professionalism and training of hoplite armies from the Peloponnesian War (late fifth century) onwards, and this reached its peak in the new professional armies established by the Macedonian king Philip II in the mid-fourth century. The Greek cities responded to the new Macedonian threat with further changes of their own, so that we see hoplite armies and institutions in the mid-fourth century which may not be at all typical of those of the preceding three centuries.

Furthermore, the ascent of Macedon and of the Hellenistic world to which Macedonian arms gave rise led to the establishment of a new genre of tactical writing, which provides a vast new source of evidence for phalanx tactics. Tactical writing has its origins, so far as we know from surviving examples, in the early fourth century with the works of Aeneas 'the Tactician', who wrote a number of works on military matters, of which just one – *On the Defence of Fortified Positions*

– has survived. Xenophon, as well as his historical works, also wrote military treatises, of which we now have *The Cavalry Commander* and *On Horsemanship*. Xenophon's historical novel, *The Education of Cyrus (Cyropaedia)*, is also full of detailed tactical information, as is *The Constitution of the Lacedaemonians*, which is attributed to him. But it was in the Hellenistic period that tactical writing really took off, and we have lists of the names of many authors of tactical handbooks, including the king of Epirus, Pyrrhus, who in the early third century fought a long and ultimately unsuccessful campaign (due to a succession of 'Pyrrhic victories') in Italy against the rising city state of Rome. There is also the historian Polybius, on whose *History* we rely for most of our knowledge of the later Hellenistic world. The tactical works of all these authors are lost, but some at least of their words survive in later adaptations by Asclepiodotus, a philosopher and writer of the second to first centuries, and the Roman period writers Arrian and Aelian. These works give us detailed knowledge of the internal structures of the Macedonian phalanx; as this was a close-order formation of spear-armed heavy infantry – very literally a phalanx of hoplites, since both terms continued in use throughout this period – it is very tempting to apply these works retrospectively to the Classical phalanx that the Macedonians defeated and (ultimately) replaced.[1]

It would be wrong not to use the Hellenistic tacticians at all to illuminate the Classical phalanx. Comparative material on later, and earlier, infantry formations can be of great value in filling in the blanks in our knowledge, and the Macedonian phalanx was in a sense a development and perfection of the earlier Greek formation; they must have had many features in common, as they both have with other close-order infantry of other regions and periods. But all too often, modern authors will state some fact about the Classical phalanx, quoting in support a reference in the Hellenistic tacticians. While they had much in common, the Greek and the Macedonian phalanxes also had many important differences. The most obvious was their armament: the Greek phalanx was a formation of spearmen who fought individually with one-handed spears, while the Macedonian phalanx used a two-handed pike, much longer than the earlier spear, which required no individual fencing or duelling, but simply to be held resolutely in advance. The Macedonian phalanx, to an extent not true of the Classical (still less the Archaic) phalanx, relied on the mass effect of the closely packed combatants. Furthermore, the Macedonian army was, or became under Philip and Alexander, a full-time professional force, highly trained, regularly drilled and often engaged in soldiering for decades at a time (quite literally). The Greek phalanx remained in contrast a citizen militia, serving part-time for brief campaigns, and resistant both to discipline and training (as we will see in Chapter 4). While in the fourth century, in response to many pressures – including the rise of Macedon and the prevalence of long-term mercenary service

– a greater degree of professionalism arose and state-sponsored military training was instituted in many cities, the Greeks remained essentially amateur soldiers and never attained the level of organization and discipline enjoyed (or endured) by the Macedonians.[2]

There was of course one exception to these statements, the Spartans, and they pose another difficulty. The Spartans – or at least their full citizen class, the Spartiates – had, from the time of their semi-legendary lawgiver Lycurgus (perhaps seventh century), devoted themselves full-time to practice for and of military action, developing a level of professionalism, drill, weapons skill and to some extent tactical expertise that was far ahead of that of the amateur militias of the other Greek cities. Inevitably this made the Spartans very successful, and gave them a fearsome reputation that was often sufficient on its own to ensure their success. Naturally, writers with an interest in military matters praised the Spartans' skills, and Xenophon, though himself an Athenian, was a particular admirer of Sparta and fought alongside one of their kings. As a result, a lot of the details about the 'Greek' phalanx that have come down to us are really about the Spartan phalanx, and we must keep in mind at all times that the Spartans were not at all typical.

Consequently, modern accounts of the hoplite phalanx which give an overall view of how the Greek phalanx in general was organized and functioned will tend naturally to draw on the two best sources of information – about the Macedonian and Spartan phalanxes – although neither of these are at all typical of Greek phalanxes generally, and the Macedonian phalanx at least was really a very different beast. Add in the temporal differences – that we know so much more about the phalanx at the peak of its development in the fourth century than we do about its evolution across the preceding three centuries – and it is inevitable that such general accounts can give a very distorted picture. These accounts must therefore always be hedged with caveats about the limitations of earlier evidence and the dangers of generalization, even as they make use of the evidence we have for comparative purposes if nothing else. The account of the phalanx in this book, being a high-level account, will inevitably fall foul of these traps; I can only urge vigilance (of author and reader) to be aware that much of what is said of the details of the phalanx may not have the general application that we might hope for or assume.

Subunits – the Spartans

The best place to start with phalanx organization is perhaps the phalanx that is best known, that of Sparta in the late fifth and early fourth centuries. Although the Spartan army is relatively well known, many obscurities remain, and a considerable literature has grown up around attempts to reconcile the contradictory accounts in the sources (chiefly Thucydides and Xenophon). With that said, the evidence for Spartan unit organization in this period is as follows.[3]

Thucydides provides the earliest detailed breakdown of Spartan units in the account he gives of the Spartan army at the Battle of Mantinea (418), in the context of his attempt to estimate the size of the Spartan army at this battle:

> 'Such were the order and the forces of the two combatants. The Lacedaemonian army looked the largest; though as to putting down the numbers of either host, or of the contingents composing it, I could not do so with any accuracy. Owing to the secrecy of their government the number of the Lacedaemonians was not known, and men are so apt to brag about the forces of their country that the estimate of their opponents was not trusted. The following calculation, however, makes it possible to estimate the numbers of the Lacedaemonians present upon this occasion. There were seven *lochoi* in the field without counting the Sciritae (who numbered six hundred men): in each *lochos* there were four *pentecostyes*, and in the *pentecostys* four *enomotiai*. The first rank of the *enomotia* was composed of four soldiers: as to the depth, although they had not been all drawn up alike, but as each *lochagos* chose, they were generally ranged eight deep; the first rank along the whole line, exclusive of the Sciritae, consisted of four hundred and forty-eight men.' (Thucydides 5.68)

So according to Thucydides, the smallest unit was the *enomotia*, in this case 32 men strong (four files of eight men), and the other units were as follows:

Enomotia	32 men
Pentecostys	128 men (four *enomotiai*)
Lochos	512 men (four *pentecostyes*)
Total	3,584 men (seven *lochoi*)

However, Xenophon (if he is indeed the author), in the *Constitution of the Lacedaemonians*, gives a slightly different organization, referring to some point in the fourth century:

> 'The men so equipped [i.e. as hoplites] were divided into six *morai* of cavalry and infantry. The officers of each citizen [or hoplite] *mora* comprise one *polemarchos*, four *lochagoi*, eight *pentekonteres*, and sixteen *enomotarchoi*. These *morai* at the word of command form *enomotiai* sometimes in [...], sometimes in threes, and sometimes in sixes.' (Xenophon, *Constitution of the Lacedaemonians* 11.4)

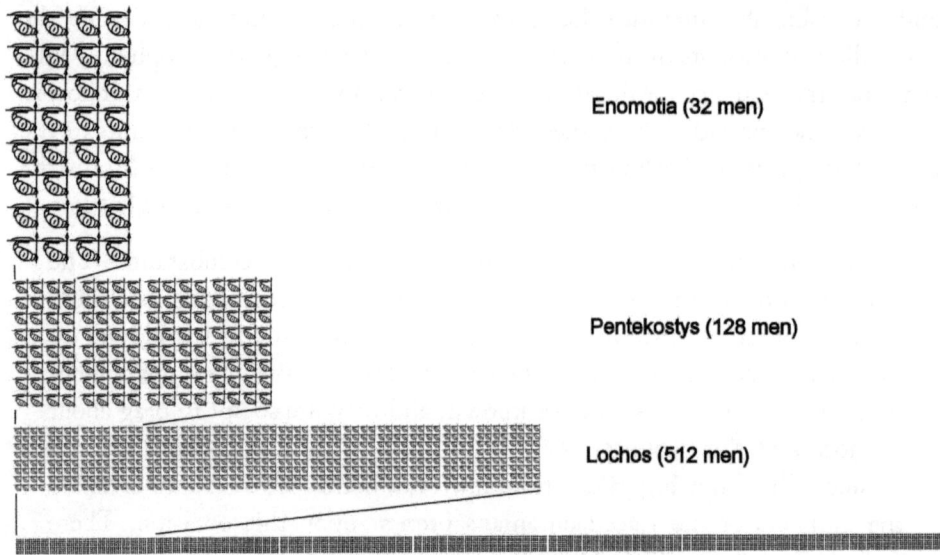

Spartan army units according to the scheme of Thucydides.

Note that Xenophon's text is suspect in this passage – a word seems to have dropped out for the first width of the *enomotia* (presumably 'ones', or perhaps 'twos'), and many have doubted the assertion that there were four *lochoi* in a *mora* (the Greek digit four, δ, would be very easily mistaken by a copyist for the Greek for two, δύο, so it could well be that originally Xenophon stated that there were two *lochoi* in a *mora*). Note also that Xenophon does not refer specifically to unit names, but only to their officers, but it seems uncontroversial to suppose that the *polemarchos* commanded the *mora*, the *lochagos* commanded the *lochos*, the *pentekonter* (or *pentekoster*, an alternative spelling) commanded the *pentekostys*, and the *enomotarchos* (or *enomatarches*, another alternative) commanded the *enomotia*. Xenophon does not specify the number of men in an *enomotia*, but at the Battle of Leuctra (371), he notes that:

> 'Coming now to the infantry, it was said that the Lacedaemonians led each *enomotia* in threes, and that this resulted in the phalanx being not more than twelve men deep.' (Xenophon, *Hellenica* 6.4.12)

Presumably 'in threes' means 'in three files', or 'three abreast', and the same can be understood for the passage above. Three files twelve men deep gives a strength of 36 men, so taking this value, and taking Xenophon for now as written (four *lochoi* per *mora*), gives the following:

Enomotia	36 men
Pentecostys	72 men (two *enomotiai*)
Lochos	144 men (two *pentecostyes*)
Mora	576 men (four *lochoi*)

If, however, we accept the emendation of 'four' *lochoi* to 'two' (which seems likely from other evidence), and assuming that all the other numbers of officers as given are correct, we would have:

Enomotia	36 men
Pentecostys	72 men (two *enomotiai*)
Lochos	288 men (four *pentecostyes*)
Mora	576 men (two *lochoi*)

This structure is broadly similar to that given by Thucydides (given a different starting size of the *enomotia*), with the major differences being the size of the *pentecostys* and *lochos*, and the absence in Thucydides of any mention of the *mora*. The latter point is the most troubling, since neither Thucydides nor Herodotus ever mention the *mora* as a Spartan (or any other army) unit, while Xenophon refers to it frequently as the basic building block of the Spartan army. Confirming the statement in the *Constitution*, in Xenophon's *Hellenica* the Spartans appear to have divided their total army into six *morai*, as shown by the following passages:

> '[T]he Lacedaemonians sent Cleombrotus, the king, across to Phocis by sea, and with him four *morai* of their own and the corresponding contingents of the allies.' (Xenophon, *Hellenica* 6.1.1)

> 'After this the ephors called out the ban of the two remaining *morai*, going up as far as those who were forty years beyond the minimum military age; they also sent out all up to the same age who belonged to the *morai* abroad; for in the original expedition to Phocis only those men who were not more than thirty-five years beyond the minimum age had served; furthermore, they ordered those who at that time had been left behind in public office to join their *morai*.' (Xenophon, *Hellenica* 6.4.17)

(The information in this passage on the functioning of Spartan recruitment will be considered in Chapter 4). To Xenophon, the most important unit is the *mora*, and six of these make up the total Spartan army, while in Thucydides, the largest division is the *lochos*, and there were at least seven of these (perhaps more, as the army at Mantinea was not the entirety of the army). There are two possibilities: either Thucydides or Xenophon is mistaken (most likely Thucydides, since

Xenophon had considerable personal knowledge of the Spartan army, though on the other hand his authorship of the *Constitution of the Lacedaemonians* is not certain), or else there was some reform of the Spartan army, perhaps around the end of the fifth century, in which the previous largest units, the *lochoi*, were grouped into *morai*. To add to the difficulty, Xenophon ceases to mention *morai* after Book 6 of the Hellenica (mentioning only *lochoi* from this point), which has led some to suppose that there was another reform in which the *mora* was again removed, and the *lochos* reinstated as the largest unit of the Spartan army.[4]

There is also a passage of Thucydides which must cast some doubt on the identification of the *lochos* as the largest unit of his time, where he refers to Spartan arrangements before the Battle of Mantinea:

> 'For when a king is in the field all commands proceed from him: he gives the word to the *polemarchoi*; they to the *lochagoi*; these to the *penteconteres*; these again to the *enomotarchoi*, and these last to the *enomotiai*. In short all orders required pass in the same way and quickly reach the troops; as almost the whole Lacedaemonian army, save for a small part, consists of officers under officers, and the care of what is to be done falls upon many.' (Thucydides 5.66.3–4)

Here Thucydides exactly matches the officer structure given by Xenophon (without giving the number of officers or the size of their units). Yet in Thucydides' scheme there is no unit for the *polemarchoi* to command, since the largest unit is the *lochos* under a *lochagos*. It is not impossible that this is correct, and that the *polemarchoi* were staff officers without specific units under their command, as is hinted at by Xenophon:

> 'The *polemarchoi* mess with the King, in order that constant intercourse may give better opportunities for taking counsel together in case of need. Three of the peers [*homoioi*] also attend the King's mess. These three take entire charge of the commissariat for the King and his staff, so that these may devote all their time to affairs of war.' (Xenophon, *Constitution of the Lacedaemonians* 13.1)

But in Xenophon's scheme, the *polemarchoi* are clearly the commanders of the *morai*. In Thucydides, the role of the *polemarchos* is unclear – for example, in the course of the Battle of Mantinea, the Spartan king made a last-minute change of plan:

> 'Agis afraid of his left being surrounded, and thinking that the Mantineans outflanked it too far, ordered the Sciritae and Brasideans to move out from their place in the ranks and make the line even with the Mantineans, and told the *polemarchoi* Hipponoidas and Aristocles to fill up the gap thus

formed, by throwing themselves into it with two *lochoi* taken from the right wing; thinking that his right would still be strong enough and to spare, and that the line fronting the Mantineans would gain in solidity.' (Thucydides 5.71.3)

Why order Hipponoidas and Aristocles to make this move rather than the commanders of the *lochoi*? It could be that Hipponoidas and Aristocles were unattached staff officers and that the order went through them to the relevant officers (as suggested by 5.66.3), or that each was commander of a larger unit (a *mora* presumably) and each was ordered to detach one of their constituent *lochoi*. Or it could be that Thucydides was mistaken about the size of units to be moved (two *lochoi*, 1,024 men by Thucydides' reckoning, would be a large force to be shifting in this way, nearly a third of the Spartan contingent). The word *lochos* is often used to designate a 'unit' in a general sense, of undefined or variable size (see below), and in the Hellenistic period it was used for the smallest unit of the Macedonian phalanx, the file of eight, ten or sixteen men.

The name of the smaller unit, the *pentekostys*, is also slightly troubling. Its name (in Greek) might suggest that it consisted of 50 men, and while it is not unknown for the traditional names of military units to not exactly match their strength (the most familiar example being the Roman 'century' of around sixty men), a strength of 128 men (in Thucydides' scheme) does seem over-large for a 'fifty'. It could be that the name of the unit instead indicates a 'fiftieth', the unit forming, roughly speaking, a fiftieth of the full Spartan army, which is a possible way around this problem.[5]

At any rate, it appears either that we have two (and possibly three) versions of Spartan organization (with, first, *lochoi* as the largest unit, then with *morai*, and then perhaps back to *lochoi* again, and with varying numbers of subunits making up each level of unit), or that we must reject either Thucydides' evidence from Mantinea or Xenophon's evidence for the later army. I think that the problem is insoluble given the state of the evidence, and that arguments either way – while compelling – can never be conclusive. Nevertheless, my inclination is to believe that the *mora* was always the largest unit, that Spartans were more conservative and did not reorganize their army as often as the alternatives would require, and that Xenophon's text is corrupt (or he was mistaken) for the number of *lochoi* and possibly also for the numbers of the other subunits. A high degree of Spartan conservatism is entirely to be expected – it was after all what many like-minded Greeks admired about their constitution. According to Herodotus (1.65.5, 'Lycurgus afterwards established their affairs of war: the *enomotiai*, the 'thirties', the common meals'), the formation of the *enomotia* (or at least the use of the name) dates all the way back to Lycurgus, while Xenophon (*Const. Lac.* 11.4, 'The men so equipped were divided into six *morai* of cavalry and infantry') includes this organization in the list of things included in 'Lycurgus' organization of the army on active service' (11.1), so it seems unlikely that the *mora* was only invented after the end of the fifth century.

In terms of numbers of men in each unit, two other issues remain. One is the number of men in the *enomotia*, which Thucydides gives as thirty-two at Mantineia and Xenophon as thirty-six at Leuctra. The suggestion has been made, and I think it is a very likely one, that the size of the *enomotia* varied according to the size of the call-up. As we will see in Chapter 4, the Spartan army was called up by age classes, the range of ages called up determining the overall size of the army for any given campaign. There were forty age classes (those aged from 20–59 inclusive), and it appears from other evidence that the age groups were distributed evenly through the phalanx (rather than being grouped into separate units). Therefore, it seems reasonable to assume that with a full call-up of the army, each *enomotia* would be forty men strong, and with a reduced call-up correspondingly smaller (so the thirty-two or thirty-six men of Mantineia and Leuctra would be made up of the first thirty-two or thirty-six age classes). This means that the size of the *enomotia*, and therefore the size of all the other units of which it was part, would not be fixed but would scale according to the size of the army as a whole.[6]

Another issue is the size of the *mora* and therefore of the full Spartan army. If it is true that there were six *morai* in the full army, and if each *mora* was around 600 men strong (640 with a full call-up, according to Xenophon's scheme), then the full Spartan army (of the fourth century) would be only 3,840 men strong. This number seems too low given the attested sizes of armies containing a Spartan component, and by comparison with the known hoplite contingents of other cities. The suggestion has therefore been made that the *mora* was (at ideal strength) double this size (1,280 men in Xenophon's scheme), which would give an ideal Spartan army of 7,680 men, comparable to those of other major cities. This could be achieved by doubling the number of subunits at any level – for example, the suggestion is that rather than Xenophon's sixteen, there were actually thirty-two *enomotiai* in a *mora*, with four in each *pentecostys*, and the other proportions unchanged. This is an attractive suggestion and it does seem that otherwise the size of the *mora*, and of the whole army, is a problem, but as with the other suggestions, it remains largely speculative. There are also other ways of making up the numbers in Spartan armies. One is to assume that the units described by Thucydides and Xenophon are those of the true Spartan citizens (the Spartiates) only, and that a full army also included (either as additional units alongside the Spartans, or possibly intermingled in Spartan units) hoplites drawn from the *perioikoi* (non-citizens) or helots (serfs).[7]

We might also note that the later Macedonian phalanx, which had a much more rigid and much better understood structure of subunits, formed each unit by doubling the number of men from the previous level, so that each unit consisted of two subunits, at every level. The Spartan army seems not to have followed this neat pattern; Thucydides' scheme has four *enomotiai* to a *pentekostys* and four *penteskotyes* to a *lochos*, while Xenophon (unamended) has four *lochoi* to a *mora*, or following the amendment, four *pentekostyes* to a *lochos*. It is not possible to come up with a scheme of doubling of strength at each level which makes the *mora* anything like large enough (320 men would be the maximum, which is smaller than any attested size and too small to give a plausible size for the full army).

One possible permutation for a full-strength army, and if we assume that the Spartans did not reorganize, might therefore be:

Enomotia	40 men
Pentecostys	160 men (four *enomotiai*)
Lochos	640 men (four *pentecostyes*, sixteen *enomotiai*)
Mora	1,280 men (two *lochoi*, eight *pentecostyes*, thirty-two *enomotiai*)

This retains Thucydides' testimony for Mantinea (but assumes that he somehow failed to mention the *morai*) but requires rejecting or altering Xenophon's version. It assumes the *pentecostys* is roughly a fiftieth of the full army (actually one forty-eighth) and allows for the larger *mora*. Other schemes are of course possible (and if the Spartans did reorganize their army on occasion, then no single scheme will be correct for the whole period anyway).

The non-appearance of the *mora* in Thucydides (and Herodotus) might therefore possibly be due to the unit not having been invented at that point, being organized (or at least named) perhaps after the Peloponnesian War. But we should remember that the appearance of technical terms in literary historians sometimes follows a surprising pattern (as we have seen, Thucydides never uses the word 'phalanx', though it seems inconceivable that it was not used by contemporaries, and it occurs frequently in Xenophon). I think it is quite plausible that the *lochos* was often the most important building block of the Spartan army, and depending on circumstances (the size of force present, and the mix of units) it might well be that recording the number of *lochoi* present made more sense than recording the *morai*, for example if there was an uneven number of *lochoi* or if *lochoi* were pulled from different *morai*. In different circumstances, as in Xenophon's accounts, *morai* might have been the more important units and so appear more frequently in the pages of Xenophon; their disappearance from Xenophon after Book 6 would then be due not to a reorganization but simply to changed circumstances which made *lochos* organization more important. The situation may be analogous to battalions and regiments in nineteenth-century AD armies, where battalions (in the British Army) might belong to a given regiment yet operate completely independently, while on other occasions (or in other armies, like the French) they would operate together and be the basic battlefield unit. On the whole, I think it most likely that Spartan organization remained unchanged and that other circumstances result in our getting an unbalanced picture of the occurrence of *morai* and *lochoi*.

To illustrate the need for caution, Xenophon records a couple of Spartan councils of war:

> '[A]s for Pausanias, he called together the *polemarchoi* and *pentekonteres*, and took counsel with them as to whether he should join battle or recover by means of a truce the bodies of Lysander and those who fell with him.' (Xenophon, *Hellenica* 3.5.22)

> 'When Agesilaus heard this, he immediately leaped up from his seat, seized his spear, and ordered the herald to summon the *polemarchoi* and *pentekonteres* and the leaders of the allies [*xenagoi*].' (Xenophon, *Hellenica* 4.5.7)

Where, we might ask, were the *lochagoi* and why were they not invited to these councils? Compare the above with this account of the standard conduct of a military ritual:

> 'And at the sacrifice are assembled *polemarchoi*, *lochagoi*, *pentekonteres*, commandants of foreign contingents [*xenon stratiarchoi*], commanders of the baggage train, and, in addition, any general from the states who chooses to be present.' (Xenophon, *Constitution of the Lacedaemonians* 13.4)

I do not believe there is any particular significance to the absence of the *lochagoi* on the first two occasions, and we cannot always count on historians (ancient or modern) to be absolutely precise about such matters.

It would appear from all the evidence quoted above that the smallest subunit of the Spartan phalanx was the *enomotia*; but in the later Macedonian armies, the smallest unit was the individual file (according to the tacticians). Xenophon offers this comment:

> 'The prevalent opinion that the Laconian infantry formation is very complicated is the very reverse of the truth. In the Laconian formation the front rank men are all officers, and each file [*stichos*] has all that it requires to make it efficient.' (Xenophon, *Constitution of the Lacedaemonians* 11.5)

Note firstly that the text of the last clause quoted is somewhat uncertain, and also that even in Xenophon's time, Spartan organization was thought complicated. But this passage would also suggest that the *enomotia* was further divided into *stichoi*, 'files', each with their own leader (the generic word *archontes*, 'those leading', is used; the 'front rank men', *protostates*, means just that and is not a rank). It is difficult to see how this would work given the way the *enomotia* could form different numbers of men abreast, and therefore with different numbers of men in the front rank, and different numbers of ranks. A fixed organization – a *stichos* of eight men including a leader, say – would be constantly disrupted according to the number of ranks and files required. It may thus be that something slightly more informal was meant at this level, perhaps no more than that the file leaders would be men of experience, as in the Hellenistic phalanxes, and at any given depth there would be men appointed to stand in the front rank. It is also likely that the rear rank men, *ouragoi* ('file closers'), were specially chosen, though evidence for them is lacking for the Spartan army.[8]

Two other Spartan units, outside of this organization of six *morai*, should be mentioned. The first is the *Hippeis*, literally horsemen, cavalry or 'knights'. It is obvious that, at least in the fourth century, they were no longer cavalry, and probably had not been for a very long time (if ever), but their name presumably represents the fact that originally they had been an aristocratic body (the aristocracy, or their nominees, did usually serve as cavalry in Greek city armies). They appear for example in Thucydides' account of Mantinea:

> 'But the Lacedaemonians, worsted in this part of the field [the left flank], with the rest of their army, and especially the centre, where the three hundred knights [*hippeis*], as they are called, fought round King Agis, fell on the older men of the Argives and the five companies so named, and on the Cleonaeans, the Orneans, and the Athenians next them, and instantly routed them.' (Thucydides 5.72.4)

Unfortunately, the name *hippeis* literally means 'cavalry' and is frequently used for cavalry, so that it is not always possible to tell whether a given body is the *Hippeis* (the 300 'knights') or just a body of cavalry (written Greek of this period having only one letter case, there is no distinction in Greek equivalent to our 'knights' and 'Knights'). To add to the confusion, Xenophon's account of the fighting at Leuctra records:

> 'But when Deinon, the *polemarchos*, Sphodrias, one of the king's tent-companions, and Cleonymus, the son of Sphodrias, had been killed, then the *hippoi*, the so-called aides of the *polemarchos*, and the others fell back under the pressure of the Theban mass, while those who were on the left wing of the Lacedaemonians, when they saw that the right wing was being pushed back, gave way.' (Xenophon, *Hellenica* 6.4.14)

The *hippoi* means 'the horses', but it is as near certain as can be that Xenophon here means the *Hippeis*, and that these are, as at Mantinea, the Spartan royal guard fighting around the king.

This body was also known to Herodotus, who records that the Spartans honoured Themistocles, architect of the naval victory at Salamis:

> 'They also gave him the finest chariot in Sparta, and with many words of praise, they sent him home with the three hundred picked men of Sparta who are called *hippeis* to escort him as far as the borders of Tegea. Themistocles was the only man of whom we know to whom the Spartans gave this escort.' (Herodotus 8.124.2–3)

As with the organization of the *morai*, Xenophon takes the institution of the *Hippeis* all the way back to the time of Lycurgus (without explicitly naming them):

> 'The Ephors [senior officials of Sparta], then, pick out three of the very best among them [the young men]. These three are called Commanders of the Guard [*hippagretai*]. Each of them enrols a hundred others, stating his reasons for preferring one and rejecting another.' (Xenophon, *Constitution of the Lacedaemonians* 4.3)

These 300 picked young men chosen by the *hippagretai* are surely the *Hippeis*.

A picked body of 300 Spartans fighting around the king is of course instantly familiar to anyone who knows anything at all about ancient Sparta, as such a body fought and died with King Leonidas at Thermopylae (480). It is possible that these particular 300 were not the *Hippeis*, however, since Herodotus notes that Leonidas 'now came to Thermopylae with the appointed three hundred he had selected, all of whom had sons' (Hdt. 7.205.2), which would be inconsistent with a pre-existing body picked from the young men. Going back even further, we find the (perhaps legendary) Battle of the Champions (mid-sixth century) fought between Argos and Sparta over a disputed tract of land, where the two armies agreed 'that three hundred of each side should fight, and whichever party won would possess the land' (Hdt. 1.82.3). Again, these may not be the *Hippeis*, but a specially selected body of 300 men is something that we will encounter again.

Nothing is known about the internal organization of the *Hippeis*; apparently they were outside and additional to the *morai* of the 'regular' army, but whether they were divided into similar subunits is not recorded. They surely were (since as Thucydides records, the whole point of Spartan organization was to have officers at many levels), but 300 does not divide naturally into the existing regular units based on an *enomotia* of forty men (at maximum strength). Entirely speculatively, and based on the method of selection described above, we might imagine there were three *lochoi* of 100 men, divided into two *pentekostyes* of fifty, each made up of two *enomotiai* of twenty-five.⁹

The second Spartan unit we encounter at Mantinea, outside of the regular organization, is the Sciritae. Thucydides describes the Spartan deployment:

> 'In this battle the left wing was composed of the Sciritae, who in a Lacedaemonian army have always that post to themselves alone; next to these were the soldiers of Brasidas from Thrace, and the *Neodamodeis* with them; then came the Lacedaemonians themselves, *lochos* after *lochos*, with the Arcadians of Heraea at their side. After these were the Maenalians, and on the right wing the Tegeans with a few of the Lacedaemonians at the extremity; their cavalry being posted upon the two wings. Such was the Lacedaemonian formation.' (Thucydides 5.67.1–2)

The importance of the various allied units and their separation in the array from the Lacedaemonians proper is apparent, but the Sciritae (or Skiritai) are a regular part of the Spartan army, though outside of the six *morai*. Thucydides, as we have seen, states that they numbered 600 men (5.68.3). Xenophon also mentions them in Spartan armies in the fourth century:

> '[T]he Lacedaemonians sent out Eudamidas, and with him emancipated Helots and men of the *Perioikoi* and the Sciritae to the total number of about two thousand.' (Xenophon *Hellenica* 5.2.24)

According to the *Constitution*:

> 'To meet the case of a hostile approach at night, he [Lycurgus] assigned the duty of acting as sentries outside the lines to the Sciritae. In these days the duty is shared by foreigners, if any happen to be present in the camp.

> ...

> 'When the King leads, provided that no enemy appears, no one precedes him except the Sciritae and the *Hippeis*.' (Xenophon, *Constitution of the Lacedaemonians* 12.3)

So while the presence of this unit in the main line at Mantinea suggests that they are hoplites, their special role in sentry duty and advanced scouting suggests that they may have been faster and more lightly equipped than the rest of the hoplites. They were drawn from the region of Sciritis (Xen., *Hell.* 6.5.24; 7.4.21) and as such were *perioikoi*, not full Spartan citizens, but were separate from the general levy of *perioikoi* and had, for whatever reason, a unique status of their own.¹⁰

To summarize, the Spartan army – the only one for which we have any detailed information and probably the only one with a complex command structure – had five levels of subunits and commanders, while the king (one of the kings, or his appointed representative) commanded the army as a whole. The phalanx was divided into six *morai*, each under a *polemarchos*, and each *mora* was divided into two *lochoi* under a *lochagos*. Each *lochos* was made up of (perhaps) four *pentekostyes*, each commanded by a *pentekoster*, and each of these contained (perhaps) four *enomotiai* each under an *enomotarchos*. Each *enomotia* contained up to forty men, depending on the number of age classes called up, with more typical values being thirty-two to thirty-six (when the older men were not required), drawn up according to need in three, four or six files. Alongside these regular forces were at least two special units, the *Hippeis* (always 300 strong) and the Sciritae (600 strong on the one occasion we know their strength). Additional units – of *perioikoi* and *neodamodeis*, 'new people' who are probably freed (emancipated, newly enfranchised) Helots – could be added to this regular core to make the full army for any given campaign, and to this would be added the Spartans' neighbors and allies, who would generally be expected to supply men in equal numbers.[11]

One final point worth making is that in the later Macedonian army, the size of the subunits was fixed (a *syntagma* was 256 men, a *chiliarchia* 1,024, and so on) and phalanxes of different sizes were formed by gathering different numbers of these fixed-sized subunits. But in the Spartan army, the size of the basic building block, the *enomotia*, was flexible according to the number of age classes called up, and the larger units correspondingly varied in size – although of course for any given operation, the force in use might have been formed from a selection of *lochoi* or *morai*. In addition, the word *lochos* was used for units of various sizes, and we can never be sure when a *lochos* is referred to whether it is a particular Spartan unit of given approximate size or a generic unit of unknown size.

Subunits – other armies

So much for the Spartans. The organization of other armies is even less clearly understood, but still some snippets of information can be gleaned. Perhaps the best known, oddly enough, is not a *polis* army at all, but the mercenary army that the Persian pretender Cyrus raised in his attempt on the throne before meeting defeat at Cunaxa (401). The march of these Ten Thousand (or so) mercenaries back to Greece was immortalized by Xenophon, who, as one of their commanders and a man intensely interested in tactical matters, has left much useful information about their organization. He never spells the matter out explicitly, however, and it may be that they had little or no formal organization to start with. He describes the army review on the march to Cilicia (in Asia Minor):

> 'Cyrus inspected the barbarians first, and they marched past with their cavalry formed in *ilai* and their infantry in *taxeis*; then he inspected the Greeks, driving past them in a chariot, the Cilician queen in a carriage.' (Xenophon, *Anabasis* 1.2.16)

Perhaps it is to be understood that the Greeks too were formed in *taxeis*, or *lochoi*, their normal unit. When the Cilician queen returned home, she was escorted by a contingent of Greeks:

> 'Cyrus sent the Cilician queen back to Cilicia by the shortest route, and he sent some of Menon's troops to escort her, Menon himself commanding them … [I]n the course of her passage over the mountains to the plain two *lochoi* of Menon's army had been lost … they numbered a hundred hoplites.' (Xenophon, *Anabasis* 1.2.20; 25–26)

Does this imply that the Greek army was formed in *lochoi* of fifty men each? I think that unlikely, and in this case *lochos* probably just means a unit (or even a group) of indeterminate size (that happened in this case to be fifty). Nevertheless, the Greek army evidently had some division into subunits, but perhaps the organization was not hierarchical. The peltasts with the army are said (for example at Xen., *Hell.* 4.3.22) to be formed in *taxeis*, again probably just generic 'units'.

At any rate, Cyrus' review should serve as a reminder of what should be obvious (but sometimes seems to be forgotten), that the Greeks were not the only ones to divide their armies into subunits and to have a formal organization, and this is by no means a feature unique to the hoplite phalanx: all regular infantry – Greek, Persian or other – were evidently organized on broadly similar lines.

The clearest statement of organization comes during the return march, when the Greeks, to protect themselves from Persian cavalry, were marching in square. They found that according to the width of the road or presence of obstructions, the ends of the square became disordered as they stretched and compressed between the sides or 'wings':

> 'When the generals [*strategoi*] came to realize these difficulties, they formed six *lochoi* of a hundred men each and put a *lochagos* at the head of each *lochos*, adding also *pentekonteres* and *enomotarchoi*. Then in case the wings drew together on the march, these *lochoi* would drop back, so as not to interfere with the wings, and for the time being would move along behind the wings; and when the flanks of the square drew apart again, they would fill up the space between the wings, by *lochoi* in case this space was rather narrow, by *pentekostyes* in case it was broader, or, if it was very broad, by *enomotiai*, the idea being to have the gap filled up in any event.' (Xenophon, *Anabasis* 3.4.21–22)

Evidently the Ten Thousand were copying Spartan organization, which is not surprising given that many of them were Spartans, while others (like Xenophon) were admirers of Spartan drill. What is perhaps surprising is that Xenophon's account makes it appear that this was an on-the-spot innovation, and that prior to this the army had no such organization (although it is hard to believe there were no subunits at all – but perhaps this did not go beyond division into *lochoi*, which were presumably larger units than those formed here). Note also that although the Spartan unit names and levels are adopted, the unit sizes are not; the *lochoi* here are smaller units, only 100 strong, and we might guess that the *pentekostyes* were therefore actually fifty-strong, two per *lochos*, and the *enomotiai* each contained twenty-five men (which would make three files eight deep, plus the *enomotarchos*). Forming 'by *lochoi*' means that each *lochos* would be in a column, one *enomotia* wide and four deep (so that the formation was as narrow as possible); 'by *pentekostyes*' would have each *pentekostys* in column, so the *lochos* would be two *enomotiai* in width and two in depth; and 'by *enomotiai*' means a phalanx with the *enomotiai* side-by-side, occupying the widest possible frontage.

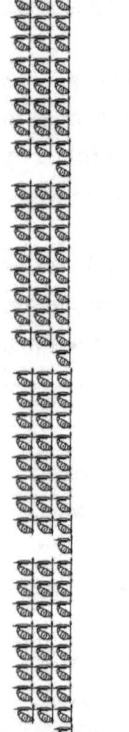

Formed by *lochos*

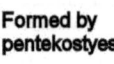

Formed by pentekostyes

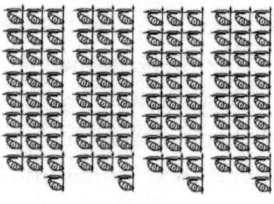

Formed by *enomotiai*

Three types of column in the army of the Ten Thousand.

A short while later, Xenophon details the arrangements for a retreat across a river in the face of the Carduchians (the local native people):

> 'And now, with the Greek baggage train and the camp followers in the very act of crossing, Xenophon wheeled his troops so that they took a position facing the Carduchians, and gave orders to the *lochagoi* that each man of them should form his own *lochos* by *enomotiai*, moving each *enomotia* by the left into the phalanx; then the *lochagoi* and *enomotarchoi* were to face toward the Carduchians and station file closers [*ouragoi*] on the side next to the river.' (Xenophon, *Anabasis* 4.3.26)

So it would appear that *enomotiai* (and their officers) were now in place throughout the army. Note also the presence of *ouragoi*, file closers. With reference to the Spartan army above, we saw that the file leaders might have been 'officers' in a sense, and the Athenians too had a concept of a special status for the front and rear rank men (see below). In this case I do not think the *ouragoi* should be thought of as a specific military rank, but rather that selected, experienced men were to be stationed in this position.[12]

Further on in the march, the Greek army formed to attack a town:

> 'Upon the arrival of the hoplites he [Xenophon] ordered each of the *lochagoi* to form his *lochos* in the way he thought it would fight most effectively ... When all preparations had been made ... the *lochagoi*, *hypolochagoi*, and those among the men who claimed to be not inferior to them in bravery were all grouped together in the line.' (Xenophon, *Anabasis* 5.2.11–13)

Hypolochagoi ('under-*lochos*-commanders') are known only from this one passage. Are these a new rank, or a general term for officers junior to the *lochagos* (*pentekonteres*, *enomotarchoi*)? I think the latter is meant. On another occasion, Xenophon gave his advice for an attack on a hill:

> '"It seems to me, fellow generals [*strategoi*], that we should station reserve *lochoi* behind our phalanx, so that we may have men to come to the aid of the phalanx if aid is needed at any point." All shared this opinion. "Well, then," said Xenophon, "do you lead on toward our adversaries, in order that we may not be standing still now that we have been seen by the enemy and have seen them; and I will come along after arranging the hindmost *lochoi* in the way you have decided upon." So while the others led on quietly, he detached the three hindmost *taxeis*, consisting of two hundred men each, and turned the first one to the right with orders to follow after the phalanx at a distance of about a *plethrum* [about 30 metres]; this *taxis* was commanded by Samolas the Achaean; the second *taxis* he posted at the

centre, to follow on in the same way; this one was under the command of Pyrrhias the Arcadian; and the last one he stationed upon the left, Phrasias the Athenian being in command of it.' (Xenophon, *Anabasis* 6.5.9–11)

Here, in the easy way Xenophon switches between *lochos* and *taxis*, we see the difficulty of seeking strict technical terminology in ancient authors, while the *taxeis* or *lochoi* of 200 men are larger than those we met earlier. Generally speaking, *lochoi* occur frequently in the pages of the *Anabasis*, and appear to be the normal name for the unit of around 100–200 men.

Finally, we see (perhaps contradictory) traces of the presence of age classes in the Ten Thousand. In an assault on a mountain, there is a suggestion that *lochoi* were grouped by age:

'Meanwhile Xenophon proceeded to climb the abandoned height with his youngest troops, ordering the rest to move on slowly in order that the hindmost *lochoi* might catch up.' (Xenophon, *Anabasis* 4.2.16)

But later, young men are clearly distributed within *lochoi*:

'Then Xenophon gave orders that the active men up to thirty years of age should move up from their several *lochoi* to the front. So he himself ran along with them, while Cleanor led the rest.' (Xenophon, *Anabasis* 7.3.3)

In summary, it appears that the hoplites of the Ten Thousand were divided into *lochoi* of 100–200 men, probably from the outset, and certainly during the return march. Subunits mirroring Spartan organization were formed only during the march (if that is what Xenophon intended to convey), but were used thereafter. It is also worth noting that the organization of the Ten Thousand suggested above (Xen., *Anab.* 3.4.22) does broadly match Xenophon's unamended text for the Spartan army (in which there are four *lochoi* to a *mora*):

	Constitution	*Anabasis*
Enomotia	36 men	25 men?
Pentecostys	72 men	50 men?
Lochos	144 men	100 men

But the problems with the *Constitution*'s scheme outlined above still stand. The drill and tactical implications of some of the manoeuvres mentioned will be considered in greater detail below.[13]

With the Spartans and the Ten Thousand we have dealt with by far the best-attested hoplite armies. The next best known is probably the Athenian, though here the evidence is scattered and slight. Xenophon records the opening moves at the Battle of the Nemea (394):

'Now the Lacedaemonians also veered to the right in leading the advance, and extended their wing so far beyond that of the enemy that only six *phylai* of the Athenians found themselves opposite the Lacedaemonians, the other four being opposite the Tegeans.' (Xenophon, *Hellenica* 4.2.19)

The *phylai* are the tribes of Athens, the traditional major groupings into which the citizen body was divided, and evidently they also formed the top layer of Athenian army organization. Athenian hoplites were drawn up in bodies, one drawn from each *phyle*, up to a total of ten, equivalent to the *morai* of the Spartan army (equivalent in that they are the largest subunit, though the regional recruitment of the Spartan *lochoi* makes them more similar to Athenian *phylai*, as we will see in Chapter 4). This division by *phylai* was evidently of some antiquity, since Herodotus records the same at the Battle of Marathon (490):

'When the presidency came round to him [Miltiades], he arrayed the Athenians for battle, with the *polemarchos* Callimachus commanding the right wing, since it was then the Athenian custom for the *polemarchos* to hold the right wing. He led, and the other *phylai* were numbered out in succession next to each other. The Plataeans were marshalled last, holding the left wing.' (Herodotus 6.111)

Note that the Athenian *polemarchos* is a more senior figure than the Spartan commander of a *mora*, and the overall command rotated around a body of ten *strategoi* (presumably one per *phyle*), as Herodotus says of the debate before the battle:

'The Athenian *strategoi* were of divided opinion, some advocating not fighting because they were too few to attack the army of the Medes; others, including Miltiades, advocating fighting. Thus they were at odds, and the inferior plan prevailed. An eleventh man had a vote, chosen by lot to be *polemarchos* of Athens, and by ancient custom the Athenians had made his vote of equal weight with the *strategoi*.' (Herodotus 6.109.1–2)

Note that the *phylai* in their military manifestation, as units of hoplites on the field, are also called *taxeis* (whether *taxis* here is a specific word for the muster of a *phyle*, or is just used as the generic word for 'unit', is a moot point), and their commander (or perhaps their administrative commander) was a *taxiarchos*. We see both *phyle* and *taxis* used in literary accounts: for example, *phyle* at Xen., *Hell.* 4.2.19 and Hdt. 6.111 quoted above, and Thuc. 6.98.4; *taxis* is mostly inferred from the title *taxiarchos*, for example Thuc. 4.4.1, while Thuc. 8.92.4 ('The hoplites in Piraeus ... among whom was Aristocrates, a *taxiarchos*, with his own *phyle*') shows the relationship. As in the Spartan army, a *taxis* was not a unit of fixed size, but could contain any number of men, perhaps typically 200 or so, while it was also used to designate the entire muster of the *phyle*, up to 1,000 or more. Used in its general sense of a 'unit', it could similarly cover many different sizes of formation.[14]

A few scattered references indicate that the *taxis* was subdivided into *lochoi*. The *Constitution of the Athenians*, traditionally attributed to Aristotle, spells this out for the fourth century at least:

> 'They [the Athenian Assembly] also elect by show of hands ten *taxiarchoi*, one for each *phyle*; these lead their fellow-tribesmen and appoint *lochagoi*.' (Aristotle, *Constitution of the Athenians* 61.3)

Xenophon records Socrates ribbing one of his companions who had taken a course in generalship:

> '"Don't you think, sirs," he said, "that our friend looks more 'majestic,' as Homer called Agamemnon, now that he has learnt generalship? ... But," he continued, "in order that any one of us who may happen to be a *taxiarchos* or a *lochagos* under you may have a better knowledge of warfare, tell us the first lesson he gave you in generalship."' (Xenophon, *Memorabilia* 3.1.4–5)

Lochoi appear, enigmatically, in Athenian comedy, where the officer Lamachos is given special orders:

> 'Herald: The *strategoi* order you forthwith to take your *lochoi* and your crests, and, despite the snow, to go and guard our borders.' (Aristophanes, *Acharnians* 1073–74)

The Greek word for the crest of a helmet, *lophos*, makes a pun with *lochos* (the joke, like most Old Comedy, falls flat in English translation – perhaps 'your crests and your companies'). Aristophanes' *Lysistrata* mentions four *lochoi* of women (Ar., *Lys.* 453–54) in a comedic scenario.

What is not revealed by any of these passages, nor by a fleeting reference in Xenophon to 'two *lochoi* of the hoplites' (Xen., *Hell.* 1.2.3), is the number of *lochoi* in each *taxis*, and it may have been the case that the number was not fixed. As the *taxis* was itself highly variable in size, then the *lochoi* too may have varied in size, or possibly *lochoi* were of a more regular size (perhaps 100 or 200 men, by comparison with the Ten Thousand and Sparta) and a different number of them would be organized in order to make up the required strength of the *taxis*, and of the total force. We are familiar from Spartan, Hellenistic and more modern armies with army organizations of a permanent nature, but the Athenian army was a militia formed on demand for a specific campaign or mission, and need have had no standing organizational structure. When a force was put together, the *strategos* – and the *taxiarchoi* of whichever *phylai* were required – could have enrolled and appointed *lochoi* and *lochagoi* according to need. As such, we may seek in vain for an answer to the question of how many *lochoi* there were in a *taxis* – the answer may simply be 'as many as were required to make up the numbers'.

As to smaller divisions below the *lochos*, for the Athenian army there is no evidence for such units. This does not mean they did not exist (since other cities did have smaller units, see below), and it may be an accident of the survival of evidence, but I think that the way Thucydides explains the Spartan system for his readers ('as almost the whole Lacedaemonian army, save for a small part, consists of officers under officers', Thuc. 5.66.4) suggests that a hierarchical structure of more than two levels, and extending down to the smallest units, was unfamiliar, and that therefore it is likely that the *lochos* was the smallest unit.[15]

However, Athenians (at least in the fourth century) were certainly aware of the idea that the file (*stichos*) should be structured, as shown in a well-known passage of Xenophon (a continuation of the passage on learning generalship quoted above):

> '["T]here are many other qualifications, some natural, some acquired, that are necessary to one who would succeed as a general [*strategos*]. It is well to understand tactics [*takitkos*] too; for there is a wide difference between right and wrong disposition of the troops, just as stones, bricks, timber and tiles flung together anyhow are useless, whereas when the materials that neither rot nor decay, that is, the stones and tiles, are placed at the bottom and the top, and the bricks and timber are put together in the middle, as in building, the result is something of great value, a house, in fact."
>
> "'Your analogy is perfect, Socrates,' said the youth; "for in war one must put the best men in the first and last place, and the worst in the centre, that they may be led by the one and driven on by the other."
>
> "'Well and good, provided that he taught you also to distinguish the good and the bad men. If not, what have you gained by your lessons? No more than you would have gained if he had ordered you to put the best money at the head and tail, and the worst in the middle, without telling you how to distinguish good from base coin.'" (Xenophon, *Memorabilia* 3.1.6–9)

This passage indicates clearly that it was understood that the file leaders (the front rank, *protostates*) and file closers (*ouragoi*) should be specially chosen; but also that it was up to the general (or other officers) to choose them, presumably on an ad hoc basis. As such, I think that this is clear evidence that the front and rear rank were special positions, but also that (in the Athenian army) they were not a special rank, as otherwise the point about knowing how to distinguish the best men would have no meaning (the front and rear ranks would be formed from the appropriate ranking officers).

Although the evidence of Onasander, a writer on generalship under the Roman Empire in the first century AD, must be used with great caution (since it is unclear which of his material is contemporary, which taken from Hellenistic theory, and which, if any, from Classical Greek), one passage on the duties of the general is of value here for the principle it illustrates:

> 'First arming the soldiers, he should draw them up in formation that they may become practised in maintaining their formation; that they may become familiar with the faces and names of one another; that each soldier may learn by whom he stands and where and after how many. In this way, by one sharp command, the whole army will immediately form in formation … For just as those who begin to learn to play a musical instrument [must learn where to place their fingers] in just this manner men unpractised and inexperienced in military formations, with great confusion and failure to find one another, will only after loss of much time take their places; but those who are well trained in formations quickly – indeed automatically, so to speak – rush to their stations, presenting a harmonious, I may say, and beautiful sight.' (Onasander, *The General* 10.1–3)

This advice should be kept in mind when reading, for example, Thucydides' account of Syracusan opposition to the Athenian siege works at Syracuse (415):

> 'The Syracusans, appalled at the rapidity with which the work advanced, determined to go out against them [the Athenians] and give battle and interrupt it; and the two armies were already in battle array, when the Syracusan generals observed that their troops found such difficulty in getting into line, and were in such disorder, that they led them back into the town.' (Thucydides 6.98.2–3)

Simply forming a phalanx in good time and good order required practice and experience (see further on this below), which in this case the Athenians had and the Syracusans did not, and the implication is that men knew, from experience, where in the line they should stand, so that the files were ordered to this extent. However, it is not evidence for there being formal ranks, and in fact (from Onasander) quite the contrary, for men knew their place by knowing the names and faces of the individuals around them, rather than because they fitted into a rigid hierarchical structure. At any rate, the Athenian army, when it was well practised, evidently had ordered files in this sense, but it is doubtful whether there was a formal army structure down to the level of the file.[16]

So the Spartans had a formal hierarchical structure of at least four levels, down to the smallest unit (the *enomotia* of thirty to forty men), together (perhaps) with a formal file structure, while the Athenians had a two-level structure, and files were perhaps ordered more by practice than by formal hierarchy. For other Greek armies, evidence is almost entirely lacking. There are scattered references to *taxeis* and *lochoi*, and it is most likely that other cities, where they had a formal structure at all, had one that resembled that of Athens.

There is, however, one intriguing scrap of evidence concerning the small city of Phlius (or Phlious), a Spartan ally in the northern Peloponnese, when in the fourth century a party of exiles tried to capture the city by surprise attack:

> 'When they had climbed up and found the posts of the guards weakly manned, they pursued the day-guards, who numbered ten (for one out of each *pempas* was regularly left behind as a day-guard); and they killed one while he was still asleep and another after he had fled for refuge to the Heraeum.' (Xenophon, *Hellenica* 7.2.6)

A *pempas* (or *pempad*) is literally a 'five', and suggests that the smallest unit of organization at Phlius was a squad of five, which itself would be either a file or a subset of a file, and in this case presumably formed part of a body of fifty (a *pentekostys*?), since taking one man from each *pempas* made ten. It is pure coincidence that we have this sort of information for Phlius (Xenophon offers no comment on this organization), and it may be pure chance that we do not have similar information for the army of Athens or Sparta.

The *pempas* as a military unit is in fact only known from one other source, that is Xenophon's semi- or wholly fictitious account of the Persian army in the *Cyropaedia*, and in fact the Persian army that Xenophon describes is known in more detail, thanks to Xenophon's writing, than any Greek army (or pretty much any other army in antiquity besides the Macedonian), and it is instructive to consider its organization. Whether Xenophon was accurately transmitting information about the real Persian army, or whether he was just describing a Utopian ideal, and to what extent if so this reflects real Greek or Persian practice, are questions to which we have no firm answers. Even so, Xenophon's account is highly informative even if only about an ideal army. Xenophon describes the Persian king Cyrus establishing competitive training for his army:[17]

> 'What he proposed was as follows: to the private soldier [*stratiotes*], that he show himself obedient to the officers, ready for hardship, eager for danger but subject to good discipline, familiar with the duties required of a soldier, neat in the care of his equipment, and ambitious about all such matters; to the *pempadarchos*, that, besides being himself like the good private, he

makes his *pempas* a model, as far as possible; to the *dekadarchos*, that he do likewise with his *dekas*, and the *lochagos* with his *lochos*; and to the *taxiarchos*, that he be unexceptionable himself and see to it that the officers under him get those whom they command to do their duty. As rewards, moreover, he offered the following: in the case of *taxiarchoi*, those who were thought to have got their *taxeis* into the best condition should be made *chiliarchoi* [and so on down the ranks]. Such, then, were the competitions appointed, and the army began to train for them ...

'Then, he had tents made for them – in number, as many as there were *taxiarchoi*; in size, large enough to each accommodate a *taxis*. A *taxis*, moreover, was composed of a hundred men. Accordingly, they lived in tents each *taxis* by itself ... He thought also that their tenting together helped them not a little to gain a perfect acquaintance with their positions. For the *taxiarchoi* had the *taxeis* under them in as perfect order as when a *taxis* was marching single file, and the *lochagoi* their *lochoi*, and the *dekadarchoi* their *dekades* and *pempadarchoi* their *pempades*. He thought, moreover, that such perfect acquaintance with their places in the line was exceedingly helpful both to prevent their being thrown into confusion and to restore order sooner in case they should be thrown into confusion; just as in the case of stones and timbers which must be fitted together, it is possible to fit them together readily, no matter in how great confusion they may chance to have been thrown down, if they have the guide-marks to make it plain in what place each of them belongs.' (Xenophon, *Cyropaedia* 2.1.22–28)

Here we have for this perhaps imaginary army a more perfect description than we have for any real army, with the following organization:

Unit	Officer	Size
Pempas	*Pempadarchos*	5
Dekas	*Dekadarchos*	10
Lochos	*Lochagos*	50?
Taxis	*Taxiarchos*	100
[*Chiliostys*]	*Chiliarchos*	1,000

The name *chiliostys* comes from Xen., *Cyrop.* 2.4.3 and 6.3.13. Xenophon adds a *muriarchos*, commander of a *murias* or Myriad (10,000 men), on top of this (for example at Xen., *Cyrop.* 8.1.14). The number of *dekades* in a *lochos* and of *lochoi* in a *taxis* is not specified, though later Xenophon refers (Xen., *Cyrop.* 2.3.21) to four *lochoi* making up a *taxis* (which would not fit easily with the scheme above, but Xenophon is not entirely consistent – at 2.4.4, the *dekas*-sized unit contains twelve men).[18]

We can see that Xenophon's scheme does not match the known organization of the Spartan or Athenian armies, and the presence of the *pempas* is all that links it to that of Phlius. Similarly, the *dekas* provides a link to another army on the fringes of the Greek world, that of the Macedonians. The Macedonian army, perhaps from the time of Philip II and certainly throughout the Hellenistic period, had a complete and formal structure of units and subunits, as is described in the Hellenistic tacticians. For the earlier period, before Philip's reforms, there is little information, although the following disputed and much-discussed passage may be relevant:

> 'Anaximenes, in the first book of the *Phillippika*, speaking about Alexander, states: "Next, after he accustomed those of the highest honour to serve as cavalry, he called them Companions, and after he had divided the majority of the infantry into *lochoi* and *dekades* and other commands, he named them Foot Companions."' (Harpocration *Lexicon* s.v. pezhetairoi: Anaximenes, *FrGrHist* 72 F 4)

Which 'Alexander' is here referred to is much debated (the options being Alexander I in the early fifth century, Alexander II in the early fourth, or Alexander III 'the Great', late fourth, and after Philip II's reforms). My own view is that on balance, Alexander I is more likely, and this passage refers to the organization of a small part of the Macedonian infantry from the fifth century. At any rate, we see here the Macedonian army with *dekades* and *lochoi* (at the least), which may conceivably have been modelled after the known decimal organization of the Persian army (of which Xenophon's idealized account is a reflection). It would be odd though if the army of Macedon (and of Phlius) had a formal organization down to a lower level than that of Athens and Sparta.[19]

One other army should be mentioned specifically – the Theban. There is almost no detail known about the Theban (or Boeotian) armies from the fourth century or earlier, but for two features. One is that the Thebans formed twenty-five ranks deep at Delium in 424 (Thuc. 4.93.4), and fifty (or rather 'not less than 50') at Leuctra in 371 (Xen., *Hell.* 6.4.12). Greek armies usually formed either eight ranks deep, or in multiples of four or eight (as we will see below), and twenty-five and fifty are not divisible by four or eight, but are by five, which might provide a hint that the Thebans, like the Phliasians, used a *pempas* as their smallest unit.

The second feature is the existence of the Theban 'Sacred Band' (*Hieros Lochos*):

> 'The sacred band, we are told, was first formed by Gorgidas, of three hundred chosen men [*epilektoi*], to whom the city furnished exercise and maintenance, and who encamped in the Cadmeia [the citadel of Thebes];

for which reason, too, they were called the city band; for citadels in those days were properly called cities.' (Plutarch, *Pelopidas* 18.1)

Plutarch goes on to note that:

'Gorgidas, then, by distributing this sacred band among the front ranks of the whole phalanx of hoplites, made the high excellence of the men inconspicuous, and did not direct their strength upon a common object, since it was dissipated and blended with that of a large body of inferior troops; but Pelopidas, after their valour had shone out at Tegyra, where they fought by themselves and about his own person, never afterwards divided or scattered them, but, treating them as a unit, put them into the forefront of the greatest conflicts.' (Plutarch, *Pelopidas* 19.3)

Gorgidas lived in the early fourth century, though Diodorus notes of the Battle of Delium (424):

'On the Boeotian side, the Thebans were drawn up on the right wing, the Orchomenians on the left, and the centre of the line was made up of the other Boeotians; the first line of the whole army was formed of what they called "charioteers and crew", a select group of three hundred.' (Diodorus 12.70.1).

The Sacred Band may thus have been older than Plutarch thought, and the archaic reference to 'charioteers and crew' suggests an institution of some antiquity. At any rate, we see here an equivalent elite unit to the Spartan *Hippeis*, and like that unit it is the magic number of 300 strong. Originally this elite unit provided the front line (so, the file leaders) of the entire phalanx, until Pelopidas separated them into a distinct unit, which was a major contributing factor to the Theban victory at Leuctra.

Other cities may also have formed elite (and professional, as we will see in Chapter 4) units of 'picked men' (*epilektoi* or *logades*). Plutarch (*Arist*.14.3) records the Athenian army's 'most zealous *lochagos*, Olympiodorus, with the three hundred picked men of his *lochos*' at Plataea (479) – note the number 300 again – and Aeschines (2.169) states that he was a member of the *epilektoi* in the mid-fourth century (see also Plut., *Phoc*. 13.2). Diodorus (11.76.2) mentions a Syracusan unit of 600 *epilektoi*, while in the Argive army at Mantinea was 'the thousand chosen Argives, which the city had for a long time caused to be trained for the wars at the public charge' (Thuc 5.67.2). Picked units of this sort, with professional training, may have become more common into the fourth century, though it is only by chance that we may hear of them.[20]

This brief survey of non-Spartan armies suggests that a level of organization – if nothing else, *lochoi* and *taxeis*, with some size of around 100 to several hundred men, and perhaps also designated files and fixed places to stand in the file – was probably common to all Greek armies. Two important questions remain, neither of which can be answered with any degree of certainty given the current state of our knowledge: to what extent was this sort of organization unique to the Greeks, and when in the development of the Greek hoplite did this organization arise?

To take the first question, little enough is known of contemporary armies for it to be possible to describe the internal structures of any non-Greek armies in any detail, but even so I think it is certain that the Persian army, at least, had a similar level of organization to Greek armies, including a number of similar units. The evidence of the *Cyropaedia* cannot be taken as proof of a historical Persian organization, but that the Persians had an organization that was broadly similar cannot be doubted. We have seen that there is evidence for a decimal unit structure with units of ten (*dekades*), 1,000 (*chiliostyes* or *chiliarchiai*) and 10,000 men (*myriades*), to give the Greek names, or in their possible Persian form, *baivarabam* for 10,000 men, *hazarabam* for the unit of 1,000, *sataba* for 100 and *dathaba* for ten men. We have also seen how Xenophon describes Cyrus' army (the historical Cyrus) being formed in *ilai* (of cavalry) and *taxeis* (of infantry) (Xen., *Anab.* 1.2.16), while Herodotus (e.g. 9.20) describes Persian cavalry charging 'by units [*kata telea*]' at Plataea (479), and Herodotus (1.103) attributes the first formal Persian organization to Cyaxares ('it was he who first organized the men of Asia in companies [*kata telea*] and posted each arm [infantry, cavalry, archers] apart'). So although we may have little information about other earlier or contemporary armies, there can be no question of the Greek army structure being in any way unique or unusual. All organized state forces, as opposed to informal tribal or pre-state forces, no doubt had similar levels of organization.[21]

The second question, as to the antiquity of these structures in Greece, is also difficult to answer with any certainty, and is of course closely tied in to the issues discussed in Chapter 1. We have already encountered Aristotle's thoughts on the matter:

> '[W]ithout orderly formation [*syntaxis*] heavy-armed infantry [*hoplitikon*] is useless, and the sciences and systems dealing with tactics [*taxis*] did not exist among the men of old times, so that their strength lay in their cavalry.' (Aristotle, *Politics* 1297b 18–22)

Aristotle uses the general word for arrangement, *syntaxis*, and it is open to question what level of organization a useful level of *syntaxis* in heavy infantry would represent; but at any rate the suggestion is that the rise in effectiveness of hoplites went hand-in-hand with the development of tactics (in the Greek sense of organization and drill) – that the one requires and implies the other (at least if the infantry are not to be 'useless').

We see a glimpse, perhaps, of the first debates about organization in Homer, such as the speech attributed to Nestor:

> 'But do thou, O King, thyself take good counsel, and hearken to another; the word whatsoever I speak, shalt thou not lightly cast aside. Separate thy men by tribes [*phylai*], by clans [*phretras*], Agamemnon, that clan may bear aid to clan and tribe to tribe. If thou do thus, and the Achaeans obey thee, thou wilt know then who among thy captains [*hegemones*] is a coward, and who among thy men, and who too is brave; for they will fight each clan for itself.' (Homer, *Iliad* 2.360–65)

Officers and men and divisions by tribe and clan would represent a bare minimum organization (it is not clear whether Nestor's advice is actually adopted), but is not very dissimilar from what is known of the structure of the Athenian hoplites, as we have seen. It is impossible to assign any date to the development of these levels of organization, but the evidence considered above (with Spartans tracing their organization to Lycurgus) suggests that the basics of army organization were of some antiquity. At the same time, we must expect there to have been developments arising from the experience of the Persian and Peloponnesian Wars and greater professionalism in the fourth century. So as discussed in Chapter 1, the likelihood is that formal organization down to a low level can be traced well back into the Archaic period for the Spartans, but that other Greek armies had a much simpler organization, perhaps just the two levels of the Athenians (or even a single level with ad hoc *taxeis*), throughout the Archaic period and well into (and indeed beyond, as we will see in Chapter 9) the Classical.

Intervals

We have seen that Greek hoplite armies, like contemporary armies of other peoples (so far as we know), were divided into at least a basic hierarchy of subunits, each with their own officers, though only in the Spartan army was this applied at multiple hierarchical levels. We must now consider how hoplite armies formed up for battle.

This question concerns the two chief measures of a phalanx (or any formation): its width or breadth from side to side (measured in files) and its depth from front to back (measured in ranks). The actual ground occupied by a phalanx was also dependent on another measure: the interval between files (usually reckoned in terms of the distance, for example, from one man's right shoulder to the next man's right shoulder, so that the interval includes the man, rather than being a gap between men). The question of the interval of the files in the hoplite phalanx has been the source of much discussion and not a little controversy, as it has become tied up with the debate over the origins and nature of the phalanx, and the dispute between the orthodox, early phalanx theory and its gradualist or revisionist late phalanx opponents, as discussed in Chapter 1. To summarize and simplify, in the orthodox view, the phalanx was always a close-order heavy formation and therefore, from the outset, adopted a close file interval which is set, in the modern debate, at 3ft per man (the reasons for which I will explore below). To the gradualists or revisionists, the phalanx was originally, and remained for some time – perhaps right through the fifth century – a more open-order formation, with file intervals of 6ft per man, not least to allow room for the integrated light infantry who are supposed to have continued to fight as part of the phalanx (at least up to the early fifth century).[22]

These file intervals are adopted from the tradition of the Hellenistic tacticians, as transmitted through Asclepiodotus, Aelian, Arrian and Polybius, who specify the file intervals of the Hellenistic Macedonian phalanx. Asclepiodotus provides the clearest exposition:

'The needs of warfare have brought forth three systems of intervals: the most open order, in which the men are spaced both in length and depth four cubits [2 metres] apart, the most compact, in which with locked shields each man is a cubit [50cm] distant on all sides from his comrades, and the intermediate, also called a compact formation, in which they are distant two cubits [1 metre] from one another on all sides.' (Asclepiodotus 4.1)

Asclepiodotus then sets out the different uses of these intervals:

'The interval of four cubits seems to be the natural one and has, therefore, no special name; the one of two cubits and especially that of one cubit are forced formations ... The former is used when we are marching the phalanx upon the enemy, the latter when the enemy is marching upon us.' (Asclepiodotus 4.3)

The cubit, the measurement used (along with feet) by ancient authors, approximates to 50cm; this is only an approximation, since there were various different sizes of cubit in antiquity. In the context of a military formation, an approximation is close enough, since the interval between men would not have been measured with precision anyway, but approximated by eye or by an outstretched arm or elbow. I will therefore work with a cubit approximating 50cm, which gives a standard file interval of 2 metres (four cubits, 6ft), close order of 1 metre (two cubits, 3ft) and closest order ('locked shields') of 50cm (one cubit, 18in).[23]

These file intervals for the Macedonian phalanx have sometimes been adopted unquestioningly for the earlier hoplite phalanx, although curiously enough, there are no indications whatever in ancient sources for what file intervals were in fact used in the Classical period or before, or indeed if there were any standard intervals at all. It does seem inherently likely that the Macedonian phalanx would have based its file intervals at least in part on those of the earlier phalanx, since Philip II, in devising Macedonian drill, most likely extended what was already familiar, rather than inventing completely out of thin air. We also have Diodorus' account of Philip's invention:

> 'Indeed he devised the compact order and the equipment of the phalanx, imitating the "locked shields" of the warriors at Troy, and was the first to organize the Macedonian phalanx.' (Diodorus 16.3.1–2)

This suggests that the Macedonian phalanx was a closer-order formation than the Greek phalanx (being inspired by supposed Homeric precedent, rather than contemporary Greek practice). The most likely interpretation is that the 'natural' spacing of four cubits and the 'close order' of two cubits were inherited from the Greek phalanx, but the one cubit 'locked shields' was a new formation devised for the Macedonians. The latter was made possible by the fact that the Macedonian phalanx, rather than duelling individually with spear and shield, relied on the mass effect of multiple lines of sarissa (pike) points extending in front of the formation, allowing the men to stand closer together (on the defensive), at the expense of reduced individual mobility.[24]

However, there is no direct evidence at all for file intervals in the hoplite phalanx. This is rather surprising, given that, although there are no tactical manuals from before the Hellenistic period, there is a relative wealth of information about other drill manoeuvres (which I will explore below) and a fair amount of information about Spartan organization, as we saw above. Xenophon at least must have been aware of what file intervals were useful in what circumstances, and Thucydides might have been expected to use this information, for example in his calculations of Spartan army size at Mantinea.

The word translated above as 'locked shields', *synaspismos*, does not occur in Classical authors. It appears in Asclepiodotus (and the other tacticians) and Diodorus in the passages quoted above, and in Arrian, and also in Plutarch, who quite commonly applies it to Romans where he is clearly not using it in a technical sense. Plutarch also uses the word in a Greek context, for example of Timoleon (who led the Sicilian Greeks in a mid-fourth century war against Carthage):

> '[H]e himself made his vanguard lock their shields in close array [*pyknosas toi synaspismoi*], ordered the trumpet to sound the charge, and fell upon the Carthaginians.' (Plutarch, *Timoleon* 27.6)

'*Pyknosis*', as a noun, is the name given by Asclepiodotus and the other tacticians to the two cubit interval, so whether Plutarch really has in mind a one cubit spacing (suggested by *synaspismoi*) or two cubits (suggested by *pyknosas*) is debatable. I think it most likely that Plutarch did not have technical meanings in mind here at all. The related verb, *synaspizō* or *synaspidoō*, occurs more frequently in several authors, but often with the sense of 'fighting alongside', not referring to any tactical formation; when Polybius uses it (12.21) for the Macedonian phalanx, it is with reference to the two cubit interval. We do at any rate see here a suggestion that a close-order formation (however close it might have been) could be adopted, and thus that a choice was available between more open or closer order.

But it is the evidence of the Classical historians, Herodotus, Thucydides and Xenophon, in which we are most interested, and they too do not offer any usage of the technical (Hellenistic) terms, but do suggest in a number of passages that a closing-up or tightening of the intervals was possible. Typical is a passage of Xenophon describing a fight between Arcadians and Spartans (Lacedaemonians) under Archidamus, who attacked first with cavalry and peltasts:

> 'The enemy, however, did not give way, but forming themselves into a compact body [*syntetagmenoi*], remained quiet. Then the Lacedaemonians attacked again. The enemy did not give way even then, but on the contrary proceeded to advance, and by this time there was a deal of shouting; Archidamus himself thereupon came to the rescue, turning off along the wagon road which runs to Cromnus and leading his men in double file, just as he chanced to have them formed. Now as soon as the two forces had come near to one another, the troops of Archidamus in column, since they were marching along a road, and the Arcadians massed together in close order [*synaspidountes*], at this juncture the Lacedaemonians were no longer able to hold out against the superior weight [*plethos*, 'crowd'] of the Arcadians … But when the Lacedaemonians as they retired along the road came out into open ground, they immediately formed themselves in line of battle against [*antiparetaxanto*] the enemy. The Arcadians on their side stood in close order [*syntetagmenoi*], just as they were, and while inferior in numbers, they were in better spirits by far, since they had attacked a foe who retreated and had killed men.' (Xenophon, *Hellenica* 7.4.22–25)

Note that Xenophon uses two words for the Arcadians' formations (*syntetagmenoi*, from *syntasso*, 'to draw up in array', and *synaspidoō*, 'have shields together'), which the translator quite freely translates as 'close order' – reasonably enough, since that is clearly the sense. Nevertheless, it seems clear that no specific technical meaning should be attached to these words, and Xenophon does not say anything about the Spartan intervals – they are forced back because they are fighting in marching column against a formed line. The impression is of the Arcadians tightening their formation by clustering more closely together to receive the initial cavalry and peltast attack, and fighting the Spartans in this formation.

We see similar instances in Herodotus, such as at Plataea (479), where a Phocian force was caught by Persian cavalry:

> 'But when the horsemen had encircled the Phocians, they rode at them as if to slay them, and drew their bows to shoot; it is likely too that some did in fact shoot. The Phocians opposed them in every possible way, drawing in together [*systrepsantes*] and closing their ranks [*pyknosantes*] to the best of their power. At this the horsemen wheeled about and rode back and away.' (Herodotus 9.18.1)

Two different words are used – *systrepho*, a very general word with meanings like 'tighten up' or 'bring together', and *pyknoō*, the verb from *pyknosis*, 'to close up'.

Thucydides too uses *systrepho* on a couple of occasions. At Plataea (431):

> 'The Thebans, finding themselves outwitted, immediately closed up [*sunestrephonto*] to repel all attacks made upon them. Twice or thrice they beat back their assailants.' (Thucydides 2.4.1–2)

An Athenian army at Amphipolis (422) was broken by a sudden attack:

> 'The Athenian right made a better stand, and though Cleon, who from the first had no thought of fighting, at once fled and was overtaken and slain by a Myrcinian peltast, his infantry forming in close order [*systraphentes*] upon the hill twice or thrice repulsed the attacks of Clearidas.' (Thucydides 5.10.9)

Clearly, these hoplite forces are forming closer order (which means that initially they were in a more open order). But I think that it is to go far beyond the evidence to assign any particular file interval to these open and closed-up formations (even if the odds are that the Hellenistic intervals of four and two cubits were used also by the Greeks), or to imagine that a specific drill is described by these 'tightenings'. The Hellenistic tacticians describe specific 'doubling' drills that an open-order formation could use to adopt close order, or drills involving closing up to the right or left (involving a quarter-turn left or right, march and face front), but there is no hint of such drills in the accounts of Classical battles, and the varying vocabulary used to describe these manoeuvres make it seem unlikely that a specific drill is meant. The Spartans, as a drilled and trained force, probably did have such drills, though they are never described, as perhaps would the picked forces, *epilektoi*, of other cities; but most Classical phalanxes were of untrained militia, and it is more likely that complex drills were not used. Rather, when we hear of a phalanx 'closing up' or 'drawing together' we should imagine a more ad hoc process of the men shuffling together, often in response to a specific threat, such as approaching cavalry or missile-armed light infantry. Infantry in such circumstances would naturally tend to bunch together for protection, and this is what the Classical authors appear to be describing. The process is very similar to the 'closing up' seen in Homer (see Chapter 1), in which formations faced with a specific threat (such as a rampaging Hector) could close up into a very tight group to meet the threat, but would then open out again for the ensuing fighting, in which more space was needed to fight effectively.

That said, it is likely that the four cubit order that seemed 'natural' to the Hellenistic tacticians would also have seemed natural for hoplites, so it is likely

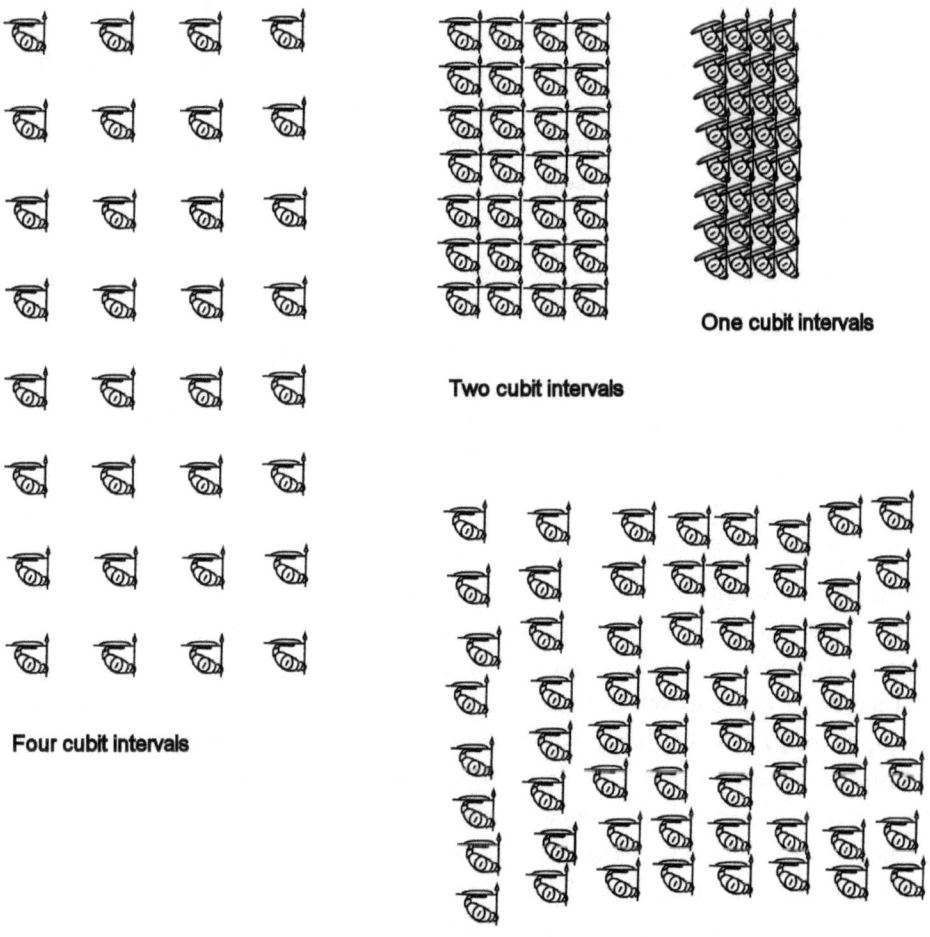

Possible file intervals used by hoplites.

that this open order (four cubits, 2 metres) was frequently adopted, especially on the march. It is also likely that some drill manoeuvres (see below), for those armies that had such drill, would have involved a process similar to the 'doubling' of the Macedonian phalanx, which would have halved these 'natural' intervals to two cubits. Furthermore, when hoplites did form in close order, as in the cases discussed above, it is likely that, armed as they were with shields roughly 1 metre in diameter, they would have used these shields as the measure of their file intervals, and so would have formed with intervals of 1 metre. This would have resulted in a formation with shields touching edge to edge (the literal meaning of *synaspismos*, 'shields together'), forming a continuous front. It is also perfectly possible that hoplites could on occasion crowd even closer together than this, overlapping their shields to a greater or lesser extent – though we lack artistic depictions, explicit descriptions or any hint of the drill that might have resulted in such a formation. Hoplites could well have crowded so closely that their shields overlapped, when occasion demanded, but the formation suggested by words like *synaspizein* was most likely that with 1 metre (two cubit) intervals.[25]

A particular passage of Thucydides is often quoted in this context – his description of the 'rightward drift' of the phalanx at Mantinea (418), already discussed in Chapter 2:

> 'All armies are alike in this: on going into action they get forced out rather on their right wing, and one and the other overlap with this their adversary's left; because fear makes each man do his best to shelter his unarmed side with the shield of the man next him on the right, thinking that the closer the shields are locked together [*aspidi ... pyknoteta tes synkleseos*] the better will he be protected.' (Thucydides 5.71.1)

'The closer the shields are locked together' could be translated more literally as 'the closer the bringing together of the shields'. This description has been interpreted in favour of four cubit, two cubit and closer than two cubit spacings, but I think that these interpretations are often over-literal. In particular, some think that hoplites held their shields (on their left arms) in such a way that half of their front was uncovered by the shield, so the entire right-hand side of their torso, as seen from the front, would be unshielded. As discussed in Chapter 2, I think this is an incorrect interpretation, both of the stance of hoplites and of the words of Thucydides. Re-enactors demonstrate clearly that the shield could be held across the front of the body, so as to cover the entire torso, and by 'his unshielded side' (literally, 'naked side') Thucydides means not the right half of the front of the torso, but the right side, as is clear from the many occasions where the unshielded side of a formation is said to be its right flank (as opposed to its front, or its left flank). So there is no reason to suppose that hoplites literally wished to shelter behind their neighbour's shield. Rather, they wished to keep their right flanks protected; that is, for the extreme right file, away from the enemy, and for the other files, in contact with their right-hand neighbours, where 'in contact' is some undefined distance – we cannot tell how close 'close enough' would be, but 'with shield edges touching' might represent an ideal degree of closeness. I think it most likely that a closed-up formation would have approximated 1 metre intervals (a shield's width), but that this passage of Thucydides sheds no particular light on this question.[26]

More fundamentally, I believe that the question of 2 metre or 1 metre intervals that has become a part of the orthodox–revisionist debate is a false dichotomy. In the early phalanx or proto-phalanx infantry formations of the early Archaic period, the formations putatively described by Homer, there is no formal drill or formation, so no fixed intervals were used. Men could line up very close together on occasion, such as to resist a dangerous attack, or they could fight in more open formation, in the normal course of events. As the phalanx acquired order, formation and drill (the 'tactics' described by Aristotle) in the course of the Archaic and into the Classical period, file intervals may have become formalized, but just as the Macedonian phalanx could use any of three intervals, hoplite phalanxes would have used at least two (though the lack of references to formal drills for changing interval suggests that they may not have been formalized even to this extent). Drilled and trained hoplites might have fought with either interval depending on the situation (tighter formations being preferred when facing cavalry or missiles, for example), or might have marched in one and fought in the other. Less-drilled hoplites would have had no fixed interval and would have stood as close together as felt appropriate – remembering that we are told by Thucydides (5.70.1) that armies typically lost cohesion on going into contact, that special steps had to be taken if they were to be kept in formation, and that the point of Thucydides' account of Mantinea is that the file interval of the (Spartan!) phalanx was not fixed by drill or determined by its officers, it was set by the individual hoplites themselves as they sought safety and protection from proximity to their neighbours. It is therefore, I believe, pointless arguing whether hoplites fought with a 1 or 2 metre frontage (or less, or more). There was probably no fixed standard other than as defined by the size of the shield, and the density of a formation would have varied according to a range of circumstances (usually aiming to match how the enemy were deployed), and in most cases is now simply unknown to us.

Depths

On the question of unit depths, we are, perhaps surprisingly, on much firmer ground. No Classical author, and no later author referring back to Classical or Archaic times, ever explicitly states the interval of files in a hoplite army. But we have numerous references to the depth (that is, the number of ranks) at which such armies formed. This suggests that depth was a much more important measure than interval, and also that there was much more opportunity for (planned) variation in depths than in intervals. If a phalanx was to avoid being outflanked by its opponents, the easiest way to do this was by adjusting the depth; greater depth meant a narrower formation, while reducing the depth allowed a formation to become wider, threatening or guarding against outflanking. The reasons for the importance of depth, and the trade-offs involved in determining unit depth and frontage, will form part of the subject of Chapter 8. For now we are concerned only with the range of depths adopted, and the means available for changing between them.[27]

Occasions where the depth in ranks of a hoplite formation are specified in the contemporary literary historians (Herodotus, Thucydides, Xenophon), plus one fourth-century orator (Isocrates), are set out below (there are a few additional references in later authors – Polyaenus, Diodorus – but they are of doubtful value and do not add anything to our knowledge, so they are excluded here):

Marathon (490)
'As the Athenians were marshalled at Marathon, it happened that their line of battle was as long as the line of the Medes. The centre, where the line was weakest, was only a few ranks deep [*epi taxias oligas*], but each wing was strong in numbers.' (Herodotus 6.111.3)

Dipaea (471)
Spartans: 'Remember the men who at Dipaea fought against the Arcadians, of whom we are told that, albeit they stood arrayed with but a single line of soldiery [*epi mias aspidas*], they raised a trophy over thousands upon thousands.' (Isocrates, *Archidamus* 99)

Delium (424):
'The Thebans formed twenty-five shields deep, the rest [Theban allies] as they pleased. Such was the strength and disposition of the Boeotian army. On the side of the Athenians, the hoplites throughout the whole army formed eight deep, being in numbers equal to the enemy.' (Thucydides 4.93.4–94.1)

Mantinea (418)
Spartans: '[A]s to the depth, although they had not been all drawn up alike, but as each *lochagos* chose, they were generally ranged eight deep.' (Thucydides 5.68.3)

Syracuse (415)
Athenians and allies: 'Half their army was drawn up eight deep in advance, half close to their tents in a hollow square, formed also eight deep ... The Syracusans, meanwhile, formed their hoplites sixteen deep.' (Thucydides 6.67.1–2)

Retreat from Syracuse (413)
Syracusans: 'Early next morning they [the Athenians] started afresh and forced their way to the hill, which had been fortified, where they found before them the enemy's infantry drawn up many shields deep to defend the fortification, the pass being narrow.' (Thucydides 7.79)

Piraeus (Munichia) (404)
Athenians: 'And the men from the city, when they came to the market-place of Hippodamus, first formed themselves in line of battle, so that they filled the road which leads to the temple of Artemis of Munichia and the sanctuary of Bendis; and they made a line not less than fifty shields in depth; then, in this formation, they advanced up the hill.' (Xenophon, *Hellenica* 2.4.11)

'As for the men from Phyle, they too filled the road, but they made a line not more than ten hoplites in depth.' (Xenophon, *Hellenica* 2.4.12)

'Now Thrasybulus and the rest of his hoplites when they saw the situation, came running to lend aid, and quickly formed in line, eight deep, in front of their comrades. And Pausanias, being hard pressed and retreating about four or five stadia to a hill, sent orders to the Lacedaemonians and to the allies to join him. There he formed an extremely deep phalanx and led the charge against the Athenians.' (Xenophon, *Hellenica* 2.4.34)

Cyrus' parade (401)
'He [Cyrus] ordered the Greeks to form their lines and take their positions just as they were accustomed to do for battle, each general marshalling his own men. So they formed the line four deep, Menon and his troops occupying the right wing, Clearchus and his troops the left, and the other generals the centre.' (Xenophon, *Anabasis* 1.2.15)

Thrace (400)
The Ten Thousand: '[W]ithin a short time the hoplites had fallen into line eight deep and the peltasts had got into position on either wing.' (Xenophon, *Anabasis* 7.1.23)

Maeander (399)
Spartans: 'When Dercylidas learned of all this, he told the *taxiarchoi* and the *lochagoi* to form their men in line, eight deep, as quickly as possible, and

to station the peltasts on either wing and likewise the cavalry.' (Xenophon, *Hellenica* 3.2.16)

Nemea (394)
Thebans, Athenians and allies: '[T]hey were negotiating about the leadership and trying to come to an agreement with one another as to the number of ranks in depth in which the whole army should be drawn up, in order to prevent the cities from making their phalanxes too deep and thus giving the enemy a chance of surrounding them.' (Xenophon, *Hellenica* 4.2.13)

'[W]hen the Athenians took position opposite the Lacedaemonians, and the Boeotians themselves got the right wing and were stationed opposite the Achaeans, they immediately said that the sacrifices were favourable and gave the order to make ready, saying that there would be a battle. And in the first place, disregarding the sixteen-rank formation, they made their phalanx exceedingly deep.' (Xenophon, *Hellenica* 4.2.18)

Corcyra (373)
Spartans: 'These latter, who were drawn up only eight deep, thinking that the outer end of the phalanx was too weak, undertook to perform an *anastrophe*.' (Xenophon, *Hellenica* 6.2.21)

Leuctra (371)
'Coming now to the infantry, it was said that the Lacedaemonians led each *enomotia* three files abreast, and that this resulted in the phalanx being not more than twelve men deep. The Thebans, however, were massed not less than fifty shields deep, calculating that if they conquered that part of the army which was around the king, all the rest of it would be easy to overcome.' (Xenophon, *Hellenica* 6.4.12)

Mantinea (370)
Spartans: 'When the phalanx had thus been doubled in depth, he [Agesilaus] proceeded into the plain with the hoplites in this formation, and then extended the army again into a line nine or ten shields deep.' (Xenophon, *Hellenica* 6.5.19)

Fourth century
Spartans: 'These *morai* at the word of command form *enomotiai* sometimes (two), sometimes three, and sometimes six abreast.' (Xenophon, *Constitution of the Lacedaemonians* 11.4)

As we can see, there is very little information in Herodotus about depths (he seems uninterested in the topic). His description of the Athenian formation at Marathon, conventionally translated and interpreted as being with fewer ranks in the centre but more ranks on the flanks, is rather less clear in the Greek. He says 'with few (or small) *taxeis*', and *taxis* – as well as being, as we have seen, the name of a unit of a few hundred men – is usually used by Herodotus to mean a position or a formation, and there is no other occasion in Herodotus where it means 'ranks' (although it is often translated as such in the English expression 'in the ranks', meaning 'in the formation'). So it is not certain that the Athenians had a shallower formation in the centre and a deeper one on the flanks, although that is a reasonable interpretation and it is difficult to see what else Herodotus might have meant.

It is also hard to know what to make of Isocrates' reference to 'a single [line of] shields' at Dipaea. As he was speaking 100 years after these events, perhaps a certain amount of mythologization had set in; a formation a single rank deep is so out of keeping with every other example we have and with everything we know (or think we know) about how hoplites fought, that I do not think much credence can be given to it. One possibility is that he meant a single rank of hoplites, with the rest of the formation being filled up with armed helots – we are reminded that at Plataea there were said to be seven helots for every Spartan hoplite (Hdt. 9.10.1), which would give a depth of eight in such a mixed formation.[28]

In Thucydides and (particularly) Xenophon, however, we find numerous clear and specific references to depths in ranks (often expressed as a depth in shields, sometimes just as, for example, 'in eights'). The number that keeps occurring is eight, and it has been proposed as a result (by modern authors, and suggested also by the Hellenistic tacticians) that eight deep was the standard Greek hoplite formation. In terms of the frequency of occurrences this is reasonable, as eight ranks do appear more frequently than any other number, though it is also apparent that there is not really any standard depth. On several occasions the number of ranks is determined individually and separately by different city contingents or even different units in the same contingent (at Delium 'Thebans formed twenty-five shields deep, the rest as they pleased', at Mantinea the Spartans 'had not been all drawn up alike, but as each *lochagos* chose', at the Nemea the allies were 'trying to come to an agreement with one another as to the number of ranks in depth in which the whole army should be drawn up'). The Spartans after Mantinea present a contrast ('When Dercylidas learned of all this, he told the *taxiarchoi* and the *lochagoi* to form their men in line, eight deep'), but it would be hard to discern a trend from these scattered examples. It seems remarkable that depth should have been left to individual contingents and even more so that it was left to individual *lochagoi*, and the fourth-century examples do perhaps indicate greater efforts at uniformity and central control, with the discussions at the Nemea producing a plan for sixteen deep, for all that the Thebans ignored the agreement and did their own thing.

The example of Cyrus' parade, where Cyrus ordered the Greeks to 'take their positions just as they were accustomed to do for battle ... So they formed the line four deep' has been taken by some to indicate that four deep was the standard depth, which seems unlikely given that this is the only time a depth of four is mentioned (aside from Diodorus 13.72.6, where a Spartan army is said to have formed four deep in 408 to surround the walls of Athens, an account which is problematic for various reasons, and does not describe a pitched battle anyway). The reasoning behind this theory has to do with the drill supposed to be used for changing depth, which I will examine below, but suffice to say for now that four deep is very unlikely to have been standard, and eight deep was, if not standard, at least much more common.[29]

We might also note that according to Xenophon, the Spartan *enomotia* could form up two, three or six abreast, and we have also seen that the strength of the *enomotia* varied according to the size of the call-up. With a twenty-five age group call-up, giving twenty-five men in the *enomotia*, three abreast would be eight deep with one man remaining. We should be clear that it must have been exceptionally rare for the number of men in every unit in an army to divide exactly into the same number of ranks deep, and some 'remainders' must have been common, indeed the norm. Such remainders would presumably have formed a partial rear rank (or more accurately, when the file closer was a position of special responsibility, they would have formed a partial central rank, with files closed up accordingly to give a rear rank all made up of file closers). The lack of uniformity of numbers in Greek units may account for delegating the depth decision to the *lochagoi* – if *lochoi* were at different strengths (due to losses, or stragglers, or different call-up responses), then it would make sense for them to deploy at (slightly) different depths so as to optimize their own formations and maintain some uniformity of frontage, which would be important when deploying from column into line, for example. A full-strength *enomotia*, theoretically forty men strong, would be twenty, thirteen to fourteen or six to seven ranks deep, deployed two, three or six abreast; a possibly more typical thirty-two-man *enomotia* would be sixteen, ten to eleven or five to six men deep. It is noteworthy that we do not see such depths (except sixteen on occasion), with eight being most common for Spartans – the twelve-man depth at Leuctra being an exception (representing thirty-six-man *enomotiai* three men abreast, and note 'not more than' twelve deep, so not all were the same depth). All in all, while eight is a number that occurs frequently, there is evidently no fixed or common depth. This is itself surprising given that, in the

Spartan army at least, the file leaders were supposedly 'officers' or at least men with greater responsibility – more on this below.[30]

We can also note the occasional very deep formations, twenty-five or fifty shields deep, particularly associated with the Thebans. The deep Athenian force at the Piraeus was constrained by roads in what was effectively a street fight, so cannot be considered typical. For the Thebans though, there is clearly a deliberate effort to experiment with deep formations, which were ultimately successful (at Leuctra), and the reasons for this success will be considered further in Chapter 8. We should note for now that the important consideration was the balance between width and depth: a deep formation was more solid and harder to break through, but easier to outflank, while a wide formation would avoid being outflanked and might outflank an opponent, but might also be broken through. In an allied army with different city contingents, a city could keep its own men safe by forming them deep and relying on other contingents to extend to cover the enemy frontage; hence the discussions about depth and the attempts (at Nemea) to reach a consensus, so that no city was shirking its responsibility (as Xenophon implies the Thebans did by forming deep, apparently as part of their own tactical plan). An army (like that of the Syracusans) that doubted its effectiveness might form deep to add solidity to their line, with depth substituting for training, experience and skill.[31]

Note also that the first definite statement we have as to depth (other than hints in Herodotus, and Isocrates' doubtful evidence) is from Delium in 424. For the first three-quarters of the fifth century, including the Persian Wars, and the entirety of the Archaic period we have no information at all about depths. Tyrtaeus is certainly describing formations with some depth, not least because the *promachoi*, the foremost, must be contrasted with men further back, and the fluid formations of Homer also have depth for the same reason. We have no clues as to what this depth may have been (and very likely it was not rigidly defined or enforced in the less well-drilled early phalanx or proto-phalanx), but at any rate we have no evidence that the Classical phalanx differed from the more fluid formations that preceded it by having greater depth. It may be that earlier formations were (on average) shallower, but there is no good reason to assume that they were.[32]

By the same token, we cannot assume that the Greek phalanx was distinguished from its Near Eastern contemporaries and predecessors by having greater depth. There are no definite statements of the depth of such armies, but for example at Plataea (479):

'[The Persian commander Mardonius] posted the Persians facing the Lacedaemonians. Seeing that the Persians by far outnumbered the

Lacedaemonians, they were arrayed in deeper ranks and their line ran opposite the Tegeans also.' (Herodotus 9.31.1–2)

'Deeper ranks' translates *epi taxis pleunas*, 'more *taxeis*', which by comparison with the *epi taxias oligas*, 'fewer *taxeis*' of the Athenians at Marathon, probably does mean 'more ranks', that is deeper. So the Persian formation on this occasion was deeper than the Greek. Later accounts, particularly Xenophon's semi-fictional account of the Persian army and its enemies in the *Cyropaedia*, assign even greater depth to 'Eastern' armies, the Lydians at Thymbrara being drawn up thirty deep and the Egyptian contingent 100 men deep (Xen., *Cyrop.* 6.3.19–20), while the Egyptian and Persian contingents at Cunaxa (401) were said to be in *plaision* ('brick') formation (Xen., *Anab.* 1.8.9), a formation which at least suggests great depth. The later Persian armies facing Alexander, as at Issus (333) and Gaugamela (331), were also said to be formed very deep – 'in such depth that they were useless' (Arr., *Anab.*2.8.8); 'in deep formation' (3.11.5). Partly this reflects the grossly inflated numbers assigned to such Eastern armies, but it is likely that Persian and allied infantry did indeed form in deep formations, so there was nothing unique about the Greek phalanx in this regard, and indeed it is likely that, in comparison with contemporary infantry, the Greek phalanx at only eight ranks deep was a relatively shallow formation.

Changing depth

We have considered so far units forming up with a given depth, but what of manoeuvres to change depth? In the case of undrilled formations, it may have been necessary to decide on a depth and form up in that depth, and it was then fixed for the duration of the battle. For drilled units (specifically, Spartans) it was different, as drills existed for changing depths on the fly:

> 'The prevalent opinion that the Laconian infantry formation is very complicated is the very reverse of the truth. In the Laconian formation the front rank men are all officers, and each file has all that it requires to make it efficient. The formation is so easy to understand that no one who knows man from man can possibly go wrong. For some have the privilege of leading; and the rest are under orders to follow. Orders to perform the *paragoge* are given verbally by the *enomotarchos* acting as a herald, and the phalanx is formed either thin or deep. Nothing whatever in these movements is difficult to understand.' (Xenophon, *Constitution* 11.5–6)

The implication of Xenophon's comments is that – even when he was writing in the fourth century – many would have thought such manoeuvres *were* complicated and difficult to understand, and that therefore this drill was not in widespread use. The Hellenistic tacticians detail drills for various forms of 'doubling', some of which alter the intervals of the phalanx (closing up and opening out the files) and some of which alter the depth. It is not always clear, even in the case of the Hellenistic phalanx, how changes in depth affected the files, since the file leader was a special position and held by a man of greater rank (and pay). If the phalanx halved its depth, say from sixteen to eight ranks deep, this would presumably require the half-file leader, the man at number nine in the sixteen-man file, to lead the second half of the file up alongside the first half (thereby, without further adjustments, halving both the intervals and the depth). This would mean that the rear rank was no longer made up of file closers, nor the front of file leaders, so may not have been commonly carried out. But Xenophon's account of the Spartan formation, with his variable widths for the *enomotia*, requires at least three depths (in which the actual number of men could also vary, depending on the strength of the *enomotia*), and is it hard to see how this fits with the special position of the file leaders.[33]

Be that as it may, the basic drill for adjusting depth seems to be the *paragoge*, left untranslated in the passage above and described in detail by Xenophon in the *Cyropaedia* (remembering that this is based on his hypothetical Persian organization, not the actual Spartan one):

'And once he saw another *taxiarchos* leading his *taxis* up from the river from the left in single file and ordering when he thought it was proper, the second *lochos* and then the third and the fourth to advance to the front; and when the *lochagoi* were in a row in front, he ordered each *lochos* to march up in double file. Thus the *dekadarchs* came to stand on the front line. Again, when he thought proper, he ordered the *lochoi* to line up four abreast; in this formation, then, the *pempadarchoi* in their turn came to stand four abreast in each *lochos*; and when they arrived at the doors of the tent, he commanded them to fall into single file again, and in this order he led the first *lochos* into the tent; the second he ordered to fall in line behind the first and follow, and, giving orders in like manner to the third and fourth, he led them inside. And when he had thus led them all in, he gave them their places at dinner in the order in which they came in. Pleased with him for his gentleness of discipline and for his painstaking, Cyrus invited this *taxis* also with its *taxiarchos* to dinner.' (Xenophon, *Cyropaedia* 2.3.21)

Various forms of the verb *parago* occur throughout this passage, and this is (one form of) the *paragoge*, the 'leading up beside', where one formation (in this case, a file) marches up alongside the one in front, used for moving from column into line (see below) and for deploying a formation at the desired depth. While

Xenophon's description is clear and precise, the manoeuvre is perhaps easier to describe visually than verbally, and the figure below is an attempt to depict the process. In the course of this drill, a *taxis* eighty men strong (of four *lochoi*, each of twenty men) starting in single file and therefore eighty men deep, deployed first to four abreast, twenty men deep, then to eight abreast, ten deep, and finally sixteen abreast and five men deep. In each case the intervals would have halved (and obviously in the earliest stages of the manoeuvre, the unit would not have been in any combat formation).[34]

This was one way to change depth – leading files or other formations up alongside those in front of them to make the line shallower. In order to make the line deeper, we hear of a rather obscure process that may have been the reverse of

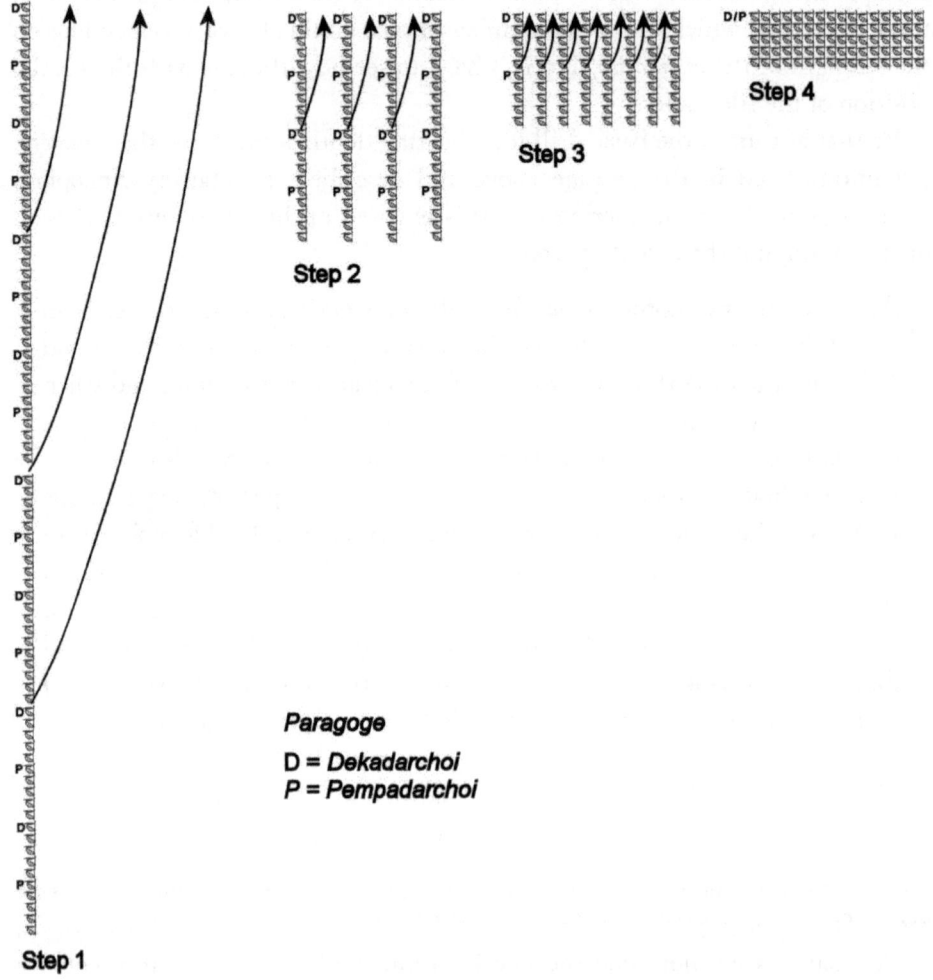

Paragoge: deploying a *lochos* from column into phalanx

the *paragoge* – the *anastrophe*. As usual, the word does not have a single technical meaning and is used to mean 'face about', 'turn around' or 'rally', and by the Hellenistic tacticians (e.g. Asclep 10.6) for a slightly different manoeuvre. But when used by Xenophon for Spartan armies, it seems to refer to a particular manoeuvre for increasing depth, as for the Spartan force before Corcyra in 373:

> 'These latter, who were drawn up only eight deep, thinking that the outer end of the phalanx was too weak, undertook to perform an *anastrophe*. But as soon as they began the backward movement, the enemy fell upon them, in the belief that they were in flight, and they did not turn back around; furthermore, those who were next to them also began to flee.' (Xenophon, *Hellenica* 6.2.21)

A similar manoeuvre was used by Agesilaus in 370 to extract his army from a difficult situation before the city of Mantinea:

> 'On the following day at daybreak he was offering sacrifices in front of the army; and seeing that troops were gathering from the city of the Mantineans on the mountains which were above the rear of his army, he decided that he must lead his men out of the valley with all possible speed. Now he feared that if he led the way himself, the enemy would fall upon his rear; accordingly, while keeping quiet and presenting his front toward the enemy, he ordered the men at the rear to perform an *anastrophe* to the right and march along behind the phalanx toward him. And in this manner he was at the same time leading them out of the narrow valley and making the phalanx continually stronger. When the phalanx had thus been doubled in depth, he proceeded into the plain with the hoplites in this formation, and then extended the army again into a line nine or ten shields deep.' (Xenophon, *Hellenica* 6.5.18–19)

Evidently in this manoeuvre, a section of the phalanx about-faces, marches to its rear, makes a quarter-turn left or right, then marches (now in column) behind the stationary part of the phalanx. (Modern translations and interpretations often introduce the word 'wheel', which only muddies the waters, as a wheel, properly speaking, involves rotating a formation on one of its corners as a pivot, whereas these are all turns; see below). Note that in the second example, the original eight-deep formation ends up nine or ten deep, with unclear consequences for the file structure of the phalanx (which evidently could not have been rigid). In the first case, the initial about-face and march to the rear was mistaken for flight by the enemy, and soon became real flight, illustrating the extreme danger of performing such fancy manoeuvres when in close proximity to the enemy; such evolutions were possible, but often the best tactic was just to keep moving straight ahead.

144 The Greek Hoplite Phalanx

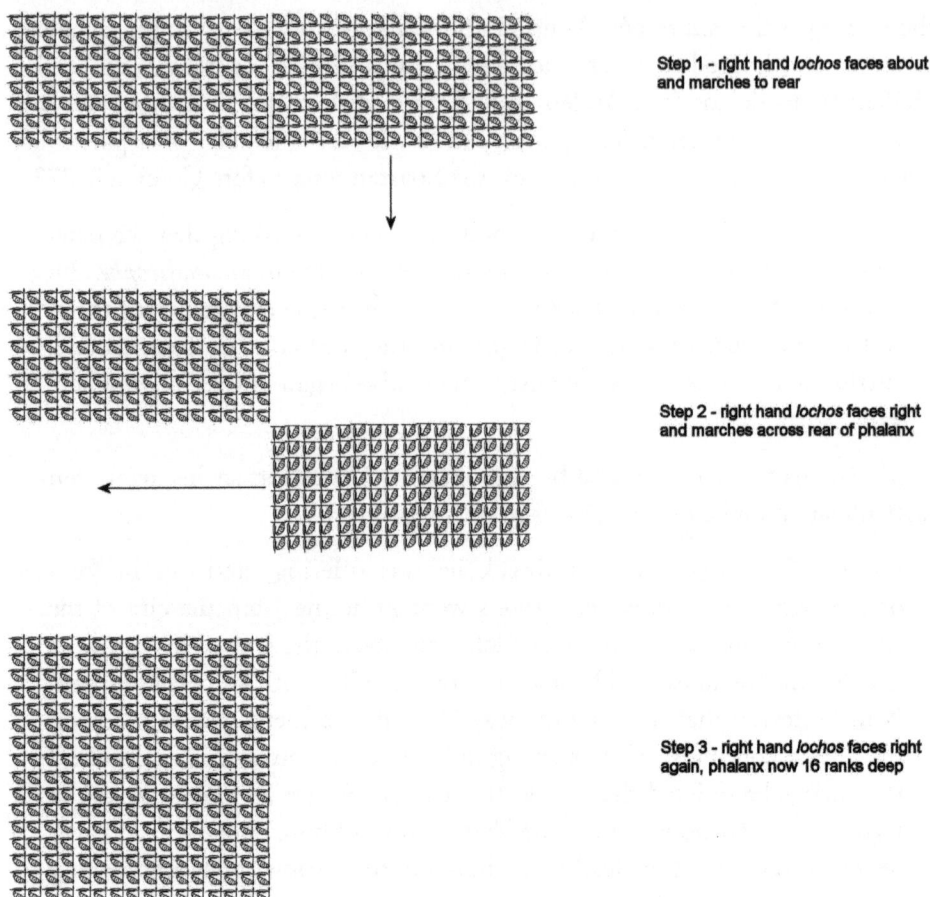

Anastrophe: reducing the width and increasing the depth of the phalanx.

Wheels, turns and countermarches

Having a unit face about and march to its rear, the first stage of the *anastrophe*, is termed a 'countermarch', or *exeligmos*, by the Hellenistic tacticians (the verb, *exelissō*, is used by Xenophon, for example *Const.Lac.* 11.9 and, for the Spartan manoeuvre at Coronea, *Hell.* 4.3.18). There were three types of countermarch available to Hellenistic phalanxes: the Laconian (Spartan), Macedonian, and Persian or Cretan. The Macedonian, we might imagine, was invented for the Macedonian phalanx, while the Persian reminds us that Greek armies were not the only ones to perform formal drill manoeuvres. The Laconian is presumably the countermarch used by the Spartan army of Xenophon's time; the individual hoplites all perform an about-face (turning 180 degrees) in position, then the file leaders (now at the back of the formation) march forward through the formation, with the rest of the file following them, to take up a new station in front of the old position. Note that simply about-facing would not be sufficient, as this would leave the file leaders at the rear of the formation, so a countermarch is necessary to restore the proper file structure. Note also that some space would have been required between the files for the men to pass through; how much space is never specified, but the ability to perform this manoeuvre would seem to require that the phalanx was not closed up as close as it could be, though a spacing of 1 metre was perhaps sufficient. Asclepiodotus (10.14) comments that this countermarch, as a way to counter an enemy appearing in the rear of the phalanx, was more imposing, as the countermarching phalanx advances on the threatening force; but as a way of withdrawing in the face of an enemy to the front it could be mistaken for flight, as happened at Corcyra.

The first step of the *anastrophe* is a turn in place, which the Hellenistic tacticians call *klisis*, or *metabole* for a complete about-face (there seems to be no single word used by Classical authors for this drill). In a turn, individual men within the unit turn 90 or 180 degrees, while the unit itself retains its original orientation. Clearly, without further manoeuvres, such as a countermarch, this would put the file leaders in the wrong position relative to their units and the rest

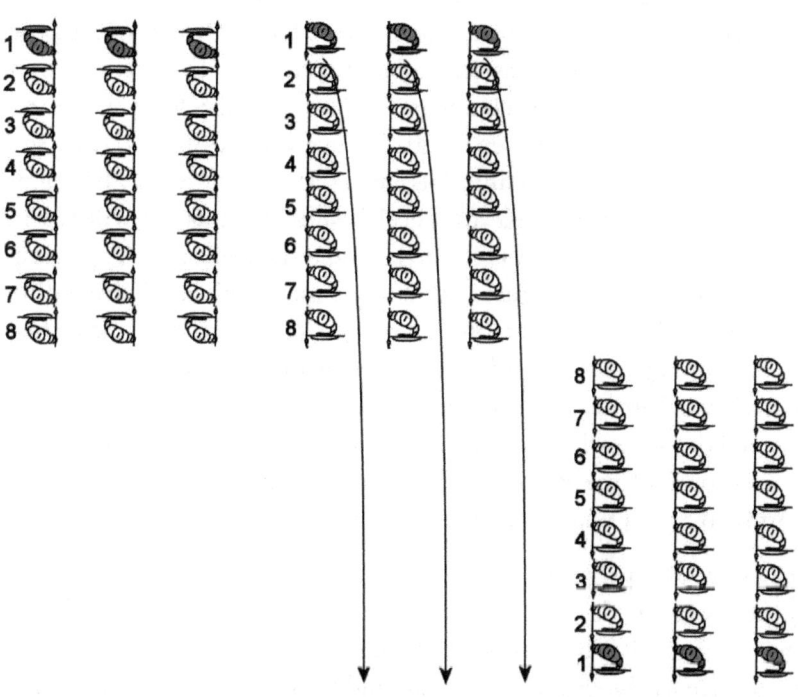

Step 1- face about

Step 1- file leaders march, rest of file falls in behind and follows

Step 3 - new position facing to rear and closer to enemy

The Laconian countermarch.

146   The Greek Hoplite Phalanx

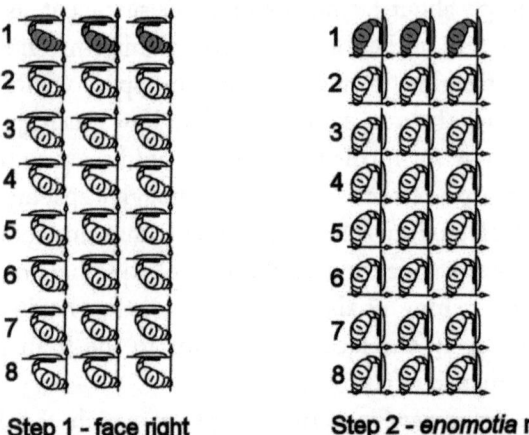

Step 1 - face right

Step 2 - *enomotia* now facing right, with file leaders in left hand file

Klisis (turning in place): changing facing to the right.

of their files. This is the quickest of manoeuvres to perform, however, as it does not require any marching, just a turn in place, and must have been well within the capability of even the most amateur and undrilled armies.

In modern authors, several turning manoeuvres are termed 'wheels' in a loose sense, though a wheel is properly, as Asclepiodotus defines it, where the unit moves 'like the body of a man in such a manner that the entire force swings on the first file leader as on a pivot' (Asclep. 10.4). Asclepiodotus calls this *epistrophe*, though again we do not find this technical usage in Classical authors. The extent to which a hoplite phalanx was able to wheel is uncertain; as we will see in Chapter 7, a common battle tactic was to outflank and then envelop an opposing phalanx, and this is sometimes depicted as being a matter of simply extending one's own phalanx to be wider than the enemy (on one or both flanks), then wheeling the overlapping parts inwards to take the enemy in the flank. However, on the few occasions where we have a clearer description of this process it appears that it was carried out somewhat differently. The best example is a fictional one, in Xenophon's account of the Battle of Thymbrara, describing the approach of the Persian army to their Lydian enemies:

> 'When they were all in sight of one another and the enemy became aware that they greatly outflanked the Persians on both sides, Croesus halted his centre – for otherwise it is impossible to execute a surrounding manoeuvre – and began to turn (*epikampto*) the wings around to encompass the Persians, thus making his own lines on either flank in form like a gamma, so as to close in and attack on all three sides at once.
>
> 'But Cyrus, although he saw this movement, did not any the more recede but led on just as before.

Organization and Drill 147

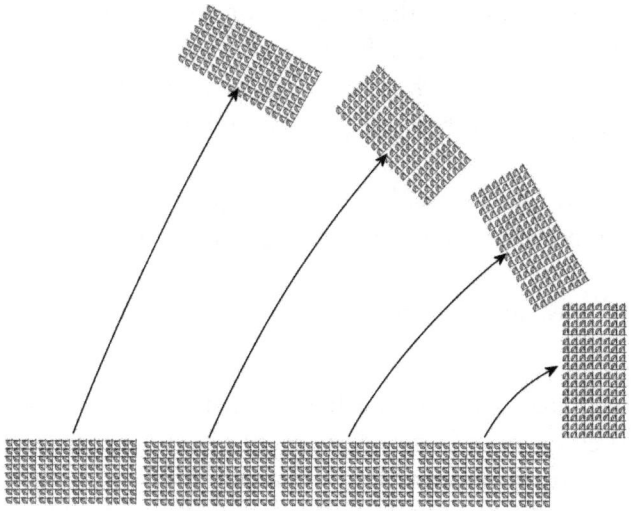

Wheel in line - individual subunits wheel and march to new location

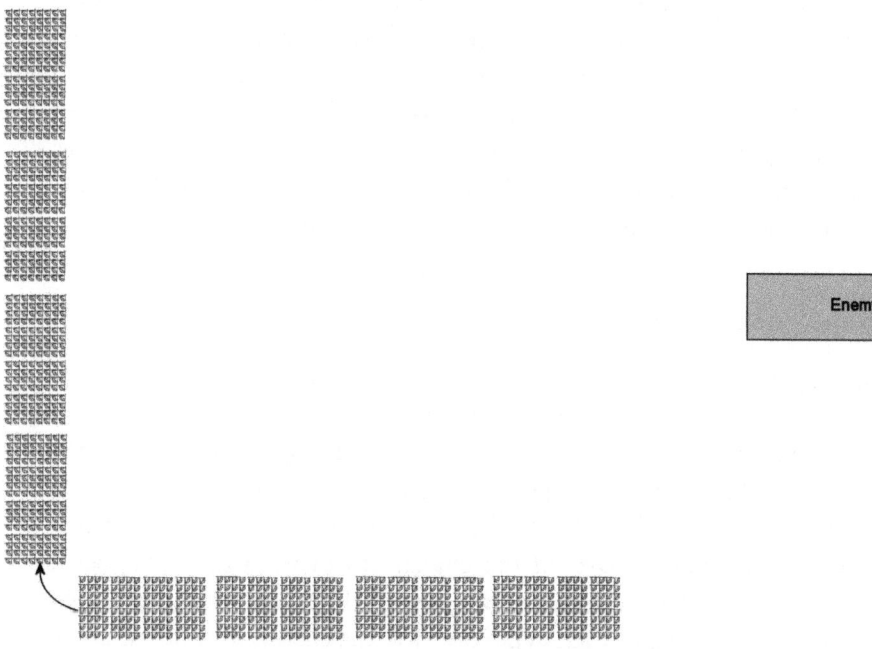

Epikampios - phalanx faces left and marches in column, wheeling right, then faces right

Kyklosis (outflanking): wheeling a phalanx by subunits (which disrupts the phalanx formation) or by forming column and wheeling in column (*epikampios*).

> "'Do you observe, Chrysantas, where the wings are drawing off to form their angle with the centre?' he asked, as he noticed at what a distance from the centre column on both sides they made their turning point, and how far they were pushing forward their wings in executing their turning movement.
>
> "'Indeed I do,' answered Chrysantas, 'and I am surprised, too; for it strikes me that they are drawing their wings a long way off from their centre.'
>
> "'Aye, by Zeus,' said Cyrus, 'and from ours, too.'" (Xenophon, *Cyropaedia* 7.1.5 f.)

Here, the Lydian flanking forces are not wheeling (pivoting on their inside corners), but rather are marching away from the centre in column, then turning in column and marching forward at 90 degrees to the main line. The manoeuvre will be completed by another 90-degree turn in place to face inwards. This is quite probably the way flanking manoeuvres (like those we will encounter in Chapter 7) were usually carried out, with lines turning into column, marching with a turn at a fixed point, then turning back into line, the whole process going by the name of *kyklosis*, 'circling' or 'outflanking'. This would have been a simpler manoeuvre to undertake than a wheel in line; while wheeling a small formation like an individual *taxis* was possible, and a longer phalanx could have been wheeled by having its constituent subunits wheel and march to their destinations, re-forming the line on arrival, this would have involved considerable disruption to the phalanx and would have been dangerous in proximity to the enemy.[35]

Column and line formations

So outflanking manoeuvres probably involved a turn into column formation. Column was also of course the normal formation for marching over any significant distance – so that an army could follow a road or pass through constrictions (like fords or passes) and so that it could avoid being broken apart by encountering a number of different types of terrain along the length of an extended line, which would have caused different parts of the phalanx to move at different speed. Armies would therefore have marched to the battlefield in column and deployed into line (that is, into phalanx) when in proximity to the enemy, so as to minimize the time spent advancing in line.

A column can take two forms. If a phalanx performs a 90-degree turn in place to left or right, then marches off in that direction, it will be in column, facing perpendicular to the original direction, and with the file leaders now forming the leftmost or rightmost file of the column, and so not in fighting formation.

Phalanx, originally facing to the left,
has formed column by facing right.
File leaders are now in the left hand file.

Column of *enomotiai*.
Files leader are at the head of each enomotia.

Two ways of forming a column, by turning the phalanx or by stacking subunits in column.

While such a manoeuvre was used – as part of the *anastrophe* or the *epikampios* flanking manoeuvre, for example – it would not be the normal way for a column to be formed, which would rather be done as Xenophon describes:

> 'The Lacedaemonians also carry out with perfect ease manoeuvres that instructors in tactics [*hoplomachoi*] think very difficult. Thus, when they march in column [*epi kerōs*], every *enomotia* of course follows in the rear of the *enomotia* in front of it.' (Xenophon, *Constitution of the Lacedaemonians* 11.8)

So rather than forming a column by turning a phalanx by 90 degrees, it is formed by stacking up units – in this case *enomotiai* – one behind the other. This would have kept the file leaders in their proper positions at the heads of the files. The method of deploying from such a column into line will be considered below.

The phalanx was of course the normal formation for fighting and the column for marching, but there are occasions where a force fights in column. The best-known example, worth quoting at length for the light it sheds on the factors a commander would have to consider when going into battle, comes from Xenophon's *Anabasis*, when the Ten Thousand faced an enemy holding the top of a high, steep ridge:

> 'At this place was a great mountain, and upon this mountain the Colchians were drawn up in line of battle. At first the Greeks formed an opposing phalanx, with the intention of advancing in this way upon the mountain, but afterwards the generals decided to gather together and take counsel as to how they could best make the contest. Xenophon accordingly said that in his opinion they should give up the line of battle and form the *lochoi* in column [*lochoi orthioi*]. "For the phalanx," he continued, "will be broken up at once; for we shall find the mountain hard to traverse at some points and easy at others; and the immediate result will be discouragement, when men who are formed in phalanx see the line broken up. Furthermore, if we advance upon them formed in a line many ranks deep, the enemy will outflank us, and will use their outflanking wing for whatever purpose they please; on the other hand, if we are formed in a line a few ranks deep, it would be nothing surprising if our line should be cut through by a multitude both of missiles and men falling upon us in a mass; and if this happens at any point, it will be bad for the whole phalanx. But it seems to me we should form the *lochoi* in column and, by leaving spaces between them, cover enough ground so that the outermost *lochoi* should get beyond the enemy's wings; in this way not only shall we outflank the enemy's line, but advancing in column our best men will be in the van of the attack, and

wherever it is good going, there each will lead forward his *lochos*. And it will not be easy for the enemy to push into the space between the columns when there are *lochoi* on this side and that, and not any easier for him to cut through a *lochos* that is advancing in column. Again, if any one of the *lochoi* is hard pressed, its neighbour will come to its aid; and if one single *lochos* can somehow climb to the summit, not a man of the enemy will stand any longer.'" (Xenophon, *Anabasis* 4.8.9–13)

The benefits of a column (or multiple columns) for traversing difficult terrain could not be more clear, as well as the trade-off between greater and lesser depth. Mutually supporting columns of this sort (Xenophon does not say how deep the columns formed) were used on a number of occasions by the Ten Thousand, generally for attacks over difficult ground (for example Xen., *Anab.* 4.2.11; 4.3.17; 5.4.22; compare Xen., *Cyrop.* 3.2.6). It seems though that outside such special circumstances – fighting on broken ground against more mobile enemies – the phalanx was still by far the preferred fighting formation.

A very deep phalanx, depending on its overall strength, might well have taken the form of a column (that is, it might have had more ranks than files and so was deeper than it was wide); the deep formations favoured by the Thebans might therefore in this sense have been fighting columns – this aspect will be examined further in Chapter 8.[36]

Deploying

An army marching in column which encounters an enemy force will need to deploy into line – into phalanx – in order to fight. Xenophon provides the details (remembering as always that this is for the unusually well-drilled Spartans – other armies would have followed a less orderly process):

'Thus, when they march in column, every *enomotia* of course follows in the rear of the *enomotia* in front of it. Suppose that at such a time an enemy in phalanx suddenly makes his appearance in front: the word is passed to the *enomotarchos* to deploy into line to the left, and so throughout the column until the phalanx stands facing the enemy. Or again, if the enemy appears in the rear while they are in this formation, each file counter-marches, in order that the best men may always be face to face with the enemy.' (Xenophon, *Constitution* 11.8)

This is the *paragoge* we encountered above, but instead of being performed by files, it is performed by *enomotiai*. It could in principle be performed by units of whatever size, according to how the column was made up; a column of *lochoi*, for

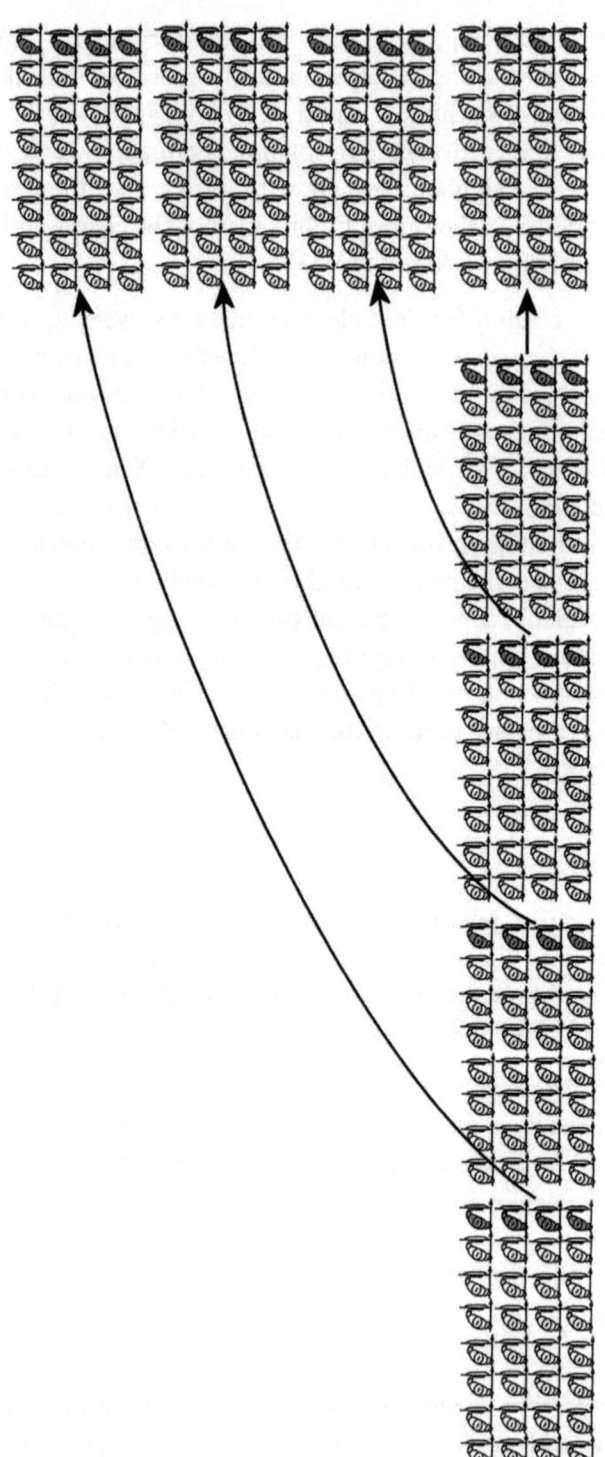

Deploying from column of *enomotiai* into phalanx by *paragoge*.

example, would deploy in the same way, by each subsequent unit in the column marching up to the left (or right) of the one in front, so as to form a phalanx.

Xenophon describes the same process in the *Cyropaedia*, though here he has *taxeis* formed in single file, thus conflating the two forms of the manoeuvre (in a way which would be completely impractical, since an army of 30,000 men, which is what he imagines, could not form in single file). With his army on parade, Cyrus wished to deploy to meet a visiting dignitary:

> '[H]e gave orders to the *taxiarchos* who was stationed first to take his stand at the head of the line, bringing up his *taxis* in single file and keeping himself to the right; he told him to transmit the same order to the second *taxiarchos* and to pass it on through all the lines. And they obeyed at once and passed the order on, and they all executed it promptly, and in a little while they were three hundred abreast on the front line, for that was the number of the *taxiarchoi*, and a hundred men deep.' (Xenophon, *Cyropaedia* 2.4.2)

This formation then advanced, but met a constriction, a narrow street, and so formed back into column:

> 'And when they had got into their places, he ordered them to follow as he himself should lead. And at once he led them off at a double quick step. But when he became aware that the street leading to the king's headquarters was too narrow to admit all his men with such a front, he ordered the first *chiliostys* [unit of 1,000 men] in their present order to follow him, the second to fall in behind the first, and so on through them all, while he himself led on without stopping to rest, and the other *chiliostyes* followed, each the one before it.' (Xenophon, *Cyropaedia* 2.4.3)

Here, the column is of stacked *chiliostyes* and is formed by the first *chiliostys* marching out in front, ten abreast, then the next following, and so on. Naturally, each successive unit, being to the side of the one it was to follow, would have had to wheel into place behind it or turn and march across.

In the *Anabasis*, we see an example of a column deploying by *lochoi* using the *paragoge*:

> 'As soon as Cheirisophus caught sight of the enemy on the pass, he halted, while still at a distance of about thirty stadia [5,400 metres], in order not to get near the enemy while his troops were marching in column [*kata keras*]; and he gave orders to the other officers also to move along their companies [*paragein tous lochous*] so as to bring the army into line of battle [*epi phalangos*].' (Xenophon, *Anabasis* 4.6.6)

154 The Greek Hoplite Phalanx

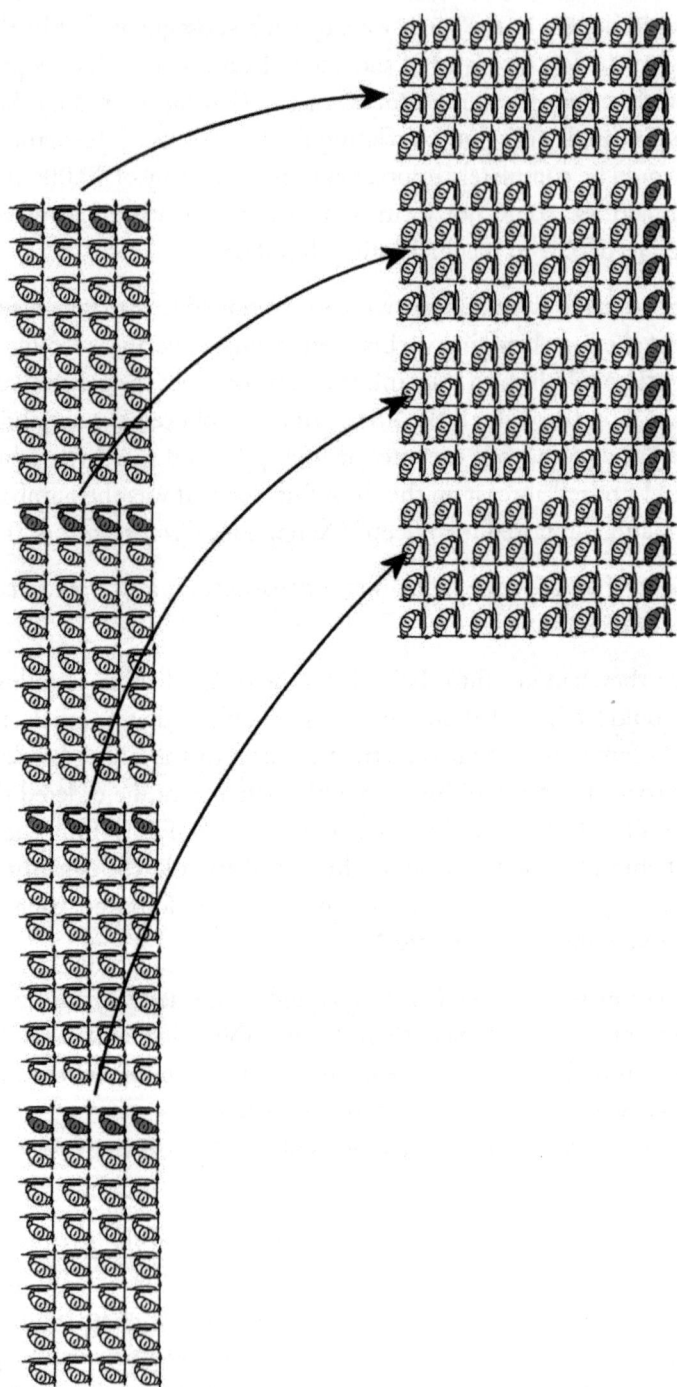

Deploying a column to meet a threat from the flank.

A column might not only need to deploy to its front, and Xenophon also describes how the Spartans could deploy to the right or left of the column:

> 'If, on the other hand, an enemy force appears on the right when they are marching in column, all that they have to do is to order each *lochos* to wheel to the right so as to front the enemy like a trireme, and thus again the *lochos* at the rear of the column is on the right.' (Xenophon, *Constitution of the Lacedaemonians* 11.10)

A similar process would be followed to the left. In this case, the formation is apparently envisaged as a column of *lochoi* (rather than the earlier column of *enomotiai*), but the principle is the same: each unit individually wheels (Xenophon uses the word *strepho*, 'turn') to face the right. As Xenophon says, this is similar to the process that a fleet of ships (triremes, the standard warship of the Classical period) would use to deploy from line ahead (a column) into line abreast (a line), by 'turning together' (to use the eighteenth-century AD naval terminology), each turning to the right to form a line at right angles to the original axis of advance.[37]

Conclusions

This brief look through the organization and drill of the phalanx has, I hope, emphasized one main point: that the most detailed information we have for both organization and drill is probably also the least typical, being that of the Spartans (or modelled on the Spartans), which was of a level of sophistication and professionalism probably not attained by other armies (outside select units) at least until the fourth century, or in some cases at all. The Spartans broke down their army into a more hierarchical structure, with more levels and more junior commanders, than other cities, which were content with perhaps two levels of units in their hierarchy. It is unclear when in the history of the development of the phalanx the internal structure of units and formal drills developed – perhaps very early on in the case of the Spartans, though other cities in the Archaic period probably had much more basic systems. We must also recall that neither internal organization nor drill were unique to the Greek phalanx, being also adopted by other infantry formations in the Mediterranean and Near East, perhaps from an early period.

The drill the Spartans used to manoeuvre their units was also more precisely defined, and other armies must have followed more ad hoc arrangements, which would have made their manoeuvres slower, involved a greater risk of disorder and also have limited them to much simpler manoeuvres, perhaps being restricted to straight-ahead marches once deployed into phalanx – only the Spartans could attempt elaborate encirclements. I will examine in Chapter 7 the ways in which this organization and drill could be used in battle conditions, but first it is time to examine the broader social and political background of the hoplite.

Chapter 4

Hoplites and Citizens

We have already seen the connection that existed between the social and economic status of citizens (and non-citizens) in the Greek cities and their military role as hoplites. In this chapter I will be looking more closely at these connections, and also at other aspects of hoplite recruitment such as numbers, training and terms of service, as well as the various types of hoplite, such as rich and poor or militia and mercenary. It is also important to keep in mind the provisos that have been pointed out in other contexts; almost all of our information comes from the Classical period, from the early fifth century onwards, and specific statements about the social and economic status of hoplites are particularly to be found in Xenophon, Plato and Aristotle, writing in the fourth century, right at the end of the period of dominance of the hoplite. For the Archaic period we have almost no firm information, and the degree to which it is valid to extrapolate backwards from the better-known fourth century is highly controversial. Similarly, the two cities about which we have most information, Athens and Sparta, are perhaps the most atypical. Athens had an unusually democratic and egalitarian constitution, and also, from the establishment of its empire in the mid-fifth century, was unusually wealthy and reliant on overseas sources of income. Sparta was unique in the professionalism and discipline of its military forces, and had a rigid, archaic constitution that emphasized the supremacy of a relatively small class of 'peers' (equals, *homoioi*) dependent for their status on dominance over a larger group of inferiors and, in particular, of a large serf class, the helots. Neither of these models were followed precisely in most Greek cities, and while of necessity we must speak mostly of the cities about which we have most information, it is important to keep in mind that there was a great deal of variety even within the limited world of the Greek cities, and that arrangements which are familiar to us due to the survival of evidence may not in fact have been typical.

The hoplite class

We saw in Chapter 1 how the development of the hoplite phalanx has been, among modern historians, tied to the rise of a 'middle class' in the Greek cities – whether that is dated as early as the seventh century or as late as the fifth.

This conception, though influenced by modern concepts of class status, does date back at least to the writings of Aristotle (late fourth century), in the well-known passage (*Politics* 1297b) quoted in Chapter 1, where the rise of the hoplite class is credited with ending the political dominance of aristocratic cavalry. Any period where cavalry was dominant in Greek warfare is now obscure to us, but in Homer at least we do see reflected an early Archaic organization of state and military which emphasized the role of the leaders, chariot-borne infantry if not actually (in this heroic setting) cavalry, fighting amongst a mass of lesser (that is, poorer) men, who made up the bulk of the infantry. Aside from ownership of horses, there was no great difference between the equipment of these groups (shield, throwing and thrusting spears, plus such armour as was available and affordable), and on the battlefield their roles are distinguished by the rich fighting in, or in front of, the front ranks while the commoners form the 'mass' that occasionally rises to prominence in Homeric battle accounts. Even in Homer we see what we might now call class conflict between leaders and led. Odysseus rebukes the common soldiers who wish to abandon the war against Troy:

> 'But when he came across some man from some locale who was making a noise, he struck him with his staff and rebuked him, saying, "What kind of spirit has possessed you? Hold your peace, and listen to better men than yourself. You are a coward and no warrior; you are nobody either in fight or council; we cannot all be kings; it is not well that there should be many masters; one man must be supreme – one king to whom the son of scheming Kronos has given the sceptre and divine laws to rule over you all."' (Homer, *Iliad* 2.197–208)

The commoners have a spokesman, one Thersites, who could answer back:

> '"Agamemnon," he cried, "what ails you now, and what more do you want? Your tents are filled with bronze and with fair women, for whenever we take a town we give you the pick of them. Would you have yet more gold, which some Trojan is to give you as a ransom for his son, when I or another Achaean has taken him prisoner? Or is it some young girl to hide and lie with? It is not well that you, the ruler of the Achaeans, should bring them into such misery. Weakling cowards, women rather than men, let us sail home, and leave this man here at Troy to stew in his own prizes of honour, and discover whether we were of any service to him or no."' (Homer, *Iliad* 2.229–38)

Thersites is presented by Homer as an upstart who should know his place (and 'the ugliest man before Troy' as well), and is beaten by Odysseus, to the general approval of the army. Nevertheless, the army before Troy is acknowledged to be mixed in terms of status and quality:

'The two Ajaxes went about everywhere on the walls cheering on the Achaeans, giving fair words to some while they spoke sharply to any one whom they saw to be remiss. "My friends," they cried, "Argives one and all – good, bad and indifferent, for there was never fight yet, in which all were of equal prowess – there is now work enough, as you very well know, for all of you. See that none of you turn in flight towards the ships, daunted by the shouting of the foe, but press forward and keep one another in heart, if it may so be that Olympian Zeus the lord of lightning will grant us to repel our foes, and drive them back towards the city."' (Homer, *Iliad* 12.265–72)

The deeds of heroes form the chief focus of Homer, but it was understood that the masses too could decide the course of a battle, as when the Lycian Sarpedon was driven back from the Greek camp:

'He therefore retired a little space from the battlement, yet without losing all his ground, for he still thought to cover himself with glory. Then he turned round and shouted to the brave Lycians saying, "Lycians, why do you thus fail me? For all my prowess I cannot break through the wall and open a way to the ships single-handed. Come close on behind me, for the more there are of us the better."' (Homer, *Iliad* 12.409–12)

Even Achilles had to admit he could not take Troy single-handed:

'He sprang forward along the line and cheered his men on as he did so. "Let not the Trojans," he cried, "keep you at arm's length, Achaeans, but go for them and fight them man for man. However valiant I may be, I cannot give chase to so many and fight all of them. Even Ares, who is an immortal, or Athena, would shrink from flinging himself into the jaws of such a fight and laying about him."' (Homer, *Iliad* 20.353–57)

The Archaic phalanx, as we have seen, may have had a similar character, a mass of soldiers of lesser status and perhaps lesser equipment, fronted by wealthy *promachoi* with the best equipment who do most of the fighting (or at least, the most prominent and best-recorded fighting).[1]

The Classical phalanx, following Aristotle's analysis of the importance of the 'middle class' (*meson*), is supposed to have been a more socially and economically homogeneous body, in some modern interpretations formed from a large class of independent middling farmers, workers of their own lands. But it is doubtful whether Aristotle's *meson* really represent a large homogenous economic class (rather than simply and practically those with more money than the poorest and less than the richest), and the idea of a phalanx of independent middling farmers is at least in part inspired by modern political ideology. It is likely that in Ancient

Greece, and even in the theories of Aristotle, the existence of the 'middle' was less important than the true fundamental distinction between different types of citizens – those who were rich and those who were poor – and the superiority of the rich arose from the fact that they did not need to work for a living (not even as hardy independent farmers), but earned enough income from their lands to live a life of leisure, something that was considered essential for full participation in political and military life. For all his talk of the importance of the 'middle', Aristotle too conceived the distinction between rich and poor, leisured and labourers, as of utmost importance:

> 'And although it is proper that the government should be drawn only from those who possess heavy armour [hoplites], yet it is not possible to define the amount of the property qualification absolutely and to say that they must possess so much, but only to consider what sort of amount is the highest that is compatible with making those who have a share in the constitution more numerous than those who have not, and to fix that limit. For those who are poor and have no share in the honours are willing to keep quiet if no one insults them or takes away any part of their substance; but this is not easy to secure, for it does not always happen that those who are in the governing class are gentlemen.' (Aristotle, *Politics* 1297b)

So the importance of the 'middle' is just that they be numerous enough to swell the ranks (literally) of the rich, and not be left out of government and so be inclined to side with the poor. In Aristotle's ideal state, there should be a distinction between those who worked for a living and those who did not need to, which was considered an established principle of great antiquity:

> 'And that it is proper for the state to be divided up into castes and for the military class to be distinct from that of the tillers of the soil does not seem to be a discovery of political philosophers of today or one made recently. In Egypt this arrangement still exists even now, as also in Crete; it is said to have been established in Egypt by the legislation of Sesostris and in Crete by that of Minos.' (Aristotle, *Politics* 1329a–1329b)

In the ideal state, the tillers of the soil would all be slaves or foreign (barbarian) subject peoples (Arist., *Pol.* 1330a), freeing all of the Greek citizens for the life of leisure, and of political and military engagement, they deserved. In the less-than-ideal states that actually existed in Classical Greece, such an arrangement was not generally possible, so various cities adopted various levels at which to set the property qualification, that is the amount of wealth each individual needed to possess in order to be allowed to fully participate in civic life.[2]

The details of the Athenian arrangement of property classes are the best known. There were four 'Solonian' property classes (named for the statesman, Solon, who defined them in the sixth century). The poorest were the *thetes*, who served as rowers in the navy or as light-armed infantry; the *zeugitai* were (broadly speaking) the hoplite class we have already encountered; *hippeis* were, as originally, those wealthy enough to provide a horse, and so serving as cavalry (in Athens' case, unlike in Sparta, as actual horse cavalry); and the wealthiest of all, the *pentacosiomedimnoi* ('500-medimnoi-men', the *medimnos* being a measure of dry goods, of varying size but approximately equivalent to 60 litres of grain, barley in this case), who would be expected to fund a trireme for the navy. Poorer men were still free, and might have some citizen rights (in Athens for example they would be allowed to vote in the Assembly and serve as jurors), but they would not be allowed to hold political office and so were not full citizens on the same level as the richer men. In Athens, the qualification for the *zeugitai* was defined by the quantity of food that a man could produce from the lands he owned and was initially set at 200 *medimnoi* of barley. This was a significant quantity, and those who met it would have been sufficiently wealthy to live a life of leisure on the income of their lands rather than having to work for a living. Even in democratic Athens then, the qualification for full citizen rights and so for hoplite service was set quite high, and the men included would have been considered wealthy.[3]

At various times, the property qualification was raised further, as after the oligarchic coup of 411, when full political rights were restricted to 5,000 men (Thuc 8.97.1); this represents about half of the total number of hoplites and 15 per cent of the number of citizens (although Thucydides implies that this body contained all those who could provide *hopla*). Generally oligarchic-minded historians approved of this arrangement:

> 'It was during the first period of this constitution that the Athenians appear to have enjoyed the best government that they ever did, at least in my time. For the fusion of the high and the low was effected with judgment, and this was what first enabled the state to raise up her head after her manifold disasters.' (Thucydides 8.97.2)

But in 404, after defeat in the Peloponnesian War led to the appointment of a Council of Thirty, the franchise was further restricted to 3,000, a ruse being required this time to exclude so many hoplite-equipped citizens:

> 'As for the Thirty, they held a review, the Three Thousand assembling in the market-place and those who were not on "the roll" in various places here and there; then they gave the order to pile arms, and while the men

were off duty and away, they sent their Lacedaemonian guardsmen and such citizens as were in sympathy with them, seized the arms of all except the Three Thousand, carried them up to the Acropolis, and deposited them in the temple. And now, when this had been accomplished, thinking that they were at length free to do whatever they pleased, they put many people to death out of personal enmity, and many also for the sake of securing their property.' (Xenophon, *Hellenica* 2.3.20–21)

The message here was clear – if too many were excluded from politics, then abuses of power would inevitably follow.

Those who met the property qualification (in Athens and in other cities), as well as having full political rights and being eligible to hold political office, were required to serve as hoplites on demand, the two things being linked. Those who met the qualification were included on the official conscription lists and were eligible to be called up as required, which was sometimes done, as we have already seen in the case of the Spartans, according to age class. Depending on the size of force needed for a particular operation, a variable number of age classes would be mobilized (with full mobilization of all age classes being reserved for emergencies) – more on these arrangements below.

This does not mean that only those who met the property qualification served as hoplites, for it is clear that the Athenian army (and in this it was probably typical) also contained some who did not meet it as well as 'resident aliens' (*metoikoi*, or metics, those originally from other cities) who did not qualify for citizenship (or a number of other rights, such as the right to own property), all of whom must have served as volunteers. Thucydides says of the Athenians at the start of the Peloponnesian War:

'Then they had an army of thirteen thousand hoplites, besides sixteen thousand more in the garrisons and on home duty at Athens. This was at first the number of men on guard in the event of an invasion: it was composed of the oldest and youngest levies and those of the metics who were hoplites.' (Thucydides 2.13.6-7)

In the Spartan army, a variable (but presumably large) proportion of the army was made up from *perioikoi*, non-citizens living in towns around Sparta, with the Spartan full citizens themselves making up about half or less of the hoplite total. Sparta could also supplement its hoplite forces with other classes – *hypomeiones*, 'inferiors', Spartans who did not meet the qualification to be full Spartiates; and *neodamodeis*, 'new people', who were perhaps enfranchised helots. Despite the generally brutal treatment of the helot population, those who served their Spartan masters loyally had a chance of being promoted in this way, and occasional mass enfranchisements of helots was a recurring feature of the Spartan army as the number of Spartiates proper steadily declined.[4]

This meant that any given hoplite phalanx could see (as was presumably the case in its Archaic antecedents) rich and poor, privileged man of leisure and manual labourer, serving alongside each other in the same formation. In Chapter 1 we saw the situation described by Plato when 'a lean, sinewy, sunburnt pauper is stationed in battle beside a rich man bred in the shade, and burdened with superfluous flesh, and sees him panting and helpless' (Plato, *Rep.* 9 556c–d). Not all poor men were considered good soldiers though, as handicraft work was not thought to make good soldiers:

> '[T]o be sure, the illiberal arts [handicrafts, artisanry], as they are called, are spoken against, and are, naturally enough, held in utter disdain in our states. For they spoil the bodies of the workmen and the foremen, forcing them to sit still and live indoors, and in some cases to spend the day at the fire. The softening of the body involves a serious weakening of the mind. Moreover, these so-called illiberal arts leave no spare time for attention to one's friends and city, so that those who follow them are reputed bad at dealing with friends and bad defenders of their country. In fact, in some of the states, and especially in those reputed warlike, it is not even lawful for any of the citizens to work at illiberal arts.' (Xenophon, *Economics* 4.2–3)

Xenophon spoke highly instead of 'the king of the Persians ... For they say that he pays close attention to husbandry and the art of war, holding that these are two of the noblest and most necessary pursuits' (Xen., *Econ.* 4.4). Xenophon has Socrates recommend husbandry (farming) as a suitable occupation for the wealthy:

> '[E]ven the wealthiest cannot hold aloof from husbandry. For the pursuit of it is in some sense a luxury as well as a means of increasing one's estate and of training the body in all that a free man should be able to do.' (Xenophon, *Economics* 5.1)

Note, however, that this is not a middling farmer making a bare living from his own land, but a landowner and master of labourers, and the comparison is with an officer, not a common soldier:

> 'Therefore nobody can be a good farmer unless he makes his labourers both eager and obedient; and the captain who leads men against an enemy must contrive to secure the same results by rewarding those who act as brave men should act and punishing the disobedient. And it is no less necessary for a farmer to encourage his labourers often, than for a general to encourage his men.' (Xenophon, *Economics* 5.15–16)

But while craftsmen and farm labourers might not make the best soldiers, they may well have been the most numerous, as the famous story told of Agesilaus illustrates, when his allies complained at the small size of the Spartan contingent:

> 'It was at this time, we are told, that Agesilaus, wishing to refute their argument from numbers, devised the following scheme. He ordered all the allies to sit down by themselves together, and the Lacedaemonians apart by themselves. Then his herald called upon the potters to stand up first, and after them the smiths, next, the carpenters in their turn, and the builders, and so on through all the handicrafts. In response, almost all the allies rose up, but not a man of the Lacedaemonians; for they were forbidden to learn or practise a manual art. Then Agesilaus said with a laugh: "You see, men, how many more soldiers than you we are sending out."' (Plutarch, *Agesilaus* 26.4–5)

The point is perhaps not so much that the Spartans were professional soldiers (which is the way this story is often interpreted today), as that they were men who did not need to work for a living at all, men of leisure, ones who devoted all their leisure time (as the best men should, it was thought) to exercising for war. The fact that the appropriate preparation for military service was thought to be gentlemanly leisure activities worked well for the Spartans, where military training and exercise were also considered the ideal forms of leisure activity, but in other cities rather less well, as we will see below when we come to consider military training.

So while the theoretical ideal was for gentleman-soldiers, the reality in most cities was that a good proportion of most hoplite armies would have been workers, agricultural or otherwise. It is difficult to quantify the proportions given the lack of evidence and uncertainties over, for example, the proportion of *perioikoi* in the Spartan army. It is probably reasonable though to see a trend from the Archaic period, when the hoplites would have been a primarily aristocratic force – for all that there were large numbers of poorer men in the early phalanx (not necessarily fighting as hoplites, but as archers or stone throwers) – through the early Classical, when more democratic constitutions led to more socially varied phalanxes, to the fourth century, when the poorer hoplites were the majority and where professional soldiering for pay came to be a common option (though mercenary service had always been possible, as we will see below). It is also perhaps not too fanciful to see a distinction in most Greek cities, at least before the rude shock of defeat by the professional armies of Macedon, between 'gentlemen' and 'players'; between rich amateurs who regarded their amateurism as a virtue and felt that military service was something that came naturally to the right sort, and the workers or professional soldiers on whom they looked down.

Hoplites and politics

As we saw in Chapter 3, some armies, particularly the Spartan, divided their armies up into the sorts of subunits with which we are familiar today (and the Greek terms are often translated with modern equivalents – company, regiment, brigade and so on). But most Greek states did not, and their armies were subdivided if at all only into rather more informal groupings (*taxeis* and *lochoi* in some combination), and in cases (such as Athens) where we have good information, the military organization of the citizens can be seen to be in fact civic groupings, with no particular relevance or suitability for military organization. In Athens, the reforms of Cleisthenes in the late sixth century divided the population into ten subdivisions, the *phylai* or 'tribes', and these were also the basic units of the Athenian phalanx as we saw in the previous chapter. But because of the complex way in which tribes were organized – each was formed from a number of different, unconnected geographical areas around Athens, so as to give them all a similar mix of inland, city or coastal regions – they make little sense as recruitment areas or as the basis for homogenous military units (compare with the regionally recruited Macedonian phalanx). Yet the Athenian army was divided, at the top level of organization, into units by *phylai*, and there were ten generals appointed over them, selected in the same way. This illustrates what appears to be a general principle of Greek armies, that their military organization mirrored their civic organization. Rather than a city's phalanx being a separate entity, it was an extension of the city's normal (peacetime) structure, naturally enough given the close link between service in the phalanx and citizenship.[5]

A related phenomenon is the way that armies in the field tended to mirror (if in altered form) the political arrangements of a Greek city. The clearest examples of this are from the Ten Thousand, the mercenary force which marched back from Persia to Greece at the end of the fifth century. Because this was a force of mixed origins, from a number of different cities, it did not exactly mirror any one city's constitution, but rather represented a generic sort of Greek city – a close analogy might be with a Greek colony, since like a colony it was united against a common enemy (the indigenous people) and did not have a pre-existing constitution. The original commanders of the Ten Thousand were captured by the Persian king, whereupon the remainder elected new commanders and held an assembly to decide on their future course of action. At this assembly, Xenophon proposed that they attempt the return to Greece:

> '"All who are in favour of this motion," he said, "will raise their hands." And every man in the assembly raised his hand. Thereupon they made their vows and struck up the paean [battle hymn].' (Xenophon, *Anabasis* 3.2.9)

Many of those present were Spartans or from other less-than-democratically governed cities, but the concept of voting in an assembly was evidently familiar to all.

The Ten Thousand were in an unusual situation, a mercenary army adrift and leaderless in hostile territory, but other more regular armies also adopted some political forms. The Athenian army besieging Syracuse in the Peloponnesian War, for example, held a council as to whether they should persevere or withdraw:

> '[The commander Nicias did not want] to have it reported to the enemy that the Athenians in full council were openly voting for retreat ... saying he was sure the Athenians would never approve of their returning without a vote of theirs [that is, of the full Athenian Assembly].' (Thucydides 7.48.1–4).

Gylippus, the Spartan mercenary commander of the Syracusans, faced greater problems. At the start of his tenure, he led the Syracusans out to defeat in battle:

> 'After this Gylippus called the soldiers together, and said that the fault was not theirs but his; he had kept their lines too much within the siege works, and had thus deprived them of the services of their cavalry and javelinmen.' (Thucydides 7.5.3)

Spartan armies themselves had something of a record of insubordination, which I will examine further below. In the Athenian army, officers were generally elected (see Chapter 5), and the relationship between officers and men was therefore a political one, rather than strictly military. This meant for example that soldiers accused of offences against military law would be accused and tried through the civil courts, rather than subject to any sort of martial law. Of course any army might prove unruly and mutinous, and any commander might have to resort to persuasion rather than just issuing orders, but it does seem a feature of Greek armies of this period that, just because they were serving as soldiers, hoplites (and their commanders) would not forget that they were citizens, and that they had political rights back home that were not wholly suspended just because they were on campaign.[6]

Hoplites and others

Citizens of Greek cities were all male, and hoplites of course were also male. There are a few individual women in antiquity who held important military commands, mostly royalty or aristocratic figures. In this period, the most well known is Artemisia, the Carian queen of Halicarnassus, who commanded a Carian contingent at the naval Battle of Salamis (480), at one point ramming an allied ship in order to avoid imminent attack from a Greek ship:

> 'Thus she happened to escape and not be destroyed, and it also turned out that the harmful thing which she had done won her exceptional esteem from Xerxes. It is said that the king, as he watched the battle, saw her ship ram the other, and one of the bystanders said, "Master, do you see how well Artemisia contends in the contest and how she has sunk an enemy ship?" When he asked if the deed was truly Artemisia's, they affirmed it, knowing reliably the marking of her ship, and they supposed that the ruined ship was an enemy. As I have said, all this happened to bring her luck, and also that no one from the Calyndian ship survived to accuse her. It is said that Xerxes replied to what was told him, "My men have become women, and my women men."' (Herodotus 8.88.1–3)

Xerxes' comment illustrates the normal attitude; a cowardly or unsuccessful soldier (or sailor) could be insulted by being called a woman, and a brave or successful woman qualified as an honorary man.

The Spartan lawgiver Lycurgus devised the military uniform of the Spartans:

> 'In the equipment that he devised for the troops in battle he included a red cloak, because he believed this garment to have least resemblance to women's clothing and to be most suitable for war.' (Xenophon, *Constitution of the Lacedaemonians* 11.3)

At the Battle of Plataea (479), the Persians sent their cavalry to challenge the Greek lines:

> 'Thereupon the horsemen rode up to the Greeks and charged them by squadrons; as they attacked, they did them much hurt, and called them women all the while.' (Herodotus 9.20)

The final comment added insult to injury. Plato considered the penalties that were appropriate for cowardice in battle, and particularly for the most heinous crime of 'shield-flinging' (*rhipsaspia*), the act of throwing away one's shield in order to run away faster:

> 'A god, it is said, once changed Kaineus the Thessalian from woman's shape to man's; but it is beyond human power to do the opposite of this; otherwise, the converse transformation – changing him from a man into a woman – would be, perhaps, the most appropriate of all penalties for a "shield-flinger".' (Plato, *Laws* 944d–e)

So a shield-flinger is no better than a woman. It is significant that the Greek for 'courage', *andreia*, literally means 'manliness'.[7]

A recurring theme throughout the accounts of the Persian Wars in Herodotus and elsewhere is that the Persians' clothing, particularly their wearing of trousers, made them effeminate, although in an interesting twist, Herodotus also records that the Athenians at Marathon (490) were:

> 'the first to endure looking at Median dress and men wearing it, for up until then just hearing the name of the Medes caused the Hellenes to panic.' (Herodotus 6.112.3)

The military reputation of the Medes (Persians) in this case was presumably enhanced by the outlandishness of their clothing. But usually the emphasis is on how the Persians, by covering their bodies, made themselves soft, pale and effeminate, unlike the bronzed, hardened Greeks with their fondness for nudity. Another theme running through Herodotus is the way the Persians were disadvantaged by their short spears, allowing the Greeks with their long spears to defeat them. Depictions of Persian spears in Persian art do not make them look particularly short, and it may not be too fanciful to see a sexual aspect to the hard, naked, masculine Greeks with their long spears thinking themselves superior to the soft, be-trousered, effeminate Persians with their short spears.[8]

Women thus provided a contrast to the masculine, military virtues of the hoplites, and for the most part their role in depictions of warfare in ancient authors is as passive victims, as spoils of war (an aspect which goes back to Homer, where the main theme of the *Iliad* is the quarrel between Achilles and Agammemnon over possession of a female captive) or as a motivation to bravery. This latter aspect is reflected in the famous story of the Spartan mother who told her son, going off to war, to return 'either with his shield or upon it' (that is, carrying his shield, not throwing it away in flight, or dead, carried on it as a stretcher; Plut., *Mor.* 240.16). Plutarch collected a whole chapter of *Sayings of Spartan Women*, cheerful stories in which wives and (particularly) mothers exhort their menfolk to be victorious or die fighting, of which one more example will suffice:

> 'Another was burying her son, when a commonplace old woman came up to her and said, "Ah the bad luck of it, you poor woman." "No, by Heaven," said she, "but good luck; for I bore him that he might die for Sparta, and this is the very thing that has come to pass for me."' (Plutarch, *Moralia* 240.8)

Such stories were a part of Sparta's legend, but no doubt also have some element of truth behind them. Aside from the particularly Spartan elements (sons being disposable, provided they are brave), a major reason for fighting would have been to defend the honour, and the lives, of women at home.

Thucydides puts into the mouth of the Athenian statesman Pericles, in his funerary oration for the Athenian dead, the following view of the role of women:

> 'On the other hand if I must say anything on the subject of female excellence to those of you who will now be in widowhood, it will be all comprised in this brief exhortation. Great will be your glory in not falling short of your natural character; and greatest will be hers who is least talked of among the men whether for good or for bad.' (Thucydides 2.45.2)

Plutarch specifically objected to this view, and composed (along with his pamphlet of Spartan women's sayings) a short work on *The Bravery of Women*, offering the view that 'not the form but the fame of a woman should be known to many' (Plut., *Mor.* 242e).

But women could on occasion play a more active, rather than just passive or exhortatory, role. Plutarch tells the intriguing tale of the female Argive poet Telesilla, when Argos was attacked by the Spartans in 510:

> 'But when Cleomenes king of the Spartans, having slain many Argives ... proceeded against the city, an impulsive daring, divinely inspired, came to the younger women to try, for their country's sake, to hold off the enemy. Under the lead of Telesilla they took up arms, and, taking their stand by the battlements, manned the walls all round, so that the enemy were amazed. The result was that Cleomenes they repulsed with great loss, and the other king, Demaratus, who managed to get inside, as Socrates says, and gained possession of the Pamphyliacum, they drove out. In this way the city was saved.' (Plutarch, *Moralia* 245d–e)

This story may not be historically accurate (Herodotus does not mention it), but if true it represents one of the very few, possibly unique, occasions where a force of women – in this case presumably armed as hoplites, having taken arms, *hopla*, from the temples (Plut., *Mor.* 223b) – actively fought against and defeated a force of men. Pausanias repeats the story, adding:

> 'Then the Lacedaemonians, realizing that to destroy the women would be an invidious success while defeat would mean a shameful disaster, gave way before the women.' (Pausanias 2.20.9)

Which smacks rather of making excuses. The city of Sinope in 370, meanwhile, is said to have disguised women as hoplites:

> 'giving them in place of shields and helmets their jars and similar bronze utensils, and marched them around the wall where the enemy were most likely to see them. But they did not allow them to throw missiles, for even a long way off a woman betrays her sex when she tries to throw.' (Aeneas Tacticus 40.4)

So women, it was thought, threw like girls. In this case they had only to look the part, but when defending against attacks on towns and cities, the women, who had much to lose from defeat (being sold into slavery was the accepted fate of women and children in a captured town, if they survived the sack), must often have taken an active part in the defence, particularly in the traditional role of throwers of tiles from roofs or of stones from the walls. The women of Argos (again), in the Peloponnesian War, also helped with the building of long walls to connect the city to the harbour and hold off the Spartans (Thuc. 5.82.6).[9]

In this Argive wall-building, the women were aided by slaves, and Pausanias also records that Telesilla, along with the Argive women, was aided by the slaves of the city (Paus. 2.20.9), and this reminds us that slaves were another great out-group that had no role in the hoplite phalanx, but could still be important in warfare more generally. Indeed, the pairing of women and slaves, particularly in defence of a city (against male, hoplite-class attackers), is quite common in ancient sources. In the abortive Theban attack on Plataea that opened the Peloponnesian War, for example, the Plataean men attacked the Theban intruders while 'the women and slaves screamed and yelled from the houses and pelted them with stones and tiles' (Thuc. 2.4.2).

Plutarch records another incident from slightly after our period, in the third century:

> '[A]t the time when Philip, son of Demetrius, was besieging their city [Chios], and had made a barbarous and insolent proclamation bidding the slaves to desert to him, their reward to be freedom and marriage with their owners, meaning thereby that he was intending to unite them with the wives of their masters. But the women, suddenly possessed of fierce and savage spirit, in company with their slaves, who were themselves equally indignant and supported the women by their presence, hastened to mount the walls, both bringing stones and missiles, and exhorting and importuning the fighting men until, finally, by their vigorous defence and the wounds inflicted on the enemy by their missiles, they repulsed Philip. And not a single slave deserted to him.' (Plutarch, *Moralia* 245b–c)

To offer freedom to slaves was about as low as one could sink, but the Chian slaves knew their place. It has been remarked how rarely attackers sought to raise the slaves of their enemies in revolt, even in the case of the Spartans, whose particular way of life (and so military dominance) was dependent on their subjection of the helots. It was only after several years of the Peloponnesian War that the Athenians made serious efforts to raise the helots against the Spartans, and only in the fourth century that Thebes finally broke Spartan power by freeing the Messenian helots and founding them a new home city. That this was so is not surprising, given that all Greek cities, whatever their differences and whatever the precise details of their constitutions, were still fundamentally dominated by a privileged class (socially and economically), and had a shared interest in maintaining that domination over outsiders (from other cities or barbarians), the poor, women and slaves. Any step which too greatly disrupted the established social order would have undermined the position of both sides, and would not normally have been justified merely to gain short-term military or political advantage. Greek hoplites from different cities arguably had more in common with each other than they did with the women, poor, slaves and foreigners who dwelt alongside them, and stressing patriotic shared interests (as for the Chian slaves) was more important than inadvertently promoting solidarity among the disadvantaged.[10]

Curiously, an exception to this rule was the Spartans themselves, as they do appear to have freed helots for military service on a number of occasions, perhaps driven by the shortage of Spartans proper. Freed helots began to be used as hoplites during the Peloponnesian War (or at least that is when evidence for the practice begins):

> 'The same summer [421] the soldiers from Thrace who had gone out with Brasidas came back, having been brought from thence after the treaty by Clearidas; and the Lacedaemonians decreed that the helots who had fought with Brasidas should be free and allowed to live where they liked, and not long afterwards settled them with the *neodamodeis* at Lepreum, which is situated on the Laconian and Elean border; Lacedaemon being at this time at enmity with Elis.' (Thucydides 5.34.1)

The *neodamodeis* ('new people') appear to be other freed helots, and here they are being used as, in effect, a form of military colonist, settled into a potentially hostile frontier area (which presumably conveniently coincided with 'where they liked'). The sending of helots with Brasidas' expeditionary force to northern Greece had itself been driven by mixed motives:

> 'The Lacedaemonians were also glad to have an excuse for sending some of the helots out of the country, for fear that the present aspect of affairs and the occupation of Pylos [by Athenian forces] might encourage them to revolt. Indeed fear of their numbers and obstinacy even persuaded the Lacedaemonians to the action which I shall now relate, their policy at all times having been governed by the necessity of taking precautions against them. The helots were invited by a proclamation to pick out those of their number who claimed to have most distinguished themselves against the enemy, in order that they might receive their freedom; the object being to

test them, as it was thought that the first to claim their freedom would be the most high spirited and the most apt to rebel. As many as two thousand were selected accordingly, who crowned themselves and went round the temples, rejoicing in their new freedom. The Spartans, however, soon afterwards did away with them, and no one ever knew how each of them perished. The Spartans now therefore gladly sent seven hundred as hoplites with Brasidas, who recruited the rest of his force by means of money in the Peloponnese.' (Thucydides 4.80.2–5)

This serves as a reminder, if one were needed, that Spartan freeing of helots was not done through any sense of shared humanity. Note that these helots are specifically armed as hoplites (presumably at state expense, since they could hardly have provided their own equipment). Brasidas' men and the *neodamodeis* formed part of the Spartan hoplite line at Mantinea (Thuc. 5.67.1). Although fighting as hoplites in the main hoplite phalanx, these men were not incorporated into the *morai* or *lochoi*, the regular Spartan units, and remained separate from them. Helots and *neodamodeis* together also formed part of the Spartan force sent to Sicily (Thuc. 7.19.3), and *neodamodeis* served also in the fourth century (Xen., *Hell*. 3.1.4) and with Agesilaus on his expedition to Asia (Xen., *Hell*. 3.4.2).[11]

The use of helots as hoplites in this way stands in contrast to the earlier, and presumably usual, practice of employing them as light infantry, in the way that poor and slave soldiers from all other cities were also employed. At Plataea (479):

'On the right wing were ten thousand Lacedaemonians; five thousand of these, who were Spartiates, had a guard of thirty-five thousand light-armed helots, seven appointed for each man.' (Herodotus 9.28.2)

'Light-armed' translates as *psiloi*, the generic Greek word for light infantry (armed in an indeterminate way, but probably with light shields, javelins and no armour). Herodotus also assigns *psiloi* to the other cities' hoplites:

'This [38,700] was the number of hoplites that mustered for war against the barbarian; as regards the number of the *psiloi*, there were in the Spartan array seven for each hoplite, that is, thirty-five thousand, and every one of these was equipped for war. The *psiloi* from the rest of Lacedaemon and Hellas were as one to every hoplite, and their number was thirty-four thousand and five hundred.' (Herodotus 9.29.1–2)[12]

There is much uncertainty about how (or if) these *psiloi* were employed in the battle, since they do not appear at all in Herodotus' account of the fighting, which characteristically recounts only the exploits of the hoplites, and it may be that their role was chiefly as servants rather than fighters (Herodotus specifically states that the Spartan helots were equipped for war, which suggests that the others were not). This is a reminder that a hoplite, as (frequently) a wealthy gentleman, would normally be accompanied by at least one servant on campaign, as Thucydides' account of the retreat of the Athenian army from Syracuse (413) makes clear:

> 'Indeed they could only be compared to a starved-out town, and that no small one, escaping; the whole multitude upon the march being not less than forty thousand men. All carried anything they could which might be of use, and the hoplites and cavalry, contrary to their wont, while under arms carried their own victuals, in some cases for want of servants, in others through not trusting them; as they had long been deserting and now did so in greater numbers than ever.' (Thucydides 7.75.5)

This should not though be taken as evidence of the general untrustworthiness of such servants, since the Athenian retreat from Syracuse was an exceptional circumstance.[13]

Greeks and non-Greeks

So far we have been looking at hoplites as a specifically Greek phenomenon, and indeed the use of other similarly armed and organized infantry is strictly outside the scope of this book. This can be justified by the fact that to the Greeks there was a very close link between ethnic origins and fighting styles, something that can be seen most clearly in the Hellenistic period, but which is also inherent in the contrasts constantly drawn between Greeks and 'barbarians' (non-Greeks) in the course of the Persian Wars and after. The Greeks felt that their way of fighting, with its emphasis on face-to-face close-quarters combat, on standing one's ground in battle and in fighting a decisive battle rather than skirmishing, were inherently different from the fighting style of other peoples, and was an essential part of their common Greekness.

As we have already seen in earlier chapters, the extent to which this was actually true in practice should not be overstated. For one thing, elements of the Greek panoply were adopted from other peoples (the shield supposedly from the Egyptians, features of shield and helmet from the Carians, who were on the margins of the Greek world), while close-order heavy infantry were already well known in several non-Greek civilizations. The concept of the decisive battle and face-to-face fighting was also perfectly familiar to the Persians, for example, despite Herodotus' much quoted satirical take on the Greek way of fighting (more on this in Chapter 7).

In addition, hoplite equipment and fighting styles – taken more narrowly as the specific combination of Argive aspis, spear, armour and close-order phalanx – was adopted by other non-Greek peoples, in particular by the Etruscans and their neighbours in Italy. This was probably initially a result of the contact with the Greek colonists of Sicily and southern Italy; Greek expansion overseas throughout the Archaic period brought large numbers of Greek settlers to Italy (*Magna Graecia*, 'Greater Greece'), and the Etruscans and very likely the Romans adopted elements of equipment and fighting styles from them (it may also be that the fighting styles were well known in Italy anyway, since close-order heavy infantry were known in the Western Mediterranean as well as the East). Indeed, some of the best-known examples of hoplite equipment, such as the Bomarzo shield, are Etruscan not Greek; the Chigi vase, with its valuable early depiction of a hoplite phalanx, was found in an Etruscan tomb; and some of the clearest descriptions of fighting in a phalanx are to be found in the pages of the historian of Rome, Dionysius of Halicarnassus, in his accounts of the early battles of Rome and the Etruscans (note though that Dionysius was writing in part to convince the Greeks that the Romans were not so different from themselves). The details of Etruscan organization and equipment are beyond the scope of this book, while that of the Romans, and the extent to which they fought in a hoplite phalanx until their adoption of their later better-known equipment of sword, large shield and javelin, are somewhat controversial. Nevertheless, it seems clear at least that there was nothing uniquely Greek about the hoplite phalanx and it represented a way of fighting common to both Greece and Italy, at least around the fifth and fourth centuries.[14]

But while it is true that the hoplite phalanx was not confined to Greece, it is also true that Greece's neighbours, where they were of different ethnic origin – and indeed Greeks themselves, where social and economic circumstances were different from the city states of the southern mainland and Ionia – did not generally adopt the full equipment of the hoplite phalanx. Within Greece, the Thessalians, for example, lived on more extensive plains in northern Greece, where the land was suitable for horse breeding and the pattern of politically separate city states did not evolve in the same way. The Thessalians were therefore noted for their aristocratic cavalry, not their hoplites (in Aristotle's terms, the middle class of hoplites never became powerful enough to overthrow the horse-borne aristocracy). Further north, the Macedonians also did not develop a city-based political system, remaining scattered farmers dominated by an aristocratic cavalry. It may be that there were Macedonian hoplites from a relatively early date, but until the reforms of Philip II in the mid-fourth century, Macedonian infantry was considered unreliable at best and probably fought, like Greek *psiloi*, as unarmoured javelineers.[15]

Further north, and west and east of Macedon, lay the lands of those on the fringes of the Greek world, the Thracians and Illyrians, divided into several subgroups and tribes. Also lacking an urban culture and relatively wealthy class which could support a force of hoplites, these tribal cultures depended on light infantry and cavalry too for their armed forces, the Thracians in particular relying on the famous Thracian peltast, a javelineer or spearman with a light shield (*pelta* or *pelte*). These infantry proved extremely effective in the rough terrain to the north of Greece, and actual Thracian peltasts – or Greek peltasts armed and fighting in the Thracian manner – became highly important in southern Greek warfare, especially during the fourth century, and also played a central role in the later Macedonian conquest of the Persian Empire. Their role in war will be considered further in Chapter 6, but for now it is enough to note that these peoples did not adopt the hoplite phalanx, most likely because their social and economic structure (rural, tribal, aristocratic) did not favour the creation of a heavy infantry force, as well as for more technical reasons around terrain and types of combat.[16]

Manpower

It is difficult to determine the total numbers of hoplites and non-hoplites that might have been available to each city, and even more so to determine what proportion of total theoretical manpower this might represent. Populations of ancient societies are difficult to define with any precision (there are few hard numbers to work from), and are also very much tied up with the questions of population dynamics and expansion that lie behind the 'hoplite debate' issues examined in Chapter 1. For individual battles, we will often have carefully broken-down contingent lists with numbers for each group, though the reliability of these may be suspect. Here we are more interested in the total manpower resources on which each city could draw, and the extent to which it used these to provide its hoplite forces.[17]

As we saw above, according to Thucydides, at the start of the Peloponnesian War, Athens 'had an army of thirteen thousand hoplites, besides sixteen thousand more in the garrisons and on home duty at Athens' (Thuc. 2.13.6–7), these last being the younger and older age classes and the metics. There were also 'twelve hundred horse including mounted archers, with sixteen hundred archers unmounted, and three hundred triremes fit for service' (Thuc. 2.13.8) – which are very modest numbers of specialist light forces, but 300 triremes would have required some 51,000 rowers if all were manned at the same time, an indication that the hoplite class was relatively small compared with the poorer rowers.

Army sizes before the Classical period are of course particularly difficult to determine. Strabo (10.1.10) records a parade in Eretria featuring 3,000 hoplites (with 600 cavalry and sixty chariots), which may be the whole of their forces; if so, this would imply that the hoplite class was a small proportion of the total population, since Eretria controlled a large territory. Later sources give the Athenian Solon 500 men for the conquest of Salamis (Plut., *Solon*. 9.2) and record a disastrous defeat of Samos where 1,000 men were killed (Plut., *Mor.* 296ab) – to be considered disastrous, this must have been a fairly high proportion of their total manpower. It appears then that up to the sixth century, hoplite forces were still modestly sized, with 3,000 or so being perhaps typical for a populous city or region.[18]

The earliest detailed figures we have for most cities are the Battle of Plataea (479), for which Herodotus lists the Greek contingents:

'On the right wing were ten thousand Lacedaemonians; five thousand of these, who were Spartans, had a guard of thirty-five thousand light-armed helots, seven appointed for each man. The Spartans chose the Tegeans for their neighbours in the battle, both to do them honour, and for their valour; there were of these fifteen hundred men-at-arms. Next to these in the line were five thousand Corinthians, at whose desire Pausanias permitted the three hundred Potidaeans from Pallene then present to stand by them. Next to these were six hundred Arcadians from Orchomenus, and after them three thousand men of Sicyon. By these one thousand Troezenians were posted, and after them two hundred men of Lepreum, then four hundred from Mycenae and Tiryns, and next to them one thousand from Phlius. By these stood three hundred men of Hermione. Next to the men of Hermione were six hundred Eretrians and Styreans; next to them, four hundred Chalcidians; next again, five hundred Ampraciots. After these stood eight hundred Leucadians and Anactorians, and next to them two hundred from Pale in Cephallenia; after them in the array, five hundred Aeginetans; by them stood three thousand men of Megara, and next to these six hundred Plataeans. At the end, and first in the line, were the Athenians who held the left wing. They were eight thousand in number, and their general was Aristides son of Lysimachus.' (Herodotus 9.29)

These city contingents might be taken to be a maximum effort, given the importance of defeating the Persian invasion, or they might represent what became normal in the later fifth century, when Sparta for example invaded Attica each year with two-thirds of its full army (Thuc. 2.10.2, 2.47.2). This would make sense if Athenian population remained fairly static during the fifth century, the 8,000 at Plataea equating then to a full force of 12,000 (field army), to compare with the 13,000 quoted by Pericles. Each component was also accompanied by equal numbers of light-armed troops, except the Spartans.

So generalizing wildly from these scattered figures, we might decide that a typical large Archaic city might muster around 3,000 hoplites, and that by the late fifth century this had risen to around 8,000–12,000 for the larger cities (much smaller numbers of course, just a few hundreds, for the smaller towns). This increase in hoplite contingent size is presumed to be due not so much to general population growth as to the extension of the franchise, or of the range of men admitted to the phalanx. These figures, if truly representative, would suggest that this was quite a late occurrence, toward the end of the sixth century at the earliest, rather than further back in the Archaic period.[19]

As usual, the Spartans are a special case when it comes to numbers. As we have seen, the Spartans made up perhaps half of their army from *perioikoi*, serving alongside (whether in separate units or possibly in the same units) the Spartans proper, and also from the time of the Peloponnesian War at least enrolled helots as hoplites. At first the numbers of these *neodamodeis* were small – just 700 for the expedition sent to Thrace under Brasidas and described above. However, by the early fourth century, attempting to build an army to oppose Epmaninondas' Boeotians, greater numbers were needed:

> 'It was also determined by the authorities to make proclamation to the helots that if any wished to take up arms and be assigned to a place in the ranks, they should be given a promise that all should be free who took part in the war. And it was said that at first more than six thousand enrolled themselves, so that they in their turn occasioned fear when they were marshalled together, and were thought to be all too numerous.' (Xenophon, *Hellenica* 6.5.28–29)

It is characteristic of the Spartans that they should be afraid of men who had enrolled for their defence. In the case of overseas expeditions, enrolling helots would have been a response to the desire not to send Spartiates far away from Sparta, but apparently by the fourth century, Sparta was suffering from the *oliganthropia* ('shortage of manpower') that plagued it for the next two centuries. Aristotle noted in the late fourth century that Sparta's poor regulation of land ownership and inheritance allowed estates to be concentrated into the hands of too few men or (even worse) women:

> 'As a result of this although the country is capable of supporting fifteen hundred cavalry and thirty thousand hoplites, they numbered not even a thousand. And the defective nature of their system of land-tenure has been proved by the actual facts of history: the state did not succeed in enduring

a single blow [that is, defeat at Leuctra in 371], but perished owing to the smallness of its population [*oliganthropia*]. They have a tradition that in the earlier reigns they used to admit foreigners to their citizenship, with the result that dearth of population did not occur in those days, although they were at war for a long period; and it is stated that at one time the Spartiates numbered as many as ten thousand. However, whether this is true or not, it is better for a state's male population to be kept up by measures to equalize property.' (Aristotle, *Politics* 2.1270a)[20]

The upshot of this is that the size of hoplite army that could be raised by a city was only indirectly related to the population of that city, since only a limited subset of the population would be admitted to the hoplite phalanx. The internal political arrangements of each city determined the size of its hoplite forces, although special measures such as enrolling foreigners, social inferiors or mercenaries could be adopted in cases of dire need.

Mercenary hoplites

So far we have mostly been looking at citizen hoplites, those serving in their city army (or city militia) through duty, as part of their obligations as citizens (see below). Such hoplites could be paid by their city, but this did not make them mercenaries as such. Mercenaries are, rather, those soldiers who serve for some foreign power or polity, purely for pay, and often as professional soldiers who might serve long-term for a particular employer or move from job to job. As Greek mercenary service is a vast subject in itself, I can offer no more than the briefest outline here. Greek hoplites, for all that they are thought of as the quintessential citizen forces, have a long history as mercenaries, going right back into the period of Greek overseas expansion in the Archaic period, as we saw in Chapter 1.[21]

There was a particularly strong tradition of mercenary service in Egypt, dating back to the first 'men of bronze' who served the Pharaoh Psammetichus (Psamtik), and whom we encountered in Chapter 2. A group of such mercenaries, on an expedition up the Nile to the great Egyptian temple at Abu Simbel, carved their names (and that of the 'Axe, son of Nobody', which they used to do the carving) into the leg of a colossal statue of Ramses II. This shows a degree of literacy and of wit, if not of respect for local culture, that reveals these men not to be the desperate poor, but men of some wealth and education. They may of course have been the leaders of a larger group; Herodotus (2.163) records that the Pharaoh Apries had a bodyguard of 30,000 Ionians and Carians. Into Classical times, Egypt (or the Persians ruling Egypt) remained favoured employers, with the

Athenian general Iphicrates spending an extended period of time on mercenary service there.[22]

Hoplites were not the only sort of men to serve as mercenaries, and other classes of soldiers (light infantry or peltasts) or sailors (since paid rowers for the fleet were also in strong demand, and remembering that slaves were rarely employed to row warships except in emergencies, as at Xen., *Hell.* 1.6.11) are commonly to be found as mercenaries. Xenophon describes how the Spartan admiral Lysander hoped to persuade the Persian prince Cyrus to win over Athenian rowers with higher rates of pay:

> '[Lysander's ambassadors] urged [Cyrus] to make the wage of each sailor an Attic drachma a day, explaining that if this were made the rate, the sailors of the Athenian fleet would desert their ships, and hence he would spend less money. He replied that their plan was a good one, but that it was not possible for him to act contrary to the King's instructions ... Lysander accordingly dropped the matter for the moment; but after dinner, when Cyrus drank his health and asked him by what act he could gratify him most, Lysander replied: "By adding an obol to the pay of each sailor." And from this time forth the wage was four obols, whereas it had previously been three.' (Xenophon, *Hellenica* 1.5.4–7)

That sailors (or more accurately, rowers) might be willing to change sides for pay in this way is explained in this case by the fact that the Athenian navy was to a large extent manned by rowers from the Athenians' overseas empire, rather than native Athenians. Despite the bad reputation that mercenaries have in general for changing sides and serving for the highest bidder, Greek mercenaries seem to have been more generally loyal, part of the reason for this being that mercenaries were often serving not as independent freelancers but as an extension of their home city's foreign policy, often with their city's express permission.[23]

There were of course exceptions to this. Mercenaries were thought to be the natural recourse of tyrants who could not rely on support from their own citizens, and the Greek cities of Sicily in particular relied on mercenaries far more than the constitutional regimes in mainland Greece. It was also thought a sensible precaution not to hire more mercenaries than there were citizen soldiers, to ensure they did not try to take over for themselves, or turn over to an enemy, the city of their employers, although the same warnings also applied to the forces of allies (Aen. Tact. 12).

There were also varied classes of mercenary. As well as the fleet rowers, light infantry – in particular peltasts (originally chiefly those from Thrace) or other specialists, such as archers – would often be hired as mercenaries. Part of the reason for this was the Greek conception of the link between ethnicity and military skill – if a force of archers was required, for example, it would be hired from a region specializing in archery (such as Crete), rather than training native soldiers to shoot.

Mercenaries of the hoplite class were among the first Greek mercenaries in the Archaic period, but domestic use of hoplite mercenaries appears not to have been particularly widespread until the fourth century. Then, the growth of mercenary service was (at least according to Athenian orators of the time, Demosthenes and Isocrates) linked to the increasing poverty at the lower ends of the hoplite class, along with incessant political upheaval in Greek cities which led to exile and property seizure, casting men adrift with no other means to make a living. This mass of mercenaries available for hire could form entire armies, like the Ten Thousand serving the Persian pretender Cyrus; such armies could contain men from a mix of backgrounds (Xenophon himself was a wealthy Athenian) and motivations, but a large proportion of them would have been veterans of the Peloponnesian War seeking employment in the economic and political upheaval of the time. In most Greek city armies, however, mercenaries did not form a major element, certainly not of the phalanx, and mercenaries were more likely to be hired as specialists (light infantry, sometimes cavalry) or for garrisons, than as replacements for citizen hoplites in the phalanx.[24]

A related question is whether mercenary hoplites would supply their own hoplite equipment, and therefore were of the hoplite class and available to serve as hoplites in their own city armies, or whether equipment was commonly supplied by the employer, in which case mercenary hoplites could be drawn from a wider and poorer pool of men, and could have included poorer farmers, artisans or labourers. There is only scattered evidence for this either way. Xenophon records that after the defeat at Cunaxa, emissaries of the Persian king came to the Greek army:

> 'For yourselves, the King demands your arms [*ta hopla*]; for he says that they belong to him, since they belonged to Cyrus, his slave.' (Xenophon, *Anabasis* 2.5.38)

The uniformity of the equipment of the Ten Thousand also suggests that they may have been centrally equipped. Diodorus records a massive weapon production programme in Syracuse under the tyrant Dionysius in the early fourth century:

> 'At once, therefore, he gathered skilled workmen, commandeering them from the cities under his control and attracting them by high wages from Italy and Greece as well as Carthaginian territory. For his purpose was to make weapons [*hopla*] in great numbers and every kind of missile ... After collecting many skilled workmen, he divided them into groups in

accordance with their skills, and appointed over them the most conspicuous citizens, offering great bounties to any who created a supply of arms. As for the armour [*hopla*], he distributed among them models of each kind, because he had gathered his mercenaries from many nations; for he was eager to have every one of his soldiers armed with the weapons of his people, conceiving that by such armour his army would, for this very reason, cause great consternation, and that in battle all of his soldiers would fight to best effect in armour to which they were accustomed ... [T]here were made one hundred and forty thousand shields [*aspides*] and a like number of daggers [*encheridia*] and helmets; and in addition corselets [*thorakes*] were made ready, of every design and wrought with utmost art, more than fourteen thousand in number. These Dionysius expected to distribute to his cavalry and the commanders of the infantry, as well as to the mercenaries who were to form his bodyguard.' (Diodorus 14.41.3–5; 43.2–3)

There are some peculiarities to this account (why so many shields, or so little armour), and we might wonder why the mercenaries did not already have their own national equipment, if they were accustomed to using it. But it is difficult to say for sure whether such episodes are the exception or the rule – probably all we can conclude is that employers might provide arms, but we cannot be certain that this was normal practice, and as such it is hard to use this argument to draw firm conclusions about the economic background of the mercenaries. It does at least seem likely that both Cyrus and Dionysius (both outsiders from the normal internal political arrangements of Greek cities) sought to expand the pool of available manpower by drawing from those poorer men who would not normally be eligible or able to serve as hoplites (whether through temporary or recent hardship, or long-term inferior economic status).[25]

Perhaps the main benefit of mercenaries was that, unlike the amateur citizen soldiers, they could afford to be away on campaign (or in garrisons) for long periods. Whether 'could afford to' means economically, in this case, depends on the still contentious answer to the question of the economic background of most citizen hoplites, but assuming that at least a proportion of the citizen hoplites were farmers (or farm managers), they could not, as a practical matter, be away from their lands for extended periods, while the wealthier citizens would not have wished to be. For local campaigns of limited duration, such as against a neighbouring city, this would not have mattered, but for overseas expeditions and particularly when facing the professional armies of Philip II of Macedon from the mid-fourth century, it was essential to have men who could remain in the field (not their own fields) for extended periods. The orator Demosthenes proposed creating a special expeditionary force 'to carry on a continuous war of annoyance against Philip' (Dem. 4.19), 'Not an imposing army – on paper – of ten or twenty thousand mercenaries!' Demosthenes explained:

> 'I propose that the whole force should consist of two thousand men, but of these five hundred must be Athenians, chosen from any suitable age and serving in relays for a specified period – not a long one, but just so long as seems advisable; the rest should be mercenaries. Attached to them will be two hundred cavalry, fifty at least of them being Athenians, serving on the same terms as the infantry.' (Demosthenes 4.21)

See below for mobilization by age classes. Serving in relays would allow the Athenian core to remain in the field for long periods, while the mercenaries would be on permanent duty (at least while the money lasted). Demosthenes does not paint a flattering picture of the value of mercenary armies, but then his intent was to persuade the Athenians to fight their own wars rather than relying entirely on mercenaries. Demosthenes warned particularly against under-funded mercenary expeditions:

> 'Wherever, I believe, we send out a force composed partly or wholly of our citizens, there the gods are gracious and fortune fights on our side; but wherever you send out a general with an empty decree and the mere aspirations of this platform, your needs are not served, your enemies laugh you to scorn, your allies stand in mortal fear of such an expeditionary force … For when your general leads wretched, ill-paid mercenaries, and finds plenty of men here to lie to you about what he has done, while you pass decrees at random on the strength of these reports, what are you to expect?' (Demosthenes 4.45–46)

Despite such doubts, another reason for hiring mercenaries was their greater experience and professionalism. Although the citizen phalanx remained essentially amateur and eschewed military training, the long campaigns of the Peloponnesian War meant that many hoplites became veterans. The Athenians sent on the expedition to Syracuse, for example, were considered expert hoplites, in contrast to the inexperienced native Syracusans. Mercenaries who had experienced long periods of service could expect to have a high level of practical experience and technical skill, and this may have made them more desirable than inexperienced and unskilled part-time citizen militias. However, military experience and technical proficiency could not give the mercenary that most valuable quality of a hoplite, the willingness to stand, fight and die, as Aristotle points out:

> '[Mercenaries] are like armed men fighting against unarmed, or trained athletes against amateurs; for even in athletic contests it is not the bravest men who are the best fighters, but those who are strongest and in the best training. But professional soldiers prove cowards when the danger imposes too great a strain, and when they are at a disadvantage in numbers and equipment; for they are the first to run away, while citizen troops stand their ground and die fighting, as happened in the battle at the temple of Hermes. This is because citizens think it disgraceful to run away, and prefer death to safety so procured; whereas professional soldiers were relying from the outset on superior strength, and when they discover they are outnumbered they take to flight, fearing death more than disgrace.' (Aristotle, *Nicomachean Ethics* 3.8.8–9)

One interesting aspect of the payment and maintenance of mercenaries is revealed by Aeneas 'the Tactician', who advises that:

> 'The wealthiest citizens should be required to provide mercenaries, each according to his means ... When as many as you need are assembled, they should be divided into *lochoi*, and the most trustworthy of the citizens placed over them as *lochagoi*. Pay and maintenance the mercenaries should receive from their employers, partly at the private expense of the latter, partly from funds contributed by the state ... Reimbursement should be made in due time to those who have incurred expense for the mercenaries, after deducting the taxes due the state from each individual.' (Aeneas Tacticus 14.2–4)

Whether this suggestion was widely adopted in this way and mercenary hire subcontracted out to individuals is unclear (although this was the way navies were raised, with ships supplied by the wealthy from their own funds). It is good to know at any rate that mercenaries were not tax-deductible.

Training

This brings us to the topic of hoplite training. We have frequently seen that, with the exception of the Spartans, Greek hoplites were usually untrained citizen militias with a fairly rudimentary level of drill and weapons skill. A possibly exaggerated view of common attitudes to training is put by Xenophon into the mouth of one of Cyrus' Persians, one Pheraulas, a commoner:

> '[W]e have been initiated into a method of fighting, which, I observe, all men naturally understand, just as in the case of other creatures each understands some method of fighting which it has not learned from any

other source than from instinct: for instance, the bull knows how to fight with his horns, the horse with his hoofs, the dog with his teeth, the boar with his tusks. And all know how to protect themselves, too, against that from which they most need protection, and that, too, though they have never gone to school to any teacher ... Furthermore, even when I was a little fellow I used to seize a sword wherever I saw one, although, I declare, I had never learned, except from instinct, even how to take hold of a sword. At any rate, I used to do this, even though they tried to keep me from it – and certainly they did not teach me so to do – just as I was impelled by nature to do certain other things which my father and mother tried to keep me away from. And, by Zeus, I used to hack with a sword everything that I could without being caught at it. For this was not only instinctive, like walking and running, but I thought it was fun in addition to its being natural.' (Xenophon, *Cyropaedia* 2.3.9–11)

Pheraulas goes on to contrast the trained Persian aristocrats with his fellow commoners:

'And yet I know that these men pride themselves upon having been trained, as they say, to endure hunger and thirst and cold, but they do not know that in this we also have been trained by a better teacher than they have had; for in these branches there is no better teacher than necessity, which has given us exceedingly thorough instruction in them.' (Xenophon, *Cyropaedia* 2.3.9–13)

We can identify three areas in which training might be beneficial: use of weapons (and also of defensive equipment such as shields); physical fitness and endurance; and drill, discipline and technical formation skills. Of these three, the first, as argued by Pheraulas, was thought not to require any special instruction but to just come naturally, at least to those who have been been brought up with such weapons. This ties into the perceived Greek notion of the link between ethnicity and specialist skills – Rhodians made good slingers because traditionally they grew up using the sling, and the same with Cretans and archery. For mainland Greeks, the spear and the sword were evidently thought of as the natural weapons. This may seem strange, since it seems self-evident that even a simple weapon like a sword or spear can be used more or less skillfully, and that it would be possible to increase one's skill in their use through training.[26]

We have no evidence either way for the presence or absence of weapons training in Archaic Greece. We can well imagine that the heroes of Homer would be expected to know how to handle their weapons without needing special instruction, though whether an outstanding warrior like Achilles was (thought to be) skilled entirely because of his innate ability, or whether he also received special instruction, is not known. During the long development of the more organized phalanx, it may be that the aristocratic front-fighters could afford, and received, some sort of weapons training, but the mass of their retainers or social inferiors in the rear ranks would hardly have done so. When weapons training does appear, in the late fifth and early fourth centuries, it is evidently a new phenomenon, and one that could be regarded with some disdain by gentleman hoplites. At this time, lessons in *hoplomachia*, 'fighting with (hoplite) weapons', began to be offered by itinerant, and self-appointed, experts. Note that, at least at first, any such training would have been private tuition undertaken on the initiative of and at the expense of the individual hoplite; there was no state programme of training (though see below for the Athenian *ephebeia*). Plato records a discussion between Socrates and a group of his followers as to whether 'fighting in armour' (*hoplomachia*) was a worthwhile subject for boys to study:

> 'Nicias: "There is no difficulty about that, Socrates. For in my opinion this accomplishment is in many ways a useful thing for young men to possess. It is good for them, instead of spending their time on the ordinary things to which young men usually give their hours of leisure, to spend it on this, which not only has the necessary effect of improving their bodily health – since it is as good and strenuous as any physical exercise – but is also a form of exercise which, with riding, is particularly fitting for a free citizen; for only the men trained in the use of these warlike implements can claim to be trained in the contest whereof we are athletes and in the affairs wherein we are called upon to contend [that is, war]. Further, this accomplishment will be of some benefit also in actual battle, when it comes to fighting in line with a number of other men; but its greatest advantage will be felt when the ranks are broken, and you find you must fight man to man, either in pursuing someone who is trying to beat off your attack, or in retreating yourself and beating off the attack of another."' (Plato, *Laches* 181d–182b)

Another of Socrates' circle offers a contrary view of the value of such training:

> 'Laches: "I conceive that if there were anything in it, it would not have been overlooked by the Lacedaemonians, whose only concern in life is to seek out and practise whatever study or pursuit will give them an advantage over others in war. And if they have overlooked it, at any rate these teachers of it cannot have overlooked the obvious fact that the Lacedaemonians are more intent on such matters than any of the Greeks, and that anybody who won honour among them for this art would amass great riches elsewhere, just as a tragic poet does who has won honour among us … But I notice that

these fighters in armour regard Lacedaemon as holy ground where none may tread, and do not step on it even with the tips of their toes, but circle round it and prefer to exhibit to any other people, especially to those who would themselves admit that they were inferior to many in the arts of war.'" (Plato, *Laches* 182e–183c)

Laches aso makes the point that no teachers of *hoplomachia* are themselves noted warriors, and offers an anecdote about a trainer who tried out an experimental weapon (a scythe attached to a spear – an early polearm) with ridiculous results. Socrates then, sadly for our purposes, leads the discussion off into the general value of education. A number of points can be made about this discussion. Firstly, it is far from a given that any sort of weapons training is considered useful, and in particular the point is made that fighting in the phalanx would require less in the way of weapon skills than more open-order fighting. A major selling point of the training is that it increases overall fitness, rather than giving specific skills, and also increases confidence and smartness (in a section not quoted above). It is not certain whether the point is that the Lacedaemonians did not themselves undertake weapons training, or that they were already far in advance in this subject of these trainers (I think the sense is that the Lacedaemonians, in 'overlooking' this training, do not themselves practice it at all). Even here at the end of the fifth century, when military professionalism was on the increase and expertise was valued, it is not a given that weapons skill is even worth studying for hoplites. We cannot be certain, however, that this dim view of such instructors was held universally. Evidently *hoplomachoi* could make a living from selling their training, and there are a few known cases of them rendering presumably valuable service, such as Phalinus, in the employ of the Persian Tissaphernes in the royal army at Cunaxa (401), who 'was held in honour by him; for this Phalinus professed to be an expert in tactics [*taxeis*] and *hoplomachia*' (Xen., *Anab.* 2.1.7). (We might wonder in this case which of Tissaphernes' men Phalinus instructed.)[27]

As for the second form of training that may be of value, physical fitness and endurance, we have seen that *hoplomachia* was of some value in increasing overall fitness, and also that valid leisure pursuits for hoplite-class gentlemen included athletics. There were also some specifically military forms of athletics, particularly the 'race in armour' (*hoplitodromos*) and the 'Pyrrhic dance' (*Pyrrhike*). According to Pausanias:

> 'The race for men in armour was approved at the sixty-fifth Olympiad [520], to provide, I suppose, military training; the first winner of the race with shields was Damaretus of Heraea.' (Pausanias 5.8.10)

The link between the race and training is explicit here, though an Olympic event hardly provided training for all hoplites, and the next event Pausanias describes was for chariots, so without direct military application. Herodotus tells us that the first Greeks to run into battle were the Athenians at Marathon (490), and so a link has been proposed between the running attack and the need to counter Persian archery (by crossing the 'beaten zone' of the archers as fast as possible). However, it is likely that the value of the *hoplitodromos* was in promoting general fitness and facility with carrying the panoply rather than with any specific military purpose or tactic.[28]

Plato provides a description of the Pyrrhic dance:

'The warlike division [of dances], being distinct from the pacific, one may rightly term *"pyrrhiche"*; it represents modes of eluding all kinds of blows and shots by swervings and duckings and side-leaps upward or crouching; and also the opposite kinds of motion, which lead to active postures of offence, when it strives to represent the movements involved in shooting with bows or darts, and blows of every description.' (Plato, *Laws* 814e–815a)

Note also that it is in fact wrestling that Plato thinks has the most military value:

'[W]restling of this kind is of all motions by far the most nearly allied to military fighting; and also that it is not the latter that should be learned for the sake of the former, but, on the contrary, it is the former that should be practised for the sake of the latter.' (Plato, *Laws* 814c–d)

How closely the movements of Pyrrhic dancing matched the movements required of the hoplite in battle – or indeed how closely hoplite fighting resembled wrestling – is a somewhat controversial matter (see Chapter 8 for further discussion of this). Note that Plato, in his ideal state, would replace all sports not directly relevant to warfare with combat sports of various kinds – in which he included (*Laws* 830c–e and 833a–e) running (with and without arms, and over various distances), field manoeuvres (also with and without arms), mock battles and assaults on fortified positions, ball games (*sphairomachia*) and fighting in armour (*hoplomachia*) – for which he recommends obtaining trainers to help draw up a set of rules. Aside from the latter case, these are mostly general exercises in fitness:

'Most important of all things for war is, no doubt, general activity of the body, of hands as well as feet – activity of foot for flight and pursuit, and of hand for the stand-up fighting at close quarters which calls for sturdiness and strength.' (Plato, *Laws* 832e–833a)

Plato contrasts the training routinely undertaken by boxers and athletes with the lack of any such training, outside his ideal state, for war:

'Is the fighting force of our State [*polis*] to venture to come forward every time to fight for their lives, their children, their goods, and for the whole State, after a less thorough preparation than the competitors we have been describing?' (Plato, *Laws* 830c)

The answer, as things stood in Greece, is 'yes', as such exercises were not widely undertaken:

'Are we all aware of the reason why such choristry and such contests do not at present exist anywhere in the States [*poleis*], except to a very small extent?' (Plato, *Laws* 831b)

The 'very small extent' perhaps reflects the small bodies of professional picked forces. The reasons why such measures are not widely adopted, Plato goes on to explain (after some prompting), are excessive love of money and the internal political divisions of cities which deter the rulers from increasing the military capabilities of the ruled – so internal politics could, in Plato's view, act as a deterrent to providing military training, so that the ruled could be kept in their place.

It appears that, on the whole, rather than seeking specific training in technical skills, Greeks preferred general fitness (and strength and flexibility) training, in line with the ethos of gentlemanly athletic pursuits. There is little trace of specific military fitness training of the type instituted by Philip of Macedon, who introduced route marches with full arms and equipment for his phalanx.[29]

Such training was not totally unknown, but was perhaps the preserve of the picked units (*epilektoi*) or of mercenaries, who did display a higher degree of professionalism (not surprisingly). The fourth-century Thessalian tyrant, Jason of Pherae, boasted of the superiority of his forces:

'"[Y]ou are aware that I have men of other states as mercenaries to the number of six thousand, with whom, as I think, no city could easily contend. As for numbers," he said, "of course as great a force might march out of some other city also; but armies made up of citizens include men who are already advanced in years and others who have not yet come to their prime. Furthermore, in every city very few men train their bodies, but among my mercenaries no one serves unless he is able to endure as severe toils as I myself." And he himself – for I must tell you the truth – is exceedingly strong of body and a lover of toil besides. Indeed, he makes trial every day of the men under him, for in full armour he leads them, both on the parade-ground and whenever he is on a campaign anywhere. And whomsoever among his mercenaries he finds to be weaklings he casts out, but whomsoever he sees to be fond of toil and fond of the dangers of war he rewards, some with double pay, others with triple pay, others even with quadruple pay, and with gifts besides, as well as with care in sickness and magnificence in burial; so that all the mercenaries in his service know that martial prowess assures to them a life of greatest honour and abundance.' (Xenophon, *Hellenica* 6.1.5–6)

Note, however, that reference is made only to training in physical fitness, not to drill or weapons training. The extent to which even picked units underwent such training is unclear, but according to Xenophon, in 370 'all the Boeotians were now training themselves in the craft of arms, glorying in their victory at Leuctra' (Xen., *Hell.* 6.5.23), so only after Leuctra did the general Boeotian levy start training, and then it was in 'gymnastics with arms' – presumably physical training rather than technical skills.

As usual, the Spartans were the (partial) exception to the above observations, since they had a state programme of training, the *agoge*, that all Spartiates at least had to go through, perhaps from the age of 7, which was intended to toughen and discipline (or brutalize and desensitize, depending on one's point of view) Spartan youth. This involved separating Spartan boys from their families to live together in 'herds', with insufficient food and clothing (they were expected to steal from the helots to make up the deficiency), where they also practised hunting, dancing and singing. This was not a strictly military programme, since it too did not feature drill or weapons handling, but instead a general means of toughening Spartan boys and accustoming them to hardship and suffering – culminating in the flogging ceremonies at the temple of Artemis Orthia, which indicate the likely religious origins of the practice. This would of course have had some military benefit in producing fitter, tougher hoplites used to hardship and able to endure both privations on campaign and the terrors of battle, but like the all-male messes (*syssitia*) in which young adult Spartans lived, it was not directly tied to the army structure (it appears that the *syssitia* did not map directly onto the army organization of *enomotiai* and other small units).[30]

Finally, we come to the third form of training, the most specifically military form: drill and tactics (to the Greeks, 'tactics' meant organization and drill, rather than the use of armies on the battlefield, in the modern sense). This is the training that produces the manoeuvres examined in Chapter 3, and practised, Xenophon implies, only by the Spartans. Tactics was, however, also taught in late fourth-century Athens at least, as Xenophon shows in another anecdote of Socrates:

'[Socrates] once heard that Dionysodorus had arrived at Athens, and gave out that he was going to teach generalship. Being aware that one of his companions wished to obtain the office of general from the state, he addressed him thus: "Young man, surely it would be disgraceful for one who wishes to be a general in the state to neglect the opportunity of learning the duties, and he would deserve to be punished by the state much more than one who carved statues without having learned to be a sculptor."

'This speech persuaded the man to go and learn. When he had learnt his lesson and returned, Socrates chaffed him. "Don't you think, sirs," he said, "that our friend looks more 'majestic', as Homer called Agamemnon, now that he has learnt generalship? For just as he who has learnt to play the harp is a harpist even when he doesn't play, and he who has studied medicine is a doctor even though he doesn't practise, so our friend will be a general for ever, even if no one votes for him. But your ignoramus is neither general nor doctor, even if he gets every vote. But," he continued, "in order that any one of us who may happen to be a *taxiarchos* or *lochagos* under you may have a better knowledge of warfare, tell us the first lesson he gave you in generalship."

'"The first was like the last," he replied; "he taught me tactics – nothing else."'

(Xenophon, *Memorabilia* 3.1.1–5)

There are several interesting features to this conversation. Firstly, the lessons are taught to prospective generals, not to the mass of the hoplites (presumably the general would be expected to then drill and train the army under him). The tutor is a private teacher and students enrol on their own initiative (and one who did not enrol would 'deserve' to be punished by the state, but obviously would not actually be so punished). Finally, the lessons taught were all about tactics (organization and drill) – Socrates goes on to remark that the full art of generalship required many other skills (Xen., *Mem.* 3.1.6). Tactics provided a necessary starting point, as Socrates points out with the analogy with building a house ('just as stones, bricks, timber and tiles flung together anyhow are useless', Xen., *Mem.* 3.1.7; the analogy is repeated by Xenophon in the *Cyropaedia* – 'just as a house, without a strong foundation or without the things that make a roof, is good for nothing, so likewise a phalanx is good for nothing, unless both front and rear are composed of valiant men', Xen., *Cyrop.* 6.3.25). But tactics alone was not enough to make a good general, as Socrates points out, as skillful use of the army was also important (3.1.10, and see further Chapter 5).

Xenophon's enthusiasm for drill training at Sparta led him to devote a considerable section of the *Cyropaedia* to his idealized way of building and training an army, much of which we have already seen in Chapter 3. Xenophon presents a scene of Cyrus and his officers over dinner, discussing their progress in drilling their men. One officer (*taxiarchos*) describes his success:

"'But as for me, when you had instructed us about the arrangement of the lines and dismissed us with orders each to teach his own *taxis* what we had learned from you, why then I went and proceeded to drill one *lochos*, just as the others also did. I assigned the *lochagos* his place first and arranged next after him a young recruit, and the rest, as I thought proper. Then I took my stand out in front of them facing the *lochos*, and when it seemed to me to be the proper time, I gave the command to advance. And that young recruit, mark you, advanced – ahead of the *lochagos* and marched in front of him! And when I saw it, I said: 'Fellow, what are you doing?' 'I am advancing, as you ordered,' said he. 'Well,' said I, 'I ordered not only you, but all to advance.' When he heard this, he turned about to his comrades and said: 'Don't you hear him scolding? He orders us all to advance.' Then the men all ran past their *lochagos* and came toward me. But when the *lochagos* ordered them back to their places, they were indignant and said: 'Pray, which one are we to obey? For now the one orders us to advance, and the other will not let us.' I took this good-naturedly, however, and when I had got them in position again, I gave instructions that no one of those behind should stir before the one in front led off, but that all should have their attention on this only – to follow the man in front. But when a certain man who was about to start for Persia came up and asked me for the letter which I had written home, I bade the *lochagos* run and fetch it, for he knew where it had been placed. So he started off on a run, and that young recruit followed, as he was, breastplate and sword; and then the whole *lochos*, seeing him run, ran after. And the men came back bringing the letter. So exactly, you see, does my *lochos*, at least, carry out all your orders.'" (Xenophon, *Cyropaedia* 2.2.6–9)

I have some doubts whether many officers would really take such things so good naturedly, but this does at least illustrate both how hopeless Xenophon expected inexperienced soldiers to be at even the most basic drill, and also how he thought best to teach them – that is, by instructing the officers first, who then instruct their men, and by following the principle of each man following the leader. This principle is what we see demonstrated in the 'dinner drill' (for *paragoge*, discussed in Chapter 3), and it may seem obvious, though Xenophon is clear that it still had to be taught, and that those other than the Spartans may not have been familiar with it.

Training of this sort must presumably have been employed for the various elite, picked forces employed by several cities, but whether there was similar training for the general levy is less clear. In Athens at least, for which we have most evidence (but that is still very little), there was the institution of the *ephebeia*, or ephebate, although its nature, and the period in which it was in place, remain unclear. At least parts of the institution probably date back to the fifth century. Young Athenians (at age 18) would be gathered to swear an oath and would then embark on a two-year programme of training. It is not certain whether this training also dates back to the fifth century or whether only the oath and ritual elements are so old, with military training being introduced in the fourth century, perhaps in direct response to the threat of Macedon's increasingly professional forces. *The Constitution of the Athenians*, attributed to Aristotle, describes the training involved:

> '[T]he people elects by show of hands one of each tribe as disciplinary officer [*sophronistes*], and elects from the other citizens a marshal [*kosmetes*] over them all. These take the *epheboi* in a body, and after first making a circuit of the temples then go to Peiraeus, and some of them garrison Munichia, others the Point [strategic locations near Athens]. And the people also elects two athletic trainers [*paidotribes*] and instructors [*didaskoloi*] for them, to teach them fighting as heavy-armed soldiers [*hoplomachein*], and the use of the bow, the javelin and the sling ... They go on with this mode of life for the first year; in the following year an assembly is held in the theatre, and the cadets give a display of drill before the people, and receive a shield and spear from the state; and they then serve on patrols in the country and are quartered at the guard-posts ... When the two years are up, they now are members of the general body of citizens.' (Aristotle, *The Athenian Constitution* 42.2–5)

It seems likely that teaching drill in this way (and providing spear and shield) must be a later innovation, given the evidence for the earlier private nature of training and the provision of equipment. It may be that the more generic training – athletics and use of missile weapons – were of greater antiquity, and the more specific training in *hoplomachos* (weapons handling) and (by implication) drill came later. It may be surprising to see future hoplites trained in the use of missiles, but this was also a feature of training throughout the Hellenistic period. It is likely that general fitness and athleticism were the aim, rather than specific hoplite skills.[31]

So there certainly were forms of training for hoplites and future hoplites, but they tended to take the form of general athletic and fitness training, and were optional and provided privately. State-provided training in the specifics of hoplite fighting, and in particular in complex drill, at least before the later fourth century, seems to have been the sole preserve of the Spartans, with their specifically military way of life.

Discipline

As citizens, Greeks of the hoplite class had certain rights and could be expected to be treated, by their peers and the state, in a certain way. But as hoplites, as soldiers, they must have been subjected to some form of military discipline, since discipline of at least a basic sort is inherent to any effective military organization. There is conflicting evidence as to the extent to which hoplites were subject to what we might now think of as normal or at least typical levels of military discipline.[32]

On the one hand, the generals of many Greek cities (Athens is as usual the main example) were elected by the citizens (Aristot., *Const. Ath.* 61, and see Chapter 5), and the citizens were themselves the hoplites of the army that the general would command. Citizens could also bring lawsuits against generals, which meant that an army that felt itself badly commanded could sue its generals, once it got back home. This naturally meant that officers would need to adopt a more exhortatory style of command – something that is already familiar from the *Iliad*, though there for different reasons, as there was no democratic oversight in Homer's world, but there was a lack of rigid command structures, and the style of command in Homer is probably typical of that in warrior societies. In the example quoted above (Homer, *Iliad* 2.198–200), Odysseus reasoned with kings or chieftains he met while trying to prevent a rout, but struck common soldiers with his stick and rebuked them. In the more egalitarian phalanx, there must have been more carrot and less stick.

The Constitution of the Athenians spells out the two-way nature of command in a citizen army (at least as it was in the late fourth century):

> 'A confirmatory vote is taken in each presidency upon the satisfactoriness of their administration; and if this vote goes against any officer he is tried in the jury-court, and if convicted, the penalty or fine to be imposed on him is assessed, but if he is acquitted he resumes office. When in command of a force they have power to punish breach of discipline with imprisonment, exile, or the infliction of a fine; but a fine is not usual.' (Aristotle, *Constitution of the Athenians* 61.2)

Note the lack of corporal or capital punishment, and also that even these disciplinary measures would have to be pursued through a court, though it could be a military court in which hoplites formed the jury. Corporal punishment was felt to be appropriate only for slaves, as a speech of Demosthenes illustrates:

> 'Indeed, if you wanted to contrast the slave and the freeman, you would find the most important distinction in the fact that slaves are responsible in person for all offences, while freemen, even in the most unfortunate

circumstances, can protect their persons. For it is in the form of money that in the majority of cases the law must obtain satisfaction from them; but Androtion on the contrary exacted vengeance from their persons, as if they had been slaves.' (Demosthenes 22.55; see also 24.167)

Naturally enough, the Spartans followed different rules, and this led to tension when Spartan officers found themselves in charge of subjects or allies. Thucydides records a mutiny over arrears of pay of the fleet under the Spartan admiral Astyochus:

> 'Most of the Syracusan and Thurian sailors were freemen, and these the freest crews in the armament were likewise the boldest in setting upon Astyochus and demanding their pay. The latter answered somewhat stiffly and threatened them, and when Dorieus spoke up for his own sailors, even went so far as to lift his staff [*bakterion*] against him; upon seeing which the mass of the men, in sailor fashion, rushed in a fury to strike Astyochus.' (Thucydides 8.84.2–3)

Disorderly scenes (though typical of sailors) are not necessarily evidence of slack discipline, so much as of the breakdown of discipline, but the resistance to corporal punishment against free men is plain. Amongst the Spartans, striking with the staff or cane was apparently an accepted punishment, as were other forms of corporal punishment such as being forced to stand to attention holding one's shield (Xen., *Hell.* 3.1.9). But note that corporal punishment might prove counterproductive, as in the case of the Spartan general Mnasippus' response to insubordination:[33]

> '[Mnasippus] ordered the *lochagoi* and *taxiarchoi* to lead forth the mercenaries. And when some *lochagoi* replied that it was not easy to keep men obedient unless they were given provisions, he struck one of them with a staff and another with the spike [*styrax*] of his spear. So it was, then, that when his forces issued from the city with him they were all dispirited and hostile to him – a situation that is by no means conducive to fighting.' (Xenophon, *Hellenica* 6.2.18–19)

The contrast between Athenian and Spartan treatment of allies is noted by Plutarch too:

> 'For, well disposed as the Hellenes were toward the Athenians on account of the justice of Aristides and the reasonableness of Cimon, they were made to long for their supremacy still more by the rapacity of Pausanias and his severity. The commanders of the allies ever met with angry harshness at the hands of Pausanias, and the common men he punished with beatings,

or by compelling them to stand all day long with an iron anchor on their shoulders.' (Plutarch, *Aristides* 23.2)

In Athens there were military crimes for which a soldier could be tried – *astrateia*, or failing to report for duty; *lipotaxia*, 'desertion'; and *deilia*, 'cowardice'; as well as *rhipsaspia*, 'shield-flinging' or running away from combat – but these were civil offences requiring a formal trial, rather than being subject to summary military discipline. We should also note that, since many hoplites who escaped alive from a lost battle probably threw away their shields, the level of prosecution for *rhipsaspia* at least must have been rather low; it was more likely a means to bring down a prominent man or political opponent than a form of discipline applied to every hoplite.

The Ten Thousand present a particularly illuminating example. Being a 'city' adrift in enemy territory, they had no established legal system through which to impose discipline, and Xenophon had to head off a mutinous movement by calling an assembly (Xen., *Anab.* 5.7.3), at which future policy was discussed, any other crimes or wrongdoings were investigated and 'it was likewise resolved that the generals should undergo an inquiry with reference to their past conduct' (Xen., *Anab.* 5.8.1), in which Xenophon found himself accused:

> 'Accusations were also made against Xenophon by certain men who claimed that he had beaten them, and so brought the charge of wanton assault.' (Xenophon, *Anabasis* 5.8.1)

Xenophon defended himself, explaining that he used force on the men only in cases of need, not wantonly:

> 'I admit, soldiers, that I have indeed struck men for neglect of discipline, the men who were content to be kept safe by you who marched in due order and fought wherever there was need, while they themselves would leave the ranks and run on ahead in the desire to secure plunder and to enjoy an advantage over you. For if all of us had behaved in this way, all of us alike would have perished. Again, when a man behaved like a weakling and refused to get up, preferring to leave himself a prey to the enemy, I did indeed strike him and use violence to compel him to go on.' (Xenophon, *Anabasis* 5.8.13–14)

Xenophon explains that if a man fell behind and so endangered the army, 'I struck such a one with the fist in order that the enemy might not strike him with the lance [*longche*]' (5.8.16). The charges were dropped, and indeed Xenophon invited the men to remember occasions where he had treated them kindly. 'Then people began getting up and recalling past incidents, and in the end all was pleasant' (5.8.26).

Plato, musing in authoritarian mode on the arrangements in his ideal state, proposed a very rigid form of military discipline:

> 'Military organization is the subject of much consultation and of many appropriate laws. The main principle is this – that nobody, male or female, should ever be left without control, nor should anyone, whether at work or in play, grow habituated in mind to acting alone and on his own initiative, but he should live always, both in war and peace, with his eyes fixed constantly on his commander and following his lead; and he should be guided by him even in the smallest detail of his actions – for example, to stand at the word of command, and to march, and to exercise, to wash and eat, to wake up at night for sentry-duty and despatch-carrying, and in moments of danger to wait for the commander's signal before either pursuing or retreating before an enemy.' (Plato, *Laws* 942a–c)

Plato also recommends that 'with a view to excellence in war, they shall dance all kinds of dances', and not wear hats or socks, which would soften and spoil their extremities. This all sounds like a taut system of discipline, but the penalties are rather less than draconian. Anyone accused of cowardice or desertion is to be tried by those of his own class (hoplites before hoplites, cavalry by cavalry and so on):

> '[A]nd any man that is convicted shall be debarred from ever competing for any distinction and from ever prosecuting another for shirking service, or acting as accuser in connection with such charges; and, in addition to this, what he ought to suffer or pay shall be determined by the court.' (Plato, *Laws* 943b)

Even in cases of *rhipsaspia*, Plato (*Laws* 943a–945a) is at pains to ensure that those who lose their arms through no fault of their own or under physical duress not be wrongly accused, and for those convicted the penalty is to never again be employed as a soldier, along with a fine (and another fine for any future officer who enrols them).

For all these examples of lenient, or legally administered, discipline, we do also see examples of more harsh punishments, and not just from Spartans (though they are sometimes associated with the fleet, not the hoplite phalanx). The Athenian Alcibiades could tell his fleet that he 'made proclamation that death would be the punishment' for attempted desertion (Xen., *Hell.* 1.1.15). Making 'traitorous signals' to an enemy (which was often a sign of a plan to betray a city to enemy forces) could result in a death sentence (Lys. 13.65 – though after due process). Iphicrates, the fourth-century Athenian general, is said to have killed on the spot a sentry he found asleep – remarking 'I left him as I found him' (Front.- *Strat.*

3.12.2) – although that the same story is told of Epaminondas suggests this may well be apocryphal. The Spartan Brasidas warned his fleet and marines:

> 'to give no excuse for any one misconducting himself. Should any insist on doing so, he shall meet with the punishment he deserves, while the brave shall be honoured with the appropriate rewards of valour.' (Thucydides 2.87.9).

Nevertheless, these few examples hardly amount to evidence for strong discipline, and indeed, even in the Spartan army, what we see is not unquestioning obedience but a remarkably high level of insubordination. One was the case, quoted already in Chapter 1, of the Spartan Amompharetus at Plataea (479), who flatly refused to obey an order to withdraw his unit, and when ordered in person engaged in a stand-up row with his superior officer, to the bemusement of watching Athenian messengers (Hdt. 9.53–55). A similar situation occurred at the Battle of Mantinea (418), where the Spartan king Agis attempted a last-minute adjustment of the deployment his phalanx:

> 'However, as he gave these orders in the moment of the onset, and at short notice, it so happened that Aristocles and Hipponoidas would not move over, for which offence they were afterwards banished from Sparta, as having been guilty of cowardice.' (Thucydides 5.72.1)

The Spartan commanders simply refused to comply with an order that they did not understand or agree with. Note that their punishment came 'afterwards' (after the return of the army to Sparta, presumably), that they were accused of cowardice (based on the Athenian evidence, this might have been the only crime with which they could be charged, there being no offence of 'failing to follow orders') and that the punishment even for this crime was not death but banishment (admittedly a high price for a Spartan to pay). Before this same battle, the Spartans had demonstrated a willingness to question orders, when Agis led an initial impetuous attack on sighting the enemy:

> 'The Lacedaemonians at once advanced against them, and came on within a stone's throw or javelin's cast, when one of the older men, seeing the enemy's position to be a strong one, hallooed to Agis that he was minded to cure one evil with another; meaning that he wished to make amends for his retreat, which had been so much blamed, from Argos, by his present untimely precipitation.' (Thucydides 5.65.2)

Shouting out advice to the commanding general when about to go into action is surprising behaviour in an army so apparently professional and disciplined as the Spartan. The point about these incidents is that to Greeks (particularly hoplites), their status as citizens was more important than their status as soldiers. Greeks themselves drew strong contrasts between their own way of doing things and the harsh discipline of the Persians (and later, the Romans); Persian armies are often described going into battle 'under the lash', and Greeks frequently referred to even high-ranking Persians as 'slaves' because of their subservience to authority.[34]

That said, hoplite armies were not just a disorderly mob, by any means. Persuasion was more important than coercion, and Greeks expected to be commanded by consent not by force, but even so, officers issued commands, and expected them to be obeyed (which they usually were). The difference between Greek armies and those of other peoples is generally one of mindset rather than a greater or lesser ability to follow orders.[35]

Recruitment and mobilization

When the time came to mobilize an army for a campaign, or perhaps for an emergency such as the city coming under attack, various means were used to call up and gather the men, the best evidence as usual being for the Spartan and the Athenian ways of doing things.

In Athens, the older method of call-up, in use during the fifth century (and perhaps before, though there is no evidence), was for the ten elected *strategoi*, in consultation with the other officers of each of the ten tribes, to draw up lists (*katalogoi*), by hand, of all those who were to be called up for the coming campaign, with the lists being posted on painted boards in a public space. These lists, being the personal selections of the generals and officers, could ensure that the best and most suitable men (by whatever means that was measured) were selected for a campaign – Thucydides remarks that for the Syracusan expedition, 'the land forces had been picked from the best muster-rolls' (Thuc. 6.31.3) – but it could also obviously be open to abuse, and to a tendency to pick the same men repeatedly. The comic playwright Aristophanes wrote of the 'damned *taxiarchoi*':

> 'Once back again in Athens, these brave fellows behave abominably; they write down these, they scratch through others, and this backwards and forwards two or three times at random.' (Aristophanes, *Peace* 1179–81)

One of the concessions won by 'the people' in another of Aristophanes' plays was that:

> 'Further, the hoplite enrolled for military service shall not get transferred to another service through favour, but shall stick to that given him at the outset.' (Aristophanes, *Knights* 1369–72)

There seem to have been a number of official complaints made about the drawing up of the *katalogoi*. The orator Lysias presented one such case:

> 'The year before last, after I had arrived in the city, I had not yet been in residence for two months when I was enrolled as a soldier. On learning what had been done, I at once suspected that I had been enrolled for some improper reason. So I went to the *strategos*, and pointed out that I had already served in the army; but I met with most unfair treatment. I was grossly insulted but, although indignant, I kept quiet.' (Lysias, *For the Soldier* 9.4)

Despite the fact that the generals took 'oaths to enroll only those who had not served in the field' (Lys. 9.15) – presumably with the intention of sharing service evenly between all those eligible – these oaths were allegedly violated on occasion.

Perhaps because of these problems, or because the system of *katalogoi* was laborious to administer, Athens in the fourth century appears to have adopted a system similar to that already long in use by the Spartans. The *Constitution of the Athenians* describes how this worked:

> '[T]he Heroes giving their names to the Tribes are ten in number and those of the years of military age forty-two, and the *epheboi* used formerly when being enrolled to be inscribed on whitened tablets, and above them the Archon in whose term of office they were enrolled and the Name-hero of those that had been Arbitrators the year before, but now they are inscribed on a copper pillar and this is set up in front of the Council-chamber at the side of the list of Name-heroes … The Name-heroes also are employed to regulate military service; when soldiers of a certain age are being sent on an expedition, a notice is posted stating the years that they are to serve, indicated by the Archon and Name-hero of the earliest and latest.' (Aristotle, *Constitution of the Athenians* 53.4)

What this means is that each Athenian tribe had a 'Name-hero' or 'Eponymous hero' (a mythical figure associated with the tribe, or *phyle*), and that similar heroes' names were also assigned to each of the yearly age classes, cycling around once the full list of forty-two had been used. Then, as in the Spartan system, various sizes of force could be created by calling up a subset of the years (by selecting a list of yearly eponymous heroes – *en tois eponymois* – and/or a subset of the tribes – *en tois meresin*). The orator Aeschines describes his service:

> 'My first experience in the field was in what is called "division service" [*en tois meresi*], when I was with the other men of my age and the mercenary troops of Alcibiades, who convoyed the provision train to Phleius. We fell

into danger near the place known as the Nemean ravine, and I so fought as to win the praise of my officers. I also served on the other expeditions in succession, whether we were called out by age-groups or by divisions.' (Aeschines 2.168)

This mixed system was the one in use in the late fourth century, when, to oppose the Macedonians in the Lamian War, the Assembly decreed 'that all Athenians up to the age of forty should be enrolled, that three tribes should guard Attica, and that the other seven should be ready for campaigns beyond the frontiers'. (Diod. 18.10.2)

Conscription by *katalogos* may have been laborious to administer, but it could produce a high-quality army, since it meant that the most suitable men could be selected. Thucydides has Nicias contrast the Athenian (and allied) army at Syracuse produced by this selection with the mass levy of the Syracusans who (whether because they had no equivalent system or because of the dire need) mobilized all available manpower:

'Where we have Argives, Mantineans, Athenians, and the first of the islanders in the ranks together, it were strange indeed, with so many and so brave companions in arms, if we did not feel confident of victory; especially when we have mass-levies [*pandemei*] opposed to our picked troops [*apolektoi*], and what is more, Siceliots, who may disdain us but will not stand against us, their skill not being at all commensurate to their rashness.' (Thucydides 6.68.2)[36]

A similar system of conscription had probably long been in use in Sparta. Sending out two-thirds of the total available levy was evidently the usual practice, and a similar requirement was placed on Sparta's allies, which might imply that they in turn used a similar method of conscription. Thucydides relates the following from early in the Peloponnesian War:

'Immediately after the affair at Plataea, Lacedaemon sent round orders to the cities in Peloponnese and the rest of her confederacy to prepare troops and the provisions requisite for a foreign campaign, in order to invade Attica. The several states were ready at the time appointed and assembled at the Isthmus; the contingent of each city being two-thirds of its whole force.' (Thucydides 2.10.1–2)

Spartan forces of two-thirds of the total appear several times in the fifth century (at Thuc. 2.47.2 and 3.15.1). Xenophon writes of the preparations for the Battle of Leuctra (371):

'[T]he Lacedaemonians sent Cleombrotus, the king, across to Phocis by sea, and with him four *morai* of their own and the corresponding contingents of the allies.' (Xenophon, *Hellenica* 6.1.1)

Four *morai* out of six (see Chapter 3) is again two-thirds of the available total. Later, the call-up was extended:

'After this the ephors called out the ban of the two remaining *morai*, going up as far as those who were forty years beyond the minimum military age; they also sent out all up to the same age who belonged to the *morai* abroad; for in the original expedition to Phocis only those men who were not more than thirty-five years beyond the minimum age had served.' (Xenophon, *Hellenica* 6.4.17)

Conscription by age class in this way allowed the Spartans to form armies of varying sizes (and therefore also made up of subunits of varying size), according to need.

Rewards of service

As we have seen, hoplites were in general drawn from the wealthier elements of society and were expected to provide their own equipment, and also to provide for their own training, if any (except in Sparta, of course). We have also seen that there were two other classes of hoplite – the elite picked units and mercenaries – who perhaps from earliest times would have been paid, making them professional soldiers, distinct from the amateur militias that constituted most hoplite armies. The Theban Sacred Band, for example, was 'of three hundred chosen men, to whom the city furnished exercise and maintenance, and who encamped in the Cadmeia' (Plut., *Pel.* 18) – the point of their encamping in the Cadmeia (the citadel of Thebes) presumably being that they were permanently on active service. Argos similarly maintained a unit of 1,000 men 'which the city had for a long time caused to be trained for the wars at the public charge' (Thuc. 5.67.2), and the Arcadians maintained a unit of *Eparitoi* which they funded using the sacred treasures of Olympia (Xen., *Hell.* 7.4.33).

Small professional forces of this sort obviously provided considerable military advantages (the Sacred Band was instrumental in the Theban overthrow of Spartan power). But republics are always fearful of standing armies, and just as there was considerable distrust of mercenary forces, professional forces could and did have a detrimental effect on the internal politics of a city. The 1,000 Argives ending up allying themselves with the Spartan enemy in order to overthrow the Argive democracy and instal a narrow constitution more to the liking of the

Spartans (Thuc. 5.81.2). The Arcadians, having decided that using the sacred treasures to pay an army was a misuse of funds, stopped the pay of the *Eparitoi*:

> 'When, accordingly, a vote had been passed in the Arcadian assembly not to make use of the sacred treasures any longer, those who could not belong to the *Eparitoi* without pay speedily began to melt away, while those who could, spurred on one another and began to enroll themselves in the *Eparitoi*, in order that they might not be in the power of that body, but rather that it might be in their power.' (Xenophon, *Hellenica* 7.4.35)

This is an interesting example of the interplay of money, hoplites and the ever-changing political dynamics of a Greek state. The provision of pay by the state had allowed poorer, more democratically inclined men to serve, but when the pay was stopped, the richer, more oligarchic citizens who could afford to serve without pay enrolled instead and threatened to force Arcadia to join the Spartan cause.

In Athens, the fifth and fourth centuries saw a steady expansion of those who were paid for public service (political and judicial), in line with Athens' democratic ethos, and this extended also to the hoplites. Pay could come from two sources – from the city's central finances (as was usual for the fleet) or from wealthy benefactors, as the orator Lysias records:

> 'Now, when the townsmen had assembled together before their setting out, as I knew that some among them, though true and ardent patriots, lacked means for expenses of service, I said that the well-to-do ought to provide what was necessary for those in needy circumstances. Not only did I recommend this to the others, but I myself gave thirty drachmae each to two men; not as being a person of great possessions, but to set a good example to the others.' (Lysias 16.14)

Demosthenes calculates the cost of a military expedition as 'more than two hundred talents, if you include the private expenses of the troops' (Dem. 19.84). This indicates that the expenses of forces on campaign could be met by a mixture of public and private funds, in what was apparently a fairly ad hoc fashion. Note also that this presumably relates to expenses (money for food and essentials while on campaign), rather than the provision of a wage, of the sort that mercenaries would have received. Aristotle notes that:

> 'Also people have a way of being reluctant to serve when there is a war if they do not get rations and are poor men; but if somebody provides food they want to fight.' (Aristotle, *Politics* 1297b)

However, the *Constitution of the Athenians* notes that during the Peloponnesian War:

'[T]he people being locked up in the city, and becoming accustomed to earning pay on their military campaigns, came partly of their own will and partly against their will to the decision to administer the government themselves. Also Pericles first made service in the jury-courts a paid office, as a popular countermeasure against Cimon's wealth.' (Aristotle, *Constitution of the Athenians* 27.2)

The wealthy Cimon had provided handouts for his supporters. Xenophon recognized the difficulties that Athens had maintaining a large military force from a not predominantly wealthy population, and how this led to Athens' exploitation of its subject cities and consequent unpopularity. He wrote a pamphlet to suggest 'what regulations I think should be introduced into the state in order that every Athenian may receive sufficient maintenance at the public expense' (Xenophon, *Ways and Means* 4.33).[37]

Whether or not men were paid by the state, they had to eat. Aristophanes presents a comic picture of an unfortunate recruit:

'The departure is set for to-morrow, and some citizen has brought no provisions, because he didn't know he had to go; he stops in front of the statue of Pandion [where the *katalogos* is posted], reads his name, is dumbfounded and starts away at a run, weeping bitter tears.' (Aristophanes, *Peace* 1181–84)

Hoplites were expected to bring an initial supply of food with them, but laying on markets at which the men were able to buy their provisions was the normal way for Greeks to provision their armies, once the initial supplies were used up (see further in Chapter 6). The central authorities might be expected to provide all the other items that might be needful, such as spare equipment, and tools and materials for making repairs, but again depending on circumstances, this might have ended up being contingent on the initiative of individual hoplites.

So, depending on local circumstances, the poorer men might hope to receive ration money, if nothing else, for hoplite service, while the richer would generally be expected to pay their own way, with exceptions such as Athens and the permanently maintained picked units. Of course mercenary service for a full wage was also an option throughout the period.[38]

In addition to monetary rewards, the Greeks were an exceptionally competitive people, and rewards of other sorts were also available; prizes for valour could be offered for particularly distinguished service, and in some cases an individual might be designated 'man of the match' (as we might say) after a battle. Herodotus records both his own and the official verdicts on the Battle of Plataea (479):

'Among the barbarians, the best fighters were the Persian infantry and the cavalry of the Sacae, and of men, it is said, the bravest was Mardonius. Among the Greeks, the Tegeans and Athenians conducted themselves nobly, but the Lacedaemonians excelled all in valour. Of this my only clear proof is (for all these conquered the foes opposed to them) the fact that the Lacedaemonians fought with the strongest part of the army, and overcame it. According to my judgment, the bravest man by far was Aristodemus, who had been reviled and dishonoured for being the only man of the three hundred that came alive from Thermopylae; next after him in valour were Posidonius, Philocyon, and Amompharetus. Nevertheless, when there was a general discussion about who had borne himself most bravely, those Spartans who were there judged that Aristodemus, who plainly wished to die because of the reproach hanging over him and so rushed out and left the battle line behind, had achieved great deeds, but that Posidonius, who had no wish to die, proved himself a courageous fighter, and so in this way he was the better man.' (Herodotus 9.71.1–3)

Note that the Spartans at this period disapproved of those who rushed out of line to perform 'deeds of valour' (in the old-fashioned style), reckoning it braver to stand and fight in the line. The reward for bravery took the form of a wreath, similar to those awarded for victory at the Olympics or other athletic festivals (not, note, a financial reward). The orator Aeschines records his own modest achievements:

'I fought in the battle of Mantinea, not without honour to myself or credit to the city. I took part in the expeditions to Euboea, and at the battle of Tamynae as a member of the picked corps I so bore myself in danger that I received a wreath of honour then and there, and another at the hands of the people on my arrival home; for I brought the news of the Athenian victory, and Temenides, *taxiarchos* of the tribe Pandionis, who was despatched with me from camp, told here how I had borne myself in the face of the danger that befell us.' (Aeschines 2.169)

As well as prizes for bravery in combat, prizes could be awarded for training, as an incentive to excellence. Xenophon records Agesilaus' efforts in this regard:

'[W]hen spring was just coming on, he gathered his whole army at Ephesus; and desiring to train the army, he offered prizes both to the heavy-armed divisions [*taxeis*], for the division which should be in the best physical condition, and to the cavalry divisions, for the one which should show the best horsemanship; and he also offered prizes to peltasts and bowmen, for all who should prove themselves best in their respective duties. Thereupon one might have seen all the gymnasia full of men exercising, the hippodrome full of riders, and the javelin-men and bowmen practising.' (Xenophon, *Hellenica* 3.4.16)

Notice how the hoplites are rewarded not for drill, weapons handling or any technical expertise, but just for being in good physical condition.

Another aspect of financial reward for hoplite service should be considered: the chance to seize plunder from captured enemy cities or defeated armies, and in some cases to seize land for settlement from the capture of enemy territory. Plunder as an incentive for a city fighting wars has, I believe, been often overstated, as the expenses of maintaining and paying an army (and often a fleet, for amphibious operations) in the field would greatly outweigh what could be seized in movable property (even including slaves). Where we hear of invading armies looting or ravaging the territory of their enemies, the aim is evidently not so much enrichment of the invader as impoverishment of the invaded, and the emphasis is on destruction, not looting, and in particular agricultural destruction (destroying crops, vines and olive trees).[39]

For individuals, though, the chance for a little personal enrichment through plunder might well have been one of the incentives for fighting. Xenophon's Cyrus articulates the principle, saying 'it is a law established for all time among all men that when a city is taken in war, the persons and the property of the inhabitants thereof belong to the captors' (Xen., *Cyrop.* 7.5.73), and while not every campaign involved the capture of cities, there must have been plentiful opportunities for this sort of gain. However, soldiers could not simply seize whatever they were able – there were rules for the collection and distribution of plunder, and the commander of an army had to account for what he had seized to his home city upon his return. Armies might be accompanied by *laphyropolai*, 'booty-dealers', whose job was to sell on plundered goods (fences, in other words; see Xen., *Hell.* 4.1.26 and Xen., *Anab.* 7.7.56, along with other examples, all relating to Spartan or mercenary armies, though it seems fair to suppose that other armies functioned in a similar way). A single Spartan example from Agesilaus' campaign against Acarnania is representative of a great many:

'Agesilaus ... offered sacrifice in the morning and accomplished before evening a march of one hundred and sixty *stadia* to the lake on whose banks were almost all the cattle of the Acarnanians, and he captured herds of cattle and droves of horses in large numbers besides all sorts of other stock and great numbers of slaves. And after effecting this capture and remaining there through the ensuing day, he made public sale of the booty.' (Xenophon, *Hellenica* 4.6.6)

The takings of this sale would have been returned to the city, with a proportion making its way back to the soldiers, either in the form of pay and subsistence or as special prizes for valour (*aristeia*). It is not absolutely certain that such awards (other than *aristeia* itself) would have been additional to rather than in lieu of regular pay, but I think it is likely that it was. There were also some cases where soldiers were allowed to plunder in a private capacity, as Xenophon records for the Ten Thousand:

> 'Whenever the army remained in camp and rested, individuals were permitted to go out after plunder, and in that case kept what they got; but whenever the entire army set out, if an individual went off by himself and got anything, it was decreed to be public property.' (Xenophon, *Anabasis* 6.6.2)

So there were opportunities for personal enrichment, but they were probably more limited than the frequent references to looting and plundering might lead us to believe. The degree of attraction of such opportunities would of course depend greatly on the circumstances of the hoplite. A wealthy landowner serving for a short seasonal campaign was less likely to be attracted by the chance to seize some movable wealth than would a poor man, or one seeking a living from soldiering as a mercenary. Thucydides specifically contrasts the 'olden days', where men expected to make a living from plunder, with more civilized times:

> 'For in early times the Hellenes and the barbarians of the coast and islands, as communication by sea became more common, were tempted to turn pirates, under the conduct of their most powerful men; the motives being to serve their own cupidity and to support the needy. They would fall upon a town unprotected by walls, and consisting of a mere collection of villages, and would plunder it; indeed, this came to be the main source of their livelihood, no disgrace being yet attached to such an achievement, but even some glory ... The same rapine prevailed also by land.' (Thucydides 1.5.1–3)

Thucydides goes on to contrast the more civilized modern nature of Athens, for example, with more backward areas on the edges of the Greek world where such practices continued.[40]

In addition to movable wealth, successful warfare and conquest could lead to the seizure of land, and again while this would usually have been at the state level (and so go by the name of empire building), land could also be alloted to individuals. This was an extension of the common Greek practice of colonization or settlement, where cities would send out some of their surplus population around the coasts of the Eastern Mediterranean and Black Sea to found colonies – which must usually have involved the forcible displacement of whatever peoples were already living there. Herodotus records the sixth-century expansion of Cyrene in North Africa:

> '[T]he Pythian priestess warned all Greeks by an oracle to cross the sea and live in Libya with the Cyrenaeans; for the Cyrenaeans invited them, promising a distribution of land; and this was the oracle: "Whoever goes to beloved Libya after the fields are divided, I say shall be sorry afterward." So a great multitude gathered at Cyrene, and cut out great tracts of land from the territory of the neighbouring Libyans. Robbed of their lands and treated violently by the Cyrenaeans, these then sent to Egypt together with their king, whose name was Adicran, and put their affairs in the hands of Apries, the king of that country. Apries mustered a great force of Egyptians and sent it against Cyrene; the Cyrenaeans marched out to Irasa and the Thestes spring, and there fought with the Egyptians and beat them; for the Egyptians had as yet had no experience of Greeks, and despised their enemy.' (Herodotus 4.159)

Such distributions of land became a major factor in the settlement of Greek and Macedonian soldiers and their families in captured Persian territory in the late fourth century. During the main period of Archaic Greek overseas expansion, the chance to seize land may well have been a major factor too, but in the Classical period, the opportunities to gain land and the degree to which it would have been an incentive to service should not be overstated. Again, it would have depended on the circumstances of the individual – a well-off Athenian landowner would be less tempted by a parcel of North African land than would an impoverished struggling smallholder or labourer.[41]

Psychology and motivation

The motivations for some hoplites are clear: mercenary hoplites fought for a living, and as a way of life. They earned a living wage from selling their skills, and appear to have done so (so far as it is possible to judge) willingly. Fighting could bring wealth and status, at least to some, and at worst it could stave off destitution. In the Archaic period, mercenary service seems to have been a development of the overseas adventuring of an earlier age – a chance for status, adventure and enrichment, in the company no doubt of like-minded individuals. A drinking song which may have its origins in this age sets the tone:

> 'I have great wealth: a spear, a sword, and the fine leather shield which protects one's skin. For with this I plough, with this I harvest, with this I trample the sweet wine from the vines, with this I am called master of serfs.

Those who dare not hold a spear, a sword, or the fine leather shield which protects one's skin, all cower at my knee and prostrate themselves, calling me "Master" and "Great King".' (Athenaeus 695f–696a)

The seventh-century poet Archilochus expressed similar sentiments:

'In my spear is my kneaded barley bread, in my spear is Ismaric wine, and I drink it leaning on my spear.' (Archilochus fr.2)

The poet Alcaeus had a brother, Antimenidas, who served under a Babylonian king and for whom he wrote a 'welcome home' verse:

'You have returned from the ends of the earth, Antimenidas, with the gold-bound ivory hilt of that sword with which, as you fought for the Babylonians who dwell in the houses of long bricks, you did a great deed, preserving them all from evil by killing a fighter who lacked only a palm of standing five royal cubits high.' (Alcaeus, fr.133)

This is the same Alcaeus who wrote the description of the great hall hung with weapons and armour that we encountered at the start of Chapter 2. We can perhaps see here the hall of an aristocratic leader and his retainers (a *promachos* and other ranks) of the sort who might, like Alcaeus' brother, seek enrichment overseas. These are not, on the whole, poor men scratching out a living by selling their swords, but wealthy men seeking adventure, honour and enrichment.

For mercenaries of the Classical period, the attractions will have been similar, though perhaps the opportunities for enrichment, rather than subsistence, rather less. Land was also the reward for colonists, who should be seen in the same light as mercenaries, as men willing to travel and fight to earn a living. How typical such overseas service was is difficult to judge; mercenaries are always present in the Classical narratives, but always take second place – then and among modern historians – to the citizen soldiers fighting for their city.[42]

Financial rewards were thus a factor in encouraging some hoplite service, but probably not a major one for most non-mercenary hoplites. What more can be said about the 'will to fight' of the typical citizen hoplite? Direct evidence is somewhat lacking, and many who have considered the subject rely instead on modern theories of combat motivation, which centre on the concept of the 'primary group', the small circle of comrades that fought together, and how maintaining status and respect within this small group was the most important factor in retaining a will to combat. Such theories obviously run the risk of being anachronistic, and lessons learned from late twentieth-century AD Western armies can only be applied with caution to any period of the ancient world. Nevertheless, the importance of one's immediate comrades in combat was also recognized in antiquity. Homer (*Iliad* 2.362–363) quotes Nestor as recommending that armies be organized around tribal groupings as we saw above, which would have the effect of ensuring that men fought alongside those whom they knew socially.[43]

This practice was indeed followed by Classical armies, although the tribal groupings of Athens, for example, were largely bureaucratic and do not carry any strong implication of familial relationships – greater cohesion was probably at the *demos* or deme level, demes being smaller local groupings that united neighbours and relatives in the civic life of the city. The great virtue of the Theban Sacred Band, with its pairs of *eromenoi* and *erastes*, was said to be that fighting alongside one's beloved would ensure devotion to duty and will to excel (see below), which can be seem as an extreme form of the primary group. More broadly, city militias would have seen family members, neighbours and people from the same city districts fighting alongside each other, which could be expected to create a sense of group solidarity. Hoplites would have personally known the hoplites fighting alongside, in front of and behind them, and the lack of formal military units in non-Spartan armies would have meant that these sorts of relationships remained intact in the field, rather than being broken up by the needs of army organization, as peacetime groupings so often were in later periods of history.[44]

What we do not see particularly is the deliberate fostering of a sense of larger unit cohesion, as is familiar from later military forces such as the Macedonians and Romans. Macedonian armies had named, permanently formed units which could develop a sense of *esprit de corps* over time (particularly with success in battle), while Romans, with their named and numbered legions and legionary symbols (chiefly the legionary standard or eagle), carried this even further, fostering a strong sense of loyalty to the unit among their increasingly professional armies. If there was any equivalent in hoplite armies it has left little trace in the sources, outside of the chosen units, the *epilektoi* or the Sacred Band of Thebes, for example. Spartans certainly took professional pride in their skills, while Athenians could be similarly proud of their experience, but we do not see obvious evidence of Spartan attachment to their individual *morai*, or of others to their particular *taxeis* or *lochoi*. We do not know whether standards were carried into battle (except perhaps as signal flags, as in the Macedonian army), nor were there named units, except inasmuch as Athenian *taxeis*, for example, would be named for the tribes from which they were recruited.[45]

Hoplites undoubtedly could feel considerable patriotic loyalty to their home city. This is of course not to say that all armies were always of one mind in matters of politics, but the fact that hoplites were (generally) drawn from a relatively restricted economic class meant that they could have a shared interest in and commitment to maintaining their privileged status, and also to furthering the interests of their home city. Pericles' funeral oration for Athenian war dead, as recorded by Thucydides, presents this patriotic view:

> 'So died these men as became Athenians. You, their survivors, must determine to have as unaltering a resolution in the field, though you may pray that it may have a happier issue. And not contented with ideas derived only from words of the advantages which are bound up with the defence of your country, though these would furnish a valuable text to a speaker even before an audience so alive to them as the present, you must yourselves realize the power of Athens, and feed your eyes upon her from day to day, till love of her fills your hearts; and then when all her greatness shall break upon you, you must reflect that it was by courage, sense of duty, and a keen feeling of honour in action that men were enabled to win all this, and that no personal failure in an enterprise could make them consent to deprive their country of their valour, but they laid it at her feet as the most glorious contribution that they could offer.' (Thucydides 2.42.1)

'Country' in this case translates *polis*, city. Aeneas defines the things a man might go out to fight for as 'their land [*chora*] and city [*polis*] and fatherland [*patris*]' (Aen. Tact., Pref.1), while the things men most wish to protect are 'shrines and fatherland, parents and children, and all else' (Aen. Tact., Pref.2). Pan-Hellenic feelings closer to what we might identify as nationalism might be roused against a common enemy, such as the Persians:

> 'On, you men of Hellas! Free your native land. Free your children, your wives, the temples of your fathers' gods, and the tombs of your ancestors. Now you are fighting for all you have.' (Aeschylus, *Persians* 402–405)

But in most cases Greeks fought Greeks, and even though love of *patris* and *polis* could provide a unifying motivation, the politics of the times could also cause divisions. Aeneas recommends, in the defence of a city against siege, picking out the most prudent and most physically fit citizens for guard duty:

> 'They must be both loyal and satisfied with the existing order, since it is a great thing to have such a group acting like a fortress against the revolutionary designs of the other party.' (Aeneas 'the Tactician', 1.5).

Loyalty to party could trump loyalty to country, or rather it could provide a different conception of where the good of the country lay. But such divisions would be highly dangerous in the phalanx, where group solidarity was everything, so the ideology of combat required every man to stand firm beside his comrades for the good of all. Political differences could be put aside when the safety and survival of the group (the phalanx) depended on this solidarity.

This ideal goes back to some of the earliest evidence we have in the poems of Tyrtaeus:

> 'This is courage, this is the finest possession of men, the noblest prize that a young man can win. This is the common good for the city and for all the people, when a man stands firm and remains unmoved in the front rank and forgets all thought of disgraceful flight, steeling his spirit and heart to endure, and with words encourages the man standing beside him. This is the man good in war.' (Tyrtaeus fr.12, 13–20)

Group solidarity was maintained not by iron discipline or, necessarily, patriotism alone, but by this shared sense of duty to the common interest, and the shame that went with letting the side (or one's neighbour) down. Ideally, citizens would fight to the death, according to Aristotle's conception of the difference between mercenary and citizen soldiers quoted earlier – 'citizens think it disgraceful to run away, and prefer death to safety so procured' (Aristotle, *Ethics* 1116b 9). Of course, in practice it was exceedingly rare for citizens to fight to the death, notable exceptions at the start and end of the Classical period being the 300 Spartans at Thermopylae (480), and the 300 Thebans at Chaeronea (338); but the ideal of standing one's ground and avoiding flight would hold men in place, it was hoped, for longer (for the importance of this, see Chapter 8).

Religious reinforcement was available to back up peer pressure and the avoidance of shame. Athenian ephebes (though perhaps only from the fourth century) took an oath that 'I shall not abandon the man beside me, wherever I stand in line'. Note again the reference to 'the man beside me', as in Tyrtaeus' verse. Spartans may have sworn an oath to follow their officers:

> 'I shall not desert my *taxiarchos* or my *enomotarches* whether he is alive or dead, and I shall not leave unless the *hegemones* lead us away, and I shall do whatever the *strategoi* may command, and I shall bury on the spot those of my fellow-fighters who die, and I shall leave no-one unburied.' (Tod II 204 ll. 25–31)

This oath, or something like it, may be what gave the *enomotia* ('sworn band') its name. The emphasis on following orders may be specifically Spartan, but the importance of not leaving the ranks, and of the proper treatment of the dead (on which see further in chapter 7), are probably typically Greek.[46]

It helped in ensuring this group solidarity that 'the man beside me' may have been (as we saw above) the hoplite's friend or neighbour, family member or lover. This last case is best known from the Theban Sacred Band mentioned above:

> '[S]ome say that this band was composed of lovers [*erastai*] and beloved [*eromenoi*]. And a pleasantry of Pammenes is cited, in which he said that

Homer's Nestor was no tactician when he urged the Greeks to form in companies by clans and tribes, "That clan might give assistance unto clan, and tribes to tribes", since he should have stationed lover by beloved. For tribesmen and clansmen make little account of tribesmen and clansmen in times of danger; whereas, a band that is held together by the friendship between lovers is indissoluble and not to be broken, since the lovers are ashamed to play the coward before their beloved, and the beloved before their lovers, and both stand firm in danger to protect each other.' (Plutarch, *Pelopidas* 18.1–2)

This incarnation of the Sacred Band, founded around 378, appears to have replaced an earlier elite body of 300 men, recorded by Diodorus (12.70.1) as consisting of pairs of 'charioteers and chariot-fighters' – an apparent reference to a Homeric style of combat (though these men fought as infantry) – and it is tempting to see this body as being formed from similar relationships as the later Sacred Band. The division into *erastai* and *eromenoi*, in the 'standard model' of Greek homosexuality, implies pairs of young adult men (the *erastai*) with adolescent or at least pre-18-year-old boys (*eromenoi*), though the fact that these men were serving together in a military unit (to which admission would not usually be made before the age of 20) suggests that this pederastic model does not apply in this case, with both partners being full adults and perhaps close or equal in age.

According to Xenophon (Xen., *Sym*. 8.34), the Eleans had a similar elite unit (perhaps the 'Three Hundred' or 'Four Hundred' of Xen., *Hell*. 7.4.13). That this was also the common practice in Sparta, though with slightly different motivations, is implied by Xenophon's further comment:

'In contrast to [the Thebans and Eleans], the Lacedaemonians, who hold that if a person so much as feels a carnal concupiscence he will never come to any good end, cause the objects of their love [*eromenoi*] to be so consummately brave that even when arrayed with foreigners and even when not stationed in the same line with their lovers [*erastai*] they just as surely feel ashamed to desert their comrades.' (Xenophon, *Symposium* 8.35)

(I will not attempt to improve on the translator's 'carnal concupiscence'.) Xenophon suggests that Spartan relationships were Platonic, though the fact that the vocabulary is the same as in normal homosexual relationships might suggest otherwise. Plato himself spelled out his ideal military unit, formed on similar lines:

'So that if we could somewise contrive to have a city or an army composed of *erastai* and *eromenoi* they could not be better citizens of their country than by thus refraining from all that is base in a mutual rivalry for honour; and such men as these, when fighting side by side, one might almost consider able to make even a little band victorious over all the world. For an *erastes* would surely choose to have all the rest of the host rather than his *eromenos* see him forsaking his station or flinging away his arms; sooner than this, he would prefer to die many deaths.' (Plato, *Symposium* 178e–179a)

Plato was writing around the same time as the formation of the Sacred Band. How common military homosexuality of this sort was in the phalanx generally, rather than in elite units, and to what extent and in what way the relationships were sexual, remains somewhat obscure – Greek writers tended to avoid explicit statements on the matter. But the implication is at any rate that romantic, perhaps sexual, relationships between hoplites were an accepted and valued way of reinforcing the prohibition on leaving one's place in the line.[47]

We must also keep in mind that some, or many (it is impossible to give statistics) enjoyed war, and they fought through choice, for pleasure. As in more modern literature, passages are available from ancient sources emphasizing the horrors of war (frequently quoted in this context is Pindar, fr.110, 'Sweet is war to the untried, but anyone who has experienced it dreads its approach exceedingly in his heart'); but equally, there are passages extolling the joys of war, and sometimes quite explicitly the pleasures of killing. No doubt in Ancient Greece, just as today, levels of aggression and of predisposition to violence varied greatly between individuals. Nevertheless, there must have been hoplites who went to war with great willingness (provided they were victorious of course, since in all ages there has been a stronger aversion to defeat and death than there is to victory and killing).

Men who, in peacetime, would favour peace and security and the rule of law (seeking redress for wrongs in the law courts rather than through feud or violence, for example), could relish both the chance to prove their courage – literally, their manliness – in combat, and to kill the enemies of their people. The well-known epitaph of Python of Megara provides a vivid example:

'This memorial is set over the body of a very good man. Python from Megara slew seven men and broke off seven spear points in their bodies ... This man, who saved three Athenians *taxeis* ... having brought sorrow to no one among all men who dwell on the earth, went down to the underworld felicitated in the eyes of all.' (IG I^3 1353, Tod 41)

There was no conflict or irony in the contrast (so evident now) between slaying seven men and bringing sorrow to no one. Xenophon expresses similar sentiments:

'For, you know, when states defeat their foes in a battle, words fail one to describe the joy they feel in the rout of the enemy, in the pursuit, in the slaughter of the enemy. What transports of triumphant pride! What a halo of glory about them! What comfort to think that they have exalted their city! Everyone is crying: "I had a share in the plan, I killed most"; and it's hard to find where they don't revel in falsehood, claiming to have killed more than all that were really slain. So glorious it seems to them to have won a great victory!' (Xenophon, *Hiero* 2.11–16)

Whatever shared interests and fellow feeling there might have been between hoplites of different cities, it did not prevent men taking delight in killing their enemies.[48]

Xenophon gives an account of the career of Clearchus, the original commander of the Ten Thousand:

'Now such conduct as this, in my opinion, reveals a man fond of war. When he may enjoy peace without dishonour or harm, he chooses war; when he may live in idleness, he prefers toil, provided it be the toil of war; when he may keep his money without risk, he elects to diminish it by carrying on war. As for Clearchus, just as one spends upon a loved one or upon any other pleasure, so he wanted to spend upon war.' (Xenophon, *Anabasis* 2.6.6)

Clearchus comes across as a very troubled figure, but there may have been many men like this in the hoplite phalanx, just as there could have been many there only through a sense of duty rather than through choice, and many who would rather not have been there at all. Some of the lovers of war were able to make careers as mercenaries, and it must have been a similar spirit that drove the raiders and adventurers of an earlier age. It is important to remember that not every man in the phalanx, by any means, was there unwillingly.[49]

Finally, though by no means of least importance, it was by fighting hand-to-hand in the phalanx, standing firm beside one's comrades and refusing to give ground, that a man demonstrated his courage (*andreia*), his honour (*timē*) and his virtue (*aretē*). That manliness and worth went hand-in-hand with success in warfare (as both cause and effect), goes back to Homer and is a common theme throughout Archaic and Classical Greece. In the Funeral Oration of Pericles, courage was the greatest virtue, that could wipe clean other vices:

'For there is justice in the claim that steadfastness [*andragathia*, manliness] in his country's battles should be as a cloak to cover a man's other imperfections; since the good action has blotted out the bad, and his merit as a citizen more than outweighed his demerits as an individual.' (Thucydides 2.42.3)

Courage, and victory in combat, was the route to all other rewards in life, as Xenophon could remind the Ten Thousand on their long march home:

> 'And whoever among you desires to see his friends again, let him remember to show himself a brave man; for in no other way can he accomplish this desire. Again, whoever is desirous of saving his life, let him strive for victory; for it is the victors that slay and the defeated that are slain. Or if anyone longs for wealth, let him also strive to conquer; for conquerors not only keep their own possessions, but gain the possessions of the conquered.' (Xenophon, *Anabasis* 3.2.39)

For the hoplite, though, it was not just any sort of fighting that was required, but the particular steadfast courage required to fight in the phalanx. Fighting bravely did not mean recklessly attacking the enemy, it meant holding one's place in the line and standing one's ground. As reported by Plato, the philosopher Socrates, himself a veteran of several of Athens' great battles, equated keeping one's place in the line with his own vocation as a philosopher:

> 'For thus it is, men of Athens, in truth; wherever a man stations himself, thinking it is best to be there, or is stationed by his commander, there he must, as it seems to me, remain and run his risks, considering neither death nor any other thing more than disgrace. So I should have done a terrible thing, if, when the commanders whom you chose to command me stationed me, both at Potidaea and at Amphipolis and at Delium, I remained where they stationed me, like anybody else, and ran the risk of death, but when the god gave me a station, as I believed and understood, with orders to spend my life in philosophy and in examining myself and others, then I were to desert my post through fear of death or anything else whatsoever.' (Plato, *Apology*, 28d–29a)

This definition of courage occurs frequently, among orators and philosophers as well as historians:

> 'But as each man of you would be ashamed to desert the post to which he had been assigned in war, so now you should be ashamed to desert the post to which the laws have called you, sentinels, guarding the democracy this day.' (Aeschines, *Against Ctesiphon* 3.7)

> 'Laches: "On my word, Socrates, that is nothing difficult: anyone who is willing to stay at his post and face the enemy, and does not run away, you may be sure, is courageous."' (Plato, *Laches* 190e)[50]

So for a Greek citizen, war was the chance to prove and (importantly) to demonstrate his worth as a man, and for a hoplite the clearest demonstration of his quality was his ability to stand his ground and fight hand-to-hand in the phalanx.

Chapter 5

Command and Control

Chains of command

In Chapter 3 we looked at the internal organization of the phalanx of various cities (chiefly Athens and Sparta) and noted that only the Spartan phalanx had a complex internal structure down to a low level, the phalanx of other states being divided more simply into probably just two levels (variously named, but usually *taxeis* and *lochoi*). Naturally enough, the command structure of a phalanx matched its internal organization, since the point of the various subunits was that each had an officer appointed over it, to control its movement on the march and on the battlefield.[1]

The command structure of the Spartan phalanx therefore matches the structure examined before, with all the accompanying uncertainties due to the differences in the accounts of Thucydides and Xenophon, and the possibility (unlikely in my view) that the Spartans went through a series of reforms, changing their phalanx structure a number of times in the fifth and fourth centuries. Whatever the precise details of the sizes and structure of the various Spartan units, the overall intent of the chain of command was described by Thucydides:

> 'For when a king is in the field all commands proceed from him: he gives the word to the *polemarchoi*; they to the *lochagoi*; these to the *pentecounteres*; these again to the *enomotarchoi*, and these last to the *enomotiai*. In short all orders required pass in the same way and quickly reach the troops; as almost the whole Lacedaemonian army, save for a small part, consists of officers under officers, and the care of what is to be done falls upon many.' (Thucydides 5.66.3–4)

So whatever uncertainty there may be over the size and number of the various subunits, what is clear at least is that the phalanx was divided into small subunits (*enomotiai* – generally taken to be forty men strong at full strength) with a five-level hierarchy of officers above them, culminating in the king himself. Xenophon makes a similar observation about the existence of a chain of command:

> 'The prevalent opinion that the Laconian infantry formation is very complicated is the very reverse of the truth. In the Laconian formation the

front rank men are all officers, and each file has all that it requires to make it efficient.' (Xenophon, *Constitution of the Lacedaemonians* 11.5)

Xenophon's 'officers' (*archontes*, the same word used by Thucydides for 'officers under officers', *archontes archontōn*), in order to constitute the whole of the front rank as he implies they did, would need to be the equivalent of the *dekadarchos*, commander of ten men, or 'file leader', which he describes in Cyrus' army and which we see also in the later Macedonian army. Such a rank is not mentioned by Xenophon or Thucydides, but it may well have existed in the Spartan army, if not in any other (the reason for doubt, as we saw earlier, is that if the file leader was a fixed rank, then this would make it difficult for the *enomotia* to draw up in different breadths and depths, as Xenophon says it did, while keeping the file leaders in the front rank). At any rate, both Xenophon and Thucydides describe a multi-level hierarchical chain of command. This comes as no surprise to us as we would expect armies to have a chain of command of this sort, but note that Thucydides singles it out as a point of difference about the Spartan army, one worthy of particular explanation, while Xenophon remarks that most people consider the Spartan army very (and thus unusually) complicated. This command structure consists of:

King
– *Polemarchos*
– *Lochagos*
– *Pentecronteres*
– *Enomotarchos*
– *Dekadarchos?*

The chain of command of non-Spartan armies (or of armies not based on the Spartan model, as the Ten Thousand apparently were) seems to have been much more rudimentary, matching their more streamlined organizational structure, and can be summarized as:

– One or more *Strategoi* (with or without a *Polemarchos*)
– *Taxiarchos*
– *Lochagos*

This is following the Athenian pattern, where the original *polemarchos* was supplemented or later replaced by ten elected *strategoi*, one for each tribe, following the reforms of Cleisthenes in 501 (Aristot., *Const. Ath.* 22.2). Note that the *strategoi* and *taxiarchoi* under them were elected officials (elected by show of hands in the Assembly – Aristot., *Const. Ath.* 61.1–3), which may have been a feature specific to the more democratic Athenian constitution. The *lochagoi* on the other hand were chosen by the *taxiarchoi* (Aristot., *Const. Ath.* 61.3). A system of electing officers naturally enough could lead to complaints that the wrong men were appointed for the wrong reasons. Xenophon records an anecdote of Socrates:

> 'Once on seeing Nicomachides returning from the elections, he asked, "Who have been chosen *strategoi*, Nicomachides?"
>
> "Isn't it like the Athenians?" replied he; "they haven't chosen me after all the hard work I have done, since I was called up, as *lochagos* or *taxiarchos*, though I have been so often wounded in action" (and here he uncovered and showed his scars); "yet they have chosen Antisthenes, who has never served as a hoplite nor distinguished himself in the cavalry and understands nothing but money-making."' (Xenophon, *Memorabilia* 3.4.1)

Socrates characteristically goes on to argue that businessmen could make good generals (see further below).[2]

The Spartan system of appointing or promoting officers is if anything more obscure than the Athenian. Overall command of the land army was held by one of the two kings, or if necessary (as at Plataea, 479) by a regent. Smaller forces (such as that defeated on Sphacteria in 425 or Lechaion in 391) would of course be commanded by more junior officers appropriate to the size of force. Senior officers were probably of aristocratic origin, but exactly how they were appointed and whether by birth or on merit is not clear. No names of Spartan officers below the rank of *polemarchos* are given by Thucydides or Xenophon. Although Herodotus (9.53.2) calls the insubordinate Amompharetos at Plataea a *lochagos*, this has been disputed. The *polemarchoi* at least seem to have had a degree of independence from the kings, judging by the insubordination of the two *polemarchoi* at Mantinea (Thuc. 5.72.1) and the fact they were only punished afterwards, not in the field, while Plutarch, discussing the Spartan practice of dining in common messes, relates that:

> 'For a long time this custom of eating at common mess-tables was rigidly observed. For instance, when King Agis, on returning from an expedition in which he had been victorious over the Athenians, wished to sup at home with his wife, and sent for his rations, the *polemarchoi* refused to send them to him; and when on the following day his anger led him to omit the customary sacrifice, they laid a fine upon him.' (Plutarch, *Lycurgus* 12.3)

Spartan kings were in a slightly strange position, being anything but absolute monarchs, and the Spartan state – meaning in practice probably the ephors (*ephoroi*), the council of five elected by the assembly – kept constant check on them. Agis in particular had to be put under special measures after a failed campaign against Argos:

> 'Meanwhile the Lacedaemonians, upon their return from Argos after concluding the four months' truce, vehemently blamed Agis for not having subdued Argos, after an opportunity such as they thought they had never had before; for it was no easy matter to bring so many and so good allies together. But when the news arrived of the capture of Orchomenos, they became more angry than ever, and, departing from all precedent, in the heat of the moment had almost decided to raze his house, and to fine him ten thousand drachmae. Agis however entreated them to do none of these things, promising to atone for his fault by good service in the field, failing which they might then do to him whatever they pleased; and they accordingly abstained from razing his house or fining him as they had threatened to do, and now made a law, hitherto unknown at Lacedaemon, attaching to him ten Spartans as counsellors, without whose consent he should have no power to lead an army out of the city.' (Thucydides 5.63)

Agis' position was unusual, but other kings also had to endure oversight from the central authorities. Agesilaus had a committee of thirty Spartiates sent out to him in Asia (Xen., *Hell.* 3.4.2), as did fellow king Agesipolis (Xen., *Hell.* 5.3.8), though Agis was able to assign commands to these apparently according to his own desires (Xen., *Hell.* 3.4.20).[3]

Generals or commanders who were unsuccessful could face severe sanctions on their return to their city. Often, however, this took the form of lawsuits, or of politically motivated action, rather than any form of military discipline, and could result from disagreement over political aims – a general in the field, with a long line of communication to the home city, could not always consult with the home government over political and diplomatic decisions, and would be required to act on his own initiative. An Argive general, Thrasylus, on his own initiative, avoided an impending battle with the Spartans in 418 by concluding a truce with the Spartan king Agis, who also agreed without consulting any other authority (Thuc. 5.59–60). Both men were subsequently blamed by their fellows for having let slip an opportunity for victory, Thrasylus being stoned by a mob and only escaping by seeking sanctuary in a temple. The Athenian generals who initially made peace with Syracuse were banished and fined (on a charge of taking bribes) upon their return to Athens (Thuc. 4.65.3). Thucydides himself was banished from Athens for his part in the defeat at Amphipolis (Thuc. 5.26.5 – which gave him leisure, as he recounts, 'to observe affairs somewhat particularly', including from the Peloponnesian side). Generals on their return to Athens had to account for their actions to the Assembly. According to Plutarch, the general Paches, obviously not expecting the hearing to go well, 'while he was giving the official account of his generalship, drew his sword in the very court-room and slew himself' (Plut., *Nic.* 6.2). Lysicles, the Athenian commander at the defeat by Macedon at Chaeronea (338), was condemned to death on the accusation of the orator Lycurgus:

> 'You were general, Lysicles. A thousand citizens have perished and two thousand were taken captive. A trophy stands over your city's defeat, and all of Greece is enslaved. All of this happened under your leadership and command, and yet you dare to live and to look on the sun and even to intrude into the market, a living monument of our country's shame and disgrace.' (Diodorus 16.88.2)

This could lead to a certain hesitancy on the part of Athenian generals; most famously Nicias, commander of the ultimately disastrous Syracusan expedition, who failed to withdraw while there was still time, saying 'he was sure the Athenians would never approve of their returning without a vote of theirs', and was fearful that the soldiers would accuse him of taking bribes to retreat (Thuc. 7.48.3–5).[4]

The Spartans did not have such democratic oversight over the actions of generals, but could still judge and censure even their kings if they disagreed with their actions, as we saw above. For a Spartan facing defeat, an honourable death in battle might be considered preferable to facing the consequences – the Spartan governor Anaxibius preferred to die fighting than seek safety in flight after he led his forces into an ambush (Xen., *Hell.* 4.8.38–39). Spartans of course had a tradition of preferring death to dishonour, though the number of senior commanders who met this fate, after the famous example of Leonidas at Thermopylae, is actually rather low. Merely to be killed in battle was not glorious in itself, unless it was in some higher cause. The Spartan hero Lysander was killed in a skirmish before the walls of Haliartus when he made a somewhat reckless attack, but Xenophon obviously felt he should have waited for the reinforcements that were coming from Sparta (Xen., *Hell.* 3.5.18).

We have seen that officers might expect to select the officers serving under them. At higher levels of command, Xenophon tends to refer to 'the Lacedaimonians' sending out officers on campaign with various forces, without clarifying whether this means the ephors, the assembly or some combination, nor what group of men the commanders were chosen from (Xen., *Hell.* 4.8.32 notes that Anaxibius 'inasmuch as the ephors had become friends of his, succeeded in having himself sent out to Abydus as governor' – having the ear of the ephors must have helped with such appointments, though on other occasions the kings or local commanders appoint governors on their own initiative). These appointments are, however, mostly of higher-ranking positions, local governors or commanders of modest expeditionary forces. There is no similar information on how the various lower-ranking officers were appointed, or on whether they held their positions permanently or were appointed afresh for each new expedition, as at Athens.[5]

Particular armies might have needed to make field appointments on the spot of officers at any rank. The Ten Thousand, after many of their original officers had been captured by the Persians, called a council. The surviving officers – *strategoi*, *hypostrategoi* ('under-generals' – it is unclear what this means exactly) and *lochagoi* – were invited from each of the *taxeis* of the army, and were addressed by Xenophon:

> 'I think you would do the army a great service if you should see to it that *strategoi* and *lochagoi* are appointed as speedily as possible to take the places of those who are lost. For without leaders [*archontōn*] nothing fine or useful can be accomplished in any field, to put it broadly, and certainly not in warfare.' (Xenophon, *Anabasis* 3.1.38)

These officers then were selected by their superiors, based on their knowledge of who was suitable from the men they commanded. This was perhaps the usual way in which appointments of junior officers were made.

In the early Athenian army, as described by Herodotus, command had a collegiate and democratic nature. Before Marathon (490):

> 'The Athenian generals were of divided opinion, some advocating not fighting because they were too few to attack the army of the Medes; others, including Miltiades [one of the *strategoi*], advocating fighting. Thus they were at odds, and the inferior plan prevailed. An eleventh man had a vote, chosen by lot to be *polemarchos* of Athens, and by ancient custom the Athenians had made his vote of equal weight with the *strategoi*. Callimachus of Aphidnae was *polemarchos* at this time.' (Herodotus 6.109.1–2)

Miltiades did his best to persuade Callimachus to cast his vote with those who favoured an attack:

> 'By saying this Miltiades won over Callimachus. The *polemarchos*' vote was counted in, and the decision to attack was resolved upon. Thereafter the generals who had voted to fight turned the presidency [*prytaneia*] over to Miltiades as each one's day came in turn. He accepted the office but did not make an attack until it was his own day to preside.' (Herodotus 6.110.1)

A system in which ten *strategoi* and a *polemarchos* took it in turns to command and voted on the course of action does not seem one destined to provide the best in military efficiency, but is characteristic of the way that military institutions reflected the political arrangements of the home city. The Athenian army of the early fifth century was explicitly organized (in the reforms of Cleisthenes at the end of the sixth century) to reduce aristocratic control of the government. The Spartan arrangement appears more conventional, with the king having total command over the army through a formal chain of command, although there were checks and balances on the actions of the king, as we have seen.[6]

The democratic, egalitarian approach of some cities (chiefly Athens) meant that credit for victory could not be claimed solely by the commanding general; the glory of victory belonged to the city and the citizens as a whole, and the general was no more than a contributor to it. The orator Aeschines records the memorialization in Athens of the victory at Marathon:

> 'And now pass on in imagination to the Stoa Poecile; for the memorials of all our noble deeds stand dedicated in the Agora. What is it then, fellow citizens, to which I refer? The battle of Marathon is pictured there. Who then was the general? If you were asked this question you would all answer, "Miltiades". But his name is not written there. Why? Did he not ask for this reward? He did ask, but the people refused it; and instead of his name they permitted that he should be painted in the front rank, urging on his men.' (Aeschines, *Against Ctesiphon* 186)

It is not certain that 'in the front rank' means that literally here, rather than 'in front' or 'among the foremost'. Pausanias, commander of the allied armies at Plataea (479), was afterwards suspected at Sparta of having despotic ambitions – 'his contempt of the laws and imitation of the barbarians' as Thucydides (1.132.1) puts it. Chief among his misdeeds was that:

> '[H]e had taken upon himself to have inscribed on the tripod at Delphi, which was dedicated by the Hellenes as the first-fruits of the spoil of the Medes, the following couplet: "The Mede defeated, great Pausanias raised this monument, that Phoebus might be praised." At the time the Lacedaemonians had at once erased the couplet, and inscribed the names of the cities that had aided in the overthrow of the barbarian and dedicated the offering.' (Thucydides 1.132.2–3)

Cities (and their citizens), not generals (even if they were regents), won wars. In the fourth century, the Athenian general Iphicrates was said to be the first to have enemy spoils dedicated in his own name rather than that of the city (though Iphicrates' position as a mercenary general meant that his relationship with Athens was not so clear-cut as a general acting solely in his official capacity).[7]

Later in the fourth century, the Athenian general Leosthenes was remembered for his role in leading Athens against the Macedonians in the Lamian War. His funeral oration stresses his role even above that of the city and the citizens:

> 'While praise is due to Athens for her policy, for choosing as she did a course not only ranking with her past achievements but even surpassing them in pride and honour, and to the fallen also for their gallantry in battle, for proving worthy of their forbears' valour, to Leosthenes the *strategos* it is doubly due; the city's guide in framing her decision, he was besides the citizens' commander in the field.' (Hyperides, *Funeral Oration* 3)

It is tempting to see in this a progressive increase in the importance of the *strategos* relative to the city and citizens, but perhaps such a progression is illusory, since the contribution of other 'great men' (Miltiades, Themistocles, Pericles) had been valued, but there were limits on how far they could be celebrated. A funeral oration for the man himself might be expected to be rather more fulsome.

The general's staff

Outside the formal command structure of the army units themselves, a general might also have a staff of officers and officials to assist with the day-to-day handling of the army. Once again, most of the information we have about such institutions comes from the fourth century and applies to the Spartan army, though something analogous must also have been in place for the armies of other cities. Xenophon calls these men 'those about the tent' or 'tent companions', and writes in the *Constitution*:

> 'The *polemarchoi* mess with the King, in order that constant intercourse may give better opportunities for taking counsel together in case of need. Three of the peers [*homoioi*, Spartiates] also attend the King's mess. These three take entire charge of the commissariat for the King and his staff, so that these may devote all their time to affairs of war.' (Xenophon, *Constitution of the Lacedaemonians* 13.1)

So the *polemarchoi*, commanders of the *morai*, messed with the king along with three officials in charge of supplies and so on. In addition to these were two Pythii (Xen., *Const. Lac.* 15.5; Hdt., 6.57.3) responsible for communicating with the Delphic Oracle, a reminder of the central role of religion in Greek warfare.

This religious role of the general was of genuine importance, as was the necessity of obtaining favourable omens before acting. Incidents such as that at Plataea where the Spartans were unable to attack the Persians until they made favourable sacrifices (Hdt. 9.61–62) occur frequently enough that it seems that, while such delays might on occasion be used as an excuse to avoid an unwanted course, it could also hold back a commander from much-needed action. When the Spartan commander Dercylidas approached the town of Cebren in 397:

> '[T]he man who commanded the garrison in Cebren, a very strong place, thinking that if he succeeded in keeping the city for Pharnabazus he would receive honours at his hands, refused to admit Dercylidas. Thereupon the latter, in anger, made preparations for attack. And when the sacrifices that he offered did not prove favourable on the first day, he sacrificed again on the following day. And when these sacrifices also did not prove favourable, he tried again on the third day; and for four days he kept persistently on with his sacrificing, though greatly disturbed by the delay; for he was in haste to make himself master of all Aeolis before Pharnabazus came to the rescue.' (Xenophon, *Hellenica* 3.1.17)

In the event, Dercylidas' sacrifices turned favourable just as he learned that the garrison was prepared to hand over the town to him (Xen., *Hell.* 3.1.19), which may perhaps suggest that religious scruples were not the only reason for the delay.[8]

So beyond the inner circle were other officials, including priests for conducting sacrifices and perhaps surgeons responsible for dealing with the wounded. Xenophon reports that the Ten Thousand 'appointed eight surgeons, for the wounded were many' (Xen., *Anab.* 3.4.30). Presumably these were selected on an ad hoc basis from the ranks, and in any case eight men does not sound a large number for many wounded in an army of 10,000. In most armies, much reliance for medical care must have been placed on the men helping each other, and on the hoplites' servants and attendants. Xenophon also records an extended junior staff in the Spartan army:

> 'The troops that are to support these [the leading *morai* in the army] are marshalled by the senior member of the King's staff. The staff consists of all peers who are members of the royal mess, seers, doctors, fluteplayers, commanding officers and any volunteers who happen to be present. Thus nothing that has to be done causes any difficulty, for everything is duly provided for.' (Xenophon, *Constitution of the Lacedaemonians* 13.7)

Accompanying a Spartan commander on overseas expeditions – the commander of which would be called *navarch*, 'admiral' – was a second-in-command and staff officer, the *epistoleus* or *epistoliaphoros* ('secretary'), responsible for official communications and dispatches. The Laconic character of the messages he sent is perhaps indicated by an example recorded by Xenophon:

> 'Meanwhile a letter dispatched to Lacedaemon by Hippocrates, *epistoleus* under Mindarus, was intercepted and taken to Athens; it ran as follows:

"The ships are gone. Mindarus is dead. The men are starving. We know not what to do.'" (Xenophon, *Hellenica* 1.1.23)[9]

Such messages might have been encrypted using one of the various methods employed by the Greeks, the most common of which was perhaps to write the message on a strip of paper wrapped around a stick (*scytale*), as described by Plutarch:

> 'The dispatch-scroll [*scytale*] is of the following character. When the ephors send out an admiral or a general, they make two round pieces of wood exactly alike in length and thickness, so that each corresponds to the other in its dimensions, and keep one themselves, while they give the other to their envoy. These pieces of wood they call '*scytalae*'. Whenever, then, they wish to send some secret and important message, they make a scroll of parchment long and narrow, like a leathern strap, and wind it round their '*scytale*', leaving no vacant space thereon, but covering its surface all round with the parchment. After doing this, they write what they wish on the parchment, just as it lies wrapped about the '*scytale*'; and when they have written their message, they take the parchment off and send it, without the piece of wood, to the commander. He, when he has received it, cannot otherwise get any meaning out of it – since the letters have no connection, but are disarranged – unless he takes his own '*scytale*' and winds the strip of parchment about it, so that, when its spiral course is restored perfectly, and that which follows is joined to that which precedes, he reads around the staff, and so discovers the continuity of the message. And the parchment, like the staff, is called '*scytale*', as the thing measured bears the name of the measure.' (Plutarch, *Lysander* 19.5–7)

This would not, it has to be said, provide a very high degree of secrecy since it is easy to decode; possibly the point was to confirm the identity of the sender (by checking that the sticks used are of the same dimensions) rather than to hide the content of the message.[10]

Officer training

The same general points about the lack of training in Greek armies that were made in Chapter 4 apply to officers and generals. There does not appear to have been any specific centrally organized training for officers, though individuals might purchase it for themselves from self-appointed trainers. In Sparta, the Spartiates at least were all subject to the *agoge*, the Spartan system for training (or more accurately, instilling toughness and discipline in) their male children.

Plato notes that the *hoplomachoi*, itinerant trainers, would avoid Sparta, in his account because their training was useless, but also presumably because there was nothing new they could teach Spartans. The skills for which the Spartan army was noted – the drill manoeuvres which allowed it to deploy from column to line and to change facing – would have been learned by all ranks and presumably there were practice drills in which the rank-and-file learned what to do, and the officers at every level learned what commands to give, but no details of this training have come down to us.

In Athens we hear of teachers of generalship such as Dionysodorus, as we saw in the previous chapter, Socrates recommending such training to one of his companions:

> 'Young man, surely it would be disgraceful for one who wishes to be a general in the state to neglect the opportunity of learning the duties, and he would deserve to be punished by the state much more than one who carved statues without having learned to be a sculptor. For in the dangerous times of war the whole state is in the general's hands, and great good may come from his success and great evil from his failure. Therefore anyone who exerts himself to gain the votes, but neglects to learn the business, deserves punishment.' (Xenophon, *Memorabilia* 3.1.1–3)

The young man duly enrolled on the course. Upon his return, he reported that the trainer 'taught me tactics – nothing else'. As noted before, 'tactics' means what we would call organization and drill, rather than battlefield manoeuvres. This was not enough for a general, Socrates explained:

> 'But then that is only a small part of generalship. For a general must also be capable of furnishing military equipment and providing supplies for the men; he must be resourceful, active, careful, hardy and quick-witted; he must be both gentle and brutal, at once straightforward and designing, capable of both caution and surprise, lavish and rapacious, generous and mean, skilful in defence and attack; and there are many other qualifications, some natural, some acquired, that are necessary to one who would succeed as a general.' (Xenophon, *Memorabilia* 3.1.6–7)

Socrates was particularly keen that the would-be general be able to tell the 'good men from the bad', specifically in the context of placing the best men in the front and rear rank (Xen., *Mem*. 3.1.9–10 and see also Chapter 3), but given what we have observed above about commanders choosing their junior officers, it was obviously important to be able to recognize talent and ability in others. In addition, Socrates saw that there were important technical skills which must be learned beyond low-level tactics:

"'But,' said Socrates, 'did he teach you only the disposition of an army, or did he include where and how to use each formation?'

"'Not at all.'

"'And yet there are many situations that call for a modification of tactics [*tattein*] and strategy [*agein*].'

"'I assure you he didn't explain that.'

"'Then pray go back and ask him. If he knows and has a conscience, he will be ashamed to send you home ill-taught, after taking your money.'"
(Xenophon, *Memorabilia* 3.1.11)

It is interesting to see this emphasis on the broader skills of generalship seen so clearly by Socrates (and Xenophon) when it does not seem to have been recognized more widely (and it may be that Xenophon's own strong views on this subject are not typical of those of his contemporaries). Presumably it was hoped that these broader, less technical skills would be acquired in the field through experience. Xenophon tells a very similar story in the *Cyropaedia*, when the young Cyrus discusses his military education with his father:

'I remember well when I came to you for money to pay to the man who professed to have taught me to be a general; and you, while you gave it me, asked a question something like this: "Of course," you said, "the man to whom you are taking the pay has given you instruction in domestic economy as a part of the duties of a general, has he not? At any rate, the soldiers need provisions no whit less than the servants in your house." And when I told you the truth and said that he had given me no instruction whatever in this subject, you asked me further whether he had said anything to me about health or strength, inasmuch as it would be requisite for the general to take thought for these matters as well as for the conduct of his campaign. And when I said "no" to this also, you asked me once more whether he had taught me any arts that would be the best helps in the business of war. And when I said "no" to this as well, you put this further question, whether he had put me through any training so that I might be able to inspire my soldiers with enthusiasm, adding that in every project enthusiasm or faintheartedness made all the difference in the world. And when I shook my head in response to this likewise, you questioned me again whether he had given me any lessons to teach me how best to secure obedience on the part of an army. And when this also appeared not to have been discussed at all, you finally asked me what in the world he had been teaching me that he professed to have been teaching me generalship. And thereupon I answered, "tactics". And you laughed and went through it all, explaining point by point, as you asked of what conceivable use tactics could be to an army, without provisions and health, and of what use it could be without the knowledge of the arts invented for warfare and without obedience. And when you had made it clear to me that tactics was only a small part of generalship, I asked you if you could teach me any of those things, and you bade me go and talk with the men who were reputed to be masters of military science and find out how each one of those problems was to be met. Thereupon I joined myself to those who I heard were most proficient in those branches.' (Xenophon, *Cyropaedia* 1.6.12–15)

Xenophon's low opinion of generalship courses may not have been universally shared (after all, such courses presumably continued to be offered and attended), but his emphasis on the superiority of practical experience was probably valid.[11]

Hoplites and officers

From armies in other places and periods we are used to officers having certain badges of rank and status, and being clearly distinguishable from the men under their command. In a hoplite army, with a generally more egalitarian ethos, and where officers were often elected from among the broad citizen population, we might not expect such clear distinctions, and indeed it seems to be the case that officers were equipped, supplied and generally lived in a fashion similar to the men they commanded. Aristophanes gives a comic scene of the *taxiarchos* Lamachos – presumably a *taxiarchos*, since he commanded more than one *lochos* (Arist., *Acharn.* 1074) – equipping himself with a haversack filled with salt, fish and onions and a bedroll strapped to his shield (Arist., *Acharn.* 1099–1101; 1136–37), as any hoplite would do, and the fact that this was to be carried by his slave or servant does not make him any different from the mass of hoplites, at least the better-off sort, all of whom would be attended by slaves or servants on campaign. Lamachos also wore a triple-plumed helmet (of which Aristophanes makes fun), presumably as a mark of his rank; but these would probably not have been official insignia, just an individual affectation and mark of his personal wealth. This was similar to the way that the Athenian commander Alcibiades was noted (and criticized) for his extravagant arms and armour – Plutarch records he 'had a golden shield made for himself, bearing no ancestral device', but this is in the context of criticism of his 'great luxuriousness of life, with wanton drunkenness and lewdness, with effeminacy in dress' (Plut., *Alcib.* 16.1–2), so it is hardly likely that these were badges of rank. It is possible that Spartan officers wore rank insignia, as there are depictions of helmets with transverse crests (running side to side of the helmet, rather than the usual front to back) which might be an indication of rank, if not just a personal choice. At any rate, senior Spartans carried a staff (*bakterion*), with which to beat wrongdoers (at least, those below citizen status), and 'a rough cloak and staff' is used by Athenians to refer to the Spartan commander Gylippus at Syracuse (Plut., *Nicias* 19.4).[12]

But officers were not treated completely like the rank and file. They received higher pay, and were expected to have greater responsibilities as well as slightly greater privileges. Xenophon made a speech to the assembled officers of the Ten Thousand:

> 'For you are *strategoi*, you are *taxiarchoi* and *lochagoi*; while peace lasted, you had the advantage of them alike in pay and in standing; now, therefore, when a state of war exists, it is right to expect that you should be superior to the common soldiers, and that you should plan for them and toil for them whenever there be need.' (Xenophon, *Anabasis* 3.1.37)

Xenophon records a pay offer received by the Ten Thousand upon their return, with messengers reporting that:

> '[The Spartans] wanted this army; also that he said the pay would be a *daric* per month for every man, twice as much for the *lochagoi*, and four times as much for the *strategoi*.' (Xenophon, *Anabasis* 7.6.1)

A *daric* was a Persian gold coin worth about twenty-five drachmas, so the rate of pay is about average (a drachma a day being typical). The proportions may also be typical. For non-mercenary armies that still received pay, like the Athenian, the proportion was perhaps similar, though there is little direct evidence, and it is possible that wealthy Athenian officers would waive their salary anyway.[13]

In terms of rations and overall living style on campaign, there was again a general equality – which is often a good idea as an example of sharing the hardships of the common soldiers, quite apart from being inevitable when officers and men are broadly of the same social class. Spartan kings did have double rations, 'not that they might eat enough for two', Xenophon helpfully explains, 'but that they might have the wherewithal to honour anyone whom they chose' (Xen, *Const. Lac.* 15.4); honouring subordinates by giving them food from one's table is a common occurrence in the *Cyropaedia* too (Xen., *Cyrop.* 2.1.30).

Qualities of the general

Sharing the soldiers' hardships went hand-in-hand with being generally approachable – Greek generals were certainly not 'Château Generals' in the style of some later militaries. Xenophon presents the Greek (or at least his own) ideal in the person of Agesilaus, in contrast with the aloof style of a Persian king:

'I will next point out the contrast between his behaviour and the imposture of the Persian king. In the first place the Persian thought his dignity required that he should be seldom seen: Agesilaus delighted to be constantly visible, believing that, whereas secrecy was becoming to an ugly career, the light shed lustre on a life of noble purpose. In the second place, the one prided himself on being difficult of approach: the other was glad to make himself accessible to all. And the one affected tardiness in negotiation: the other was best pleased when he could dismiss his suitors quickly with their requests granted.' (Xenophon, *Agesilaus* 9.1–2)

Needless to say, not every general or officer shared this style of leadership, and there are examples of those with very different styles. Xenophon's pen portraits of the original leaders of the Ten Thousand provide some valuable illustrations of good and bad command styles, very different from that of Agesilaus, and are worth quoting at length. Clearchus, as we saw in the last chapter, was a lover of war more than of his fellow men:

'[F]or he was gloomy in appearance and harsh in voice, and he used to punish severely, sometimes in anger, so that on occasion he would be sorry afterwards. Yet he also punished on principle, for he believed there was no good in an army that went without punishment; in fact, he used to say, it was reported, that a soldier must fear his commander more than the enemy if he were to perform guard duty or keep from harming his friends or without making excuses advance against the enemy. In the midst of dangers, therefore, the troops were ready to obey him implicitly and would choose no other to command them; for they said that at such times his gloominess appeared to be brightness, and his severity seemed to be resolution against the enemy, so that it appeared to betoken safety and to be no longer severity. But when they had got past the danger and could go off to serve under another commander, many would desert him; for there was no attractiveness about him, but he was always severe and rough, so that the soldiers had the same feeling toward him that boys have toward a schoolmaster. For this reason, also, he never had men following him out of friendship and good-will, but such as were under him because they had been put in his hands by a government or by their own need or were under the compulsion of any other necessity, yielded him implicit obedience. And as soon as they began in his service to overcome the enemy, from that moment there were weighty reasons which made his soldiers efficient; for they had the feeling of confidence in the face of the enemy, and their fear of punishment at his hands kept them in a fine state of discipline. Such he was as a commander, but being commanded by others was not especially to his liking, so people said.' (Xenophon, *Anabasis* 2.6.9 15)

Proxenus the Boeotian presents a contrasting, and rather sorry, figure:

> 'As a leader, he was qualified to command gentlemen [literally, 'the great and good'], but he was not capable of inspiring his soldiers with either respect for himself or fear; on the contrary, he really stood in greater awe of his men than they, whom he commanded, did of him, and it was manifest that he was more afraid of incurring the hatred of his soldiers than they were of disobeying him. His idea was that, for a man to be, and to be thought, fit to command, it was enough that he should praise the one who did right and withhold praise from the one who did wrong. Consequently all among his associates who were gentlemen were attached to him, but the unprincipled would plot against him in the thought that he was easy to deal with.' (Xenophon, *Anabasis* 2.6.19–20)

Positive reinforcement was evidently not a command style of which Xenophon approved (nor have many militaries throughout history). Finally, in a mercenary army, there were some with purely mercenary motivations:

> 'Menon the Thessalian was manifestly eager for enormous wealth – eager for command in order to get more wealth and eager for honour in order to increase his gains ... As for making his soldiers obedient, he managed that by bearing a share in their wrongdoing. He expected, indeed, to gain honour and attention by showing that he had the ability and would have the readiness to do the most wrongs; and he set it down as a kindness, whenever anyone broke off with him, that he had not, while still on terms with such a one, destroyed him.' (Xenophon, *Anabasis* 2.6.26–27)

We saw above how some writers at least clearly recognized that there was more to generalship than just tactics (in the Greek sense of organizing and drilling an army). Xenophon in particular stresses the importance of army management, of the technical skills of providing supplies and ensuring that the army is properly equipped and has all that it needs (in terms of spare equipment, as well as food and drink). Generals and officers also of course needed to show leadership and to be able to inspire confidence amongst the men they commanded. Socrates (via one of Xenophon's dialogues) stresses such skills when arguing for the suitability of a man skilled in business as an officer, with a list of virtues he felt were common to both businessmen and generals:

> '"Come then, let us review the duties of each that we may know whether they are the same or different."
>
> '"By all means."

"'Is it not the duty of both to make their subordinates willing and obedient?'"

"'Decidedly.'"

"'And to put the right man in the right place?'"

"'That is so.'"

"'I suppose, moreover, that both should punish the bad and reward the good.'"

"'Yes, certainly.'"

"'Of course both will do well to win the goodwill of those under them?'"

"'That is so.'"

"'Do you think that it is to the interest of both to attract allies and helpers?'"

"'Yes, certainly.'"

"'And should not both be able to keep what they have got?'"

"'They should indeed.'"

"'And should not both be strenuous and industrious in their own work?'"

"'All these are common to both; but fighting is not.'"

"'But surely both are bound to find enemies?'"

"'Oh yes, they are.'"

"'Then is it not important for both to get the better of them?'"

"'Undoubtedly; but you don't say how business capacity will help when it comes to fighting.'"

"'That is just where it will be most helpful. For the good business man, through his knowledge that nothing profits or pays like a victory in the field, and nothing is so utterly unprofitable and entails such heavy loss as a defeat, will be eager to seek and furnish all aids to victory, careful to consider and avoid what leads to defeat, prompt to engage the enemy if he sees he is strong enough to win, and, above all, will avoid an engagement when he is not ready.'"

(Xenophon, *Memorabilia* 3.4.7–11)

People skills and good sense and judgement were thus just as important as tactics and technical skills, and less likely to be learned from a trainer. Because of this emphasis on people skills, and because of the lack of specific insignia for officers or clear visually defined ranks, leadership at all levels was personal, and based on mutual familiarity and knowledge between officers and men, as individuals. Xenophon stresses the importance of knowing subordinates by name – Cyrus' subordinates 'remarked to one another what a good memory Cyrus had and how he called every one by name as he assigned them their places and gave them their instructions' (Xen., *Cyrop.* 5.3.46). Xenophon continues:

'Now Cyrus made a study of this; for he thought it passing strange that, while every mechanic knows the names of the tools of his trade and the physician knows the names of all the instruments and medicines he uses, the general should be so foolish as not to know the names of the officers under him; and yet he must employ them as his instruments not only whenever he wishes to capture a place or defend one, but also whenever he wishes to inspire courage or fear. And whenever Cyrus wished to honour any one, it seemed to him proper to address him by name.' (Xenophon. *Cyropaedia* 5.3.47)

That this same principle held in the Spartan army is implied when Xenophon writes: 'The formation is so easy to understand that no one who knows man from man can possibly go wrong' (Xen., *Const. Lac.* 11.6). These are not armies in which a *taxiarchos* might show up and start issuing orders to some unfamiliar *taxis* over which he has been given authority by a distant central authority. Command and leadership are based on personal familiarity within fairly small and evidently stable groups.

In battle, the main role of any officer may have been to inspire confidence in his men through his own dynamic leadership, especially in those armies that lacked any complex tactical drills. Xenophon again has Cyrus represent the ideal:

'Cyrus shouted, "Bravest of men, now let each press on and distinguish himself and pass the word to the others to come on faster." And they passed it on; and under the impulse of their enthusiasm, courage, and eagerness to close with the enemy some broke into a run, and the whole phalanx also followed at a run. And even Cyrus himself, forgetting to proceed at a walk, led them on at a run and shouted as he ran: "Who will follow? Who is brave? Who will be the first to lay low his man?" And those who heard him shouted with the same words, and the cry passed through all the ranks as he had started it: "Who will follow? Who is brave?"' (Xenophon, *Cyropaedia* 3.3.61–62)

An enthusiastic advance of this sort, at the run and with much shouting, was favoured by many armies, as we will see in Chapter 7, though Thucydides (5.70) contrasts such an approach with the more measured advance of the Spartans ('slowly and to the music of many flute-players'). It is perhaps surprising to see Xenophon here approving of the non-Spartan way.

As well as shouting words of encouragement during the advance, a general was also expected to inspire his men with a speech, shortly before, but not immediately before, battle was joined. Pre-battle speeches are a problematic feature of ancient Greek historiography, since historians traditionally include lengthy pre-battle speeches before any major battle account, largely as an exercise in rhetoric, since it is very doubtful whether such speeches were ever written down, and difficult to see how a general could deliver a set-piece speech to an army thousands or tens of thousands of men strong. But we need not doubt that generals did make speeches, or at least address the troops, before battle. It appears that there were two forms such speeches could take: firstly, more formal speeches delivered to an assembly of the army in camp, some time before the battle; and secondly, encouraging remarks made to individuals and units by the general passing down the line, just before battle was joined. Thucydides again contrasts Spartan practice with that of other Greeks at Mantinea (418):

> 'The armies being now on the eve of engaging, each contingent received some words of encouragement from its own commander ... The Lacedaemonians meanwhile, man to man, and with their war-songs in the ranks, exhorted each brave comrade to remember what he had learnt before; well aware that the long training of action was of more saving virtue than any brief verbal exhortation, though never so well delivered.' (Thucydides 5.69.1–2)

Xenophon has Cyrus too adopt such an approach with his army:

> 'Now I should be ashamed indeed to suggest to you how you ought to conduct yourselves at such a time; for I know that you understand what you have to do, that you have practised it, and have been continually hearing of it just as I have, so that you might properly even teach others.' (Xenophon, *Cyropaedia* 3.3.34)

When Cyrus' Assyrian enemies were seen being made the audience of a speech, Cyrus remarks that:

> '[N]o speech of admonition can be so fine that it will all at once make those who hear it good men if they are not good already; it would surely not make archers good if they had not had previous practice in shooting; neither could it make lancers good, nor horsemen; it cannot even make men able to endure bodily labour, unless they have been trained to it before.' (Xenophon, *Cyropaedia* 3.3.50)

A speech is thus no substitute for training and experience (which the Spartans had in plenty), but it was still an expected part of the general's role in preparing the army for battle.[14]

Another aspect of the general's role is his religious duty, as we saw above in the context of the pre-battle sacrifice and quest for favourable omens. Xenophon stresses the importance of religious observance before battle, 'for in the performance of such service the God-fearing have less fear of men' (Xen., *Cyrop.*

3.3.58). This involved the singing of the *paean* or battle hymn on the approach to battle (see Chapter 7), and also performing the proper sacrifices and rituals before battle. Again, the Spartan practice was particular to them, consisting of a last-minute sacrifice: 'When a goat is sacrificed, the enemy being near enough to see, custom ordains that all the flute players present are to play and every Lacedaemonian is to wear a wreath' (Xen., *Const. Lac.* 13.8). Xenophon mentions this practice at the Battle of the Nemea – 'when the armies were now not so much as a stadium [180 metres] apart, the Lacedaemonians sacrificed the goat to Artemis the Huntress, as is their custom' (Xen., *Hell.* 4.2.20).[15]

Campaigning

We have seen the central importance placed on logistics and managerial skills by Xenophon – an army had to be fed, watered and supplied with proper equipment above all, if it was to have any chance of success in battle. But the general must also be able to conduct campaigns on the strategic and operational level (that is, not in direct contact with the enemy, where tactical considerations, which I will examine below, were paramount).

Precisely how a general might plan out his campaigns is not clear. More modern armies rely on large staffs to control the movement of armies, and on detailed and accurate maps to direct them to the right place. As we saw above, the general staff of a hoplite army was modest in size, consisting chiefly of the senior commanders of the main tactical units – there does not seem to have been a large military bureaucracy to control strategic moves, which were instead directly organized by the junior commanders. As for maps, such things were known in the Greek world – the Spartans saw 'a bronze tablet on which the map of all the earth was engraved, and all the sea and all the rivers' (Hdt. 5.49.1). A map on such a scale would obviously have been pretty useless for practical campaigning, and we do not hear of larger-scale maps that might have been more useful, though they may have existed.[16]

More likely, generals relied on local knowledge, either their own or obtained from local inhabitants, pressed or enticed into service. One well-known incident came when the Persians of King Xerxes were blocked by the Greek army at Thermopylae:

> 'The king was at a loss as to how to deal with the present difficulty. Epialtes son of Eurydemus, a Malian, thinking he would get a great reward from the king, came to speak with him and told him of the path leading over the mountain to Thermopylae.' (Herodotus 7.213.1)

It seems unlikely that nobody else in Xerxes' army, which included contingents recruited from northern Greece, knew of this path, or that Xerxes was not able to send out scouts (from his allegedly million-strong army with large cavalry and light infantry contingents) to find the path for themselves. Nevertheless, reliance on local guides is a common feature of ancient campaigning, in this case perhaps because finding the correct path at night required a local guide; compare with the Roman general Cato's attempt to force the same path at Thermopylae three centuries later – 'They climbed the heights, but their guide, who was a prisoner of war, lost the way, and wandered about in impracticable and precipitous places until he had filled the soldiers with dreadful dejection and fear' (Plut., *Cat. Ma.* 13.1.2). In this case of course Greek armies operating in Greece had things rather easier, but on expeditions overseas would have had the same reliance on prisoners and coerced locals.

The lack of detailed maps, accurate timekeeping or any easy means to transmit orders over a distance meant that it was very difficult to coordinate the movements of dispersed forces. This meant that complex campaign plans were subject to even more friction than is normal in military operations. In 418, the Spartan king Agis mustered three forces to act against Argos – the Spartan muster including helots, under Agis himself, the Tegeans and other allies from Arcadia, and the rest of the allies including Boeotians, who mustered at Phlius (Thuc. 5.57.1–2). These musterings were presumably arranged by messengers sent (by the secretary) to the respective cities. The dispersed forces engaged in a complex series of night-time marches:

'Reinforced by the Mantineans with their allies, and by three thousand Elean heavy infantry, [the Argives] advanced and fell in with the Lacedaemonians at Methydrium in Arcadia. Each party took up its position upon a hill, and the Argives prepared to engage the Lacedaemonians while they were alone; but Agis eluded them by breaking up his camp in the night, and proceeded to join the rest of the allies at Phlius. The Argives discovering this at daybreak, marched first to Argos and then to the Nemean road, by which they expected the Lacedaemonians and their allies would come down. However, Agis, instead of taking this road as they expected, gave the Lacedaemonians, Arcadians, and Epidaurians their orders, and went along another difficult road, and descended into the plain of Argos. The Corinthians, Pellenians, and Phliasians marched by another steep road; while the Boeotians, Megarians, and Sicyonians had instructions to come down by the Nemean road where the Argives were posted, in order that if the enemy advanced into the plain against the troops of Agis, they might fall upon his rear with their cavalry. These dispositions concluded,

Agis invaded the plain and began to ravage Saminthus and other places.' (Thucydides 5.58.1–5)

The upshot of all this manoeuvring was that the Argive army found itself facing the isolated Spartan contingent – something that the Argives considered a golden opportunity for a victorious battle – while the Spartan allies, separated from the Spartans, nevertheless threatened the Argive rear around the plain, which the Spartans likewise considered a winning position. In the event, battle was (probably wisely) avoided by both sides, in favour of a truce and arbitration in the issue under dispute. It is apparent that neither side really had a clear idea of the movements and intentions of the other, and the fact that some on both sides felt they had the enemy exactly where they wanted them, while the commanding generals themselves preferred to avoid battle altogether, indicates how little matters had gone to plan on either side.

Plans for separated forces to act simultaneously could easily go wrong, as in the Athenian plan to attack Boeotia and take Delium in 424. Various elements of the plan were put in place:

> '[A]ll these events were to take place simultaneously upon a day appointed, in order that the Boeotians might be unable to unite to oppose them at Delium, being everywhere detained by disturbances at home.' (Thucydides 4.76.4)

However, the simplest of errors stymied the plan:

> 'A mistake, however, was made in the days on which they were each to start; and Demosthenes sailing first to Siphae, with the Acarnanians and many of the allies from those parts on board, failed to effect anything, through the plot having been betrayed by Nicomachus, a Phocian from Phanotis, who told the Lacedaemonians, and they the Boeotians.' (Thucydides 4.89.1)

Similarly, when Pausanias' campaign of 395 failed, leading to the death of Lysander, Pausanias 'was charged with having arrived at Haliartus later than Lysander, though he had agreed to reach there on the same day' (Xen., *Hell.* 3.5.25). Errors over the calendar, and the eternal problem of unreliable allies in faction-riven cities, could foil a plan, and widely separated forces would be unable to communicate any change of plan with each other. It was no doubt partly for this reason that armies generally tended to stick together and march in a single column to their destination. The concept, familiar from later eras, of 'marching divided and fighting concentrated' would not have applied when communications between separate marching columns were so limited.[17]

Battlefield command

The most important role of the general and of the subordinate commanders and officers was arguably to command the army in battle. Here we will be looking at the nature of battlefield command, and at how, or if, it changed across the lifetime of the phalanx.

In Homeric descriptions of fighting, the role of the senior figures is to fight in the front rank, to seek personal glory by fighting bravely and defeating their opposite numbers in single combat, but also on occasion to cajole and encourage the masses (the *phalanges*) among which they fought, sometimes ordering changes of formation, or at least changes of density, such as when organizing the defence against a rampaging opposing hero. This represents the earliest form of battlefield command, which was perhaps that still in place for much of the Archaic period, at least. The general or officer is distinguished by his social status rather than by his military rank (being invariably an aristocrat), and his importance is every bit as much in fighting in the front rank, or in front of the formation, as in any sort of command role.[18]

Gradually, as the phalanx formation became more formalized, and alongside social and political changes in the Greek cities, the role of the officer changed. Individual prowess no longer had such a prominent place, as phalanx fighting became more a matter of the encounter of homogenous masses than of the duelling of individuals, and there was less scope for individual glory in the more ordered ranks. At the same time, the limited tactics available to the early phalanx meant that there was little opportunity for the general to act as 'battle manager'; there were, generally speaking, no complex tactical manoeuvres to be devised or executed, or clever deployments. Army contingents tended to draw up (see Chapter 7) in traditional or honorary positions in the line, particularly in armies which consisted of a large number of forces from allied cities, and even such a basic tactical decision as how many ranks deep to form up was in the hands of the individual contingents, not the overall commander. Battle largely consisted of both sides advancing into contact and fighting until one side ran away, so there was little need or scope for a general to issue orders or commit forces.

This situation changed only very gradually, if at all, in the fifth century, and even in the fourth, where we see somewhat greater tactical nuance, the opportunities for tactical generalship were quite limited. It is tempting to see, with the greater complexity of hoplite battles in the fourth century, a corresponding increase in the importance of the general as battle manager rather than as warrior, and to equate this with the development, in Homeric terms, of a style of command more akin to that of Odysseus (wily, a master of tricks and stratagems) than of Achilles (brave, a fearless warrior). To an extent this is true, but this development should not be overstated. Even in the fourth century, the opportunities for clever tactics in hoplite battle were limited, as much by the limitations of the amateur armies involved as by limited Greek conceptions of tactics, while the transition from Achilles-style to Odysseus-style command should also not be pushed too far. Even in Homer, both styles of leadership were important and had their place, and while there may have been a tension between these two models, this was as old as (or older than) the phalanx itself, and the importance of the general as strategist and trickster was important already in the fifth century or earlier, albeit perhaps less so in the limited circumstances of the pitched battle. While the 'agonal' nature of warfare in the Archaic period (on which see Chapter 7) supposedly limited the opportunities for clever tricks, it is doubtful in practice to what extent this was really the case.[19]

To a great extent, leadership by example and fighting in the front rank are mandated not just by the lack of drill and complex tactics available to some forces, but also by the lack of formal authority that the war leader might have had, the power to punish and coerce rather than just to exhort and lead. The Roman historian Tacitus makes a similar point about the leadership of German tribes:

'They choose their kings by birth, their generals for merit. These kings have not unlimited or arbitrary power, and the generals do more by example than by authority. If they are energetic, if they are conspicuous, if they fight in the front, they lead because they are admired. But to reprimand, to imprison, even to flog, is permitted to the priests alone, and that not as a punishment, or at the general's bidding, but, as it were, by the mandate of the god whom they believe to inspire the warrior.' (Tacitus, *Germania* 7.1–2)

As we have seen, hoplite generals were often in a rather similar position; they were chosen for merit (or for merit at winning elections, at least), and while there was formal discipline it was of a much less rigorous nature than in the militaries of more recent societies, with punishment administered not by the god, as for the Germans, but by the mandate of the law, by which a general was bound. In these circumstances, leadership by example was bound to be more effective than leadership by coercion.[20]

In Homer, the advice of Nestor to divide the army into clan groups, which we have already seen in the context of army organization, also had a command element to it: 'thou wilt know then who among thy captains [*hegemones*] is a coward, and who among thy men' (Homer, *Iliad* 362–365). The Trojans too could be organized under leaders, perhaps on a more ad hoc basis – 'the men divided and arrayed themselves, and marshalled in five companies they followed after the leaders [*hegemones*]' (Hom., *Iliad* 12.86–87) – while Achilles also organized his men into groups under five named *hegemones* (Hom., *Iliad* 16.165–

199). *Hegemones* in this sense are clearly more than just aristocratic warriors; they obviously have a leadership (if not, strictly speaking, a command) role over the men placed under them, and it is just as important that they encourage and cajole their men in action as it is that they fight in the forefront themselves.[21]

Such activities in the course of combat would require a looser and more open formation than the presumed close order of the fully developed phalanx, in which the ability of the commander to move from front to rear to provide encouragement would (we must imagine) be greatly limited, and the commander's place would therefore be much more fixed in (or near) the front rank. Xenophon notes that the front rank of the Spartan phalanx were all officers; to what extent this was literally true may be doubted, but it is at least likely that the higher-ranking officers – *lochagoi*, *taxiarchoi* and the *polemarchoi* and kings themselves – would also have taken up their position in the front rank. Direct evidence for this is lacking, but there is no indication that officers would have stationed themselves outside the phalanx, which would have required them to stand either behind or to one side of the unit they commanded, with the latter requiring that there be gaps between units in which the officers would stand. It seems more likely that higher-ranking officers took their place in the front rank and also at the head of a file, as the lower-ranking officers did. This would have followed on naturally from the fact that aristocrats in the early phalanx would have stationed themselves at the front of the formation, as the highest-status position and the one that offered the best opportunity for conspicuous gallantry.[22]

The first hoplite battle for which any detailed account exists (brief and inadequate though it is) is Marathon (490). Here, the nominal overall commander of the army (the *polemarchos* Callimachus) and one of the ten *strategoi* were both killed in the fighting (Hdt. 6.114), though this was the fight about the ships after the rout of the Persian army, so does not prove that these officers fought at the front of the phalanx in the pitched battle stage of the fight. Following the rout, no doubt both sides lost order and the fighting became more open and confused (and more similar to the fighting in Homer). Earlier, Callimachus had commanded the right wing of the Athenian phalanx ('since it was then the Athenian custom for the *polemarchos* to hold the right wing', Hdt. 6.111.1), though precisely where he was positioned is not made clear.

Leading from and fighting at the front seems to have been common during the Peloponnesian War, judging by the number of senior commanders in all armies killed or wounded in combat. Yet at Amphipolis (422), when the Spartan commander Brasidas was mortally wounded and carried from the field by those around him, this was not in the course of direct combat but after the Athenian left was routed ('it was in full retreat and Brasidas was passing on to attack the right', Thuc. 5.10.8). It is not clear who was fighting whom or how Brasidas came to receive his wound. The Athenian commander in the same battle, Cleon, was also killed, although this (according to Thucydides) was not while bravely fighting at the front, as 'Cleon, who from the first had no thought of fighting, at once fled and was overtaken and slain by a Myrcinian peltast' (Thuc. 5.10.9).[23]

If officers did fight in the front rank, they must of course have been fighting on foot. There are examples of officers on horseback – Xenophon, for example, rode a horse in several actions in the *Anabasis*, including the occasion where he was criticized for doing so by some of the hoplites, as it meant he was not sharing in their hardships (Xen., *Anab.* 3.4.46–49). This does not, however, indicate that Xenophon would have commanded the phalanx in battle from horseback (the incident in question here was a forced march to the summit of a hill). Undoubtedly, senior commanders and kings could ride a horse, if they had one, on campaign, or in special circumstances such as when leading cavalry – as Agesilaus did on occasion (Plut., *Ages.* 16.5) – but this is by no means evidence that he would ride when in command of a phalanx in pitched battle.[24]

While the relatively high casualties among senior commanders might suggest that, right up to the level of the Spartan king, they would be expected to fight in the front rank of the phalanx, there is evidence for hoplites being stationed in front of the king, and so it may be that the general fought near the front of the formation, but not necessarily actually in the front rank. Xenophon's somewhat cryptic account of the positioning of the Spartan king for battle suggests that his protection was considered important:

> 'When the King leads, provided that no enemy appears, no one precedes him except the Sciritae and the *hippeis*. But if ever they think there will be fighting, he takes the lead of the first *mora* and wheels to the right, until he is between two *morai* and two *polemarchoi*.' (Xenophon, *Constitution of the Lacedaemonians* 13.6)

(Totalling 900 men, the Sciritae and *hippeis* are rather a large 'apart from'.) Presumably, the intent of this was to place additional forces to the king's right so that he was not on the very end of the line (and so in danger of flank attack). He might still have been in the front rank. Similarly, at Mantinea there are men 'around' the king:

> 'But the Lacedaemonians, worsted in this part of the field, with the rest of their army, and especially the centre, where the three hundred *hippeis*, as they are called, fought round King Agis [and defeated the enemy].' (Thucydides 5.72.4)

It is not clear then that Agis was necessarily right in the front rank, though as in other cases, the contingent where the king or commander was positioned could be expected to be successful (both because it contained the best men and because it would be inspired by his presence). Agesilaus also practised such personal leadership:

> 'As for courage, he seems to me to have afforded clear proofs of that by always engaging himself to fight against the strongest enemies of his state and of Greece, and by always placing himself in the forefront of the struggle.' (Xenophon, *Agesilaus* 6.1)

Plutarch records the following action at Coronea (394):

> '[A] battle ensued which was fierce at all points in the line, but fiercest where the king himself stood surrounded by his fifty volunteers, whose opportune and emulous valour seems to have saved his life. For they fought with the utmost fury and exposed their lives in his behalf, and though they were not able to keep him from being wounded, but many blows of spears and swords pierced his armour and reached his person, they did succeed in dragging him off alive, and standing in close array in front of him, they slew many foes, while many of their own number fell.' (Plutarch, *Agesilaus* 18.3)

It appears from this as if Agesilaus fought in the front rank until forced to withdraw behind his bodyguard after being wounded. Similar is the case of Cleombrotus, mortally wounded in the fighting at Leuctra and dragged away by his bodyguard (Xen., *Hell.* 6.4.13), or of Archidamus, who 'speedily received a wound straight through his thigh and speedily those who fought in front of [*pro*] him kept falling' (Xen., *Hell.* 7.4.23).

This personal leadership extended to other senior officers below the king of course. At Tegyra (375):

> 'Confident of victory, the *polemarchoi* of the Spartans, Gorgoleon and Theopompus, advanced against the Thebans. The onset being made on both sides particularly where the commanders themselves stood, in the first place, the Lacedaemonian *polemarchoi* clashed with Pelopidas and fell; then, when those about them were being wounded and slain, their whole army was seized with fear and opened up a lane for the Thebans, imagining that they wished to force their way through to the opposite side and get away.' (Plutarch, *Pelopidas* 17.3–4)

If the general really was incorporated into the rank-and-file of the phalanx, as is often supposed, it must have vastly limited the extent to which he could have exercised command over any part of the phalanx, or indeed even have seen what was happening in the battle with any clarity. There are suggestions that commanders must have remained at least somewhat more apart from the phalanx until the very last moment. Take the actions of Agis at Mantinea (418), for example, as described by Thucydides:

> 'Just before the battle joined, King Agis resolved upon the following manoeuvre ... Agis afraid of his left being surrounded, and thinking that the Mantineans outflanked it too far, ordered the Sciritae and Brasideans to move out from their place in the ranks and make the line even with the Mantineans, and told the *polemarchoi* Hipponoidas and Aristocles to fill up the gap thus formed, by throwing themselves into it with two *lochoi* taken from the right wing.' (Thucydides 5.72)

Thucydides' account gives no indication how Agis was able to observe the overlap of his left wing by an army that was already close at hand, in a battle line that stretched at a minimum for several hundred metres, nor does he reveal how Agis sent these very specific orders to four separate contingents and their commanders. This can hardly have been done by signal (no army would have a pre-prepared signal for 'take two *lochoi* from the right wing and fill the gap that has opened on your left'), so the only alternatives are that these orders were either passed down the line – in the way the watchword was passed (see below) – or were delivered by runners (and presumably then by word of mouth). If Agis was standing in the front rank of a dense, approximately eight-deep close-order phalanx, it is hard to see how he could have dictated orders to and dispatched four runners (or maybe only one, who visited each officer in turn, but the basic problem remains). No really satisfactory solution to these problems has been proposed, to my knowledge, and we must perhaps just accept that the mechanics of command in the phalanx (like many other aspects of its operation) remain somewhat mysterious. It does suggest to me that the idea of a monolithic dense formation with the general in the front rank must be in error, and we should perhaps envisage either a looser formation or one with intervals between units to allow a greater degree of movement by individuals, at least until contact with the enemy was imminent. The general may then have stood forward of the phalanx proper, where he had a better view of both lines and could be surrounded by a small staff, including runners to deliver messages, and bodyguard. Only when the phalanx was about to come into contact with the enemy (certainly after the sacrifice of the goat, in the Spartans' case, so at a distance of less than 200 metres) might the general have fallen back into an interval in the phalanx and taken his place in line with the front rank, or as a part of the front rank, men presumably moving back or aside to make room.[25]

At Plataea, we see mounted messengers being used for battlefield communication when the Spartans began their planned withdrawal:

'[The Athenians] sent a horseman of their own to see whether the Spartans were attempting to march or whether they were not intending to depart, and to ask Pausanias what the Athenians should do.' (Herodotus 9.54.2)

This messenger witnessed the row between Pausanias and Amompharetus, who would not obey the order to withdraw:

'[Pausanius] asked the man to tell the Athenians of his present condition, and begged them to join themselves to the Lacedaemonians and, as for departure, to do as they did.' (Herodotus 9.55.2).

Plataea is a slightly unusual case in that the army was more widely dispersed (and larger) than usual for a hoplite battle, which may have necessitated the use of a horseman as messenger. It is not clear if the messenger was one of a number of such retained for this purpose, or just happened to be an available man with a horse (presumably an aristocratic Athenian serving as a cavalryman). At any rate, we see here how information was obtained, and important battlefield commands transmitted, by this nameless messenger. Pausanias for his part went in person to see why Amompharetus would not withdraw and to try to persuade him to do so, and the order to withdraw Herodotus describes with the general-purpose word *parangello*, 'give orders' – he 'gave orders to the Lacedaemonians to take up their arms likewise and follow the others who had gone ahead' (Hdt. 9.53.1). Exactly how he gave these orders (shouted, messengers, signal) is not revealed.

Thucydides described the way orders were transmitted in the Spartan army, in the passage quoted above for the information it gives on Spartan organization: the king 'gives the word' to the *polemarchoi*, they to the *lochagoi*, they to the *pentecostyes* and they to the *enomotarchoi* – 'In short all orders required pass [*parangelseis*] in the same way and quickly reach the troops' (Thuc. 5.66.4). Again we see the word *parangello*, and the same word is used by Thucydides to describe the sending of orders by Agis at Mantinea discussed above. It seems highly unlikely that passing orders down the line from man to man is meant in this case, and we must surely understand a runner (or possibly a mounted messenger) carried the orders.

Xenophon gives many instances of the transmission of orders (often by himself), and considerable variety of means is evident. At the storming of Drilae (Xen., *Anab*. 5.2.6–20), the advance force first 'sent to Xenophon' news of the presence of the stronghold (presumably, 'sent a messenger' is to be understood). Xenophon 'ordered [*keleuo*] the hoplites to halt there under arms' and went forward to reconnoitre with the *lochagoi*. This small group of officers surveyed the situation and 'the capture of the place, in the opinion of the *lochagoi*, was feasible, and Xenophon fell in with their opinion'. So Xenophon 'sent the *lochagoi* to bring over the hoplites', and upon their arrival 'he ordered each of the *lochagoi* to form his *lochos* in the way he thought it would fight most effectively'. While the *lochagoi* made these arrangements, Xenophon 'passed word [*parengeille*] to all the peltasts to advance with hand on the thong, so that they could discharge their javelins when the signal should be given, to the bowmen to have their arrows upon the string, ready to shoot upon the signal, and to the slingers to have their bags full of stones; and he despatched the proper persons to look after all these things'. Who the 'proper [or suitable] persons' may be is not revealed, but presumably we may understand a group of officers or messengers, perhaps the officers of the peltasts and archers (by comparison with the way the *lochagoi* of the hoplites came to Xenophon to receive their orders, which they then took back to the hoplites). The order given was a specific tactical instruction which would come into effect when a signal was given (see more on this below). Once the army was deployed and ready, 'they struck up the *paean* and the trumpet [*salpinx*] sounded, and then, at the same moment, they raised the war cry to Enyalius, the hoplites charged forward on the run, and the missiles began to fly all together'. Evidently the trumpet gave the signal for which the peltasts and archers had been told to prepare, and the same signal also served to start the attack of the hoplites. The light forces succeeded in forcing their way into the town, but Xenophon, seeing the threat of another enemy force approaching (how did he see this if he was engaged in the fighting?) and 'taking his stand at the gates, kept out as many as he could of the hoplites'. Then those who had broken in began rushing out again, and 'when those who were tumbling out were questioned, they said that there was a citadel within', from which the defenders had launched a counterattack. At this point, 'Xenophon ordered Tolmides the herald [*kerux*] to proclaim that whoever wanted to get any plunder should go in' (as a ruse to reverse the rout of those inside).

In this short, eventful passage we see a number of ways in which a commander could obtain intelligence about the enemy and issue orders to his own men. Xenophon relies greatly on direct observation, both before the battle had started (going forward to reconnoitre the town) and during the action (spotting the approaching enemy forces), but for events outside his immediate view he could receive messages from those better placed, or he could question passing men. The planning of the attack was conducted in the company of the *lochagoi*, and orders were given directly to them, though they had discretion to organize their men as they thought best. Other forces were given specific instructions, carried to them by appropriate persons, the execution of which was to be triggered by a trumpet signal. In action, Xenophon could place himself directly in the way of his men (in the gate) and, we may assume, make his wishes known verbally, but he also had a herald or crier to hand who could (presumably with the benefit of a particularly loud voice) give specific instructions. The vocabulary suggests a distinction between 'passing orders' (*parengellein*) – presumably by messenger – and 'ordering' (*keleuein*), directly by word of mouth. We must assume that Xenophon would be surrounded by a staff of at least the herald and trumpeter, as well as some 'appropriate persons', and that these would remain by him even in action. Tolmides, the herald in this case, had also acted as herald for Clearchus (the original commander of the Ten Thousand), and was 'the best herald of his time' (Xen., *Anab.* 2.2.20), though it is not clear if he was involved in the transmission of any other orders in this episode.

Mention of a herald reminds us of Xenophon's comments on the command system of the Spartan army:

> 'Orders to perform the *paragoge* are given verbally by the *enomotarchos* acting as a herald [*kerux*], and the line is formed either thin or deep.' (Xenophon, *Constitution of the Lacedaemonians* 11.6)

So we are to understand that the *enomatarchos* shouted the orders, like a herald, to the (up to) forty men in an *enomotia*. Despite Xenophon's assertion that 'nothing whatever in these movements is difficult to understand' (11.6), there are still some obscurities. We can imagine a shouted order being transmitted to a unit of forty men who, eight deep, would have a frontage of only five men – a very compact body – but it is less clear how the orders were transmitted down from the king or *polemarchos* through the intervening levels of command. In the *Cyropaedia* (assuming Xenophon's imaginary Persians reflect Spartan practice), there is a rather similar disconnect. Cyrus made a rousing speech to various officers of his army and gave them their orders:

> 'When he had said this he sent them away to their several *taxeis* with orders to issue, as they marched, the same directions each to his own *dekadarchoi* (for the *dekadarchoi* were in the front so as to hear); and they were to bid the *dekadarchoi* each one to announce it to his *dekas*.' (Xenophon, *Cyropaedia* 4.2.27)

Note here the two 'ordering' words together – 'they were to bid ... each one to announce', *keleuein, parangellein*. The importance of verbal orders, and of being in a position to hear them, is particularly stressed, but this is at the very lowest level of command, the *dekas* (the ten, or file). Xenophon stresses the difficulty of shouted orders even at this level:

> 'Moreover, the men shout words of encouragement [*parakeleuontai*] to the *enomotarchos*, for it is impossible for each *enomotarchos* to make his voice travel along the whole of his *enomotia* to the far end. The *polemarchos* is responsible for seeing that all is done properly.' (Xenophon, *Constitution of the Lacedaemonians* 13.9)

The meaning and the text of the passage are obscure, and it is not immediately obvious why the *enomotarchos* cannot make himself heard by so small a body as the *enomotia* (nor why 'encouragement' from the ranks would help, though presumably the intended meaning is that the men pass orders and encouragement along between themselves). The gap in the chain of command between *enomotarchos* at the lowest level and *polemarchos* at the highest is striking, and there is no hint of how orders were transmitted between the intervening levels.[26]

The shouting of encouragement or orders between the men could also extend to them shouting helpful suggestions to their generals, something that was not viewed as insubordination, and the general might well accept the proferred advice, as we saw in the previous chapter. At Mantinea (418), 'one of the older men' shouted a suggestion to the Spartan king Agis that he should abandon his planned attack (Thuc. 5.65.2), while at Nemea (394) 'when the first *polemarchos* was about to attack them [the Argives] in front, it is said that someone shouted out to let their front ranks pass by' (Xen., *Hell.* 4.2.22), which they did.[27]

In Xenophon's imagining of Cyrus' command methodology, orders to the higher-ranking officers are given by gathering them together before action and addressing them communally, as Xenophon did at Drilae with the *lochagoi*. After sighting the enemy the day before the battle, and while still at some distance (enemy numbers and dispositions being reported by scouts):

> '[Cyrus] summoned together the commanders of the cavalry, the infantry, and the chariot corps, and also the officers in charge of the engines, of the baggage train, and of the wagons, and they came.' (Xenophon, *Cyropaedia* 6.3.8)

This group of senior officers then discussed what had been reported and interrogated the prisoners, and Cyrus sent out a force of cavalry with instructions to chase off some enemy scouts. Cyrus then gave specific instructions to each contingent, of which a sample will suffice:

> '"And then do you, Arsamas," said he, "and you Chrysantas take charge of the right wing, as you always have done, and the rest of you *myriarchoi* take the posts you now have. When the race is on, it is not the time for any chariot to change horses. So instruct your *taxiarchoi* and *lochagoi* to form a phalanx with each separate *lochos* two deep [or 'in twos']."' (Xenophon, *Cyropaedia* 6.3.21)

> 'And you, Pharnuchus and Asiadatas, keep each of you the regiment [*chiliostys*, unit of 1,000] of cavalry under your command out of the phalanx and take your stand by yourselves behind the carriages, and then come to me with the rest of the officers. You must be just as fully ready, though in

the rear, as if you were to be the first to have to join battle.' (Xenophon, *Cyropaedia* 6.3.32)

Conferences of this sort in which detailed instructions are given to contingent commanders must have been the normal practice (though in most cases, tactics being simple, there would be little to instruct). Note that the cavalry commanders are to attend with their junior officers, presumably so that further orders can be given directly to them by Cyrus, by word of mouth. The method of transmitting orders, when necessary, to particular larger subunits which may be at a distance from the general (as to the two Spartan units at Mantinea) remains obscure, and we must again assume the use of messengers to deliver a spoken order, which would then be translated into specific drill commands and shouted to the men by the junior officers.

The first-century AD author Onasander, who collected a lot of Greek and Hellenistic material but perhaps in this case also reflects Classical Greek practice, sets out the way orders should be transmitted in greater detail:

> 'Time is lost in passing orders down the line [*parangellein*], and confusion arises, as all the soldiers question each other at the same time ... But one should communicate his orders to his higher officers and they should repeat them to the officers next below them, who in turn pass them to their subordinates, and so on to the lowest, the higher officers in each case telling the orders to those below them.' (Onasander, *The General* 25)

Obviously it is only at the lowest level of officer – the *enomotarchoi* in the Spartan example – that the orders will be transmitted directly to the men, the regular hoplites, all the intermediate levels just passing on their orders to other, lower ranks of officer.

Reliance on orders transmitted by word of mouth had its obvious downsides, in particular that often in the heat of battle it might have been impossible to hear the orders. Thucydides makes this point when recounting the hardships of the Spartan hoplites isolated on Sphacteria:

> '[T]hey themselves were helpless for offence, being prevented from using their eyes to see what was before them, and unable to hear the words of command for the hubbub raised by the enemy; danger encompassed them on every side, and there was no hope of any means of defence or safety.' (Thucydides 4.34.3)

This point is also made by the later Hellenistic tacticians, who recommend using visual (flag) and audible (trumpet) signals to supplement shouted orders (for example Asclepiodotus 12.10), but this practice does not seem to have been adopted in any rigorous fashion by Classical Greek armies.

Xenophon mentions the use of signals given by trumpet, a method of command that is more familiar to us from later eras (the military bugle forming an essential method of command in pre-industrial modern armies). Apparently the trumpet was not used, like the bugle, to deliver specific orders, but only orders of the most general kind, and the usual practice appears to have been to establish a desired course of action verbally (by shouted orders, or messengers, as appropriate), and have the trumpet act as an executive signal for the cautionary verbal order to be carried out (thus ensuring that implementation of the order is simultaneous). Xenophon provides an example in a drill display of the Ten Thousand:

> 'When he [Cyrus] had driven past them all, he halted his chariot in front of the centre of the phalanx, and sending his interpreter Pigres to the *strategoi* of the Greeks, gave orders that the troops should advance arms and the phalanx move forward in a body. The *strategoi* transmitted these orders to the soldiers, and when the trumpet sounded, they advanced arms and charged.' (Xenophon, *Anabasis* 1.2.17)

We see the familiar pattern of the commanding general (Cyrus in this case) making his wishes known directly (via an interpreter) to the *strategoi*, who pass on the orders to the men (presumably down the chain of command, if there was a chain of command at this stage); then the execution of the orders is triggered by a trumpet signal. Note also that this model example of controlled drill does not last long, the phalanx rapidly getting out of the control of its officers, even here on the parade ground (though this uncontrolled advance had the desired effect of terrifying the barbarian onlookers). In an engagement in Bithynia, the Ten Thousand advanced against the enemy:

> 'The orders had been to keep their spears on the right shoulder until a signal should be given with the trumpet; then, lowering them for the attack, to follow on slowly, nobody to break into a run. And now the watchword was passed along, "Zeus Saviour, Heracles Leader".' (Xenophon, *Anabasis* 6.5.25)

Again, specific orders are sent out (presumably by word of mouth), and the trumpet is used as the signal to execute them. In the event, the peltasts on the flanks 'raised the battle-cry and proceeded to charge upon the enemy without waiting for any order' (6.5.26), though the hoplites showed better discipline and charged only on the trumpet signal being given (6.5.27).

There is some evidence that a very limited range of specific orders could be given by different notes or phrases on the trumpet – perhaps only 'advance' and 'retire'. On one occasion, the Ten Thousand deceived a force of Carduchian skirmishers that had been pressing closely on their rear as they tried to withdraw across a river. First the Greeks charged toward the enemy to make them withdraw:

> 'At that instant the Greek trumpeter sounded his signal; and while the enemy began to flee much faster than before, the Greeks turned about and set out on their own flight through the river at top speed.' (Xenophon, *Anabasis* 4.3.32)

Here the Carduchians had come to associate the trumpet signal with a charge, and the Greeks were able to use this to their advantage by adopting the opposite course of action. The 'retire' signal appears in Thucydides' account of the fighting at Amphipolis. Cleon the Athenian commander went ahead of his army to reconnoitre the town:

> '[H]aving done so, being unwilling to venture upon the decisive step of a battle before his reinforcements came up, and fancying that he would have time to retire, [he] bid the retreat be sounded and sent orders [*parengeile*] to the men to effect it by moving on the left wing in the direction of Eion, which was indeed the only way practicable.' (Thucydides 5.10.3)

It is not absolutely explicit that the signal was a trumpet signal, and on this occasion the Athenian forces fell into disorder, because – it has been suggested – Cleon gave the 'retreat' signal before he had verbally passed on the details of how the retreat was to be executed, so that the executive order preceded the cautionary. Xenophon is more explicit regarding the *strategoi* of a force of hoplites fearing an attack on their camp, who 'immediately sounded the recall with the trumpet and set out on the return journey' (Xen., *Anab*. 4.4.22).[28]

There were other uses for the trumpet outside of battle – it was used to sound the reveille, to summon the men to arms and to call for silence in meetings – but there is no evidence for a complex set of specific sounds associated with orders, other than the simple 'charge' and 'retire'. There is reason to suppose that Hellenistic armies did use more sophisticated trumpet signals, but there is no evidence for such signals in Classical Greek armies.[29]

Similarly, while there are occasional examples of visual signals being used in battle in the Classical period, there is no sign of the systematic use of unit standards that apparently became common in the Hellenistic phalanx. We do see visual signals used to send messages outside of pitched battle, as at Potidaea (432):

> 'Meanwhile the auxiliaries of the Potidaeans from Olynthus, which is about seven miles off, and in sight of Potidaea, when the battle began and the signals were raised, advanced a little way to render assistance; and the Macedonian horse formed against them to prevent it. But on victory speedily declaring for the Athenians and the signals being taken down, they retired back within the wall; and the Macedonians returned to the Athenians.' (Thucydides 1.63)

Was this a general distress signal, some sort of 'SOS' or a prearranged signal, the meaning of which had been agreed beforehand? This calls to mind the signal given to the Persian fleet from traitors in Athens in 490 by flashing the sun off a shield, the meaning of which could only have been prearranged.

Diodorus records an ill omen in the Boeotian army shortly before the Battle of Leuctra (371):

> 'For as the clerk [*grammateus*] advanced with a spear and a ribbon attached to it and signalled the orders from headquarters [literally 'from the *hegemones*'], a breeze came up and, as it happened, the ribbon was torn from the spear and wrapped itself around a slab that stood over a grave.' (Diodorus 15.52.5)

What sort of orders these might have been, and precisely how they were signalled, is not revealed, nor why the orders could not have been transmitted verbally. This was in the context of Epaminondas leading out his army, so perhaps these signals were a general 'prepare to march' order. Epaminondas, incidentally, won retrospective praise (after the subsequent victory) for ignoring the omen and 'was considered to have excelled in military shrewdness' (Diod. 15.52.7).

Xenophon observed the Persian use of a standard to mark the location of the king – 'the royal standard, a kind of golden eagle on a shield, raised aloft upon a pole' (Xen., *Anab*.1.10.12, and compare Xen., *Cyrop*. 7.1.4) – and he was sufficiently impressed by the value of this in providing a rallying point that he used the idea widely in his fictional army of Cyrus:

> 'And all the officers had banners over their tents; and just as in the cities well-informed officials know the residences of most of the inhabitants and especially those of the most prominent citizens, so also in camp the aides under Cyrus were acquainted with the location of the various officers and were familiar with the banner of each one; and so if Cyrus wanted one of his officers, they did not have to search for him but would run to him by the shortest way.' (Xenophon, *Cyropaedia* 8.5.13)

There is no sign of this being extended to mark the location of units in battle, still less to use the standards to give orders.[30]

The purpose of the passing of the watchword (*sunthema*), which we encountered above, is not clear. The procedure occurs a number of times in Xenophon, for example at Cunaxa where Cyrus was talking to Xenophon in front of the army:

> 'While saying this he heard a noise running through the ranks, and asked what the noise was. Xenophon replied that the watchword was now passing along for the second time. And Cyrus wondered who had given it out,

and asked what the watchword was. Xenophon replied "Zeus Saviour and Victory". And upon hearing this Cyrus said, "Well, I accept it, and so let it be." After he had said these words he rode back to his own position.' (Xenophon, *Anabasis* 1.8.16–17)

Evidently the Greeks' own commanders decided on this watchword (this may not have been a Persian practice, though Xenophon has his fictional Cyrus follow the same procedure, Xen., *Cyrop.* 3.3.58, 7.1.10). A short phrase, as here of a religious nature, was passed by word of mouth (*parangellein*) down the line and back again. The purpose of watchwords and countersigns (*parasunthema*) was in theory to establish the identity of combatants and the veracity of any orders they might be transmitting. Thucydides 7.44.4–5 describes the Athenian difficulties in a night attack at Syracuse, when by constantly asking for the watchword to identify men looming out of the moonlight, they ended up causing more confusion and giving the watchword away to the enemy. But no doubt the practice acquired a more symbolic and religious significance over time. Onasander perhaps reflects a change in practice after the Classical period:

> '[The General] should give the countersign not by the voice but by some gesture, as a wave of the hand or the clash of weapons, or dipping a spear, or a side-wave of his sword, in order that when confusion arises the soldiers may not have to trust to the spoken watchword alone – for the enemy hear this so often that they are able to get it – but also to the countersign. This is most useful in the case of allies who speak a different language, for, unable to speak or to understand a foreign tongue, they differentiate between friends and enemies by this countersign' (Onasander, *The General* 26)

This is a reminder that in similarly equipped armies without uniforms, telling friend from foe was not a trivial matter, as much among forces with the same language and physical appearance as those who speak different languages, and 'friendly fire' incidents on the Greek battlefield were not unknown. Where orders are passed by word of mouth, the watchword and countersign will have served as a way of confirming that the order did indeed originate with one's own side, though it is difficult to see in practice how so subtle a sign as 'a side-wave' of the general's sword could be seen and acted upon by an army in combat.[31]

Overall, it is clear that while orders of any level of complexity could be passed by word of mouth or in conference with senior officers, the scope for sending out specific detailed orders, or orders which deviated at all from the previously determined plan, was very limited. In part this reflected the fact that, other than the Spartans, the small-unit organization to allow drills and manoeuvres to be executed was simply not available, nor the chain of command that could set such evolutions in motion, so there was little point having the ability to send complex orders. A phalanx could be formed in variable depth and at a desired location, but battlefield control then came down largely to ordering it to advance and hoping for the best. Where more complex manoeuvres were attempted, largely by Spartan armies, they did not always go well, and the common Spartan tactic of outflanking on one wing would have involved following a set and predetermined series of drill manoeuvres which did not require much real-time input from the commander. The command and control capabilities of Greek armies were rudimentary, reflecting their nature as largely untrained and lightly drilled militias.

Despite this, it was felt that the officers, and particularly the general, were highly important to the effective functioning of an army. After the defeat at Cunaxa (401), the commanders of the Ten Thousand were tricked and killed by the Persians. Xenophon gives their reasoning:

> 'You observe that our enemies did not muster up courage to begin hostilities against us until they had seized our generals; for they believed that so long as we had our commanders and were obedient to them, we were able to worst them in war, but when they had got possession of our commanders, they believed that the want of leadership and of discipline would be the ruin of us. Therefore our present commanders must show themselves far more vigilant than their predecessors, and the men in the ranks must be far more orderly and more obedient to their commanders now than they used to be.' (Xenophon, *Anabasis* 3.2.29–30)

This reveals an assumption on the part of the Persians, at least as Xenophon saw it, that 'decapitating the leadership' of the army would destroy its effectiveness (a practice very much enshrined in the tactics of some modern armies), though in the event, discipline was maintained and new, even more effective commanders (including Xenophon himself) were appointed.

There was perhaps a conception that an army could be defeated by destroying its high command. Theban tactics at Leuctra concentrated on the Spartan king:

> 'The Thebans, however, were massed not less than fifty shields deep, calculating that if they conquered that part of the army which was around the king, all the rest of it would be easy to overcome.' (Xenophon, *Hellenica* 6.4.12)

However, we should not be too ready to see this as a case of command decapitation, since the objective was rather to defeat the Spartans themselves so that their less-willing allies would then give way (the famous 'crushing the head of the snake' tactic):

'In order to encourage the Thebans to make a vigorous attack on the Lacedaemonians, Epaminondas produced a large snake, and crushed its head in front of the army. "If you crush the head," he said, "you see how impotent is the rest of the body. So let us crush the head of the confederacy, that is the Laconians, and the power of their allies will become insignificant." The Thebans appreciated the force of his argument. They attacked and routed the Laconian phalanx; after which, the whole army of the allies immediately gave way and fled.' (Polyaenus 2.3.15)

Here, the head of the snake was the Spartan contingent as a whole, not just the king. The relative importance of the general was debated in antiquity, Plutarch recording one aspect of the debate:

'For if, as Iphicrates analyzed the matter, the light-armed troops are like the hands, the cavalry like the feet, the phalanx itself like chest and breastplate, and the general like the head, then he [the general], in taking undue risks and being over bold, would seem to neglect not himself, but all, inasmuch as their safety depends on him, and their destruction too. Therefore Callicratidas, although otherwise he was a great man, did not make a good answer to the seer who begged him to be careful, since the sacrificial omens foretold his death; "Sparta," said he, "does not depend upon one man."' (Plutarch, *Pelopidas* 2.1)

This is a debate that would come to the fore in the Hellenistic and Roman periods, when the person of the king came to be even more important to the state, and the concept arose of the general as battle manager, directing tactics, feeding in reserves and making high-level decisions. In this period though, the number of things the general could manage was more limited – something that in itself was a cause of the relatively simple tactics employed, there being no point retaining a reserve, for example, if there was nobody to decide when to commit it or to issue the orders to do so.

Chapter 6

Hoplites at War

The hoplite, as the heavy infantryman of Classical Greece, was probably the most important warfighting component of the Greek city armies, and is certainly presented that way to us in the extant works of the ancient historians (who were all themselves of the hoplite class). Nevertheless, he was not the only soldier in Classical Greece, but rather was part of an integrated group that also contained, in broad terms, cavalry, light infantry and sailors (or rowers). It is tempting to minimize the role of these other arms (except perhaps the sailors and rowers) because this matches the picture presented to us in the ancient sources, and it was long the fashion among historians to do so and to emphasize the contribution of the hoplite far above that of the other arms and social classes. As a reaction against this tendency, many modern authors will now stress the importance of the 'others', and note that the hoplite was (ideally) only one component of a combined-arms system.[1]

But another possible distortion can creep in instead, which is to see a steady progression from the putative very hoplite-centred warfare of the Archaic period, through the increasing importance of light infantry, in particular, in the fifth and early fourth centuries, a process culminating in the true combined-arms armies of the Macedonians with their integrated forces of heavy infantry, light infantry and cavalry. It may be that some such progression does indeed match historical reality, but we must also be acutely aware of the fact that there are practically no detailed records of Greek warfare before the fifth century, that for much of the fifth century itself we have only Herodotus' account of the (highly atypical) Persian Wars, and that detailed accounts of battles and campaigns only really become available with the writings of Thucydides and Xenophon in the late fifth and early fourth centuries – a time at which we already see light infantry in particular playing an important role in warfare. It may be, therefore, that the progression we seem to see from highly stylized ('agonal', from the Greek *agon*, 'contest') warfare between exclusively heavy infantry forces at the start of the fifth century, to combined-arms forces with a greater reliance on light infantry in the fourth, reflects not changes in practice, but simply changes in the quality and availability of evidence. If we had writers like Thucydides and Xenophon to record the events of the late sixth century, we may find that warfare was not so very different at this date than it was in the early fourth century, and the sense of progress from rigid, formalized agonal warfare to something more resembling 'combined arms' is largely illusory. The Macedonians for their part certainly had balanced armies with integrated forces of light infantry and cavalry, but they also had a phalanx that was in many ways even heavier and more specialized (for close-quarters combat) than that of the Greeks, so it may well be an error to see the Macedonians as a culmination of a process that had been going on in Greek warfare for two centuries. With these caveats in mind, this chapter will look at the role of the hoplite alongside other arms in waging war in Classical Greece.[2]

Light infantry

Hoplites are always listed as the most important component of Greek armies, to the extent that sometimes it is hard to discern whether other arms were present at all in a given battle or campaign, or what their role might have been. Yet even in the Persian Wars, as Herodotus makes clear, large forces of light infantry accompanied the hoplites into battle. The last stand of the Spartan hoplites at Thermopylae is a model of the spirit of self-sacrifice, and of determination to stand one's ground, that is central to the ideal of the hoplite. Nevertheless, Herodotus tells us that when Xerxes allowed observers to inspect the aftermath of the battle, as well as the bodies of fallen Lacedaemonians and Thespians, 'helots were also there for them to see' (Hdt. 8.25.1), though there is little indication in Herodotus' account of what role they played in the fighting.

Similarly at Plataea, a 'maximum effort' by all the non-Medizing Greek cities saw large light infantry forces accompanying the hoplites – Herodotus states that 'as regards the number of the light-armed men, there were in the Spartan array seven for each hoplite, that is, thirty-five thousand, and every one of these was equipped for war. The light-armed from the rest of Lacedaemon and Hellas were as one to every hoplite, and their number was thirty-four thousand and five hundred' (Hdt. 9.29.1–2). The account of the fighting that follows makes no mention of what these light-armed troops did, or even if they fought at all, yet as at Thermopylae they sustained casualties, as shown by the tombs that were built to hold the dead:

> 'The Lacedaemonians made three tombs; there they buried their *irenes* [men aged 20–30] ... in the second the rest of the Spartans, and in the third the helots. This, then is how the Lacedaemonians buried their dead. The Tegeans, however, buried all theirs together in a place apart, and the Athenians did similarly with their own dead.' (Herodotus 9.85.1–2)

The light infantry of a more egalitarian city such as Athens could be poorer citizens, and deserving of being buried with the hoplites, unlike the serf class helots, but in neither case were their exploits considered worth recording.

Because the light-armed fighters would be drawn from a broader social class than the hoplites, they could be available in very large numbers, although the actual number might not be recorded. Thucydides describes the full Athenian levy for the invasion of Megara in 431:

> 'And this was the greatest army that ever the Athenians had together in one place before, the city being now in her strength and the plague not yet amongst them. For the Athenians themselves were no less than ten thousand hoplites, besides the three thousand at Potidaea; and the foreigners that dwelt amongst them [*xenoi*] and accompanied them in this invasion were no fewer than three thousand hoplites more, besides other great numbers of light armed soldiers.' (Thucydides 2.31.2)

'Great numbers' or equivalent seems to have been a common designation – compare with Xenophon's estimate of the manpower of Thessaly available to its ruler:

> '[H]e had more than eight thousand horsemen, including the allies, his hoplites were reckoned at not fewer than twenty thousand, and there were peltasts enough to be set in array against the whole world; for it is a task even to enumerate the cities which furnished them.' (Xenophon, *Hellenica* 6.1.19)

Hyperbole aside, light infantry could be plentiful and historians do usually make a better attempt at quantifying them, but their military role is often downplayed. Partly this may be the result of the prejudice of hoplite-class historians belittling the efforts and contribution of their social inferiors, and partly down to the concept that, when push came to shove, it was the efforts of the steadfast hoplites that won battles and wars, not any of the other arms (more on this idea below).

We should also note Herodotus' comment that the helots at Plataea were 'equipped for war'. As we have seen in Chapter 4, servants and attendants would frequently accompany hoplites, at least the wealthier sort, on campaign, and would be involved particularly with transport (carrying the shield, for example) and logistics (obtaining supplies). But such men would not necessarily be expected to fight (either in pitched battle or in any sort of irregular operation), except as a last resort. Light infantry 'equipped for war' would have had weapons of some sort, at the least, and perhaps a shield and maybe a little armour (possibly a helmet, if only a felt *pilos*), and would have been used in combat. Greek light infantry came in two basic types – *psiloi* or generic light infantry who may have carried no more than their offensive weapons, being perhaps unshielded and unarmoured (they are sometimes called *gumnetes* or 'naked men'), and *peltastai* or peltasts, men equipped with and named after a light shield, the *pelte* or *pelta*, a smaller shield than the hoplite's heavy *aspis*, of a flatter section (without the pronounced rim),

Hoplites at War 257

Greek light infantry: a) Thracian peltast (Harvard University, Robinson Collection), b) *psilos*, with possibly plundered sword, c) peltast, note handles of shield (Vienna Kunsthistorisches Museum), d) Scythian archer. (*British Museum*)

of varying shape (the traditional Thracian *pelta* was crescent-shaped) and made of light materials (perhaps of wicker or leather, though light woods might also have been used, and sometimes there may have been a bronze covering).[3]

For offensive weapons, *psiloi* and peltasts alike would often have carried javelins, presumably a small bunch of light javelins rather than the two throwing spears of early hoplite infantry. Javelins seem to have been the usual armament of all peltasts (at least into the fourth century). The javelin appears to have been the default weapon of the poorer classes in Greece, and those states, like Aetolia, that lacked hoplite forces will have relied instead on large forces of *psiloi* equipped as javelineers. The javelin was cheaply made and required no special expertise to use, and as thrown javelins could be retrieved and thrown back, there were few difficulties with ammunition supply. Other more specialized weapons were available: archers were used by some cities, Athens in particular making use of specialist forces of archers (and archers also formed part of the complement of the ships of the navy), but bows and arrows were more expensive to produce and required greater skill to use, so never took on the importance that they had in the Persian army. Slings, whether used to hurl ordinary stones or specially made lead bullets, were also cheap to produce and (if stones were used) offered few problems of ammunition supply. However, sling use required more skill than javelins, and slingers tended to be drawn from those regions (particularly Rhodes) that had a particular cultural tradition of the practice. As a last resort, anyone could also throw stones picked up off the ground; stone throwing occurs surprisingly frequently, even among hoplites, and would have been especially appropriate in the defence of a fortified city, for example.[4]

Part of the reason for the lack of accurate numbers for light infantry on many occasions, aside from prejudice, was no doubt that light infantry were not formally recruited and organized in the way the hoplites and cavalry were, so no official records would have been available. As we have seen, Athens (and perhaps other cities) posted lists of the individual names of hoplites called up for a particular campaign, and grouped them into the familiar units (*taxeis* and *lochoi* if nothing else), but light infantry would not have been individually conscripted in this way and there is little evidence that they were organized into formal units (at least before the fourth century). As such, it may have been the case that even the commanding general did not know with any certainty how many light infantry he was fielding. This is not to say that light infantry would fight with no organization, since being divided up into bodies was probably essential for any force to operate effectively, but that such units may have been more ad hoc – for example, on Sphacteria, when the Athenians under Demosthenes landed their light forces on the island, 'Demosthenes had divided them into companies [*lochoi*] of two hundred, more or less' (Thuc. 4.32.3).

As shown by this example, another source of light infantry when on campaign was likely the rowers of the ships (usually triremes) which served as transports and escorts. The rowers were drawn from the same lower social classes as the light infantry, and while there is not always direct evidence that the rowers were used as light infantry, there are a few occasions (such as Sphacteria) when it is made explicit that they were. It seems inherently likely that this source of manpower would not have been neglected on campaign, especially overseas, when special arrangements had to be made for the transport of hoplites (and cavalry horses), and it is not likely that similar transports would have been laid on for light infantry. For the fighting on Sphacteria, the Athenian commander Demosthenes:

> 'drew up under the fortification and enclosed in a stockade the galleys remaining to him of those which had been left him, arming the sailors taken out of them with poor shields made most of them of osier, it being impossible to procure arms in such a desert place, and even these having been obtained from a thirty-oared Messenian privateer and a boat belonging to some Messenians who happened to have come to them.' (Thucydides 4.9.1)

Note that nothing is said of offensive weapons. When these rowers went into action, they were armed 'with the arms they carried' (Thuc. 4.32.2), presumably javelins for some or all, or just stones (plentiful on the island). The fact that weapons were available and only shields were required in addition suggests that sailors were normally armed but not shielded, so could fight as *psiloi* but not as peltasts.[5]

On other occasions, rather than relying on makeshift shields and luck, equipment was prepared in advance. In 409, an Athenian force was fitted out:

> 'Thrasyllus took the ships which had been voted him, equipped five thousand of his sailors so that he might employ them as peltasts also, and set sail at the beginning of the summer for Samos.' (Xenophon, *Hellenica* 1.2.1)

The role of light infantry was to undertake all those tasks for which the hoplites were not well suited, and to support the hoplites when it came to pitched battle. Usually, both sides in a campaign would be fielding light forces, and so in pitched battle these might face off against each other in the space between the two opposing phalanxes as a prelude to the main fighting. Again, ancient sources tend to be dismissive of the importance of such encounters, as in the first clash between Athenians and Syracusans in 415:

> 'First, the stone-throwers, slingers, and archers of either army began skirmishing, and routed or were routed by one another, as might be expected between light troops.' (Thucydides 6.69.2)

It was the indecisive nature of such fighting that both aroused a degree of scorn and prevented light forces from playing a central part in pitched battles between hoplite armies. Engaging at a distance, with missile weapons, in generally low-intensity combat (with low rates of lethality), and able to run away from the enemy in order to establish a safe distance rather than standing their ground, as hoplites did (I will return to these concepts later), light infantry would find it difficult to win a decisive victory over similarly equipped opponents. Light infantry (we assume) fought in a way closer to the 'Homeric' style discussed in Chapter 1, moving forward or backward toward the enemy as personal preference and situation dictated, rather than holding a strict formation, whereas hoplites stood their ground and fought at close quarters until one side gave way, at which point the rout would be final and decisive. This meant that light infantry could not, generally speaking, win or lose a battle on their own, and is one reason why their contribution to battle was not considered of great importance.

Where light infantry came into their own was in the many types of fighting outside of pitched battle: skirmishing, raiding, plundering, and also in siege warfare of various kinds. Clearly, missile weapons are of great value in siege warfare, to clear or defend the walls. In the various low-intensity forms of warfare involved in general campaigning, light infantry were also well suited, especially over difficult terrain, where heavily encumbered hoplites were not able to operate freely or to maintain their close formation. Yet we should not overstate the inability of hoplites to engage in such forms of fighting. As we will see below, hoplites were perfectly capable of fighting in more varied circumstances than just the pitched battle, and were not utterly helpless outside of the formed phalanx, despite the claims to that effect by some modern authors.

While light infantry would not have a decisive impact on a pitched battle between two hoplite phalanxes, where a force of light infantry on one side faced a force of hoplites without light infantry support on the other, the odds could often be very much in favour of the light infantry. Because the light infantry could stand off at a distance and pelt the hoplites with missiles, and the hoplites, encumbered by their equipment, could not easily catch the light infantry to engage them hand-to-hand, light infantry could engage hoplites more or less without risk to themselves. They could not hold ground, since they would have to fall back before any advance by the hoplites, nor could they decisively defeat the hoplites in a short engagement, but given no shortage of time, plentiful space in which to operate and sufficient supplies of ammunition and ideally a numerical advantage, light infantry could expect to defeat an unsupported force of hoplites by simply shooting them down from a distance and retreating before any advance. From the late fourth century, we encounter a number of well-known engagements in which hoplite forces were defeated by light infantry – the best-known perhaps being Sphacteria, where a force of Spartan hoplites was shot down and the survivors forced to surrender by the makeshift Athenian light forces we encountered above (see Thuc. 4.1–41 and especially 4.34.1–3).

In the fourth century, in another famous encounter, the Athenian general Iphicrates with a force of mercenary peltasts destroyed a Spartan *mora* at Lechaeum. The Spartan hoplites intended to march past enemy-held Corinth:

'But those in the city of the Corinthians, both Callias, the son of Hipponicus, commander of the Athenian hoplites, and Iphicrates, leader of the peltasts, when they descried the Lacedaemonians and saw that they were not only few in number, but also unaccompanied by either peltasts or cavalry, thought that it was safe to attack them with their force of peltasts.' (Xenophon, *Hellenica* 4.5.13)

Accordingly, the peltasts came out to challenge their heavily armed opponents:

'Now when the Lacedaemonians were being attacked with javelins, and several men had been wounded and several others slain, they directed the shield-bearers to take up these wounded men and carry them back to Lechaeum; and these were the only men in the *mora* who were really saved. Then the *polemarchos* ordered the first ten year-classes to drive off their assailants. But when they pursued, they caught no one, since they were hoplites pursuing peltasts at the distance of a javelin's cast; for Iphicrates had given orders to the peltasts to retire before the hoplites got near them.' (Xenophon, *Hellenica* 4.5.14–15)

Note the use of the younger age classes (who presumably formed the front ranks and were considered fitter) to try to catch the peltasts; the Spartan commander tried again with the youngest fifteen year-classes (4.5.16), but with no better success. This pattern of failed attempts to catch the peltasts, and taking losses from javelins when the pursuers fell back on the main body, continued for some time. Eventually, having lost nearly half their men, the Spartans broke into flight when the Athenian hoplites advanced against them – note that they took heavy losses from the peltasts, but in the end it was only the threat of enemy hoplites that caused them to flee.[6]

These are the best-known examples, but in fact defeats of hoplite forces by light infantry are more common than we might suppose. Demosthenes, the victorious Athenian commander on Sphacteria, had himself learned the hard way the danger posed to hoplites by light infantry in an earlier campaign in Aetolia:

'Led on by his advisers and trusting in his fortune, as he had met with no opposition, without waiting for his Locrian reinforcements, who were to have supplied him with the light-armed javelinmen in which he was most deficient, he advanced and stormed Aegitium ... Meanwhile the Aetolians had gathered to the rescue, and now attacked the Athenians and their allies, running down from the hills on every side and darting their javelins, falling back when the Athenian army advanced, and coming on as it retired; and for a long while the battle was of this character, alternate advance and retreat, in both of which operations the Athenians had the worst. Still as long as their archers had arrows left and were able to use them, they held out, the light-armed Aetolians retiring before the arrows; but after the captain of the archers [*toxarchos*] had been killed and his men scattered, the soldiers, wearied out with the constant repetition of the same exertions and hard pressed by the Aetolians with their javelins, at last turned and fled.' (Thucydides 3.97–98)

Note that the small force of Athenian archers served to hold the Aetolians at bay for a while, but the Aetolians were in greater numbers and the Athenians lacked a large supporting light infantry force of their own.

Later in the Peloponnesian War, a larger Athenian force in northern Greece met a similar fate in its attack on the Chalcidian town of Spartolus:

'[T]he Chalcidian hoplites, and some auxiliaries [*epikouroi*] with them, were beaten and retreated into Spartolus; but the Chalcidian horse and light troops defeated the horse and light troops of the Athenians. The Chalcidians had already a few peltasts from Crusis, and presently after the battle were joined by some others from Olynthus; upon seeing whom the light troops from Spartolus, emboldened by this accession and by their previous success, with the help of the Chalcidian horse and the reinforcement just arrived again attacked the Athenians, who retired upon the two *taxeis* which they had left with their baggage. Whenever the Athenians advanced, their adversary gave way, pressing them with missiles the instant they began to retire. The Chalcidian horse also, riding up and charging them just as they pleased, at last caused a panic amongst them and routed and pursued them to a great distance.' (Thucydides 2.79.3–6)

So we see the familiar pattern that the light infantry (and cavalry) are not able to influence the battle between the hoplites, but once the Athenians' own supporting light infantry are driven off, their hoplites find themselves defenceless against the enemy light forces.

Xenophon and the Ten Thousand encountered similar situations many times in their retreat through Persian territory – they faced mainly lighter-armed opponents skilled in the use of missiles. For example, early in their retreat after crossing the River Zapatas, their rearguard was threatened by 'about two hundred horsemen and by bowmen and slingers – exceedingly active and nimble troops – to the number of four hundred' (Xen., *Anab.* 3.3.6). This was a very modest force, yet it caused difficulty to the Greeks:

'And the Greek rearguard, while suffering severely, could not retaliate at all; for the Cretan bowmen not only had a shorter range than the Persians, but besides, since they had no armour, they were shut in within the lines of the hoplites; and the Greek javelin-men could not throw far enough to reach the enemy's slingers. Xenophon consequently decided that they must pursue the Persians, and this they did, with such of the hoplites and peltasts as were guarding the rear with him; but in their pursuit they failed to catch a single man of the enemy.' (Xenophon, *Anabasis* 3.3.7–8)

Without cavalry support, and with an inadequately equipped and short-ranged force of light infantry, the Greeks could not defend themselves effectively against their tormentors. But they could also not be defeated, as they continued their march, although taking losses, another reminder of the general inability of light troops to force a decisive conclusion in normal circumstances. Xenophon's solution on this occasion was to find the Rhodians in the army (who presumably had an innate knowledge of use of the sling) and equip them as slingers, and also to gather together all available horses (used as transports or mounts for officers) and equip a small force of cavalry (Xen., *Anab.* 3.3.12–20). Next day, the cavalry charged out to drive off the threatening Persians, while the Rhodian slingers kept them at a safe distance (Xen., *Anab.* 3.4.4; 3.4.15). The importance of supporting light infantry and cavalry for a force of hoplites facing a missile-equipped opponent is clear.

That hoplites would be unable to catch peltasts was not a given, however, and the balance of power could be more evenly matched between the two than these examples suggest. Iphicrates' own peltasts had earlier experience of this.

'Iphicrates and his troops invaded many districts of Arcadia also, where they plundered and made attacks upon the walled towns; for the hoplites of the Arcadians did not come out from their walls at all to meet them; such fear they had conceived of the peltasts. But the peltasts in their turn were so afraid of the Lacedaemonians that they did not approach within a javelin's cast of the hoplites; for it had once happened that the younger men among the Lacedaemonians, pursuing even from so great a distance as that, overtook and killed some of them. But while the Lacedaemonians felt contempt for the peltasts, they felt even greater contempt for their own allies; for once, when the Mantineans went out against peltasts who had sallied forth from the wall that extends to Lechaeum, they had given way under the javelins of the peltasts and some of them had been killed as they fled; so that the Lacedaemonians were even so unkind as to make game of their allies, saying that they feared the peltasts just as children fear hobgoblins.' (Xenophon, *Hellenica* 4.4.16–17)[7]

Even so, the ability of light infantry to overcome heavy – in the right circumstances – was so familiar to Aristotle in the late fourth century that he could even assume their superiority over the hoplites:

'[W]here there is a large multitude of this class [poor light infantry], when party strife occurs the oligarchs [hoplites and cavalry] often get the worst of the struggle; and a remedy for this must be adopted from military commanders, who combine with their cavalry and heavy infantry forces a contingent of light infantry. And this is the way in which the common people get the better over the well-to-do in outbreaks of party strife: being unencumbered they fight easily against cavalry and heavy infantry.'
(Aristotle, *Politics* 1321a)

Aristotle's proposed solution is for the hoplites to recruit light infantry of their own, but also 'for the men of military age to be separated into a division of older and one of younger men, and to have their own sons while still young trained in the exercises of light and unarmed troops' (*ibid.*). Recall that the Athenian *ephebeia* involved educating the young recruits in javelin and bow use, and also that the common solution for hoplite forces facing light infantry was for the youngest age classes to be tasked with running out ahead of the formation to attempt to catch up with the enemy (see examples above).

Because examples in literary histories of such defeats of heavy infantry by light only appear from the late fifth century, there is a tendency to see them as a new development, as the power of light infantry was gradually discovered. The growth in mercenary service of light infantry and particularly peltasts in the fourth century may have increased the experience and confidence levels of light forces and so made such outcomes more frequent, but we must beware of being unduly influenced by the availability of evidence. With almost no literary sources for wars before the mid-fifth century, other than the Persian Wars, the absence of evidence for light infantry's power over heavy in the early fifth century and before cannot be taken as evidence of absence. It may well be that light infantry had always posed the threat it did to heavy ever since light and heavy forces became formally separated at some time during the Archaic development of the phalanx. That this did not matter in many cases was because wars were primarily decided – as they continued to be into the fourth century – by the pitched battle encounter of very large forces of opposed hoplites in phalanx formation. Defeats, however spectacular, of isolated forces of hoplites by light-armed troops might have taken place throughout the period of existence of the hoplite phalanx, but given that Greek cities were contending chiefly against Greek cities, that each (for internal social, economic and political reasons) fielded a hoplite phalanx and that pitched battles between these phalanxes were the means by which disputes were chiefly settled, then light infantry never advanced beyond their peripheral, supporting role in warfare. I will return to this question below.[8]

Cavalry

We have already encountered cavalry in their role alongside light infantry in sometimes inflicting embarrassing defeats on hoplite forces. Greek cavalry was generally fairly lightly equipped. Xenophon, who wrote two treatises specifically aimed at the cavalry (*Horsemanship* and *The Cavalry Commander*), recommended a full set of equipment including helmet, cuirass and 'the so-called arm' (apparently a piece of armour for the left arm, as cavalry did not generally carry shields), with further armour for the horse (Xen., *Horse.* 12.1–10). But depictions of cavalry in art suggest they were much more lightly equipped (since there were, generally speaking, no regulations for equipment that must be carried). Often cavalry are depicted unarmoured, with just tunic, cloak and hat. In terms of offensive armament, cavalry – like light infantry – used javelins (at least they are recommended to do so by Xenophon – 'in place of the spear with a long shaft, seeing that it is both weak and awkward to manage, we recommend rather the two Persian javelins of cornel wood', Xen., *Horse.* 12.12). Also like light infantry, the primary tactic was harassment from a distance. Xenophon notes that 'we recommend throwing the javelin at the longest range possible. For this gives a man more time to turn his horse and to grasp the other javelin' (Xen., *Horse* 12.13); an indication that cavalry would, after throwing a javelin, turn around and ride away while readying for the next throw. With only two javelins, however, this method of fighting could not have been maintained for very long, and Xenophon is not clear what happened once the javelins were expended, other than that cavalry should also carry a sword. He recommends the *machaira*, apparently in this case the same as the curved *kopis*, in preference to the standard *xiphos* sword (Xen., *Horse.* 12.11). At any rate, except against broken or fleeing enemies, such cavalry did not charge into contact with the lance the way later Macedonian cavalry appear to have done, but either took on more of a skirmishing role or were used against enemy light infantry.[9]

Cavalry, like light infantry, had very much a secondary role (or according to many surviving accounts, no role at all) in pitched battles between hoplite armies. What they were useful for was

Greek light cavalry, with two javelins and *petasos* hat.

scouting, raiding and harrying enemy forces, and in the latter role were well suited for defending friendly territory against invasion, as the threat of cavalry attack could prevent infantry armies from scattering across the countryside to engage in plunder and agricultural vandalism. The Athenian cavalry played a central role in defending Attica in this way during the Peloponnesian War, to the extent that it could be claimed that Spartan invasions were completely dependent on the support of allied Boeotian cavalry to keep the Athenian cavalry at bay (Thuc. 4.95.2). The Spartans for their part did not traditionally use cavalry, their own *hippeis*, 'knights', having transformed into an elite body of infantry at an early stage, as we have seen, until they took what was for them the unusual step of raising a force of forty cavalry (and some archers) in 425, to counter the threat of Athenian raids. The Spartan way of raising cavalry is particularly noteworthy, as cavalry were generally drawn from the wealthiest, but as Xenophon records (with reference to the poor performance of the Spartan cavalry at Leuctra), wealthy Spartans only provided the horse, which was ridden by 'weak and unambitious' riders appointed to the task (Xen., *Hell.* 6.4.11).

Despite their potential value, a low opinion of the cavalry was common. In Athens, Lysias records that a man could be charged with cowardice 'because, when it was his duty to share the danger with the infantry, he chose to serve in the cavalry' (Lysias 14.7). Cavalry service was regarded as a soft option, as Lysias records in another case:

> 'I had been enrolled by Orthobulus for service in the cavalry: I saw that it was everyone's opinion that, whereas the cavalry were assured of safety, the infantry would have to face danger; so, while others mounted on horseback illegally, without having passed the scrutiny, I went up to Orthobulus and told him to strike me off the roll, as I thought it shameful, while the majority were to face danger, to take the field with precaution for my own security.' (Lysias 16.13)

The underlying reason for this was that, as with light infantry, cavalry were able to seek safety in flight (or tactical withdrawal) from the enemy, while hoplites were required to stand their ground (and would likely be cut down if they did try to run).[10]

So cavalry were not highly regarded, and their main use was in irregular actions outside of pitched battles. Yet even in pitched battle, their intervention could be valuable. For example, at Solygeia (425), Athenian and Corinthian hoplite forces engaged with no decisive result at first:

> 'After holding on for a long while without either giving way, the Athenians aided by their cavalry, of which the enemy had none, at length routed the

Corinthians, who retired to the hill and halting remained quiet there, without coming down again.' (Thucydides 4.44.1)

On this occasion the Athenians had gone to the trouble of bringing cavalry with them in horse transports, while the Corinthians, despite being in their own territory, appear – like the Spartans – not to have had any native cavalry. Exactly what form the aid provided by the Athenian cavalry took, Thucydides does not record; perhaps they harried the flanks and rear of the Corinthian hoplites, since generally speaking, hoplites who stood their ground firmly had little to fear from cavalry engaging them frontally. The ability of cavalry to attack infantry will form part of the subject of the next chapter, but hoplites who stood their ground could expect to repulse cavalry with relative ease. This meant that cavalry would only be useful in battle as an adjunct to a hoplite phalanx, rather than as a force in their own right.

That cavalry did not pose a great threat to hoplites that stood their ground was something that later writers rather take for granted, but which was clearly not yet firmly established at the time of the Persian Wars, when the Persian cavalry was feared. Nevertheless, despite the Persians' strength in cavalry, this arm features little in the invasions of either Darius or Xerxes. That cavalry were mysteriously absent from the initial Persian defeat at Marathon is well known, and has been the subject of much debate and speculation. The Persians apparently had cavalry with them, yet they do not feature at all in Herodotus' description of the battle. It may be, as modern authors have suggested, that the Persian cavalry had already been re-embarked on the Persian fleet at the time the battle began, and indeed that the Greeks timed their advance precisely to take advantage of this. According to Herodotus, the Persians thought the Athenians 'absolutely crazy, since they saw how few of them there were and that they ran up so fast without either cavalry or archers' (Hdt. 6.112.2), since Persian armies were combined-arms forces with a balanced mix of cavalry, light and heavy infantry. At any rate, the fighting as described by Herodotus was a pure infantry concern.[11]

In Xerxes' invasion a decade later, the initial fighting in the narrows of Thermopylae was not well suited for cavalry, but the broad plains of Plataea, where the decisive battle of the campaign was fought, provided ideal cavalry country, and the Persian cavalry played a significant role in the build-up to the battle. In the initial skirmishing:

'Mardonius sent against [the Greeks] his entire cavalry ... Thereupon the horsemen rode up to the Greeks and charged them by squadrons; as they attacked, they did them much hurt.' (Herodotus 9.20)

Charging 'by squadrons [*kata telea*]' suggests that the Persian cavalry (who like Greeks were primarily javelin-throwers) used a succession of charges and withdrawals, riding up, hurling javelins (and insults) then falling back out of range of retribution, a standard operating procedure for cavalry across the ages in the face of steadfast infantry. Only if the infantry broke formation or tried to run away might the cavalry charge in amongst them and cut them down with their swords.

On this occasion, the Megarian contingent bore the brunt of the cavalry attacks and sent a messenger to the Greek commander, Pausanias, to report that if they were not relieved, they would abandon their position. They were then replaced by the Athenian 300 picked men, presumably an elite force (Hdt. 9.21.3) who, crucially, were supported by a force of archers. The Persians continued to charge by squadrons, but the archers gave the Athenians the ability to hit back, and the horse of the Persian commander Masistius was shot from under him, leaving him to be killed on the ground. The Persian cavalry 'knew nothing of this, for they had not seen him fall from his horse, or die. They wheeled about and rode back without perceiving what was done' (Hdt. 9.22.3). Once the cavalry realized their commander was missing, they massed to try to recover his body and attacked 'not by squadrons as before, but all together' (Hdt. 9.23.1), but in the ensuing melee the Persians were driven off when Greek reinforcements arrived. The Greeks 'were greatly encouraged that they withstood and drove off the charging horsemen', as Herodotus tells us (9.25.1), as this greatly increased their confidence in their ability to withstand cavalry attack. Yet over the following days, Persian cavalry continued to harass the Greek army in its position in the plain: 'The horsemen rode at them and shot arrows and javelins among the whole Greek army to its great hurt, since they were mounted archers and difficult to deal with' (9.49.2), leading to the Greek decision to withdraw to higher ground (and the insubordination of Amompharetus, who refused to do so). It was Persian cavalry who caught up with and attacked Pausanias' retreating Spartans (9.60.1), even though the Spartan hoplites were accompanied by a supposedly vast force of light infantry. In the ensuing fighting, when the Spartans and Athenians turned on and defeated their tormentors, there is strangely no direct mention in Herodotus of the Persian cavalry, as if having harassed and halted the Greek formations, they were powerless to take any further action against them, and it was infantry fighting that decided the battle. The cavalry subsequently served only to protect the Persian survivors as they fled (9.68).[12]

That cavalry fought, like light infantry, with repeated alternating advances and withdrawls set them apart from the steadfast hoplite, who was required and expected to stand his ground. Plato records a Socratic dialogue on the nature of courage:

'Socrates: Let us take that man to be courageous who, as you describe him yourself, stays at his post and fights the enemy.'

'Laches: I, for one, agree to that.'

'Socrates: Yes, and I do too. But what of this other kind of man, who fights the enemy while fleeing, and not staying?'

'Laches: How fleeing?'

'Socrates: Well, as the Scythians are said to fight, as much fleeing as pursuing; and as you know Homer says in praise of Aeneas' horses, that they knew "how to pursue and to flee in fright full swiftly this way and that way" and he glorifies Aeneas himself for this very knowledge of fright, calling him "prompter of fright".'

'Laches: And very properly too, Socrates; for he was speaking of chariots; and so are you speaking of the mode of the Scythian horsemen. That is the way of cavalry fighting but with hoplites it is as I state it.'

(Plato, *Laches* 191a–b)

Socrates and Laches agree that cavalry who fight this way display courage too, and that there are different types of courage appropriate to the circumstances, so it is not simply the case that Greeks looked down on those who 'fought by flight' (Socrates records the curious assertion that the Spartan hoplites fought this way at Plataea, as we will see in the next chapter). According to this account, a hoplite stood his ground because that was the form of fighting appropriate to the phalanx, rather than because it was necessarily a superior form of courage. Herodotus is happy to count the Persian infantry and the Sacae (Scythian) cavalry as the most courageous on the Persian side at Plataea (9.71.1).

Another important role for cavalry might be in keeping light infantry away from the vulnerable phalanx, since cavalry, with their greater speed, could charge down and catch light infantry before they could run to safety. However, the mere presence of cavalry was not in itself a guarantee of safety, as Xenophon records of the Spartan hoplites' defeat by light infantry at Lechaeum. At first the hoplites tried to catch the peltasts themselves, but were shot down as they returned to the formation:

'And now that the best men had already been killed, the horsemen joined them, and with the horsemen they again undertook a pursuit. But when the peltasts turned to flight, at that moment the horsemen managed their attack badly; for they did not chase the enemy until they had killed some of them, but both in the pursuit and in the turning backward kept an even front with the hoplites.' (Xenophon, *Hellenica* 4.5.16)

By tying themselves too closely to the hoplites they were escorting, the cavalry surrendered their advantage of speed, on the face of it a remarkable failure, but we must remember that Spartan cavalry were not well regarded and were doubtless not enthusiastic about riding ahead against peltasts on their own.

In the early fourth century, the Spartan king Agesilaus campaigned in Asia Minor and frequently faced forces of Persian cavalry. Here, the nature of the ground was crucial, for cavalry – more so than hoplites – required open unencumbered terrain to operate successfully, so that the Persian commander Tissaphernes advanced aggressively to catch the Spartans in open ground before they reached unsuitable cavalry country (Xen., *Hell.* 3.4.12). Agesilaus had a small accompanying force of cavalry of his own, but the morning after this force was defeated in a skirmish with the Persians, the sacrifices turned out unfavourable; 'this sign having presented itself, he turned and marched to the sea' (Xen., *Hell.* 3.4.15), with some relief, we may suppose. Agesilaus knew what was required:

> 'Perceiving that, unless he obtained an adequate cavalry force, he would not be able to campaign in the plains, he resolved that this must be provided, so that he might not have to carry on a skulking warfare. Accordingly he assigned the richest men of all the cities in that region to the duty of raising horses; and by proclaiming that whoever supplied a horse and arms and a competent man would not have to serve himself, he caused these arrangements to be carried out with all the expedition that was to be expected when men were eagerly looking for substitutes to die in their stead.' (Xenophon, *Hellenica* 3.4.15)

This is not a dissimilar arrangement to the way Spartan cavalry was raised at home, though in this case the provision of professional riders worked better than relying on the wealthy. This improved force soon proved its worth when the Persian cavalry attacked a scattered force of Greeks out pillaging:

> 'Therefore, after offering sacrifice, [Agesilaus] at once led his phalanx against the opposing line of horsemen, ordering the first ten year-classes of the hoplites to run to close quarters with the enemy, and bidding the peltasts lead the way at a double-quick. He also sent word to his cavalry to attack, in the assurance that he and the whole army were following them. Now the Persians met the attack of the cavalry; but when the whole formidable array together was upon them, they gave way, and some of them were struck down at once in crossing the river, while the rest fled on.' (Xenophon, *Hellenica* 3.4.23–24)

The Greeks plundered the Persian camp, capturing among other things a number of camels, which they took back with them to Greece. On this occasion we see once again the use of the younger men in a more mobile role, facing cavalry this time, and the importance of a combined-arms force with hoplites, light infantry and cavalry, when facing a mobile opponent. Note also that, rather than hoplites requiring a flat and open plain on which to operate, they actively avoided such terrain when facing cavalry and kept instead to the hills.[13]

Greek cavalry were not just used outside pitched battle and to drive off light infantry – they also had a role in hoplite versus hoplite battles, albeit a small one, which I will examine in the next chapter. It was the Macedonians, under Philip II and his son Alexander (the Great), who discovered a major role for cavalry in pitched battle, leading to the overthrow of the power of the Greek city states and their hoplites and the dominance of Macedon, which will form part of the subject of Chapter 9.

Rations and logistics

The proper home of a hoplite phalanx was the field of pitched battle, as we will see in the next chapter, and the phalanx was the dominant force on the battlefield, whatever the importance of light infantry and cavalry in other operations. But in order to get to the battlefield, a hoplite phalanx, like any other army, had to march, eat, pass defended obstacles, negotiate terrain and arrive in good fighting order on the chosen field. In other words, it had to conduct successful campaigns. All aspects of operational and strategic-level warfare are of course too large a subject for a single book, let alone chapter, so I will here attempt no more than to sketch some aspects of the tasks facing hoplites on campaign, my intention being simply to draw attention to all those forms of warfare outside battle of which they had to be capable.[14]

As we saw in Chapter 4, hoplite armies may or may not have received pay, depending on the arrangements in force at the particular time and place, but on the whole they were expected to provide their own provisions (though this too could vary). The way this was done was that when the hoplite was first called up (his name included on the muster lists for the coming campaign, for example), he would also be instructed to provide a certain amount of rations (expressed as rations sufficient for a given number of days). This initial supply would not be expected to cover the entire duration of the campaign, and for the rest of the time the commander was expected to provide markets at which the soldier could buy his own food. Note that supply was outsourced in this way to third parties, rather than the commander directly providing food, although there were of course exceptions when particular stockpiles of food might be made available to an army, sometimes in an effort to buy its goodwill, and often in the form of supplies plundered from enemy territory.

The *Cyropaedia*, as usual, provides Xenophon's ideal picture of how an army should be provisioned. Cyrus instructed his men to bring along food sufficient for twenty days, along with wine, though 'only enough to last till we accustom ourselves to drinking water'. Greeks generally drank water mixed with wine, which is safer given the unknown and at the time unknowable microbes lurking in the water, and Xenophon envisages the army gradually weaning itself off wine (Xen., *Cyrop.* 6.2.25–26). After this twenty-day period, merchants accompanying the army would be allowed to sell provisions to the men (Xen., *Cyrop.* 6.2.38–39) – and not before, presumably so that nobody was tempted to skimp on rations.

In practice, things might not always work out smoothly. Xenophon records supply difficulties for the Ten Thousand in Cyrus the Younger's army on the march into Persia:

> 'As for the troops, their supply of grain gave out, and it was not possible to buy any except in the Lydian market attached to the barbarian army of Cyrus [at greatly inflated prices] … The soldiers therefore managed to subsist by eating meat.' (Xenophon, *Anabasis* 1.5.6)

Note that meat was little eaten (and would have posed great practical difficulties on a summer campaign). The experience of the Ten Thousand after they arrived back at the friendly city of Byzantium was no better, as the Spartan admiral Anaxibius was supposed to provide for them:

> 'Anaxibius would not give them pay, but made proclamation that the troops were to take their arms and their baggage and go forth from the city, saying that he was going to send them back home and at the same time to make an enumeration of them. At that the soldiers were angry, for they had no money with which to procure provisions for the journey, and they set about packing up with reluctance.' (Xenophon, *Anabasis* 7.1.7)

Once outside the city, Anaxibius shut the gates against them:

> 'Then Anaxibius called together the *strategoi* and *lochagoi* and said: "Get your provisions from the Thracian villages; there is an abundance there of barley and wheat and other supplies; when you have got them, proceed to the Chersonese, and there Cyniscus will take you into his pay."' (Xenophon, *Anabasis* 7.1.13)

Hearing this, the men mutinied and threatened to take the city by force. Evidently being told to live off the land was not sufficient – the men expected ration money to pay for supplies at market in the proper way.

Hoplites would therefore have needed expenses money or travel money to live on while they were on campaign (even, paradoxically, if they were paid for their service), which might be a problem for some. This also meant that armies on campaign would be carrying around a significant quantity of cash with them, and not, so far as we know, in centrally guarded strongboxes but among each man's individual possessions. This explains why a fortified camp could be of such importance (see below), along with slaves or attendants to guard it while the hoplite was out fighting, and why an army might expect to benefit greatly from plundering a defeated enemy's camp. After the defeat of the Athenian expedition to Syracuse, the 6,000 men who surrendered were carrying enough silver coins to fill the bowls of four upturned shields (Thuc. 7.82.3).

Armies could subsist off the land if they had to by seizing stockpiles of food or perhaps by harvesting food directly from enemy fields themselves, though this latter course surely cannot have been the norm, since it required precise and limited timing to catch the harvest at the right moment (and before it had been harvested by its owners), while gathering and threshing grain must have been enormously time-consuming. In the limited campaigns that characterized the early Peloponnesian War, where a Spartan army made a strictly time-limited invasion of Attica each summer, this may have happened more often. The invasion of Attica in 425, for example, had to be called short:

> '[H]aving made their invasion early in the season, and while the corn was still green, most of their troops were short of provisions: the weather also was unusually bad for the time of year, and greatly distressed their army. Many reasons thus combined to hasten their departure and to make this invasion a very short one; indeed they only stayed fifteen days in Attica.' (Thucydides 4.6.1–2)

If an army was to carry provisions on the march, even if only for a short period, it would need transportation of some sort, since twenty days' provisions would be too much for one man to carry on his back along with his other equipment. Pack animals and wagons were both used – 400 wagons carried emergency rations for the Ten Thousand on their march to Cunaxa (Xen., *Anab.* 1.10.18). Wagons can carry a greater weight but require smooth going, ideally a road, while pack animals carry less but are more mobile. In both cases, there is the additional complication that the animals themselves, whether pulling carts or carrying packs, had to eat, so provisions had to be provided for them too, or foraged from the country, so simply increasing the number of animals did not increase the range of an army or solve its supply problems. There are a number of occasions where armies deliberately travelled light by reducing their dependence on wagons in particular, as Xenophon advised the Ten Thousand:

> 'I think we should burn up the wagons which we have, so that our cattle may not be our captains, but we can take whatever route may be best for the army. Secondly, we should burn up our tents also; for these, again, are

a bother to carry, and no help at all either for fighting or for obtaining provisions.' (Xenophon, *Anabasis* 3.2.27)

'Our cattle may not be our captains' translates more literally as 'our yoked animals may not be our generals' – 'so that we might not be led by donkeys', we might say. Even this reduction was not sufficient for the Ten Thousand in its long overland march in enemy territory, and a short while later they resolved on even more drastic steps, reducing the baggage animals further and releasing the captives they had taken:

> 'For the baggage animals and the captives, numerous as they were, made the march slow, and the large number of men who had charge of them were thus taken out of the fighting line; besides, with so many people to feed it was necessary to procure and to carry twice the amount of provisions.' (Xenophon, *Anabasis* 4.1.13)

In the fourth century, one of the key reforms of the Macedonian army under Philip II was to increase its mobility by similarly reducing its, particularly wheeled, baggage.[15]

In terms of what was eaten, aside from harvested crops or, if desperate, meat, the meals provided by the common messes (*syssitia*) in Sparta perhaps give a picture of a common military diet:

> 'They met in companies of fifteen, a few more or less, and each one of the mess-mates contributed monthly a bushel of barley-meal, eight gallons of wine, five pounds of cheese, two and a half pounds of figs, and in addition to this, a very small sum of money for such relishes as flesh and fish. Besides this, whenever any one made a sacrifice of first fruits, or brought home game from the hunt, he sent a portion to his mess.' (Plutarch, *Lycurgus* 12.2)

As well as food, an army on the march requires water, with even greater urgency, and availability of water must have moulded the course of many campaigns, for all that our sources often skip over this aspect. The preliminaries to the Battle of Plataea (479) were dominated by the water needs of both armies. The Greeks decided 'they would march down to Plataea, for they saw that the ground there was generally more suited for encampment than that at Erythrae, and chiefly because it was better watered' (Hdt. 9.25.2). The attempted Greek withdrawal before this battle was triggered when the Persians 'spoiled and blocked the Gargaphian spring, from which the entire Greek army drew its water', the Greeks also being at this time 'barred from the [River] Asopus, not being able to draw water from that river because of the horsemen and the arrows' (Hdt.

9.49.2–3) – an important role for cavalry, as we have seen, being to deny less mobile armies the opportunity to spread out to collect food or water. Similarly, the Athenian general Nicias, informing home of his difficulties before Syracuse in the Peloponnesian War, complained that 'expeditions for fuel and for forage, and the distance from which water has to be fetched, cause our sailors to be cut off by the Syracusan cavalry' (Thuc. 7.13.2). Availability of water placed a limit on the movements, actions and size of all armies of this period, and Greek armies were no exception. Campaigning in a hot dry climate in summer must have made problems posed by the availability of water particularly acute.[16]

Marching

A hoplite phalanx is a battle formation (as I will examine in more detail in the next chapter), but before hoplites could get to battle, they would have to march, and as with soldiers of all other eras, hoplites would have spent far more of their time marching than they would fighting. Our sources have far more to say about hoplites in battle (though that is still little enough) than they do about hoplites on the march. As is often the case, the best evidence comes from the fourth century and from Xenophon, so we cannot be certain how typical what he describes is of practice in the fifth century or earlier.

What we do see in Xenophon is that Greek armies knew the importance of a fixed marching order and formation, and of maintaining good march discipline:

> '[A]n army in disorder is a confused mass, an easy prey to enemies, a disgusting sight to friends and utterly useless – donkey, trooper, carrier, light-armed, horseman, wagon, huddled together. For how are they to march in such a plight, when they hamper one another, some walking while others run, some running while others halt, wagon colliding with horseman, donkey with wagon, carrier with trooper? If there is fighting to be done, how can they fight in such a state? For the units that must escape by flight when attacked are enough to trample underfoot the hoplites.'
> (Xenophon, *Economics* 8.4–6)

Xenophon also recounts how the Greek practice was to order the march according to the capabilities of the different arms:

> 'And Xenophon said: "Well, now, consider this point, whether, if we are to make a night march, the Greek practice is not the better: in our marches by day, you know, that part of the army takes the lead which is suited to the nature of the ground in each case, whether it be hoplites or peltasts or cavalry; but by night it is the practice of the Greeks that the slowest arm

should lead the way; for thus the various parts of the army are least likely to become separated, and men are least likely to drop away from one another without knowing it; and it often happens that scattered divisions fall in with one another and in their ignorance inflict and suffer harm.'" (Xenophon, *Anabasis* 7.3.37–38)

In the *Cyropaedia*, Xenophon elaborates on the ideal order of march, spreading the baggage train out 'many lines of wagons and pack-animals abreast' where the nature of the ground allowed it (which would make for a shorter line, front to back, and so faster progress), or where the country was more narrow, flanking the baggage with columns of hoplites (Xen., *Cyrop.* 6.3.2–3). Xenophon also records that each *taxis* marched alongside its own baggage, guided by a flag or standard (*semeia*), so that equipment was always ready to hand when it was needed (Xen., *Cyrop.* 6.3.4).

This was the ideal, but the march of the younger Cyrus to Cunaxa (401) showed that it did not always work out so smoothly.

'But since the King did not appear at the trench and try to prevent the passage of Cyrus' army, both Cyrus and the rest concluded that he had given up the idea of fighting. Hence on the following day Cyrus proceeded more carelessly; and on the third day he was making the march seated in his chariot and with only a small body of troops drawn up in line in front of him, while the greater part of the army was proceeding in disorder and many of the soldiers' arms and accoutrements were being carried in wagons and on pack-animals.' (Xenophon, *Anabasis* 1.7.19–20)

Protecting the baggage within a column of hoplites was a common practice, to guard against surprise attacks. The ultimate extension of this, in open country, was to march in a vast hollow square with the baggage (and any other forces that needed protecting) in the centre. This is first attested under the Spartan commander Brasidas in the fifth century, in terms that suggest that this was not an innovation:

'At daybreak Brasidas, perceiving that the Macedonians had gone on, and that the Illyrians and Arrhabaeus were on the point of attacking him, formed his hoplites into a square, with the light troops in the center, and himself also prepared to retreat. Posting his youngest soldiers to dash out wherever the enemy should attack them, he himself with three hundred picked men in the rear intended to face about during the retreat and beat off the most forward of their assailants.' (Thucydides 4.125.2–3)

Note the preparation of younger men to charge out against attacking light infantry and the special role for the picked men. A similar formation was used for the Athenian retreat from Syracuse:

'The army marched in a hollow square, the division under Nicias leading, and that of Demosthenes following, the hoplites being outside and the baggage-carriers and the bulk of the army in the middle.' (Thucydides 7.78.2)

Agesilaus also used this formation in his campaigns in Asia and northern Greece (Xen., *Hell.* 4.3.4). Xenophon provides a detailed account of some of the difficulties of marching in square:

'Then it was that the Greeks found out that a square is a poor formation when an enemy is following. For if the wings draw together, either because a road is unusually narrow or because mountains or a bridge make it necessary, it is inevitable that the hoplites should be squeezed out of line and should march with difficulty, inasmuch as they are crowded together and are likewise in confusion; the result is that, being in disorder, they are of little service. Furthermore, when the wings draw apart again, those who were lately squeezed out are inevitably scattered, the space between the wings is left unoccupied, and the men affected are out of spirits when an enemy is close behind them. Again, as often as the army had to pass over a bridge or make any other crossing, every man would hurry, in the desire to be the first one across, and that gave the enemy a fine chance to make an attack.' (Xenophon, *Anabasis* 3.4.19–20)

The solution, as we saw in Chapter 3, was to improve the internal organization of the units of hoplites and appoint junior commanders, so that they could use formal drills to expand and contract their frontage in an orderly fashion (Xen., *Anab.* 3.4.20–23).

The importance of maintaining good order is repeatedly stressed, yet hoplites would not have marched in the same close order in which they fought (which would have been too difficult to maintain over varied terrain and for long distances), but in a more open order. The later Macedonian phalanx probably operated with its most open order, four cubits (2 metres) per man, when marching, closing up to close order only to fight, and it is likely that hoplites did the same (assuming they marched in formation at all), although the intervals used and the drills for switching between them may have been more loosely specified.

In terms of distances covered, this could of course vary greatly according to the size of force, terrain and (perhaps above all) need of haste, but some impressive distances are mentioned. The Spartans, having failed to respond (for religious reasons) to the Athenians' request for aid at Marathon (490), did eventually send a contingent after the battle was over:

'After the full moon two thousand Lacedaemonians came to Athens, making such great haste to reach it that they were in Attica on the third day after leaving Sparta. Although they came too late for the battle, they desired to see the Medes, so they went to Marathon and saw them.' (Herodotus, 6.120.1)

According to Isocrates (Isoc. 4 87), this was a distance of 1,200 stadia, about 216 kilometres, in 'three days and as many nights'.[17]

Herodotus reckoned on an average speed of around 26km per day to cover the distance from the sea to the Persian capital at Susa in three months along the Persian royal road (Hdt. 5.52–54), and while this sort of continuous marching would not have been practical for an army, the distance covered in a day is more achievable than such extreme exertions as the Spartan march to Marathon. On the approach to Cunaxa (401), Cyrus' army covered around 30km per day, with rest days on average about every other day (rest days would be necessary to rest and feed baggage animals as well as to gather supplies), so a maximum daily distance of around 26km and an average of about half that may be a reasonable estimate for normal army movements.[18]

Camping

When not marching, foraging or fighting, an army would spend its time encamped. Unlike the Romans, who are famous for their orderly and regimented camps, Greeks had a rather more haphazard approach, but we should not imagine that Greek armies did not build properly fortified and laid-out camps, up to a point. The difference in approaches is pointed out by Polybius (who is talking of later Hellenistic Greek practice, but we can probably assume that earlier Greek camps were similar):

> 'The Greeks in encamping think it of primary importance to adapt the camp to the natural advantages of the ground, first because they shirk the labour of entrenching, and next because they think artificial defences are not equal in value to the fortifications which nature provides unaided on the spot. So that as regards the plan of the camp as a whole they are obliged to adopt all kinds of shapes to suit the nature of the ground, and they often have to shift the parts of the army to unsuitable situations, the consequence being that everyone is quite uncertain whereabouts in the camp his own place or the place of his unit is. The Romans on the contrary prefer to submit to the fatigue of entrenching and other defensive work for the sake of the convenience of having a single type of camp which never varies and is familiar to all.' (Polybius 6.42)

Xenophon has little to say on the Spartan methods of building camps, noting only that 'seeing that the angles of a square are useless, [Lycurgus] introduced the circular form of camp, except where there was a secure hill or wall, or a river afforded protection in the rear' (Xen., *Const. Lac.* 12). This reinforces Polybius' point about the use of natural defences and suggests that non-Spartan camps might adopt square or rectangular shapes, despite the problem of the corners (whether this means that the corners provided an awkward space on the inside, or that the corners were a weak point against attack, is not clear). Xenophon also reveals that Spartans posted sentries facing inward to watch the helots, which was a practice particular to them, though sentries facing outwards in the normal way were common to all Greek armies.[19]

Once again, Greeks were not unique in their use of fortified camps. Earlier Assyrian and contemporary Persian camps were defended by a ditch:

> '[T]he Assyrians and their allies drew a ditch around their camp, as even to this day the barbarian kings do whenever they go into camp; and they throw up such entrenchments with ease because of the multitude of hands at their command. They take this precaution because they know that cavalry troops – especially barbarian cavalry – are at night prone to confusion and hard to manage.' (Xenophon, *Cyropaedia* 3.3.26)

Persians fleeing the defeat at Plataea sought refuge inside the 'wooden walls' of the camp (Hdt. 9.65.10), which had been built using the cut-down trees of the Persians' Theban allies (9.15.2), further strengthened with towers (9.70.1). According to Herodotus, the Spartans were unable to make any headway against these defences as they lacked any sort of siegecraft, and the wall was only breached when the Athenians arrived, with 'valour and constant effort' (9.70.2) – and, we must assume, ladders or makeshift rams.

Although they did not like digging ditches, Greek armies defended themselves where necessary with a similar makeshift palisade of felled trees. Xenophon records this of the Thebans, writing that 'wherever the Thebans encamped they at once threw down in front of their lines the greatest possible quantity of the trees which they cut down, and in this way guarded themselves' (Xen., *Hell.* 6.5.30). A ditch might be added by an army that felt vulnerable, such as the Spartan force of Praxitas before Corinth in 392, which 'thought themselves to be few in number, and therefore made a stockade and as good a trench as they could in front of them' (Xen., *Hell.* 4.4.9).

Inside the camp, men might sleep in tents or on the ground in the open. The Ten Thousand crossed the Euphrates on rafts made by stuffing their tent covers with hay (Xen., *Anab.* 1.5.10), but Xenophon later recommended (*Anab.* 3.2.27) that they should also burn their tents, as we saw above, which would only be practical in a warm dry Mediterranean or Near Eastern summer. Soldiers in more permanent camps, like the Athenians besieging Syracuse, may have used more solidly constructed huts, such as those burned by the Syracusans at Catana (Thuc. 6.75.2), though these could also have been tents as the same word is used as by Xenophon.[20]

Raids, sieges and defensive positions

The stereotypical view of Greek hoplite armies is that they required flat open ground on which to operate, and that they deliberately sought out such ground on which to fight. This is the picture of Greek warfare that Herodotus puts in the mouth of an adviser to Xerxes:

> 'When they have declared war against each other, they come down to the fairest and most level ground that they can find and fight there.' (Herodotus 7.9.2)

Yet Greece is a generally mountainous country in which the flat plains are more the exception than the rule, and Greek armies fought from Italy to Persia on terrain of all kinds, so it cannot be the case that Greek armies were incapable of operating on steep, rough or broken ground. The effect of terrain on pitched battles will be considered further in the next chapter, but the broken terrain of Greece also offered opportunities for all sorts of combat other than pitched battles on level terrain. Histories, ancient and modern, are dominated by the great set-piece battles like Marathon, Plataea, Mantinea or Delium, but Greek cities could and did fight each other in many and varied ways.[21]

It was possible for armies to avoid battle completely, either (as the attacker) by engaging in raids and skirmishes, or (as the defender) by holding defensive terrain or fortified positions, setting ambushes or simply staying within the safety of the city walls and refusing to come out to fight. The standard view of Greek warfare is embodied in the Spartan annual invasions of Attica in the opening years of the Peloponnesian War: each summer, the Spartan army would march into Athenian territory, hoping to tempt the Athenian army to come out of their city and give battle, or if they failed to do so, to ravage Athenian territory. The Athenians, contrary to (what is assumed to be) usual practice, remained within their walls, including specially constructed 'Long Walls' linking the city to the harbour at Piraeus which, by permitting the importing of food by sea, allowed the Athenians to feed themselves despite the destruction wrought on their agriculture.

It is thought that this 'Periclean strategy' marked a departure from the previous practice of giving battle in the open (as suggested by Herodotus above). This notion has come under attack from several directions in modern scholarship. It appears that armies would routinely only offer or accept battle if they felt they had a good chance of winning the ensuing clash (not surprisingly), and that a city that felt itself outmatched (in number or quality of hoplites) would routinely refuse an offer of battle. While there was some loss of face and honour in refusing battle this way, that could be more than outweighed by the pleasures of remaining alive, and it appears that an outmatched opponent had less to lose from refusing a suicidal resistance, as the Athenian envoys told the Melians when they had suggested that it would be a sign of baseness and cowardice not to resist Athenian imperialism:

> 'Not if you are well advised, the contest not being an equal one, with honor as the prize and shame as the penalty, but a question of self-preservation and of not resisting those who are far stronger than you are.' (Thucydides 5.101)

Doubt has also been cast on the amount of damage that an invading army could actually do to the defender's agriculture, since it would take a considerable commitment of time and effort to destroy olive groves or vineyards, and while ripe crops could be trampled (or pre-emptively harvested), no lasting damage would be caused by this. We also, as usual, do not have the evidence for the period before the Peloponnesian War for how unusual it was to avoid offers of battle. The Greek lack of capacity for siege warfare, which made it possible to avoid battle by remaining within the walls, can also be overstated, since Greek armies could and did take defended fortified cities (and means other than direct assault were available, most commonly taking advantage of the endless internal political divisions from which every city suffered). Protracted formal sieges, such as that of Plataea, were possible, as were *coups de main* taking advantage of internal

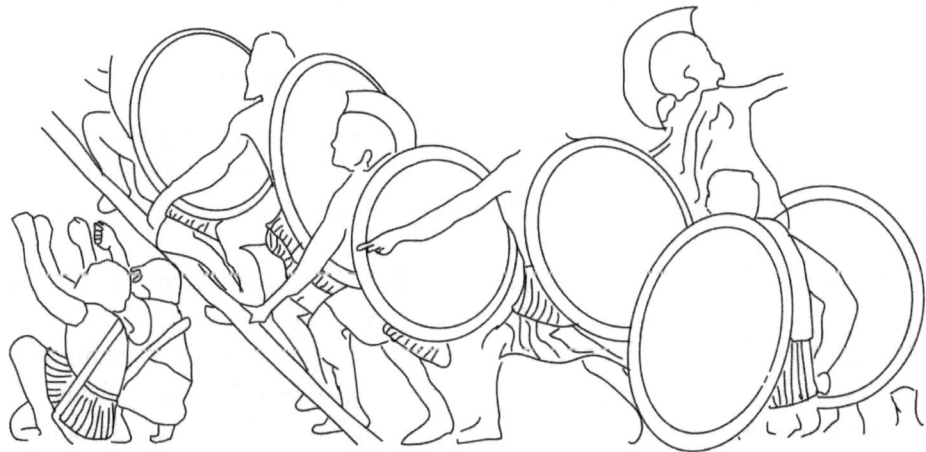

Hoplites attack a city using ladders, with covering fire from archers, Nereid Monument. (*British Museum*)

treachery, and the Spartan reluctance to engage in siege warfare may be more to do with their own peculiar internal arrangements (and reluctance for their army to be absent for extended periods) and not be typical of other cities. Siege warfare as such is outside the scope of this book, although the main workforce and assault troops in a siege would of course be hoplites.[22]

At the same time, where circumstances were right, both sides felt they had a reasonable chance of winning and the loss of agricultural produce (or rich estates) was too high a price to pay, it is clear that a common Greek practice was indeed to accept an offer of battle and fight a (reasonably) fair and open battle on (fairly) level ground. Yet other options were available, and were sometimes adopted by both attackers and defenders.

One option was that both sides could simply not offer battle, and instead engage in mutual raids and forays. On a grand scale, this was how the Peloponnesian War developed, with the Spartans stymied by Athenian refusal to fight in Attica, and the Athenians countering with seaborne raids around the Peloponnese, playing on the Spartans' fear of their own helot population. Between smaller cities which were closer neighbours, this was probably a common form of warfare, and that it is not reported often in our sources may simply be because historians did not find this sort of warfare interesting. Thucydides records one such campaign between neighbours:

> 'During this winter [419/418] hostilities went on between the Argives and Epidaurians, without any pitched battle taking place, but only forays and ambuscades, in which the losses were small and fell now on one side and now on the other. At the close of the winter, towards the beginning of spring, the Argives went with scaling-ladders to Epidaurus, expecting to find it left unguarded on account of the war and to be able to take it by assault, but returned unsuccessful. And the winter ended, and with it the thirteenth year of the war ended also.' (Thucydides 5.56.4)

Hoplites would have played just as much a part in such fighting as in pitched battle, since it was not only light infantry that could mount 'forays and ambuscades', and assaults (or in this case an attempted *coup de main*) against a town were very much the realm of the hoplite.

Another option is that the defender or weaker party might attempt to defend its frontiers against an attacker, rather than fighting a battle on the plains. This was quite commonly done in Greek warfare, and we often hear of attempts to hold passes (since in mountainous country, passes represent the obvious defensive chokepoints), along with some evidence of attempts to establish fortified frontiers as, perhaps, around Attica. Xenophon records Socrates recommending the defence of Attica's frontiers to Pericles (Xen., *Mem.* 3.5.25–27; note that it is 'active young

Athenians, more lightly armed' that he recommends using). Simply holding the passes could be enough to deter a (perhaps not very enthusiastic) attacker, as in the Spartan campaign of 376 against the Theban-Athenian alliance:

> 'The Lacedaemonians, however, when spring was just beginning, again called out the ban and directed Cleombrotus to take command. Now when he arrived at Cithaeron with the army, his peltasts went on ahead for the purpose of occupying in advance the heights above the road. But some of the Thebans and Athenians who were already in possession of the summit allowed the peltasts to pursue their ascent for a time, but when they were close upon them, rose from their concealment, pursued them, and killed about forty. After this had happened, Cleombrotus, in the belief that it was impossible to cross over the mountain into the country of the Thebans, led back and disbanded his army.' (Xenophon, *Hellenica* 5.4.59)[23]

But we hear of such defensive efforts as failed attempts as much as successful defences. Cleombrotus himself had enjoyed better results a short time earlier:

> 'Now the road which leads through Eleutherae was guarded by Chabrias with peltasts of the Athenians; but Cleombrotus climbed the mountain by the road leading to Plataea. And at the summit of the pass his peltasts, who were leading the advance, found the men who had been released from the prison, about one hundred and fifty in number, on guard. And the peltasts killed them all, except for one or another who may have escaped; whereupon Cleombrotus descended to Plataea, which was still friendly.' (Xenophon, *Hellenica* 5.4.14)

There are numerous problems with holding frontiers, particularly for relatively small militia armies. One is that the enemy can simply find some other route into the disputed territory, as happened in the Leuctra campaign:

> 'Cleombrotus did not enter Boeotia from Phocis at the point where the Thebans expected him to enter and where they were keeping guard at a narrow pass; but proceeding by way of Thisbae along a mountainous and unexpected route, he arrived at Creusis.' (Xenophon, *Hellenica* 6.4.3)

Armies could also be transported around an obstacle by sea (more on this below), thus circumventing the defences completely. Even if the held position is attacked frontally, it is fairly easy to turn any defended pass by finding a smaller alternative route through the hills to either side – as indeed happened in the most famous example of an attempt to block a pass, the defence of Thermopylae against Xerxes' invasion in 480. Here, it was only possible to hold the narrows at all so long as the Greek fleet could prevent the Persian ships from simply carrying a landing force behind the defences, and without naval support it is doubtful whether holding the pass would have been attempted. Also, the intention was probably never to hold the pass indefinitely against the Persian army but rather to delay its advance – ideally for long enough to give the Greek fleet a chance to defeat its opposite number. After a few failed attempts to force the pass by frontal assault, the Persians did what any attacker would do in the circumstances and found another route round, at which the Greeks had to abandon the position, except for the Spartan and Tegean rearguard. Greek history is full of such attempts to hold passes, and frequently they end the same way: three times just at Thermopylae, with the attackers (Persians, Gauls and Romans) using the same route to bypass the position on at least two occasions (which suggests that Greek generals were not good at learning from history). The number of areas narrow and secure enough to be held for a long period against an attacker is very small (in Greece, Thermopylae, Tempe and the Isthmus are probably the only examples), and as even these can be turned or circumvented, it is clear that holding passes was not a secure way to prevent an enemy invasion of a city's territory. Such frontier defences are of more use in delaying an enemy (perhaps to allow for naval or diplomatic manoeuvres elsewhere to bear fruit) than they are in permanently preventing passage.[24]

Away from such passes, any system of frontier defence through fortification would require, firstly, an enormous outlay on building such fortifications (Greek barrier defences such as city walls tended to be built on a smaller scale, defending population centres rather than entire territories), and also permanently stationed defenders, which in turn means professional soldiers, since any barrier defence that is not manned can never do any more than inconvenience an attacker. Linear defences of frontiers are uncommon in the ancient world, and where they are known (the Roman *limes*, the Chinese Great Wall), they require a large professional standing army and serve only to dissuade or delay mobile 'barbarian' enemies, rather than large and well-equipped 'civilized' armies. As the Greeks, at least before the fourth century, lacked large numbers of professional soldiers, and were faced by regular armies rather than barbarian raiders, it is not surprising that linear or frontier defences were not widely relied upon.

Such fortified positions as existed – chiefly fortified towns and cities, with some frontier forts or walls – would have to be attacked using the techniques of siegecraft (as we have seen, the Spartans were regarded as being particularly deficient in this area, and the Athenians unusually skilled). Even if an army did give battle, defeat did not mean the loss of the city, since the defenders could retreat within their walls and continue to resist. This would prevent an enemy from simply marching into the defeated town, but whether it meant the town was safe depended greatly on circumstances. Not every city had a wall, but even if it did not, the houses themselves could form a viable defensive position. This was the case with Sparta after their defeat at Leuctra, where the gaps between houses were blocked to make an impromptu defence, as Aeneas 'the Tactician' records in his handbook *On the Defence of Fortified Positions* (2.2), one of very few surviving Greek military handbooks from the Classical period. Perhaps for such reasons, even an unwalled city might be safe, as Xenophon records of the Spartan king Agis, who devastated the territory of Elis:

> 'When Agis reached the city he did some harm to the suburbs and the gymnasia, which were beautiful, but as for the city itself (for it was unwalled) the Lacedaemonians thought that he was unwilling, rather than unable, to capture it.' (Xenophon, *Hellenica* 3.2.27)

If a city was walled, then a prolonged siege was an option. This might involve building a wall around the city to prevent the citizens or any relieving force coming and going or gaining access to their agricultural lands. Perhaps the best-known example is the siege of Plataea carried out by the Spartans (who apparently had acquired some skill at siegecraft by this date), early in the Peloponnesian War, under their king Archidamus:

> 'First he enclosed the town with a palisade formed of the fruit-trees which they cut down, to prevent further egress from Plataea; next day they threw up a mound against the city, hoping that the largeness of the force employed would ensure the speedy reduction of the place.' (Thucydides 2.75.1)

Thucydides doesn't specify precisely who acted as the workforce – hoplites or light infantry (helots) – and we might assume that both did; at any rate, the construction work fell to the Spartans' allies, with Spartan officers (*xenagoi*) overseeing them (Thuc. 2.75.3). Thucydides describes the ingenious methods used by both sides – the Spartans building a mound of timbers packed with earth against the wall, the Plataeans countering by raising the height of their own walls, building a secondary wall inside the main wall and mining under the mound and taking earth from inside it so that it constantly subsided. The Spartans did also use 'engines' (*mechanas*), siege machines of unspecified form but presumably rams since one 'shook down a good part' of the wall (Thuc. 2.76.4). The Plataeans countered by lassoing the rams or dropping heavy beams on them – techniques recommended in Aeneas' handbook (Aen. Tact. 32). Finally the Spartans tried to set fire to the town using piles of wood with pitch and sulphur applied, but were stymied by a sudden downpour. Eventually the Spartans gave up on these methods, and replaced their wooden palisade with a proper wall of circumvallation made of mud bricks, built and manned by their allies (Thuc. 2.78.1), to starve the Plataeans out, which took the best part of a further two years.[25]

In 381, the Spartan king Agesilaus besieged the small town of Phlius. When the Phliasians refused to give in to Agesilaus' demands, 'he invaded their land and quickly built a wall of circumvallation around the city and besieged them' (Xen., *Hell.* 5.3.16). The idea was to starve out the defenders, which again took a considerable time:

> 'Agesilaus had already gone beyond the time for which the food-supply in Phlius was said to suffice; for self-restraint in appetite differs so much from unrestrained indulgence that the Phliasians, by voting to consume half as much food as before and carrying out this decision, held out under siege for twice as long a time as was to have been expected.' (Xenophon, *Hellenica* 5.3.21)

Even so, the Phliasians eventually had to submit, after enduring a year and eight months of siege (Xen., *Hell.* 5.3.25). Surrounding and starving out a city in this way was a common technique, and the two greatest sieges of the Classical period – that of Syracuse by the Athenians and of Athens itself by the Spartans at the end of the Peloponnesian War – took this form. In the case of Syracuse, bodged Athenian circumvallation and their loss of naval supremacy eventually allowed the Syracusans to break the siege, while Athens was eventually forced by hunger and sickness to surrender.

However, a complicating factor in the siege of Phlius was internal division (Agesilaus' attack was in support of pro-Spartan Phliasian exiles). Given the divided loyalties and internal strife so common in highly politicized Greek cities, the hope of treachery, or of support from some element of the population, often allowed sieges to be concluded more rapidly. In the case of Phlius:

> '[Their leader Delphion] was able to shut up and keep under guard those whom he distrusted, and had the power to compel the masses of the people to go to their posts and by putting sentinels over them to keep these people faithful.' (Xenophon, *Hellenica* 5.3.22)

The need to defend against internal treachery constantly plagued the defender – large parts of Aeneas' handbook deal not with technical engineering matters of walls or siege engines, but with methods for ensuring the loyalty of the population and preventing opportunities to betray the city to an attacker. Plataea, for example, which withstood siege for so long at the start of the Peloponnesian War, fell to the Thebans through treachery:

> '[T]he gates were opened to them by a Plataean called Naucleides, who, with his party, had invited them in, meaning to put to death the citizens of the opposite party, bring over the city to Thebes, and thus obtain power for themselves.' (Thucydides 2.22)

Because of this threat, or opportunity, the occasions where a lengthy siege was required are relatively few, as either treachery could lead to the capture of a city or the threat of treachery could encourage the defenders to make terms.[26]

If a city did not fall to treachery, and was not to be surrounded and starved out, then it could be taken by direct assault, surprise attack or escalade. Such attacks were by no means guaranteed to succeed, of course, but they meant that despite the lack of formal siegecraft, Greek armies were still able to assault and take fortified positions. There is no doubt that Greeks of this period did not develop the sort of technical engineering siegecraft that was to become so prominent under Philip II of Macedon and throughout the Hellenistic period. Classical Greeks lacked any sort of artillery (the torsion catapult used by Hellenistic armies was developed in Sicily in the fourth century), and also the array of siege machines (covered rams, towers and variants on them) which Hellenistic engineers developed (although simple battering rams for attacking gates and walls were known). Hoplite armies also, we may surmise, lacked the skill or enthusiasm for siege engineering that was so characteristic of Hellenistic and later Roman armies, and crucially they lacked the professionalism to allow such skills to be developed or the long duration of service necessary to see a siege through to completion (with obvious exceptions as noted above). Hoplite armies were temporary armies, campaigning for a single short season before returning home, so with the exception of some circumvallation sieges, lengthy engineering works were not usual for them. Surprise attacks, escalades and (above all) treachery offered much quicker returns, for much lower outlay in time, treasure and blood.

Amphibious operations

As noted above, one way to circumvent an enemy's defences was to go by sea, and in general, Greece being a sea-girt land, Greek armies would often find themselves transported by sea to reach their destination. The origins of this practice of course (like so much else) go back to mythical history and the Trojan War, where the 'thousand ships' launched to retrieve Helen bore the Greek army to Troy, and where most of the fighting described by the *Iliad* took place around the Greek ships drawn up on the beach and the palisade that was constructed around them. This mythical history perhaps reflects the actual Greek late Bronze Age and Dark Age identity as seaborne raiders, who launched piratical raids against their neighbours, their enemies and anyone who happened to live within sight of the sea in the Eastern Mediterranean. The Greeks of this period appear to have been the Vikings of their time, and may perhaps be identified with the Sea Peoples who so troubled Egypt and the Levant. In more settled Archaic times, Greeks were still active seaborne traders, raiders and colonizers (like the

Vikings), spreading Greek settlements around the Mediterranean coasts from Italy to the Black Sea and bringing back to Greece the spoils (slaves, livestock and treasure) seized there. In this period, the fighting men would themselves have rowed and sailed the boats in which they travelled, and the crew of their small ships (typically carrying fifty men) would have formed the parties that raided so many coasts.[27]

This maritime history continued into the Classical period, though with obvious variations. The expansion of the franchise – and so of the pool of manpower from which to draw heavy infantry, as discussed in Chapter 4 – meant that a city's hoplites would be numbered in the thousands rather than hundreds, and could no longer be carried in a few small ships. The founding of overseas colonies by many of the main Greek cities meant that the period of raiding, followed by that of colonization, now gave way to a period of empire building, trade and politics, as cities sought to expand their influence with the overseas colonies (their own or their rivals') in an extension of the conflicts of the mainland. Finally, shipbuilding techniques – driven on probably by contact with the Persian fleet at the end of the sixth century in Ionia – meant that ships became much larger than the fifty-man raiders of earlier times, and the principle method of naval warfare, which earlier had probably consisted mostly of chaotic boarding actions between opposing crews, now became the use of highly technical manoeuvres by skilled rowers with a view to ramming and sinking (or at least swamping) enemy vessels.[28]

Even in this changed naval environment, the hoplite had an important role. Ships needed heavy infantry, both to board enemy vessels and to prevent the boarding of their own, if a failed ramming attack should bring ships into proximity. Consequently, the main warship of the day, the trireme, would routinely carry a complement of hoplite marines (*epibatai*, literally 'passengers') – perhaps as few as ten in the Athenian navy of the late fifth century (along with a small number of archers). These would be stationed on the open upper deck of the trireme, with the rowers below in the body of the ship. Other cities' navies may have carried more hoplites per ship, since unlike the more expert Athenians with their skilled rowers, they preferred to stick with 'old-fashioned' naval warfare based on boarding rather than using the new tactics of ramming. We should note, incidentally, that the use of hoplites as marines in this way, fighting aboard ships and engaging in boarding actions of enemy ships, should be enough to dispel once and for all the strange notion that the hoplite was too heavily equipped to fight in anything other than a phalanx, and then able to do nothing more than face stolidly ahead and push. The hoplite marine was clearly a flexible and manoeuvrable fighter able to fight effectively in what must have been exceptionally difficult conditions for any sort of infantry, and there is no good evidence that there was any particular change of equipment involved for fighting aboard ship.²⁹

The marines would have been drawn from the hoplite class in the usual way, though these could be supplemented by mercenaries and by making use of those below the hoplite class equipped as hoplites, as in the Athenian expedition to Syracuse:

> '[T]he Athenians weighed from Corcyra, and proceeded to cross to Sicily with an armament now consisting of one hundred and thirty-four triremes in all (besides two Rhodian fifty-oars) of which one hundred were Athenian vessels – sixty fast ships [*tacheiai*] and forty troopships [*stratiotides*] – and the remainder from Chios and the other allies; five thousand and one hundred heavy infantry in all, that is to say, fifteen hundred Athenian citizens from the rolls at Athens and seven hundred *thetes* shipped as marines [*epibatai*], and the rest allied troops, some of them Athenian subjects, and besides these five hundred Argives, and two hundred and fifty Mantineans serving for hire; four hundred and eighty archers in all, eighty of whom were Cretans, seven hundred slingers from Rhodes, one hundred and twenty light-armed exiles from Megara, and one horse-transport carrying thirty horses.' (Thucydides 6.43)

The *thetes* in Athens were the large class ('labourers') not obliged to perform hoplite service (though they could evidently do so as volunteers), as we saw in Chapter 4.

When an expedition was mounted overseas, triremes would form the bulk of the naval assets, accompanied perhaps by merchant ships (which were sailed rather than rowed) to carry supplies and equipment. This means that the bulk of the manpower on arrival would come from the ships' crews – the hoplites being drawn from the marines, while the rowers might provide light infantry (as we saw above). Amphibious operations of this sort, with armies delivered overseas on the decks of triremes, are very common in the fifth and fourth centuries.³⁰

As shown by the example from Thucydides, a couple of variants on transport by trireme appear in the Peloponnesian War (though they may have been introduced earlier). One is the use of the horse transport, which may have been based on a trireme hull but with some rowers replaced by space to stable horses. These allowed a small force of cavalry to accompany an expedition. The second innovation was the specialist troop transport. The exact configuration of these remains a mystery, although we may see their origin in the adaptation of regular triremes by the Athenian Cimon in the mid-fifth century:

'Learning that the generals of the [Persian] King were lurking about Pamphylia with a great army and many ships, and wishing to make them afraid to enter at all the sea to the west of the Chelidonian isles, he set sail from Cnidus and Triopium with two hundred triremes. These vessels had been from the beginning very well constructed for speed and manoeuvring by Themistocles; but Cimon now made them broader, and put bridges [*diabasis*, 'gangways'?] between their decks, in order that with their numerous hoplites they might be more effective in their onsets.' (Plutarch, *Cimon* 12.2)

The general idea may have been, like the horse transport, to replace some of the rowers on a trireme with space for soldiers to sit – perhaps special decks, or perhaps they were expected to just sit on the rowers' benches – or alternatively the hoplites may all have been accommodated on the enlarged upper deck. It seems likely that the difference between triremes and troopships was not solely one of manning of oars and that there was a difference in construction, as shown by Thucydides' comments on the Athenian preparations for the Syracusan expedition:

'The fleet had been elaborately equipped at great cost to the captains and the state; the treasury giving a drachma a day to each seaman, and providing empty ships, sixty fast ships and forty troop ships [*hoplitagogos*], and manning these with the best crews obtainable.' (Thucydides 6.31.3)

Note the use of a different word for troopships here, while 'fast ships' are triremes equipped for speed and ramming combat, rather than carrying a lot of marines (compare for example with a later occasion where 'fifteen ships were manned by the Megarians and the other allies, more properly transports [*stratiotidon*] than fast ships', Xen., *Hell*. 1.1.36).[31]

The downside to this arrangement of course was that the ship, with a reduced number of rowers and/or greater weight, would be slow and unable to fight effectively should need arise on the voyage. Therefore, from the late fifth century we hear of hoplites travelling 'self-rowed' (*auteretai*) – for example, the Athenians 'sent to Mytilene ... a thousand Athenian hoplites who handled the oars themselves' (Thuc. 3.18.3–4), or the recommendation that they 'must send to Sicily a force of hoplites who will themselves handle the oars and will take the field immediately on landing' (Thuc. 6.91.4). Instead of just being passengers on a troopship, these hoplites would take to the oars and row, which means they could travel in normal triremes. Since rowing required some skill and practice, this was not presumably an option for every hoplite, but in cities with a wide hoplite franchise, the difference between poorer hoplites and better-off rowers need not have been great (as shown by the Athenian *thetes* serving as *epibatai* above), so the duty need not have been considered demeaning, and the men might already have experience as rowers. Self-rowing was also of course the norm in earlier periods, when warrior bands would have travelled in their own small ships to seek fortune or adventure, but rowing one's own small open boat on raids was a different proposition than descending into the cramped and smelly confines of a specialized weapon of war like the trireme. The labour and practice required to become proficient at rowing was also too much for some leisured citizens; the Ionians who at the start of the fifth century initially agreed to train as rowers, after just a week, 'untried as they were in such labour and worn out by hard work and by the sun', mutinied, 'raised tents on the island where they stayed in the shade' and refused to get back on their ships (Hdt. 6.12). Rowing just well enough to complete a voyage would have required less effort than training for combat, as the Ionians were attempting to do. The comparison with the later Macedonian phalanx, whose soldiers in the second century enthusiastically trained as rowers (Pol. 5.2.4–6), is revealing.

The numbers given above for the Sicilian expedition by Thucydides – 5,100 hoplites and 1,300 light infantry, divided between sixty Athenian fast triremes, forty Athenian troopships and thirty-four allied triremes of unknown configuration – gives on average about forty-eight soldiers per ship (in addition to their rowers and deckhands). As we do not know how many men would be carried on the fast ships nor how many allied ships were fast, we cannot be certain how many were carried on the troopships. Nevertheless, assuming twelve marines and eight archers on each fast ship (which would allow the 700 *thetes* and 480 archers to all be on the Athenian ships and is close to the normal complement of marines), and that all the allies were fast, this would leave 114 men on the troopships, which may be regarded as a maximum figure. We might guess that the complement of a fast ship might be between ten and thirty marines, and that a troopship could carry up to 100 men, with considerable variability based on local circumstances. It is difficult to imagine that such numbers could have been accommodated in any approximation to comfort, and this was for a long voyage from Athens to Sicily, but ancient fleets sailed close to shore and put in on a beach every night to eat, camp, resupply and perform other essential tasks, so a certain degree of crowding during the voyage would have been tolerable.

We have little detail of the nature of ship-to-ship fighting by marines, other than that ships intending to fight this way (the 'old-fashioned' way, as opposed to the 'modern' tactics of speed and ramming) would contact prow-to-prow with enemy ships (if no better angle was available; fast ships would avoid such contact), presumably grapple themselves together, and then the marines would fight as if on land – although the difficulties of crossing from one ship to another must have been considerable. Grapples, boarding planks and ropes must surely all have been utilized, though we lack detailed accounts. Ships moored on beaches would have been accessed by a plank or ladder (*apobathra*) propped between the stern and the beach, so a similar arrangement could have provided a usable, if precarious, way to cross between ships. Missile weapons would seem on the face of it to be valuable in such fighting, but although archers were carried, hoplites – so far as we know – fought with their usual weapons of spear and shield, though javelins may have been widely used.

For fighting on land, hoplites would normally disembark and fight in the normal way (in a phalanx if appropriate), though they did also on occasion perform opposed amphibious landings. Polyaenus (3.9.63) provides an account of an amphibious landing performed by Iphicrates, where the ships backed toward the beach and dropped anchor, and the men jumped from the stern, fully armed, into the sea (when it was shallow enough), thereby arriving at the beach already in formation. Other examples of such landings are to be found in the fighting around Pylos in 425, where an Athenian force on the mainland had to resist Spartan attacks from land and sea. The bulk of the Athenians defended the landward side, while a force of sixty hoplites and 'a few' archers protected the rocky shore against Spartan landings (Thuc. 4.9.2). The Spartans had forty-three ships (type and number of marines unspecified), but they could only send in a few at a time due to the narrow approach and were hanging back for fear of wrecking their ships on the rocky shore until one ship's captain (trierarch) encouraged them to run their ships aground in order to force a landing (4.9.4):

> 'Not content with this exhortation, he forced his own steersman to run his ship ashore, and stepping on to the gangway, was endeavouring to land, when he was cut down by the Athenians, and after receiving many wounds fainted away. Falling into the bows, his shield slipped off his arm into the sea, and being thrown ashore was picked up by the Athenians, and afterwards used for the trophy which they set up for this attack. The rest also did their best, but were not able to land, owing to the difficulty of the ground and the unflinching tenacity of the Athenians.' (Thucydides 4.12.1–2)

The trierarch, Brasidas, survived to have a distinguished if brief career leading a campaign in northern Greece, while his shield could conceivably be the very shield now on display in the Agora Museum in Athens. At any rate, while in this case the defending Athenians (although on land, not on their preferred element, the sea, as Thucydides points out) successfully repelled the landing, it does at least show that hoplites could if required fight their way from ship to shore, an activity as far as it is possible to get from the restricted fighting in the phalanx of which they are sometimes claimed to be solely capable.

So hoplites could certainly fight in a greater variety of ways than in the phalanx in pitched battle, yet as so often there were value judgements attached to various types of fighting, and the technical question of whether a given soldier, with his equipment, was suited for or capable of a certain type of fighting, was overlaid with a range of opinions as to whether such fighting was really suitable for a hoplite. Brasidas and his men exemplified Spartan hoplite courage and determination at Pylos (for all that they were unsuccessful), yet Plato reflected on Athens' tribute of seven youths imposed on them by Minos in the mythical past, forced on them as they had no navy:

> 'And indeed it would have profited them to lose seventy times seven children rather than to become marines instead of staunch hoplites; for marines are habituated to jumping ashore frequently and running back at full speed to their ships, and they think no shame of not dying boldly at their posts when the enemy attack; and excuses are readily made for them, as a matter of course, when they fling away their arms and betake themselves to what they describe as "no dishonourable flight". These "exploits" are the usual result of employing naval soldiery, and they merit, not "infinite praise", but precisely the opposite, for one ought never to habituate men to base habits, and least of all the noblest section of the citizens.' (Plato, *Laws* 4.706)

Plato refers to marines as 'naval hoplite service' (*nautikos hopliteia*) rather than the more technical *epibatai*. We might reflect that (as Patton pointed out) nobody ever won a war by dying for his country, boldly or otherwise, but even so this is a reminder that while hoplites were certainly capable of fighting in many different ways, the hoplite ethos still placed fighting in pitched battle in the phalanx – with its particular requirements of steadfast stubbornness – at the top of the hierarchy of honour, as will be explored in the next chapter.

Pillaging, devastation and economic warfare

We have already encountered a number of occasions where invading armies engaged in the destruction of the enemy's agricultural capability, in particular by destroying crops and cutting down trees. In the early Archaic period, when warfare probably mostly took the form of plundering expeditions, such activities – and in particular the seizure of cattle, slaves and women – would have been the primary objective of warfare. By the Classical period, war aims had become more complex (see below), and agricultural devastation was used as a threat and a ploy (to force an unwilling enemy to offer battle), as well as a form of economic warfare to damage the enemy city in the long term, though a certain amount of enrichment through plunder might also be expected.[32]

A defender could attempt to forestall this by moving behind the walls of the city everything that could be moved, though this would not be done willingly, as Thucydides recounts of Pericles' strategy at the start of the Peloponnesian War:

> 'The Athenians listened to his advice, and began to carry in their wives and children from the country, and all their household furniture, even to the woodwork of their houses which they took down. Their sheep and cattle they sent over to Euboea and the adjacent islands. But they found it hard to move, as most of them had been always used to live in the country.' (Thucydides 2.14)

Abandoning the country in this way meant giving up, if only temporarily, a way of life, and while movable goods could be saved, income from agricultural lands could not. Livestock could be moved, onto an island as in the case of the Athenians, to a remote headland (Xen., *Hell*. 4.5.1) or mountain area (Xen., *Hell*. 4.6.4), or even brought within the walls of the city, though Aeneas (10.1) advises against this, and it is easy to see why. Crops could also be pre-emptively harvested and the grain stored (or eaten). But trees, in particular olive trees (and other fruit trees and vines), had to be abandoned, and we have already encountered occasions where the cutting down of trees, whether for practical purposes for use in siegeworks or just as an act of malice, played a large part in the calculations of an attacker. Modern accounts have tended to downplay the amount of physical damage that could be done to agricultural land, but the stress ancient accounts place on the fate of trees suggests that these modern revisions go too far. At any rate, the threat of agricultural devastation was used to influence the actions of an enemy, and not just to force their hoplites to come out and give battle. A complex interplay of motives and capabilities was in play in such calculations, as shown for example by a campaign of Agesilaus in 389, with two *morai* of Spartans and the equivalent in allies, against the Acarnanians:

> '[W]hen Agesilaus arrived at the borders of the enemy's country, he sent to the general assembly of the Acarnanians at Stratus and said that unless they discontinued their alliance with the Boeotians and Athenians and chose his people and the Achaeans as allies, he would lay waste their whole territory, one portion after another, and would not spare any portion of it. Then, upon their refusing to obey him, he proceeded to do so, continually devastating the land as he went and hence advancing not more than ten or twelve stadia [about 2km] a day. The Acarnanians, therefore, deeming it safe on account of the slow progress of the army, brought down their cattle from the mountains and continued to till the greater part of their land. But when it seemed to Agesilaus that they were now very bold, on the fifteenth

or sixteenth day from the time when he entered the country, he offered sacrifice in the morning and accomplished before evening a march of one hundred and sixty stadia [29km] to the lake on whose banks were almost all the cattle of the Acarnanians, and he captured herds of cattle and droves of horses in large numbers besides all sorts of other stock and great numbers of slaves. And after effecting this capture and remaining there through the ensuing day, he made public sale of the booty.' (Xenophon, *Hellenica* 4.6.4–6)

It was on this occasion that Agesilaus had to use special manoeuvres to get his army away from vengeful Acarnanian light infantry, as we saw earlier. It is also far from clear who the public sale was made to, deep in Acarnanian territory. The Achaean allies were dissatisfied with this outcome as no permanent gain had been made, and asked Agesilaus 'to stay long enough to prevent the Acarnanians from sowing their seed', but Agesilaus replied that 'I shall again lead an expedition hither next summer; and the more these people sow, the more they will desire peace' (Xen., *Hell.* 4.6.13). In the event, the threat of a further invasion next year was sufficient, the Acarnanians 'thinking that inasmuch as their cities were in the interior they would be just as truly besieged by the people who destroyed their corn as if they were besieged by an army encamped around them' (Xen., *Hell.* 4.7.1), and they agreed to the Spartan terms. Note the importance of the possibility (or impossibility in this case) of supply by sea.

This campaign perhaps represents an unusually effective use of the threat and practice of agricultural devastation. But even so, while a city or region could not be brought to its knees by a single such campaign, repeated attacks could be expected to wear it down, remembering that the objective was to alter the enemy's foreign policy, not to destroy them or annihilate their people. In all eras, economic warfare has been a blunt and slow instrument, but that it usually does not give rapid or spectacular results does not mean that it is ineffective, nor that its only purpose was to provoke a particular military response (accepting pitched battle).

Agricultural warfare could be made more effective if armies (or at least detachments) were stationed permanently in or beside enemy territory, so that the threat to farmers could be maintained rather than being limited to the short campaigning season, with the need to march to enemy territory and back again. This became the dominant strategy of the Peloponnesian War, with Athenian outposts on Pylos and Cythera and a Spartan position at Decelea. Such outposts allowed longer-term harrassing of agriculture, and also acted as focal points or objectives for escaping slaves (the agricultural workforce consisted largely of slaves, so providing an incentive for their desertion could be effective). This practice, known as *epiteichismos*, was frequently employed from the late fifth century, but its effects were not dramatic or decisive; it just increased the overall effectiveness of such economic warfare (and so the pressure on the enemy to come to terms).[33]

Devastation could also have a more symbolic aspect, by showing that an enemy was unwilling or unable to come out of their city to fight, and many examples of small-scale devastation that could have had little effect economically are of this kind, as an act of deliberate defiance or the assertion of dominance over a weaker opponent. When Agesilaus heard of the defeat of the Spartan *mora* by peltasts at Lechaeum, his response was to march his main army against Corinth:

> 'He did not throw down the trophy [erected by the peltasts], but by cutting down and burning any fruit-tree that was still left, he showed that no one wanted to come out against him. When he had done this, he encamped near Lechaeum.' (Xenophon, *Hellenica* 4.5.10)

Asserting Spartan dominance and saving face were clearly the objective here, not economic damage. When Agesilaus later withdraw precipitately from Arcadian territory, he was still content:

> 'For he seemed to have brought the state [Sparta] some relief from its former despondency, inasmuch as he had invaded Arcadia and, though he laid waste the land, none had been willing to fight with him.' (Xenophon, *Hellenica* 6.5.21)

So attacks on agriculture certainly could have an economic dimension, intended to influence rather than to subjugate their target, and could also be more symbolic, an assertion of dominance. Whether the actual agricultural or economic damage caused was very great is perhaps beside the point.[34]

Tricks, surprise and ambushes

Writing in the second century, the historian Polybius contrasted the 'treacherous dealings' of the Macedonian king Philip V with the higher standards prevalent in the 'good old days':

> 'The ancients, as we know, were far removed from such malpractices. For so far were they from plotting mischief against their friends with the purpose of aggrandizing their own power, that they would not even consent to get the better of their enemies by fraud, regarding no success as brilliant or secure unless they crushed the spirit of their adversaries in open battle. For this reason they entered into a convention among themselves to use against

each other neither secret missiles nor those discharged from a distance, and considered that it was only a hand-to-hand battle at close quarters which was truly decisive. Hence they preceded war by a declaration, and when they intended to do battle gave notice of the fact and of the spot to which they would proceed and array their army. But at the present they say it is a sign of poor generalship to do anything openly in war.' (Polybius 13.3.2–6)

The idea that 'the ancients', which to Polybius meant the Archaic and early Classical Greeks, fought wars according to such honourable principles – even to the extent of outlawing missile weapons – was popular in antiquity and has continued to attract many adherents to this day. In recent years, however, there has been a reaction against this view, and many historians now point out that nostalgia may have influenced writers like Polybius more than cold hard facts, and that treachery, double dealing and generally seeking the maximum advantage had always been an essential part of Greek warfare. As usual, such denials should not be taken too far, since there clearly was a notion of the open, decisive battle between heavy infantry forces, which will be the subject of the next chapter. However, the idea of 'agonal' war – that war was conducted according to rules requiring open, honourable behaviour and that battles were formal, prearranged contests of honour – has rightly been criticized in view of the many counterexamples in ancient literature.[35]

Going back to the mythical origins of Greek warfare in Homer, there are two aspects of kingship and the conduct of war, epitomized by Achilles and Odysseus. Achilles was the ultimate warrior, excelling in face-to-face fighting at close range (if that indeed was the nature of Homeric combat, but at any rate the Homeric equivalent), while Odysseus was the trickster, wily and full of schemes. It was Achilles who was celebrated as the great hero of the *Iliad* (though his behaviour, sulking in his tent because of a perceived slight to his honour, is not simply or straightforwardly heroic to our eyes), but it was Odysseus who came up with the scheme that led to the fall of Troy (a wooden horse, perhaps a mythic echo of an early siege machine such as was used by the Assyrians), and Odysseus was in turn the hero of his own epic, the *Odyssey*. It is not clear that in Archaic and Classical Greece the approach of Achilles was to be preferred to that of Odysseus; clearly both were valued and necessary, for all that there was often a tension between them. This dual nature to warfare was also reflected in the existence of two war gods: Ares, representing brute force and aggression, and Athena, goddess of wisdom and representative of the more cerebral approach to warfare. As usual, the most detailed accounts of warfare we have come from the Persian and particularly the Peloponnesian War and after, which has led to the appearance of an increase in trickery, but this appearance may be deceptive, and it may be that warfare did not really change very much, in this respect, from the time of Odysseus (or of Homer) to the fourth century. Indeed, given that Archaic Greeks would have read of Odysseus' tricks (and those of the other heroes) in Homer and other mythological writings, and chose to have such scenes depicted in poetry and pottery, it is unlikely that they they would have avoided implementing such measures in their own wars.

Ambushes and surprise attacks on enemy forces are already apparent in the sixth century. Herodotus tells a story of the Aeginetans and Argives ambushing an Athenian invasion – 'when the Athenians disembarked on the land of Aegina, the Argives came to aid the Aeginetans, crossing over from Epidaurus to the island secretly. They then fell upon the Athenians unaware and cut them off from their ships' (Hdt. 5.86.4) – and the Athenian Miltiades in the mid-sixth century 'made war first on the people of Lampsacus, but the Lampsacenes laid an ambush and took him prisoner' (Hdt. 6.37). Such ambushes continued a tradition going back to Homer and are well illustrated in Archaic art. Surprise attacks were also known before the fifth century. In the mid-sixth century, the tyrant Pisistratus was victorious over another Athenian force: 'The Athenians of the city had by this time had breakfast, and after breakfast some were dicing and some were sleeping: they were attacked by Pisistratus' men and put to flight' (Hdt. 1.63.1). The Battle of Sepeia in the early fifth century saw a cunning plan enacted by the Spartans against the Argives:

> 'When [the Argives] had come near Tiryns and were at the place called Hesipeia [Sepeia], they encamped opposite the Lacedaemonians, leaving only a little space between the armies. There the Argives had no fear of fair fighting, but rather of being captured by a trick ... Therefore they resolved to defend themselves by making use of the enemies' herald, and they performed their resolve in this way: whenever the Spartan herald signalled anything to the Lacedaemonians, the Argives did the same thing. When Cleomenes saw that the Argives did whatever was signalled by his herald, he commanded that when the herald cried the signal for breakfast, they should then put on their armour and attack the Argives. The Lacedaemonians performed this command, and when they assaulted the Argives they caught them at breakfast in obedience to the herald's signal; they killed many of them.' (Herodotus 6.77–78)

An unlikely story perhaps, but it is at least very far from being an example of fair and open battle.[36]

Needless to say, many more examples of surprise and trickery are available from the Peloponnesian War and the fourth century. The most-often cited exponents of such fighting are the Athenian Demosthenes and the Spartan Agesilaus.

Demosthenes, for example, launched a classic dawn attack on the Ambraciot army at Idomene in 426:

'At dawn he fell upon the Ambraciots while they were still abed, ignorant of what had passed, and fully thinking that it was their own countrymen – Demosthenes having purposely put the Messenians in front with orders to address them in the Doric dialect, and thus to inspire confidence in the sentinels, who would not be able to see them, as it was still night. In this way he routed their army as soon as he attacked it, slaying most of them where they were, the rest breaking away in flight over the hills.' (Thucydides 3.112.2–5)

The Doric dialect was that of the Ambraciots' Spartan allies, as opposed to the Ionic dialect of the Athenians (Thucydides frequently draws attention to the different dialects of the opposing sides in the Peloponnesian War, which sometimes had tactical significance). Catching an army in bed or at breakfast was not the only way to take it by surprise; surprise attacks were also possible in pitched battle, as Demosthenes managed at Olpae, as we will see in in the next chapter. Demosthenes also led the Athenians in the series of irregular actions that led to the surrender of the Spartan garrison on Sphacteria (425), which began with a similar surprise attack on a Spartan outpost on the island:

'The advanced post thus attacked by the Athenians was at once put to the sword, the men being scarcely out of bed and still arming, the landing having taken them by surprise, as they fancied the ships were only sailing as usual to their stations for the night.' (Thucydides 4.32.1)

Needless to say, there is no sign of Demosthenes earning any opprobrium or censure for these tricks.

Another surprise attack was launched by the Athenian democrats gathering at Phyle (403) to end the tyranny of the Thirty:

'Now by this time about seven hundred men were gathered at Phyle, and during the night Thrasybulus marched down with them; and about three or four stadia from the guardsmen he had his troops ground their arms and keep quiet. Then when it was drawing towards day and the enemy were already getting up and going away from their camp whithersoever each one had to go, and the grooms were keeping up a hubbub as they curried their horses, at this moment Thrasybulus and his men picked up their arms and charged on the run. They struck down some of the enemy and turned them all to flight, pursuing them for six or seven stadia [about 1,100 metres]; and they killed more than one hundred and twenty of the hoplites, and among the cavalry Nicostratus, nicknamed "the beautiful", and two more besides, catching them while still in their beds.' (Xenophon, *Hellenica*.2.4.5–6)

The Spartan king Agesilaus was especially admired by Xenophon for his trickery. When he outwitted the Persian Tissaphernes, Xenophon remarked that:

> 'This achievement also was thought to be a proof of sound generalship, that when war was declared and trickery in consequence became righteous and fair dealing, he showed Tissaphernes to be a child at deception.' (Xenophon, *Agesilaus* 1.17)

Xenophon goes on to further praise Agesilaus' cunning:

> 'As for the enemy, though they were forced to hate, he gave them no chance to disparage him. For he contrived that his allies always had the better of them, by the use of deception when occasion offered, by anticipating their action if speed was necessary, by hiding when it suited his purpose, and by practising all the opposite methods when dealing with enemies to those which he applied when dealing with friends.' (Xenophon, *Agesilaus* 6.5)

It seems unlikely that a dramatic change in attitudes to trickery took place just during the course of the fifth century, enough for trickery to be considered 'righteous and fair dealing', and the probability must be that this represents a continuation of a side to Greek warfare that had always existed. The ethos of Odysseus was thus probably always as strong in Greek warfare as that of Achilles.

War aims and outcomes

Classical Greece was, broadly speaking, a land of walled cities, and the relative weakness of Greek armies at offensive siege warfare meant that it was difficult for an attacking army to capture an enemy city. Ideally (from the attacker's point of view), the defenders would come out to fight a battle in defence of their farmland and their honour, and once beaten, the issue of the war could be settled as the defenders would now accept the terms that the attackers imposed on them.

Many Greek wars were disputes over land, or attempts to impose a particular foreign policy on another city (to force them to join, or to leave, a particular alliance). Land disputes were often concerned with marginal border lands, sometimes lands which had been disputed between neighbours for generations. It is unlikely that the prime motivation for such conflicts was economic, since such marginal lands cannot have contributed greatly to the economy of either side. This is a controversial point, since some have argued that such lands were brought under greater cultivation through this period, and that therefore there was economic advantage to be sought in claiming them. Even so, the economic gain must have been so slight, and economic justifications (for marginal gains) so lacking in the reasons given for wars in the sources, that it must be doubted if such gain was really the prime motivation, rather than perhaps a contributory factor. Greek wars seem rather to have been primarily about establishing status and dominance (often framed in terms of honour), and political motivations (such as directing a neighbour's foreign policy) or economic motivations (gaining agricultural land, plunder or tribute) were more likely used as justifications for the desired assertion of dominance than as real motivations for war. Thucydides is explicit in assigning such motivation to Sparta in the Peloponnesian War: Sparta needed to assert its dominance over the rising power of Athens. He frames this in defensive terms, that the Spartans feared Athens' growing power and sought to assert their dominance for their own safety, but as so often, defensive motivations are often easily applied to aggressive intentions. Athens for its part had spent the previous fifty years building an empire which brought in tremendous wealth, but its aim in doing so was to challenge the superiority of Sparta, not as an end in itself.[37]

The combination of relatively invulnerable cities and wars fought between neighbours to establish status, along with the shared cultural values of the combatants, meant that Greek war aims were frequently limited, and wars of annihilation or conquest relatively rare. Of course there are exceptions to every rule, and there certainly are examples of Greek cities capturing and destroying opposing cities or annexing their territory. Spartan power was after all entirely based on their annexation of Messenian territory and subjection of the population as helots, while there is also the example of Plataea, which after its two-year siege and capture at the start of the Peloponnesian War was destroyed at the behest of the Thebans (who built a religious sanctuary on the ruins). Going back to Homeric times, a common epithet for a king is 'sacker of cities', and it is certainly not the case that Greeks were always restrained in their war aims or generous in their treatment of a defeated enemy.

Yet it is also true, perhaps especially in the Archaic period but also in the Classical, that on the whole, war aims were of a more limited nature and that once honour was satisfied and status established, aims could be considered met. Although there was clearly a surge in Greek population and population dynamism in the early Archaic era, this was generally directed outward, away from the existing Greek settled lands, and resulted in a wave of emigrations and colonization of 'barbarian' lands around the Mediterranean littoral, rather than of empire building within Greece itself. In the Classical period, population dynamics seem to have stabilized, and in some cities – particularly Sparta – shortage of manpower (*oliganthropia*, at least shortage of men of the right sort) was more of a problem than what we might anachronistically call *lebensraum*. That said, there was sufficient surplus population to drive a wave of migration into the formerly Persian lands opened up by the conquests of Alexander's Macedonians in the late fourth century, and orators such as Isocrates would advocate precisely this policy during the fourth century, opening up the East to Greek settlement. However, this was driven by the desire to relieve Greece of the burden of these surplus people, and the objective, in a spirit of pan-Hellenism, was to send them off overseas, not to find room for them in Greece at the expense of political enemies.

Indeed, given that Greek cities co-existed alongside each other in a state of perpetual conflict for several centuries, a certain equilibrium must have been reached. The ubiquity of warfare in ancient Greece is understood both from the frequent wars described in the ancient sources and, more theoretically, from a comment of Plato:

> 'For "peace", as the term is commonly employed, is nothing more than a name, the truth being that every *polis* is, by a law of nature, engaged perpetually in an informal war with every other *polis*.' (Plato, *Laws* 626a)

Cities were on occasion destroyed – an example being Thespiae in Boeotia, which suffered heavy losses in a succession of battles across several generations, falling alongside Leonidas' Spartans at Thermopylae (480), and in greater numbers (Hdt. 7.222, 7.226), suffering heavy losses at Delium (424, Thuc. 4.96.3) and again at Nemea (Xen., *Hell.* 4.2.22). Indeed, the Thespians seem to have made a habit of fighting bravely to the last man. Evidently there was a long-running feud between Thebes and Thespiae; the Thebans took advantage of Thespiae's weakness to destroy their walls (Thuc. 4.133.1), and (probably) after defeating the Spartans at Leuctra (371), the Thebans finally finished the job, expelling the residents (Diod. 15.46.6; Xen., *Hell.* 6.3.10). Even so, it took a century of enmity with Thebes and several catastrophic defeats in battle before Thespiae was destroyed (and the city was refounded shortly after, joining Alexander the Great to take revenge on the Thebans).

But most wars ended once the dispute in hand – whether on a point of honour, boundary lands or political alignment – was settled. Even the Macedonians, despite achieving military dominance by defeating Thebes and Athens at Chaeronea, and then taking the unusual step of destroying Thebes after a further revolt, never completely subjugated Greece, which retained a varying degree of political independence throughout the Hellenistic period. Only Rome finally put an end to independent Greek cities, and that only after twice defeating the Macedonians, and then defeating another Greek army and destroying a further major Greek city, Corinth. Usually, a war could be settled by coming to terms following a decisive or at least symbolically important military success, and this was most likely to be achieved in a single pitched battle between the armies of the opposing sides.[38]

Low-intensity and high-intensity combat

For all these examples given above of surprise and ambush, and for all the many and varied types of fighting that a hoplite might be called on to perform, there is no doubt that the 'fair and open' pitched battle had a special place, both in the ideals of hoplite behaviour and in practical military terms. Pitched battles, that is formal-seeming encounters between armies – almost exclusively hoplite armies formed in phalanx formation, which, if not following a set of rules, at least regulated by a number of established norms – are indeed common throughout Greek history.[39]

For the Greeks, as we have seen in Chapter 4, fighting hand-to-hand in close order was not just a tactical or strategic choice, it was also a moral one: it was a sign of the greater virtue (*aretē*) of the citizen, as opposed to the less steadfast ways of the lower social orders or of tribal or barbarian enemies. The speech which Thucydides puts in the mouth of the Spartan Brasidas, trying to extract his army from Macedonian territory in the face of hordes of hostile light infantry, makes this point:

> 'Where an enemy seems strong but is really weak, a true knowledge of the facts makes his adversary the bolder, just as a serious antagonist is encountered most confidently by those who do not know him. Thus the present enemy might terrify an inexperienced imagination, they are formidable in outward bulk, their loud yelling is unbearable, and the brandishing of their weapons in the air has a threatening appearance. But when it comes to real fighting with an opponent who stands his ground, they are not what they seemed; they have no regular order [*taxis*] that they should be ashamed of deserting their positions when hard pressed; flight and attack are with them equally honourable, and afford no test of courage; their independent mode of fighting never leaving any one who wants to run away without a fair excuse for so doing. In short, they think frightening you at a secure distance a surer game than meeting you hand-to-hand; otherwise they would have done the one and not the other. You can thus plainly see that the terrors with which they were at first invested are in fact trifling enough, though to the eye and ear very prominent. Stand your ground therefore when they advance, and again wait your opportunity to retire in good order, and you will reach a place of safety all the sooner, and will know for ever afterwards that rabble

such as these, to those who sustain their first attack, do but show off their courage by threats of the terrible things that they are going to do, at a distance, but with those who give way to them are quick enough to display their heroism in pursuit when they can do so without danger.' (Thucydides 4.126.4–5)

Only hand-to-hand fighting provided a 'test of courage' (*andreia*), so that as well as the tactical advantage afforded by their close formation, they could also be confident (or so Brasidas wished to reassure them) that they were better men.

We should not, however, be too quick to decide that a preference for pitched battle necessarily reflected a particular cultural restriction or formalized way of waging war. There are strong practical reasons to prefer a formal contest on the battlefield rather than the potentially more chaotic type of fighting that would arise if either side engaged at will, or indeed avoided a battlefield encounter at all. In such informal fighting, outcomes might be more down to chance, and the stronger side might be particularly disadvantaged. This applies to night attacks, attacks on camps and many other sorts of encounters – they might give one side an advantage, but they would also be subject to much greater uncertainty. Agesilaus' campaign against Mantinea well illustrates the point:

> 'On the next day Agesilaus encamped at a distance of about twenty stadia [3.6km] from Mantinea. But the Arcadians from Tegea, a very large force of hoplites, made their appearance; they were skirting the mountains between Mantinea and Tegea, desiring to effect a junction with the Mantineans, for the Argives, who came with them, were not in full force. And there were some who tried to persuade Agesilaus to attack these troops separately; he, however, fearing that while he was marching against them the Mantineans might issue forth from their city and attack him in flank and rear, judged it best to allow the two hostile forces to come together and, in case they wished to fight, to conduct the battle in regular fashion and in the open.' (Xenophon, *Hellenica* 6.5.16)

Agesilaus was not one to avoid an unfair advantage if it was available, as we have seen, but in this case (as commander of the more highly rated army) he preferred an open battle – where the fighting qualities of his men would be decisive – to the potential uncertainty and confusion of a piecemeal engagement. Similar considerations must have applied to many battles.

Another factor leading to a preference for open battle is that it provided the defeated enemy with no excuses. If a defeat could be explained away as the consequence of 'not fighting fair', then it would be psychologically less crushing for the defeated and less impressive to third parties. If an army accepted open battle and lost, it did not mean the end of the war or that the defeated side would necessarily submit to the victors' demands, but it did at least mean that they were unequivocally shown to be the worse side (on this occasion). This concept is of course by no means unique to Greek warfare, and is common to many cultures.

This leads on to another modern concept that has gained a surprising amount of traction in recent years, that the Greeks' preference for pitched battle was unique to them at the time, and shared only by those cultures (termed 'Western') that inherited this preference, along with other values, such as a liking for democratic forms of government and the rule of law, and a dislike for autocratic rule (which the Greeks considered akin to slavery, though not to be confused of course with actual slavery, of which Greeks were enthusiastic advocates). This idea is on the face of it ludicrous, since the concept of pitched battle is self-evidently shared by many martial cultures, and even those cultures which might not have a system of formal rules or norms around such battle could still adopt a certain formalism, such as meeting to fight at a mutually convenient place and time.[40]

The Greeks' main external enemies throughout the period we are considering here were of course the Persians, who are considered the epitome of non-Western (presumably Eastern, though the adjective is rarely applied) warfare, with a preference for ranged weapons rather than hand-to-hand fighting, and for forms of warfare other than the pitched battle, including raiding, ambushes and in general what we might term low-intensity warfare, that is warfare that is prolonged temporally and spatially, but produces few casualties and no decisive results. The grounds for making such a distinction between Greeks and Persians goes back to the Greeks themselves, from Herodotus' famous and often-quoted anecdote of the Persians ridiculing the Greek practice of selecting the fairest piece of land on which to fight and there virtually annihilating each other (Hdt. 7.9.2), to the common distinction made between the Greeks, fighting face-to-face with the spear, and the Persians, fighting at a distance with the bow. To the Greeks, these were not merely technical, tactical differences, but wrapped in a range of value judgements which made Greeks better, braver, truer, freer and manlier men. But the propagandistic nature of these claims should alert us to be cautious of accepting them at face value. Although Persian warfare is strictly outside the scope of this book, recent studies of Persian armies stress that their dependence on the bow was not an image that they had about themselves, and the Persians appear to have considered themselves to be spear users and fighters at close-quarters just as much as the Greeks. The most cursory look at Persian-Greek interactions reveals a succession of pitched battles – Marathon, Thermopylae, Plataea, Mycale, not to mention the battles against Alexander the Great at Granicus, Issus and Gaugamela – in which Persian armies massed on the battlefield to fight formal engagements against similarly deployed opponents, together with at least one battle (Cunaxa) pitting Persians against Persians. That we do not see more Persian internal battles, or battles against other opponents, is partly down to the usual imbalance of our surviving sources, and partly to the fact that the Persian Empire – as a geographically extensive imperial power – had fewer external enemies to fight than the chronically divided Greeks, with their heavily populated cities in a tiny geographic area in a state of almost constant war with each other. But as we have seen above, Greeks had no problem at all with irregular or low-intensity forms of fighting when it suited them, nor with trickery, surprise and ambush. It is by no means clear that their desire to fight formal, open battles on the occasions that they did so was not a preference fully shared by the Persians, when it suited them to do so. Indeed, facing Alexander's invasion, it was a Greek (Memnon of Rhodes) who suggested the Persians should avoid battle and carry out a scorched earth retreat, and the Persian commanders who preferred to stand and fight (at the Granicus), the commander, Arsites, stating 'that he would not suffer one house to be burned belonging to his subjects' (Arrian, *Anab.* 1.12.9). The latter was a view that in the mouth of a Greek might be taken as evidence of their willingness to undergo pitched battle in order to protect their agricultural land, in the supposedly Western way.[41]

I do not believe that there is anything unique, still less anything 'Western', about the Greek preference for fighting pitched battles – whether amongst themselves or against external opponents. Whether this was indeed a strong preference is hard to determine, given that it is difficult to compile meaningful statistics of battle compared with low-intensity warfare, since battles are by their nature one-off episodes, and other forms of fighting more numerous and protracted. However, we can accept – based on all the evidence and on their own expressed views – that the Greeks did hold the pitched battle in special regard, and viewed close-quarters fighting as morally inherently superior to other forms. I believe that the other cultures (Etruscans, Romans, Carthaginians) who fielded similar armies and fought in a similar style did so because this feature is common to many cultures, not because they copied it from the Greeks. But we should still seek to explain the Greeks' preference, and also those few occasions where the arrangements were even more formal, such as the 'Battle of Champions' and the possible censure of missile weapons.[42]

Explanations for this preference are not hard to find. As we have seen, the hoplites of the Greek cities were, on the whole, a socially, economically and politically privileged group, although the exact nature and degree of their privilege varied from city to city. Because of the close linkage seen in Chapter 4 between military service and political privilege, the hoplite class would have been keen to maintain and maximize their special military status, and the way to do this was to limit fighting, so far as possible (and while still advantageous to do so), to the types of fighting at which heavy armed infantry excelled, and in which light infantry and cavalry were at a disadvantage.

Some modern historians have expressed surprise that the Greeks, living in a mountainous landscape, should adopt heavy infantry as their preferred military arm, but the reasons for this are also apparent. Broadly speaking, military functions are related to both geography and social class. Plains produce cavalry, since in purely practical terms they provide good grazing for horses and also open ground on which the speed and manoeuvrability of cavalry are seen to full advantage. Mountains and forests produce light infantry, since the difficult ground precludes the use of cavalry or the raising of horses, and also undermines the effectiveness of slow-moving, closely ordered infantry formations. Cities produce heavy infantry, since they provide dense populations that can be made accustomed to manoeuvring *en masse*, and in which a shared sense of civic duty can encourage the steadfast fighting and willingness to face danger essential to the effectiveness of heavy infantry and (as Aristotle points out) its ability to withstand cavalry. On the other axis, the poorer classes produce light infantry, especially in rural areas, since light infantry equipment is cheap and readily available, and the skills required are easily and naturally acquired, especially among rural warrior populations. The wealthy produce cavalry, since they can afford horses and the land on which to raise them. The middling classes produce heavy infantry, with more expensive equipment but without the special needs of horsemen.

On the outskirts of the Classical Greek world, in areas such as Thessaly and Macedon (extensive plains) or Aetolia (mountains and forests), armies developed based around cavalry or specializing in light infantry. In the heart of the Greek world, cities in constrained plains (in terms of area), dominated by politically privileged middling classes, produced heavy infantry armies. The type of fighting at which heavy infantry excel is the formal pitched battle on good terrain, in which the infantry are able to maintain their close order (to repel cavalry) and to close and fight at close-quarters (rather than being shot down by light infantry). As we saw at the start of this chapter, light infantry could pose a serious threat to heavy infantry, especially on difficult terrain. Many an expensively mounted and armoured knight in later periods of history similarly found that they could be brought down by a peasant (or a yeoman) with a bow. Formal rules outlawing the use of missile weapons, so far as such things existed, and informal rules and norms that tended to minimize the tactical role of missiles, should be seen in this light – they were a way of excluding the socially and politically disadvantaged from playing an important role in military affairs, or still worse from actually dominating their social betters. Naturally enough, in the cold light of practical advantage, and given the varied and constantly changing internal circumstances of many Greek cities, it was often found better to make use of the special abilities of light infantry and cavalry than to outlaw them, but inasmuch as Greek warfare did display a preference for the pitched battle, it was because the pitched battle is the type of fighting at which heavy infantry excel.

The Greeks were able to develop such norms due to the broadly homogenous nature of Greek society, for all the many variations of detail. Though democratic and oligarchic cities might have very different constitutions and citizenships, they were all variants on the common theme of a dominant heavy infantry class, and all Greek cities, whatever their differences, had more in common with each other than they did with external enemies like the (monarchical, imperial) Persians or (tribal) Thracians. This made Greek warfare naturally highly symmetrical, in that both sides in a conflict would have similar values and preferences, and therefore – alongside cultural, linguistic and ethnic homogeneity – this made it easy to develop forms of warfare that advantaged the politically dominant class. Greek armies certainly did on occasion engage in asymmetric warfare – that is, warfare where the two sides did not share such values and had no interest in finding a mutually advantageous way to settle their differences. Hoplite armies operating in the fringe areas of the Greek world would likely encounter peltasts or *psiloi*, as we saw above, while those marching through Asia (like the Ten Thousand) would have to contend with light mobile tribal forces, as well as numerous cavalry. Such asymmetric warfare can pose great difficulties for armies that specialize in high-intensity symmetric combat, as modern American and European armies (and those based on similar principles) have discovered to their cost since the end of the last major symmetric war in 1945. No wonder then that for several hundred years, Greeks frequently fought in what appears to us a formalized way, emphasizing the high-intensity combat at which they excelled.

Chapter 7

The Hoplite Battle

Reconstructing ancient battles

Hoplites could fight in a variety of ways, from small engagements through sieges to amphibious landings, and were by no means restricted to a single type of fighting. But their natural habitat, and the fighting for which their equipment, formation and cultural traditions were optimized, was undoubtedly the pitched battle. Even so, formal large-scale pitched battles are actually quite rare (at least, ones about which we have any sort of information, which is of course only a subset). Very few Archaic battles are known even by name or combatants, and none are known in any detail. The Peloponnesian War, well documented though it is, contains single-figure numbers of large pitched battles of phalanx against phalanx (though there are many other engagements which either are not known in any detail at all or were primarily not of hoplite against hoplite). There are a similar number of battles known from the fourth century, and a handful from the Persian Wars and the rest of the fifth century. Although we have a picture of regular invasions of enemy territory by the constantly feuding Greek cities, leading to formal offers of battle decided by a decisive clash somewhere on the plain, battles are more unusual than we might expect, and where they do occur it is often in the course of more complex campaigning rather than a simple offer and acceptance of battle.

For a very large number of battles, all that we know about them is, at most, the broad geographical area where they occurred, the year and perhaps season, the combatants and a winner (and loser). A typical example might be the Battle of Alope in 431, of which the entirety of our knowledge is this account in Thucydides:

> 'About the same time the Athenians sent thirty ships to cruise round Locris and also to guard Euboea; Cleopompus, son of Clinias, being in command. Making descents from the fleet he ravaged certain places on the sea-coast, and captured Thronium and took hostages from it. He also defeated at Alope the Locrians that had assembled to resist him.' (Thucydides 2.26.1–2)

Here we have at least a date and season, the combatants, one of the commanders and the outcome, though there is no word on the numbers involved and nothing whatever on the events and course of the battle, whether because Thucydides did not know or did not think this minor engagement important enough to record in greater detail.

Another typical example where there is slightly more information is the fighting at Syracuse in 414, of which Thucydides has this to say:

> 'Gylippus [the Spartan commander aiding the Syracusans] ... constantly led out the Syracusans and their allies, and formed them in order of battle in front of the lines [of siegeworks], the Athenians forming against him. At last he thought that the moment was come, and began the attack; and a hand-to-hand fight ensued between the lines, where the Syracusan cavalry could be of no use; and the Syracusans and their allies were defeated and took up their dead under truce, while the Athenians erected a trophy. After this Gylippus called the soldiers together, and said that the fault was not theirs but his; he had kept their lines too much within the works, and had thus deprived them of the services of their cavalry and javelinmen.' (Thucydides 7.5.1–3)

Here we at least have some tactical details – the cavalry and light-armed were not involved due to the location – as well as the build-up to the battle and Gylippus' morale-boosting taking of the blame for the defeat. Yet the course of the fighting itself is entirely encompassed in just twenty-six English words, and we are left to guess how each side deployed, how and why the Syracusans were defeated, and what was the nature of the 'hand-to-hand fight' (something that has particularly troubled modern historians of ancient history, as we will see in the next chapter).[1]

The number of battles for which longer accounts are available, with more tactical detail, is remarkably small, and largely limited to the major engagements (in terms of combatants or importance) of the war. In the pages of Thucydides, just two battles (Mantinea and Delium) have really detailed accounts, and even these amount to no more than a few hundred words at best. The situation for the Persian Wars is similar: Marathon, Thermopylae and Plataea, famous as they are, are each covered in a few hundred words of Herodotus. Marathon in particular is almost totally devoid of any tactical detail at all, which has not prevented a steady stream of books and articles appearing to analyze and reconstruct the battle and offer vastly expanded accounts of the fighting, filling in far more detail than can be found in the few sentences of Herodotus.

This phenomenon has aptly been termed the 'inverted pyramid' effect, where a vast edifice of modern scholarship is built on the exceedingly narrow foundations of a few hundred words in a single ancient source. A few favourite battles have been subjected to a particular storm of scholarship, Marathon (because of its apparent importance and appealing 'East v. West' narrative) and Leuctra (because of its tactical innovation – though there is no agreement what form these tactical innovations actually took) being probably the most studied. As long ago as 1964, even before the modern flood of journal articles and book publishing, one scholar was moved to write an article appropriately titled 'On the Possibility of Reconstructing Marathon and Other Ancient Battles', in which he posed the important question of whether it is really even possible, or meaningful, to 'reconstruct' battles based on such sparse evidence. Despite his answer being largely in the negative, this has done nothing to stem the tide – on which this book also is of course carried along – but does mean that authors since then feel obliged to preface their descriptions of battles with a reference to the article and the problems it raises.[2]

The conclusion of that article was that it was probably not possible to reconstruct ancient battles, even those for which we have relatively detailed accounts, with any degree of certainty. If such an attempt was to be made, the scholar wrote, it would be best done by a more indirect approach, by studying the armies and combatants and the ways in which they operated, attempting to understand the physical realities of warfare at the time and examining the terrain over which the battle was fought. I am of course broadly in agreement with this argument, and this book is in large part an attempt to provide such an underpinning of understanding for the case of Greek hoplites, though I confess to doubts about the value or possibility of understanding the terrain or 'lie of the land'. Emphasis on terrain can be found in many modern works, going back to the seminal works of the nineteenth-century German scholars who walked and mapped the battlefields of ancient Greece, and the opinion is often offered that the best (or only) way to understand a particular ancient battle is to walk the ground over which it was fought. I have doubts about this, partly because of the question of changes to topography and ground cover – which I will examine further in the section on terrain below – and partly because of the very small number of battles in which we can state with reasonable certainty where they were fought.[3]

Great doubt must remain over the validity of reconstructing the course of a particular battle, not only because of the paucity of evidence that has come down to us in almost every case, but also because we might question how much was known about the events of a battle in antiquity, even in the days immediately after the battle took place, and amongst the combatants themselves. The first Duke of Wellington is often quoted in this context, with this observation:

> 'The history of a battle is not unlike the history of a ball. Some individuals may recollect all the little events of which the great result is the battle won or lost, but no individual can recollect the order in which, or the exact moment at which, they occurred, which makes all the difference as to their value or importance.'

In the battles of Wellington's day, there was at least a commanding general whose job it was to position himself so as to obtain an overview of the whole of the battle, so far as possible, and to write an official report on the events of the battle immediately afterward. But the commanders of Greek armies, as we have seen, often saw it as their job to fight the battle, not to watch it. Furthermore, the compact nature of ancient armies and the flat ground over which battles were mostly fought means that it must have been exceptionally difficult for anyone to obtain a clear idea of the course of events. So far as we know, no ancient commander was required to report on the events of a battle afterward (though some, such as Julius Caesar, of course did so for the purposes of self-aggrandizement), and there were no despatches or official histories (at least none that have survived). The course of a battle must therefore have been pieced together from the numerous recollections of a variety of participants, transmitted largely orally, and the accounts that came to be written down and that have survived to this day – such as those of Thucydides – must either have been reconstructed from these recollections or at second- or third-hand via intervening histories now lost. At best, we have no more than anecdotal accounts in which a few details may have been picked out because they seemed important to a few individuals who were interviewed by a later historian; and at worst, accounts which have been processed through decades of misremembrance and mythologizing.[4]

I am not, therefore, optimistic that there is much mileage in attempting to reconstruct – if by which we mean expand on and fill out with greater detail – any accounts of battles from the ancient world. The accounts we have, imperfect though they are, may be the best there can be, and all we can hope to do is properly understand the vocabulary used and the nature, more broadly, of battle in this period. This suggests an alternative approach to trying to reconstruct the events of any individual battle, and that is to create composite accounts of the nature of battle in this period more generally, by drawing details from a large number of individual battles and engagements to obtain a better overall picture of the realities and possibilities of battle. This approach has been particularly fruitful and successful over the past forty years or so, following the 'Face of Battle' technique which I will examine in more detail in the following chapter. This book will therefore follow in this tradition, and rather than attempting detailed accounts of individual battles, I will instead attempt a composite picture of hoplite battle in general, being careful, I hope, to point out the inevitable exceptions to the general rules. In the account of Greek hoplite battle that follows, I will be working thematically, but it is of course important to recognize that there were developments throughout the period in tactics and techniques, and that there was no true perfect battle that represents all the aspects I will discuss.

Numbers

In Chapter 4 I looked briefly at the manpower resources available to each city, and at the ways in which this manpower was mobilized to form an army for a particular campaign. Because of the way armies were mobilized – the hoplite component of armies at any rate – with a call-up of particular age classes or tribal divisions, the numbers made available for a given campaign, and therefore (allowing for stragglers, garrisons, casualties and other sundry losses) the numbers present at any given battle, must have been well known in antiquity, at least to the commanders of armies. Whether or how such information then made its way into the pages of a historian like Xenophon or Thucydides is less clear. Ideally, the historian would have been able to consult written records, though we cannot know how often this was actually the case. Thucydides gives us an insight into the difficulties involved in the case of Mantinea, in a passage we have already encountered in other contexts:

> 'The Lacedaemonian army looked the largest; though as to putting down the numbers of either host, or of the contingents composing it, I could not do so with any accuracy. Owing to the secrecy of their government the number of the Lacedaemonians was not known, and men are so apt to brag about the forces of their country that the estimate of their opponents was not trusted. The following calculation, however, makes it possible to estimate the numbers of the Lacedaemonians present upon this occasion. [Calculation follows based on the numbers and sizes of sub-units.]' (Thucydides 5.68)

Even these calculations, as we have seen, do not provide us with accurate figures for the Spartan army (which in theory should be 448 times eight, or 3,584 men), because of the suspicion that Thucydides has misreported the Spartan unit hierarchy, and because of the vexed question of whether the *perioikoi* were included in the count of the regular Spartan units. At least here, though, we see Thucydides' admission that numbers were often not known (in the case of the Spartans, as a deliberate policy of secrecy) or were prone to exaggeration (up or down). Yet for many battles, the historians provide us with very detailed breakdowns of the forces involved, with numbers for each individual contingent. In Chapter 4 we saw the detailed unit list that Herodotus provided for Plataea. Similar, if usually less detailed, contingent lists are provided for other major battles, for example Thucydides' reckoning of the forces at Delium:

> 'The Boeotians ... appeared over the hill, and halted in the order which they had determined on, to the number of seven thousand hoplites, more than ten thousand *psiloi*, one thousand horse, and five hundred peltasts ... On the side of the Athenians, the hoplites throughout the whole army

formed eight deep, being in numbers equal to the enemy, with the cavalry upon the two wings. *Psiloi* regularly armed there were none in the army, nor had there ever been any at Athens. Those who had joined in the invasion, though many times more numerous than those of the enemy, had mostly followed unarmed [*aoploi*, unequipped], as part of the levy in mass of the citizens and foreigners at Athens, and having started first on their way home were not present in any number.' (Thucydides 4.93.3–5, 4.94.1)

Such numbers, though not subdivided by unit, sound precise and accurate enough, though where we have another historian's account of a battle, numbers given may vary spectacularly. Diodorus, for example, records of this campaign:

'Pagondas, who commanded the Boeotians, having summoned soldiers from all the cities of Boeotia, came to Delium with a great army, since he had little less than twenty thousand infantry and about a thousand cavalry.' (Diodorus 12.69.3)

These numbers can be made to roughly match those of Thucydides if we assume that Diodrous counted all the infantry, hoplites, *psiloi* and peltasts, which in Thucydides' account come to 17,500, which is a little less than 20,000. For other battles we are not always so lucky, and the numbers given can vary irreconcilably between different accounts. Xenophon often does not give numbers at all, though he will at least give a number of units involved, as for the Spartan army for the Leuctra campaign:

'[T]he Lacedaemonians sent Cleombrotus, the king, across to Phocis by sea, and with him four *morai* of their own and the corresponding contingents of the allies.' (Xenophon, *Hellenica* 6.1.1)

This would at least allow us to calculate, in the manner of Thucydides, the size of the Spartan army if we knew precisely how strong the *morai* were, which we do not.

I am inclined to believe, therefore, that the question of numbers, for all that it is the start and indeed the end point of many modern investigations of ancient battles, may not be a fruitful path of investigation. Even in antiquity, numbers were evidently not known with any great accuracy (except, one hopes, to the army's commanders), as historians may or may not have recorded numbers correctly, and where it so happens that we have a precise number, especially in a later historian like Diodorus or Plutarch, we should treat it with the greatest suspicion.

Numbers also probably mattered comparatively little in determining the course and outcome of a battle. As we will see, ancient battles were emphatically not battles of attrition in which the 'big battalions' might be expected to grind down their less numerous foes. Because hoplites in particular fought in a single line (phalanx), usually without reserves, because the breaking and rout of a phalanx would usually be final and decisive, and because such breaking was due to many factors – of which the raw numbers are probably the least important – the numbers present on each side is perhaps one of the least important pieces of information we can have about a battle. However, some general observations can be made. One is that hoplite numbers were often only a fairly small proportion of the total, as for the Thebans at Delium – where the hoplites made up less than half of the total infantry force – or the Spartans at Plataea, although the numerous *psiloi* at both battles seem to have had no effect at all on the course of the action. Hoplites generally numbered in the low thousands; a hoplite force of 10,000 represents a very large army. Battles also tended to be fought between forces approximately equal in size – as we saw in the previous chapter, an army that saw itself to be completely outmatched would simply not fight. Finally, the numbers attributed to Persian armies are (naturally) to be trusted even less than those attributed to Greeks, with Thucydides' observation that 'men are so apt to brag about the forces of their country' (or in this case, the forces of their defeated opponents) applying particularly strongly.[5]

Rules and customs

The concept of warfare consisting chiefly of the formal, tradition-bound ('agonal') battle is, as we saw in the previous chapter, probably overstated. Greek warfare always saw a tension between the ethos of 'fair and open battle' (Achilles and Ares) and trickery and guile (Odysseus and Athena) – in which, of course, it is far from unique. We can be fairly certain at any rate that Greeks serving as mercenaries as far afield as Egypt, or Greek colonists carving out territories for themselves on the Mediterranean littoral at the expense of the local inhabitants, would not feel themselves bound by any 'rules of war'. In battles between those with much in common socially, economically, culturally and ethnically, it is more likely that unwritten rules, conventions and norms will be followed, and that shared conventions – precisely because they are shared – are more likely to be observed. Some of the more formal aspects of Greek battle can be understood in this way, in particular the request by the loser for access to their dead and the erection of a trophy by the winner, which I will examine further below. But it is likely that, just as a Spartan king like Agesilaus would happily resort to guile where it offered an advantage, such schemes and stratagems had always been a part of Greek warfare (exemplified in myth by the capture of Troy through the use of the Trojan Horse), even alongside what appear to be more formal aspects such as the offer and acceptance of battle and formal deployment. Certainly there is no evidence for actual written rules of war, and no central authority that could have imposed and policed such rules. The only possible candidate for such rules would be the supposed ban on the use of missiles described by Polybius (13.3.2–4), but whether this represents an actual historical event is disputed; even if it did, it is best seen, as discussed in Chapter 6, as a means of excluding socially disadvantaged classes from playing a significant role in battle. Other norms certainly existed and were observed in hoplite warfare, but could just as easily be ignored if some advantage was offered by doing so. We must also be careful not to include norms which may have had a strong practical value (such as the offer of battle or reluctance to engage at night) amongst such rules of war – there were often very good practical reasons for both sides to avoid the uncertainties inherent in less-regular types of warfare, and a strong incentive for both to play safe by sticking to the rules.[6]

Nevertheless, that there were customs which governed (or at least moderated) relations between cities – and also between cities and outside agents such as the Persians – is undeniable. In some cases we see these as common practices, such as the way in which the winner of a battle would mark their victory by erecting a trophy (which usually consisted of captured items of armour attached to a tree or post) and stripping the bodies of the dead, while the losers would acknowledge defeat by sending a herald under truce to apply to the victors for the return of their own dead (a request which was almost never denied). It was almost universal practice between Greek cities to follow this procedure, although we cannot tell if it was of any great antiquity, since it is known only from the early fifth century. Whether we choose to call these rules or customs is largely immaterial; they were not rules in that they were probably nowhere codified, and there were no sanctions (other than disapproval) that could be brought to bear against transgressors, but they were followed so routinely that they had the force of rules, and genuinely constrained the behaviour of armies at the time. At the same time, it is easy to see the practical benefit to both sides of such an arrangement. Between culturally and socially similar opponents, such agreements would develop naturally, and in a society where the treatment of war dead and recovery of bodies was of great importance, such customs gave the combatants on both sides the reassurance that, if the worst should happen, their bodies would at least receive due honours (Onas., *Strat*.36.1–2). They also provided a mechanism by which otherwise indecisive battles might be decided 'on points' as it were, a useful feature where the function of battle was to decide political disputes rather than to destroy the enemy, and where long campaigns and frequent battles were undesirable. This does not mean that Greek warfare was formalized and that all available advantages would not be sought by either side, and it is a false dichotomy to divide formalized warfare from total, unregulated warfare in this way. As with other periods and cultures, Greek warfare followed certain mutually agreeable and practical conventions where appropriate, while also allowing plentiful opportunities for trickery and guile.

That this arrangement could sometimes produce a tension between following conventions and seeking advantage (practical or propaganda) is well illustrated by the negotiations that took place after the Battle of Delium (424). The Boeotians were victorious in this battle, defeating an invading Athenian army and erecting a trophy on the battlefield:

> 'Meanwhile a herald came from the Athenians to ask for the dead, but was met and turned back by a Boeotian herald, who told him that he would effect nothing until the return of himself, the Boeotian herald, and who then went on to the Athenians, and told them on the part of the Boeotians that they had done wrong in transgressing the law of the Hellenes. Of what use was the universal custom protecting the temples in an invaded country if the Athenians were to fortify Delium and live there.' (Thucydides 4.97.2–3)

The 'law of the Hellenes' (*ton nomon tois Hellesin*) could also be translated 'the custom of the Hellenes', *nomos* being the usual word for 'customs' as well as 'laws', as we will see in examples below. In the above case, the Boeotians were objecting to the Athenians' occupation of a temple, that should have been held sacred and inviolate.

> 'Accordingly for the god as well as for themselves, in the name of the deities concerned, and of Apollo, the Boeotians invited them first to evacuate the temple, if they wished to take up the dead that belonged to them.' (Thucydides 4.97.4)

The Athenians sent back their own herald to argue their case, who made a number of legalistic arguments, including those regarding the fate of the temple. In the first place, they argued that 'the law of the Hellenes' gave them possession of the temples in any territory they conquered, so it was their duty to ensure that proper religious process was followed, not the Boeotians'. Also, they had exploited the temple only under military necessity:

> 'Besides, anything done under the pressure of war and danger might reasonably claim indulgence even in the eye of the god … In short, which were most impious – the Boeotians who wished to barter dead bodies for holy places, or the Athenians who refused to give up holy places to obtain what was theirs by right? … They stood where they stood by the right of the sword [spear]. All that the Boeotians had to do was to tell them to take

up their dead under a truce according to the national custom [*ta patria*].' (Thucydides 4.98.6–8)

These arguments were not put before a court of arbitration, but were used in the direct dispute between the two parties (we also, of course, do not know precisely how Thucydides knew what was argued, and must trust that he represents the two sides accurately). They clearly show the force of the 'law of the Hellenes' in moderating behaviour, but the Athenians' argument that 'anything done under the pressure of war and danger might reasonably claim indulgence even in the eye of the god' is striking – 'all's fair in love and war' was the Athenians' argument, even though they still went to the trouble of formulating reasons why they should not have to barter for their bodies. These two conflicting impulses – to follow the rules (especially where doing so inconvenienced one's enemies) or to ignore them (especially where doing so gave advantage to oneself) – are a common feature of Greek warfare, and there seems no compelling reason to suppose that such arguments were a new feature of the late fifth century, rather than being as old as Greek warfare itself. In this case, the Athenians failed to shift the Boeotians by their arguments, but the Boeotians relented after they had achieved their military objectives by recapturing the town of Delium.

We might contrast the Athenians' actions in this case with those of the Spartan king Agesilaus, who had the luxury of having just won a major victory over the Boeotians at Coronea (394):

'Now when the victory had fallen to Agesilaus and he himself had been carried, wounded, to the phalanx, some of the horsemen rode up and told him that about eighty of the enemy, still armed, had taken shelter in the temple of Athena, and asked him what they should do. And he, although he had received many wounds, nevertheless did not forget the deity, but ordered them to allow these men to go away whithersoever they wished, and would permit them [his men] to commit no wrong.' (Xenophon, *Hellenica* 4.3.20)

Xenophon approved of Agesilaus' piety, but he also praised him (as we saw in Chapter 5) for seeking whatever advantage he could in war on other occasions. It was easy to follow the rules when victory was already in the bag.

The same word – *nomos* – could be applied to such grand quasi-legal concepts and also to more local, small-scale customs. Herodotus recounts the well-known story of how Persian scouts before Thermopylae (480) observed the Spartans combing their hair. Xerxes' Greek adviser, Demaratus, explained the reason:

'These men have come to fight us for the pass, and it is for this that they are preparing. This is their custom [*nomos*]: when they are about to risk their lives, they arrange their hair.' (Herodotus 7.209.3)

The Greeks were of course not the only ones to be bound (or at least, guided) by custom – the Persian commanders before Plataea (479) disagreed about the wisdom of fighting at that time and place:

> 'Mardonius' counsel, however, was more vehement and intemperate and not at all leaning to moderation. He said that he thought that their army was much stronger than the Greeks and that they should give battle with all speed so as not to let more Greeks muster than were mustered already. As for the sacrifices of Hegesistratus, let them pay no heed to these, nor seek to wring good from them, but rather give battle after Persian custom [*nomos*].' (Herodotus 9.41.4)

This is a characteristic mix of practical arguments (they should fight before the Greeks received reinforcements) and invoking custom (they must fight as that is the Persian custom), and a reminder that a preference for pitched battle was by no means unique to the Greeks (or to 'the West'). If Mardonius had felt that practical benefit and custom did not align, he would no doubt have shaped his arguments differently.

The episode of the Persian heralds sent by Darius to Greece to demand 'earth and water' of the Athenians and Spartans, as symbols of submission, reminds us that many such customs were shared between Greeks and Persians. The Athenians and Spartans threw the heralds into a pit and a well to find their 'earth and water' there (Hdt. 7.133), but this act was contrary to the rules of religion and of warfare. Herodotus took it for granted that Athens and Sparta would be punished for it (by the gods, via some earthly instrument):

> 'What calamity befell the Athenians for dealing in this way with the heralds I cannot say, save that their land and their city were laid waste [although he attributes this to punishment for some other transgression] ... Now there was a long period after the incident I have mentioned above during which the Spartans were unable to obtain good omens from sacrifice. The Lacedaemonians were grieved and dismayed by this and frequently called assemblies, making a proclamation inviting some Lacedaemonian to give his life for Sparta. Then two Spartans of noble birth and great wealth, Sperthias son of Aneristus and Bulis son of Nicolaus, undertook of their own free will to make atonement to Xerxes for Darius' heralds who had been killed at Sparta. Thereupon the Spartans sent these men to Media for execution.' (Herodotus 7.133.2 – 7.134.3)

Xerxes magnanimously refused to kill the men:

"'You,' said he, 'made havoc of all human law by slaying heralds, but I will not do that for which I censure you, nor by putting you in turn to death will I set the Lacedaemonians free from this guilt.'" (Herodotus 7.136.2)

In the end, Sparta's guilt was assuaged, most indirectly, when the sons of these men, fifty years later in the Peloponnesian War, went as heralds to the Thracians, were handed over by them to the Athenians and were executed (Hdt. 7.137.2–3).[7]

So warfare in Greece – and with and between Greece's neighbours – was bound by laws and customs, which could, however (like all laws, in the absence of a non-divine authority to enforce them), be ignored on occasion. Greek warfare does not come across as being unusually formal in this regard, and there are many other factors in play that made such customs desirable, yet these customs, traditions, common practices and rules (some of which we will encounter below) did shape the way in which Greek warfare – and in particular its most formal manifestation, the pitched battle – took place.

Offering battle

In order for two armies to fight a battle, they had to come together on the battlefield, at the same time and place, and deploy in formation ready for the battle to start. Other types of battle could occur, such as the meeting engagement, where two armies on the march meet each other unexpectedly and fight as they are, without any formal deployment; or ambushes, where one side lies in wait for the other to march past and attacks them as they do so. Throughout military history, all these types of battle have co-existed alongside each other, taking place with greater or lesser regularity, and Classical Greece is no exception – battles of all these types can be found in the historical accounts.

But it is certainly true that the formal pitched battle appears most often in surviving accounts. Meeting engagements and ambushes would be less likely, given the small scale of most Archaic and Classical campaigns, in which armies were often fighting close neighbours in territory familiar to both. In order to fight a pitched battle, both sides had to be willing to fight – otherwise, the side that did not rate its chances highly would remain within fortifications or on difficult or advantageous terrain, and in these circumstances no battle would occur if the other side did not wish to engage in a risky attack on such a defended position. Many battles thus began with a formal-seeming 'offer' of battle, where the defending side would leave its advantageous position, its city or its camp, and advance to such a point that a regular battle could occur. This phenomenon is seen throughout ancient history, and indeed in more recent times, and although the mutual agreement required for such battles sometimes seems surprising, there were good practical reasons for it.[8]

There was no formal offer of battle as such (such as by exchange of heralds), but the simple presence of both armies on a suitable battlefield, or the presence of one and the arrival of the other, was an indication that both were willing for a pitched battle to occur, and so constituted the 'offer'. Before Potidaea (432), for example:

> '[T]he Athenians themselves broke up their camp and marched against Potidaea. After they had arrived at the isthmus, and saw the enemy preparing for battle, they formed against him, and soon afterwards engaged.' (Thucydides 1.62.4–5)

A similar process preceded the battle at Lyncestis (423), though divided into separate engagements for infantry and cavalry:

> 'The infantry on either side were upon a hill, with a plain between them, into which the horse of both armies first galloped down, and engaged in a cavalry action. After this the Lyncestian hoplites advanced from their hill to join their cavalry and offered battle; upon which Brasidas and Perdiccas also came down to meet them, and engaged and routed them with heavy loss; the survivors taking refuge upon the heights and there remaining inactive.' (Thucydides 4.124.3–4)

'Offering battle' in this case means simply putting one's army in a position where an equal battle (without advantage of terrain to either side) could occur. Note also that just as an army could avoid battle by taking up a position on a hill and not descending to the plain to fight, a defeated army might retreat onto a hill and there make a stand, in the expectation that the enemy would not risk an unequal engagement against them, even after winning the first round. This is what happened after the Battle of Solygeia (425):

> '[The defeated Corinthian army] retired to the high ground and there took up its position. The Athenians, finding that the enemy no longer offered to engage them, stripped his dead and took up their own and immediately set up a trophy.' (Thucydides 4.44.2–3)

Refusing an offer of battle, or failing to make such an offer, was of course an indication of weakness, a sign that one had no confidence in the outcome of a fair fight. After the Thebans had defeated a Spartan force at Haliartus (395), the Spartan king Pausanias arrived with reinforcements:

> 'When, however, on the next day the Athenians arrived and formed in line of battle with them [the Thebans], while Pausanias did not advance against them nor offer battle, then the elation of the Thebans increased greatly.' (Xenophon, *Hellenica* 3.5.22)

Offering or not offering battle was therefore something that had to be judged finely, weighing the risks of battle against the damage done to morale, and the boost to the confidence of the enemy, of a failure to fight. We see several variants on this in the Athenian expedition to Syracuse (415). Shortly after the Athenians arrived, nervous of the Syracusan advantage in cavalry, they took up position protected by a palisaded camp. The Syracusans advanced to meet them:

> 'At first they came close up to the Athenian army, and then, finding that they did not offer to engage, crossed the Helorine road and encamped for the night.' (Thucydides 6.66.3)

The next day, however, the Athenians did march out to give battle:

> 'The Syracusans were not at that moment expecting an immediate engagement, and some had even gone away to the town, which was close by; these now ran up as hard as they could, and though behind time, took their places here or there in the main body as fast as they joined it.' (Thucydides 6.69.1)

In later fighting around Epipolae (414), the Athenians were engaged in building siege works around Syracuse, while the Syracusans were now commanded by the Spartan Gylippus, in the example we saw earlier:

> 'Gylippus ... constantly led out the Syracusans and their allies, and formed them in order of battle in front of the [siege] lines, the Athenians forming against him. At last he thought that the moment was come, and began the attack.' (Thucydides 7.5.1–2)

The Syracusans were defeated, leading to Gylippus' apology to his army for offering battle too soon, and his working to improve their morale and prospects in a future engagement:

> 'After this he embraced the first opportunity that offered of again leading them against the enemy. Now Nicias and the Athenians were of opinion that even if the Syracusans should not wish to offer battle, it was necessary for them to prevent the building of the cross wall, as it already almost overlapped the extreme point of their own, and if it went any further it would from that moment make no difference whether they fought ever so many successful actions, or never fought at all. They accordingly came out to meet the Syracusans. Gylippus led out his heavy infantry further from the fortifications than on the former occasion, and so joined battle.' (Thucydides 7.6.1–2)

Note again the necessity of mutual agreement for a battle to be fought, in which a number of factors (here the progress of the siegeworks and the potential role of the Syracusan cavalry) all had to be weighed, and how the offer of battle took the form of advancing a sufficient distance from the fortifications so that a fair battle could take place. If the Syracusans had remained too close to their walls, the Athenians would simply not have engaged them at a disadvantage.

Other forms of battle

Formal pitched battles of this sort were not the only type of battle fought, however, and there are examples of other types of engagement, and of battles in which stratagems and guile played a greater part than in such fair and open battles. At Stratus (429), a Peloponnesian force with Chaonian allies advanced against the Acarnanian town:

> 'While they [the Chaonians] were coming on, the Stratians, becoming aware how things stood, and thinking that the defeat of this division would considerably dishearten the Hellenes behind it, occupied the environs of the town with ambuscades, and as soon as they approached engaged them at close quarters from the city and the ambuscades. A panic seizing the Chaonians, great numbers of them were slain; and as soon as they were seen to give way the rest of the barbarians turned and fled.' (Thucydides 2.81.5–6)

An ambush of course was not always successful, as when the Athenian force landed in Sicily:

> 'Two Messinese battalions [*phylai*] in garrison at Mylae laid an ambush for the party landing from the ships, but were routed with great slaughter by the Athenians and their allies.' (Thucydides 3.90.2)

An ambush could also be laid to be sprung in the course of an otherwise frontal engagement, such as that by the Athenian general Demosthenes commanding an Acarnanian force at Olpae (426/425):

> 'Demosthenes led them near to Olpae and encamped, a great ravine separating the two armies. During five days they remained inactive; on the sixth both sides formed in order of battle. The army of the Peloponnesians was the largest and outflanked their opponents; and Demosthenes fearing that his right might be surrounded, placed in ambush in a hollow way overgrown with bushes some four hundred hoplites and light troops, who were to rise up at the moment of the onset behind the projecting left wing of the enemy, and to take them in the rear.' (Thucydides 3.107.3)

On this occasion the plan worked perfectly:

> '[T]he Acarnanians from the ambuscade set upon them [the Peloponnesians] from behind, and broke them at the first attack, without their staying to resist; while the panic into which they fell caused the flight of most of their army, terrified beyond measure at seeing the division of Eurylochus and their best troops cut to pieces.' (Thucydides 3.108.1)

Demosthenes was a general particularly noted for his use of guile, being also responsible for a brutally effective surprise attack at Idomene (426/425), described in the previous chapter. Demosthenes himself had learned the use of such irregular tactics the hard way, having commanded an Athenian force that was destroyed by light infantry, as we have seen.

Surprise attacks, or at least unexpected advances, could also form an element of more formal battles between frontally deployed armies. We have already seen above how the Syracusan army was taken by surprise by the Athenians advancing to give battle after many of the Syracusans had remembered urgent errands they had to do in town. A number of major pitched battles of the period began somewhat unexpectedly (to one side) in a similar way. At the first Battle of Mantinea (418), the Spartan army had already faced the allies but battle was avoided on that occasion owing to the allies' strong position on a hill. When the Spartans returned:

> '[They] suddenly saw their adversaries close in front of them, all in complete order, and advanced from the hill. A shock like that of the present moment the Lacedaemonians do not ever remember to have experienced: there was scant time for preparation, as they instantly and hastily fell into their ranks, Agis, their king, directing everything, agreeably to the law [*nomos*].' (Thucydides 5.66.1–2)

At the Nemea (394), a Spartan army was similarly surprised:

> 'Now for a time the Lacedaemonians did not perceive that the enemy were advancing; for the place was thickly overgrown; but when the latter struck up the paean, then at length they knew, and immediately gave orders in their turn that all should make ready for battle.' (Xenophon, *Hellenica* 4.2.19)

This tells us a lot about the scouting capabilities (or customs) of Classical armies – evidently the Spartans had not placed any sort of pickets in front of their position to warn of enemy approach, although such pickets were used on other occasions. The way in which the defenders of the pass at Thermopylae (480) gained intelligence of the Persian flanking move is one example:

> 'The seer Megistias, examining the sacrifices, first told the Hellenes at Thermopylae that death was coming to them with the dawn. Then

deserters came who announced the circuit made by the Persians. These gave their signals while it was still night; a third report came from the watchers running down from the heights at dawn.' (Herodotus 7.219.1)

At least some system of signalling was in place (using fire signals presumably), and examining animal entrails was not the only means of gaining intelligence.[9]

The Theban general Epaminondas made a particular feature of the unexpected advance. The start of the Battle of Leuctra (371) came as a surprise to at least some of the Spartans, including their king Cleombrotus, in Xenophon's version of events:

'Now when Cleombrotus began to lead his army against the enemy, in the first place, before the troops under him so much as perceived that he was advancing, the horsemen had already joined battle and those of the Lacedaemonians had speedily been worsted; then in their flight they had fallen foul of their own hoplites, and, besides, the companies of the Thebans were now charging upon them.' (Xenophon, *Hellenica* 6.4.13)

At the second Battle of Mantinea (362), Epaminondas surprised the Spartans on the open field again:

'In the first place, as was natural, he formed them [the Boeotian army] in line of battle. And by doing this he seemed to make it clear that he was preparing for an engagement; but when his army had been drawn up as he wished it to be, he did not advance by the shortest route towards the enemy, but led the way towards the mountains which lie to the westward and over against Tegea, so that he gave the enemy the impression that he would not join battle on that day.' (Xenophon, *Hellenica* 7.5.21)

This was a ruse to cause the enemy 'a relaxation of their mental readiness for fighting, and likewise a relaxation of their readiness as regards their array for battle'. Epaminondas grounded arms as if about to camp and made some tactical manoeuvres (see below), then:

'[H]e gave the order to take up arms and led the advance; and his troops followed. Now as soon as the enemy saw them unexpectedly approaching, no one among them was able to keep quiet, but some began running to their posts, others forming into line, others bridling horses, and others putting on breast-plates, while all were like men who were about to suffer, rather than to inflict, harm.' (Xenophon, *Hellenica* 7.5.22)

Being taken by surprise in this way seems to be rather a feature of Spartan armies, though often their high level of training and delegated command structure allowed them to adjust in time. Sometimes the boot was on the other foot, the Spartan general Brasidas defeating an Athenian army at Amphipolis (422) by means of a surprise attack from the town while the Athenians were manoeuvring. Brasidas' calculations on this point (as recorded by Thucydides) are of interest:

> 'He did not venture to go out in regular order against the Athenians: he mistrusted his strength, and thought it inadequate to the attempt; not in numbers – these were not so unequal – but in quality, the flower of the Athenian army being in the field, with the best of the Lemnians and Imbrians. He therefore prepared to assail them by stratagem [*techne*].' (Thucydides 5.8.2)

To 'go out in regular order [*antitaxis*]' required and implied an assumption of equal strength and equal chances on both sides; if either side felt itself outmatched, it had either to avoid battle completely or to seek some unfair advantage. Thucydides gives Brasidas a speech to his men justifying this approach:

> 'But the most successful soldier will always be the man who most happily detects a blunder like this [the Athenians' bungled manoeuvres], and who carefully consulting his own means makes his attack not so much by open and regular approaches, as by seizing the opportunity of the moment; and these stratagems, which do the greatest service to our friends by most completely deceiving our enemies, have the most brilliant name in war.' (Thucydides 5.9.4–5)

That stratagems could have 'the most brilliant name in war' by the last third of the fifth century does rather suggest that they would not have been considered unacceptable a century earlier. The tension between open battle with even chances, and making one's own chances where odds were not even, must always have existed wherever armies were not perfectly matched in strength and prospects.

Deployment

Assuming open battle was desired by both sides, and that battle was offered and accepted, then the two armies would deploy into line of battle, that is phalanx formation for the hoplites, with light infantry and cavalry probably on the flanks, as we will see. In later periods where fortified camps were the norm, and where armies were very large (upward of 50,000 men), this was a quite formal process of exiting the camp and forming line within no more than a few hundred metres of the enemy, with each side completing their deployment before the signal was given for the battle to begin. Classical battles were if anything less formal than this. Armies tended not to be deployed from fortified camps (obviously there are exceptions), and one side would often be in their home city and would march out and deploy from there. There are also several examples of armies coming upon each other unexpectedly or of them stealing a march by a rapid advance before the other side was fully prepared, as we saw above. Such unexpected openings were made more likely by the apparently rudimentary level of scouting and battlefield intelligence carried out by Classical armies, and by the lack of well organized light infantry and cavalry forces which would, in later periods, skirmish in the gap between two armies or camps, so as to provide a buffer against, and early warning of, an unexpected advance. Even so, armies would ordinarily have to deploy either from a column of march into line of battle, or through the gates of a city or camp, a process which must have taken some considerable time and been susceptible to some confusion if done badly, but which literary accounts tend to refer to, if at all, as simply 'forming up' or some similar expression.[10]

Sometimes, as in a meeting engagement, this process would have been more hurried, and Xenophon provides a vivid account of the way the Greeks in Cyrus' army had to deploy rapidly before Cunaxa (401) on receiving news of the approach of the royal army (which, as it turned out, was further away than they feared):

> 'It was now about full-market time and the stopping-place where Cyrus was intending to halt had been almost reached, when Pategyas, a trusty Persian of Cyrus' staff, came into sight, riding at full speed, with his horse in a sweat, and at once shouted out to everyone he met, in the barbarian tongue and in Greek, that the King was approaching with a large army, all ready for battle. Then ensued great confusion; for the thought of the Greeks, and of all the rest in fact, was that he would fall upon them immediately, while they were in disorder ... Thereupon they proceeded in great haste to take their places ... At this critical time the King's army was advancing evenly, while the Greek force, still remaining in the same place, was forming its line from those who were still coming up.' (Xenophon, *Anabasis* 1.8.1–2; 4; 14)

Being caught by an enemy before fully forming up was a source of fear, since the disorganized side would be at a great disadvantage in battle, as we saw from the various surprise engagements above. Ideally, an army would have completed its deployment before the fighting started, and in this case there were two principal decisions that had to be taken – the position of the various contingents (usually city contingents in an army made up of several allies, though picked units might also need to be specially positioned) and the depth at which each unit was to form.

To determine position in the line it appears, in allied armies or even among the tribes of a single army such as the Athenian, that there was an established order of precedence, with the right of the line being the most prestigious position which would therefore usually be occupied by the senior, and often highest quality, unit. Sometimes this order of precedence would rotate from day to day. The Athenian army at Marathon (490) was deployed by its commander Militades (whose command was also rotated, each *strategos* present being in overall command for one day at a time):

> 'When the presidency came round to him, he arrayed the Athenians for battle, with the *polemarchos* Callimachus commanding the right wing, since it was then the Athenian custom [*nomos*] for the *polemarchos* to hold the right wing. He led, and the other tribes [*phylai*] were numbered out in succession next to each other.' (Herodotus 6.111.1)

If the right wing position was most prestigious, the left wing was second most, leading to discord at Plataea (479):

> 'During the drawing up of battle formation there arose much dispute [*othismos*] between the Tegeans and the Athenians, for each of them claimed that they should hold the second wing of the army, justifying themselves by tales of deeds new and old.' (Herodotus 9.26.1)

Note that position in the line was determined by prestige (and argument), not by tactical considerations. On this occasion, the Athenians' arguments won the day:

> '[T]he whole army shouted aloud that the Athenians were worthier to hold the wing than the Arcadians. It was in this way that the Athenians were preferred to the men of Tegea, and gained that place.' (Herodotus 9.28.2)

However, according to Herodotus, when Pausanias, the Spartan commander, heard that the Persians were planning to attack, he had a change of heart:

> 'At the message Pausanias was terrified by the Persians, and said: "Since, therefore, the battle is to begin at dawn, it is best that you Athenians should take your stand opposite the Persians, and we opposite the Boeotians and the Greeks who are posted opposite you; for you have fought with the Medes at Marathon and know them and their manner of fighting while we have no experience or knowledge of those men. We Spartans have experience of the Boeotians and Thessalians, but not one of us has experience with the Medes. No, rather let us take up our equipment and change places, you to this wing and we to the left." "We, too," the Athenians answered, "even from the moment when we saw the Persians posted opposite you, had it in mind to make that suggestion which now has first come from you. We feared, however, that we would displease you by making it. But since you

have spoken the wish yourselves, we too hear your words very gladly and are ready to do as you say."' (Herodotus 9.46.1–3)

So here tactical considerations (the Athenian familiarity, from Marathon, with Persian tactics) did take precedence over traditional notions of prestige, but only when proposed by the senior commander himself (there is, however, good reason to doubt the details of Herodotus' story here, since in the event the switch was not made).

Deployment for the Battle of Mantinea (418) also shows a mix of tradition and some practical considerations:

'In this battle the left wing was composed of the Sciritae, who in a Lacedaemonian army have always that post to themselves alone; next to these were the soldiers of Brasidas from Thrace, and the *Neodamodes* with them; then came the Lacedaemonians themselves, company after company, with the Arcadians of Heraea at their side. After these were the Maenalians, and on the right wing the Tegeans with a few of the Lacedaemonians at the extremity; their cavalry being posted upon the two wings. Such was the Lacedaemonian formation. That of their opponents was as follows: on the right were the Mantineans, the action taking place in their country: next to them the allies from Arcadia; after whom came the thousand picked men of the Argives, to whom the state had given a long course of military training at the public expense; next to them the rest of the Argives, and after them their allies, the Cleonaeans and Orneans, and lastly the Athenians on the extreme left, and their own cavalry with them.' (Thucydides 5.67.1–2)

Note that the position of the Sciritae on the left is down to tradition, while the Mantineans hold the allied right in their capacity as hosts of the battle, with the Athenians ceding first place and occupying the left. This practice of placing the best contingents on the right wing of each army meant that very often, the right wing of each army would be victorious, resulting in a disputed outcome overall unless, as at Coronea, the two victorious right wings had a rematch against one another. At the Nemea (394), the Boeotians exploited this, along with the notion of rotating the precedence order day to day, to avoid having to fight the formidable Spartans, at least in Xenophon's hostile account:

'Now the Boeotians, so long as they occupied the left wing, were not in the least eager to join battle; but when the Athenians took position opposite the Lacedaemonians, and the Boeotians themselves got the right wing and were stationed opposite the Achaeans, they immediately said that the sacrifices were favourable and gave the order to make ready, saying that there would be a battle.' (Xenophon, *Hellenica* 4.2.18)

One of the main tactical innovations of the Theban general Epaminondas was to break with the tradition of forming the best men on the right, or using notions of prestige to decide the order of deployment, and instead forming his best units on his left opposite the Spartans, so that by defeating the Spartans in a single decisive clash at the outset of the battle, he could avoid either the uncertainty of leaving the outcome to inferior allied forces, or the risk of an indecisive battle with both right wings victorious – I will look further at this innovation below.[11]

The depth to be adopted by each part of the line has already formed part of the subject of Chapter 3. Here we need only note again that, rather surprisingly to modern eyes, rather than there being an overall tactical plan determined by the commanding general, depth might either be left to individual unit commanders, as at Mantinea (418) – 'as to the depth, although they had not been all drawn up alike, but as each *lochagos* chose, they were generally ranged eight deep' (Thuc. 5.68.3) – or to individual city contingents, as at Delium (424), where 'the Thebans formed twenty-five shields deep, the rest as they pleased' (Thuc. 4.94.4), or Nemea (394):

> '[Before the battle the allies] were negotiating about the leadership and trying to come to an agreement with one another as to the number of ranks in depth in which the whole army should be drawn up, in order to prevent the states from making their phalanxes too deep and thus giving the enemy a chance of surrounding them.' (Xenophon, *Hellenica* 4.2.13)

A deeper phalanx would be safer, as it was less likely to be broken through by the enemy and there was less danger of suffering serious losses in the pursuit, but a deeper, narrower formation would occupy less frontage, so contingents to either side would have to form shallow lines to make up the space or risk being outflanked themselves. At the Nemea, the Thebans disregarded the agreed depth (sixteen ranks) and formed 'exceedingly deep' (Xen., *Hell.* 4.2.18). The Thebans had long experimented with deep deployments – at the Nemea greater depth worked, in that the Theban contingent was victorious, but the resulting narrow front of the whole army allowed the Spartans to outflank the Athenians and so roll up the whole allied line. Epaminondas' deployments at Leuctra (371) and Second Mantinea (362) used another 'exceedingly deep' Theban phalanx up to fifty ranks deep (Xen., *Hell.* 6.4.12), but by matching it against the Spartan right and also by other tactical innovations (see below) avoided falling into the same trap as at the Nemea. As to how and why, precisely, a deep formation gave an advantage in battle, this controversial topic will be considered in Chapter 8.[12]

Among the considerations to be taken into account when deciding on a depth to form up would be the numbers available to each side, the available space, and the quality and experience of the men on each side. An outnumbered force might have to form a shallower phalanx to avoid being outflanked by a larger army, though this could also work the other way, a large force deliberately forming deep against a smaller one so as to use its numerical advantage to gain the advantage of depth, rather than in an attempt to outflank; the Persians at Plataea (479) deployed this way, 'seeing that the Persians by far outnumbered the Lacedaemonians, they were arrayed in deeper ranks [*taxis*]' (Hdt. 9.31.2). In a confined space, both armies might have to form deep in order to fit all their force into the space available, as at Munichia, a battle fought in the streets of the town:

> 'And the men from the city [Athens], when they came to the market-place of Hippodamus, first formed themselves in line of battle, so that they filled the road which leads to the temple of Artemis of Munichia and the sanctuary of Bendis; and they made a line not less than fifty shields in depth; then, in this formation, they advanced up the hill. As for the men from Phyle, they too filled the road, but they made a line not more than ten hoplites in depth.' (Xenophon, *Hellenica* 2.4.11–12)

In the first fighting at Syracuse, the Athenians, as a picked force, had greater confidence than the mass levy of Syracusans, so formed up eight deep (and held back a reserve), while the Syracusans formed up sixteen deep (Thuc. 6.67.1–2).

Because the depth adopted was so often left to the initiative of junior or contingent commanders, and because there were so many different considerations in determining depth, one oft-quoted example of the use of different phalanx depths – the Athenian phalanx at Marathon (490) – must be treated with caution. In Herodotus' account:

> 'The centre, where the line was weakest, was only a few ranks [*taxeis*] deep, but each wing was strong in numbers.' (Herodotus 6.11.3)

In the ensuing fighting, the Persians broke through the Athenian centre but the Athenian wings were each victorious and able to envelop the Persians. This has been put down to a deliberate ploy of Miltiades, but Herodotus' language is more vague than the translation above suggests, and the course of later battles, where victorious wings were only very rarely able to perform the manoeuvres necessary to re-engage a victorious enemy portion of the line, suggests that the idea of a deliberate Athenian double envelopment at Marathon may be a misunderstanding. It may be that Miltiades simply had to extend his centre to prevent his overall line being so short as to be easily outflanked (Herodotus records that 'as the Athenians were marshalled at Marathon, it happened that their line of battle was as long as the line of the Medes', which was most likely by design), and any resulting envelopment of the Persian force was a happy accident.[13]

It was now, once the army was deployed, that the general might pass along the line offering short words of encouragement to each contingent as he passed. This was distinct from the more formal speech delivered to the army, or at least to a subset of the army (perhaps the officers) in camp or in the days before battle, as we saw in Chapter 5.

Pre-battle skirmishes

The most important part of the battle line was of course made up of the hoplite phalanx itself (at least according to our hoplite-class sources). But as we have seen, Greek armies often (so far as we can tell) contained large numbers of light infantry and usually had at least a modest contingent of cavalry. These forces also had to be deployed in front of or on the ends of the main hoplite phalanx, and would have their own part to play in the battle. Contests of light infantry against hoplites outside of pitched battles were considered in the previous chapter, while the role of cavalry in later stages of the battle will be looked at below. But on those occasions where such forces were stationed before the line, their fighting might serve as a prelude to the clash of the heavy infantry.

It is often difficult to discern where, or even whether, light and cavalry forces were deployed relative to the main line. At Plataea (479), there were present very large forces of Greek light infantry, as detailed by Herodotus (9.29.1–2), but despite these impressive numbers, the light-armed are afforded no role in the battle in Herodotus' account. Although the Persians used large forces of cavalry and their infantry were (to a debatable extent) equipped as archers, and although the Greeks certainly recognized the value of light infantry in these circumstances (Pausanias sent a messenger to the Athenians to request they send him a force of archers, Hdt. 9.60.3), there is little sign of light infantry in the course of the fighting, and we are left to speculate if or how they were employed.

In battles of the Peloponnesian War and later, light and mounted forces are sometimes to be found on the flanks of the main phalanx, as at Delium (424), 'the cavalry and the light troops being at the extremity of each wing' (Thuc. 4.94.4). When Thucydides records that 'the extreme wing of neither army came into action, one like the other being stopped by the water-courses in the way' (Thuc. 4.86.2), we must presumably understand that this included the light and mounted forces, prevented from fighting – ironically enough – by difficult terrain. And yet, according to Thucydides:

> 'Not quite five hundred Boeotians fell in the battle, and nearly one thousand Athenians, including Hippocrates the general, besides a great number of light troops and camp followers.' (Thucydides 4.101.2)

Evidently they were involved in some capacity, to take heavy casualties (perhaps unsuccessfully guarding the camp). At Syracuse in 415/414, the light infantry were apparently formed up in front of the phalanxes, this being the occasion when the Syracusans were caught by surprise by the Athenians' advance:

> 'First, the stone-throwers, slingers, and archers of either army began skirmishing, and routed or were routed by one another, as might be expected between light troops.' (Thucydides 6.69.2)

The light forces are not expected to have a significant impact on the course of the battle because their method of fighting does not produce a decisive result. If one side, however, was weak in light infantry, they would face the danger that heavy forces always faced, of being shot down by them without any opportunity to strike back. A large part of the role of light infantry in battle was therefore simply to neutralize the enemy's light forces by their mere presence.[14]

The somewhat peculiar position of light forces in pitched battle is revealed by Xenophon's account of Leuctra (371):

> 'Again, when both sides were arming themselves and it was already evident that there would be a battle, in the first place, after those who had provided the market and some baggage-carriers and such as did not wish to fight had set out to withdraw from the Boeotian army, the Lacedaemonian mercenaries under Hieron, the peltasts of the Phocians, and, among the horsemen, the Heracleots and Phliasians made a circuit and fell upon these people as they were departing, and not only turned them about but chased them back to the camp of the Boeotians. Thereby they made the Boeotian army much larger and more densely massed than it had been before.' (Xenophon, *Hellenica* 6.4.9)

Note that as at Delium, camp followers were considered fair game, that the Spartan light forces and cavalry were able to go around to the rear of the Boeotian army, yet they played no further positive role in the battle, and that the presence of a large number of non-combatants was said to make the Boeotian army large and more densely massed. This last comment in particular may indicate that our common image of a clear distinction between units and types of soldiers in Greek armies, with a distinct phalanx and separate, distinct formations of light troops, may be in error, and there may well have been – except for the front ranks perhaps – a much more crowd-like nature to Greek armies, with all sorts of men making up the numbers to the rear, in a fashion similar to, though on a much larger scale than, the familiar *plethos* ('mass') and *promachoi* ('fore-fighters', 'vanguard') of Homeric warfare. This could account for the strange invisibility of light infantry at Plataea; they may have been massed behind the fighting men of the phalanx proper.

Cavalry could also, like light infantry, fight against each other on the flanks, or in front of the main lines before the main infantry forces engaged. At Spartolus (429), these forces engaged first:

> '[T]he Chalcidian hoplites, and some auxiliaries with them, were beaten and retreated into Spartolus; but the Chalcidian horse and light troops defeated the horse and light troops of the Athenians.' (Thucydides 2.79.2)

These cavalry went on to play a more active role on the battle (see further below). At Lyncestis (423) there was a similar initial engagement of cavalry as we saw above (Thuc. 4.124.3). At Leuctra, an initial engagement of cavalry had an even greater impact on the course of the battle:

> '[S]ince the space between the armies was a plain, the Lacedaemonians posted their horsemen in front of their phalanx, and the Thebans in like manner posted theirs over against them … the horsemen had already joined battle and those of the Lacedaemonians had speedily been worsted; then in their flight they had fallen foul of their own hoplites.' (Xenophon, *Hellenica* 6.4.10; 13)

The disruption of the Spartan phalanx by the retreating cavalry is given by Xenophon as one of the reasons for the Spartan defeat. Cavalry in particular did not always have a peripheral role in pitched battle, and light infantry could be deadly to heavy in the right circumstances – the nature of the fighting between these forces will be considered further below.[15]

Sacrifices and religion

Once armies were committed to battle and deployed, and any initial skirmishing or cavalry action had taken place, one final step remained – seeking the approval and assistance of the gods for the ensuing battle. Religion also played a key part at an earlier phase of course, since often battle would not be joined at all unless the omens were suitable, but it is striking – particularly in the Spartan army – just how late in the process of giving battle a religious ceremony was inserted.

At Plataea (479), despite being under galling Persian archery and in a desperate tactical position, Herodotus says the Spartans and Tegeans refused to even engage until they had received favourable omens:

> 'These offered sacrifice so that they would fare better in battle with Mardonius and the army which was with him. They could get no favourable

omen from their sacrifices, and in the meanwhile many of them were killed and by far more wounded ... Since the Spartans were being hard-pressed and their sacrifices were of no avail, Pausanias lifted up his eyes to the temple of Hera at Plataea and called on the goddess, praying that they might not be disappointed in their hope.' (Herodotus 9.61.2–3)

In the event the Tegeans' patience snapped:

'While he was still in the act of praying, the men of Tegea leapt out before the rest and charged the barbarians, and immediately after Pausanias' prayer the sacrifices of the Lacedaemonians became favourable. Now they too charged the Persians.' (Herodotus 9.62.1)

This delay has sometimes been rationalized as being for tactical not religious purposes, but the consistent importance of religion in Greek battles suggests that we should not be too quick to dismiss the motives given. Accusations of manipulating religion for tactical ends were, however, also present in the ancient world, as Xenophon's account (Xen., *Hell.* 4.2.18) of Nemea (394), quoted above, illustrates, the Boeotians conveniently obtaining favourable sacrifices only when they were in the less dangerous position in the line. At this same battle, the Spartans made their typically last-minute sacrifice:

'And when the armies were now not so much as a stadium [180 metres] apart, the Lacedaemonians sacrificed the goat to Artemis Agrotera, as is their custom [*nomos*], and led the charge upon their adversaries.' (Xenophon, *Hellenica* 4.2.20)

This, if perhaps not the precise timing of it, was evidently the usual practice. At Syracuse, after the indecisive skirmishing we saw above, 'soothsayers brought forward the usual victims, and trumpeters urged on the heavy infantry to the charge' (Thuc. 6.69.2). Clearly this was an important part of the psychological preparation of an army before battle, to ensure that the men (of both sides) knew that the gods were on their side. Note also the timing of the sacrifice, after the skirmishing of the light-armed forces was over. Evidently, the battle proper was considered to start only when the phalanxes themselves were about to go into action, with the preliminary skirmishing not forming part of the real fighting.

We can perhaps detect a certain cynicism in Diodorus' account of the preparations for Leuctra:

'Both sides eagerly drew together for the decisive conflict, their armies in battle formation, while the soothsayers, having sacrificed on both sides, declared that victory was foreshadowed by the gods.' (Diodorus 15.85.1)

Yet it is entirely natural for men going into battle to seek this sort of reassurance, and much battlefield religious practice was probably more sincere than that performed in the comfort and safety of the home city.

We must also remember that such religious practices were another element of warfare largely shared by the Greeks and Persians. Before Cunaxa (401), Xenophon approached the Persian pretender Cyrus and asked if he had any orders to give: 'Cyrus pulled up his horse and bade Xenophon tell everybody that the sacrificial victims and omens were all favourable' (Xen., *Anab.* 1.8.15). At Plataea, it was not just the Greeks who were delayed by the need for divine approval:

> 'When no favourable omens for battle could be won either by the Persians themselves or by the Greeks who were with them (for they too had a diviner of their own, Hippomachus of Leucas), and the Greeks kept flocking in and their army grew, Timagenides son of Herpys, a Theban, advised Mardonius to guard the outlet of the pass over Cithaeron, telling him that the Greeks were coming in daily and that he would thereby cut off many of them.' (Herodotus 9.38.2)

In this case the clear tactical benefit of fighting before the Greek army grew any stronger was negated by the inability of either Persian or collaborating Greek priests to obtain the necessary omens.

Another element of hoplite battle with a religious aspect was the singing of the paean before battle – a paean being something between a communal singalong and a hymn. Xenophon describes an evening of music and dance laid on for the Ten Thousand in Paphlagonia, which started with all making libations and singing the paean:

> '[T]he Mantineans and some of the other Arcadians arose, arrayed in the finest arms and accoutrements they could command, and marched in time to the accompaniment of a flute playing the martial rhythm and sang the paean and danced, just as the Arcadians do in their festal processions in honour of the gods.' (Xenophon, *Anabasis* 6.1.11)

In battle, the paean was sung at the start of and possibly during the advance to contact, as at the second phase of Solygeia (425, Thuc. 4.43.3) or at Delium (424, Thuc. 4.96.1), or where the Ten Thousand began an advance against opposing light infantry (Xen., *Anab.* 4.3.29, see also 4.8.16, 5.2.13). The purpose of the paean, aside from its religious significance, is given by Aeschylus:

> '[S]ing the victory song [paean], the sacred cry of joy and goodwill, our Greek ritual of shouting in tribute, that brings courage to our friends and dissolves fear of the enemy.' (Aeschylus, *Seven Against Thebes* 270)

So the paean was a way to build group solidarity and to heighten the combatants' courage in preparation for the coming fight.[16]

Advancing and charging

> 'When they had been set in order and the sacrifices were favourable, the Athenians were sent forth and charged the foreigners at a run.' (Herodotus 6.112.1)

So Herodotus describes the opening of the battle of Marathon (490). He goes on to give more details of this charge at the run:

> 'The space between the armies was no less than eight stadia [about 1,440 metres]. The Persians saw them running to attack and prepared to receive them, thinking the Athenians absolutely crazy, since they saw how few of them there were and that they ran up so fast without either cavalry or archers. So the foreigners imagined, but when the Athenians all together fell upon the foreigners they fought in a way worthy of record. These are the first Hellenes whom we know of to use running against the enemy.' (Herodotus 6.112.1–3)

The question of whether it is really plausible that the Athenians would have run for nearly a mile, in formation and in full hoplite equipment, prior to fighting for a potentially long period, has been one of the great controversies of Marathon – and of Greek warfare generally – and has generated a literature of its own. I am inclined to believe that no sane commander would risk the fatigue and disorder inherent in such a long run (although I am sure it was physically possible), and that Herodotus has simply conflated two distinct pieces of information: that the armies first deployed eight stadia apart, and that the Athenians completed their charge at the run. The reason why they would have done this is clear enough – the Persian army contained a large proportion of archers, so a running charge over the last 100 metres or so would have minimized the time the Athenians were on the receiving end of their arrows. Herodotus' further comment may also be relevant:

> 'They are also the first to endure looking at Median dress and men wearing it, for up until then just hearing the name of the Medes caused the Hellenes to panic.' (Herodotus 6.112.3)

If the Persians ('Medes') had a particularly fearsome reputation, then the Athenians (individually or their commanders) may have wished to fire up the hoplites' courage and reduce the amount of 'thinking time' before battle – a reminder that as well as possible tactical and combat purposes, there were also good psychological reasons for a running charge, as for singing the paean, to increase the courage of one's own side and, with luck, to frighten the enemy.[17]

This being so, it is perhaps surprising that if Herodotus is to be believed, hoplite armies before this date (490) did not charge at the run, presumably instead making their advance at a steady walk, as some continued to do. Many have for this reason been inclined to doubt Herodotus on this point, though we should note (see below) that the Spartans, the arch-traditionalists, were among those who continued to advance at a walk. Hoplite armies relied on good order to a greater extent than other types of infantry, and when hoplites faced hoplites there was no necessity either to rapidly cross a 'beaten zone' of missiles or to catch a retiring enemy. It is therefore perhaps not too surprising if hoplites in the earlier period had always advanced at the walk. An innovation introduced to combat the fearsome, missile-armed Persians may have proven its worth, but not to the extent that every hoplite army advanced at the run, since many certainly did not do so, and a running charge may be more an indication of ill-discipline and likely defeat in the ensuing combat than a winning tactical choice.[18]

Running charges do certainly occur, however. The Persians themselves are said to have charged at the run at Plataea, believing in this case that the Greeks were in retreat and all that remained was to pursue them. Herodotus describes them attacking 'each at top speed, no battalion [*taxis*] having order in its ranks nor place assigned in the line' (Hdt. 9.59.2), and adds that 'they ran pell-mell and shouting' (9.60.1). This disorganized advance could help account for Greek success in the ensuing fighting, since the Greeks halted to receive their attacks, at least at first.

Running charges were also not necessarily performed at the whim of the commander, but could arise spontaneously. Xenophon describes the parade of the Ten Thousand on the march to Cunaxa (401), where Cyrus ordered the Greek phalanx to advance:

> 'The *strategoi* transmitted these orders to the soldiers, and when the trumpet sounded, they advanced arms and charged. And then, as they went on faster and faster, at length with a shout the troops broke into a run of their own accord, in the direction of the camp.' (Xenophon, *Anabasis* 1.2.17)

The English word 'charged' used in the translation has too strong a connotation of a rapid, running advance; the Greek *epeimi* has a more general sense of 'advance' or 'attack', and it was only later that they spontaneously broke into a run. This running charge caused great fear among the watching barbarians, and was to be repeated at the actual battle:

> 'At length the opposing lines were not three or four stadia apart [about 600 metres], and then the Greeks struck up the paean and began to advance against the enemy. And when, as they proceeded, a part of the phalanx billowed out, those who were thus left behind began to run; at the same moment they all set up the sort of war-cry which they raise to Enyalius, and all alike began running ... And before an arrow reached them, the barbarians broke and fled. Thereupon the Greeks pursued with all their might, but shouted meanwhile to one another not to run at a headlong pace, but to keep their ranks in the pursuit.' (Xenophon, *Anabasis* 1.8.17–19)

The result was the same as at the parade, the Persians being frightened into flight by the mere charge (as Xenophon tells it) – the ideal outcome for such a running charge, and a large part of its purpose, for all that it did not always work out this way. Note that the running charge began spontaneously as an attempt by different parts of the phalanx, advancing at slightly different speeds, to keep contact with the rest, a reminder that most Greek armies probably did not march in step. The Greeks will not have run the whole 600 metres or so; they started their advance at a slower pace and broke into a run only as they got closer, which is most likely also what happened at Marathon. The speed of the pursuit was also controlled spontaneously by shouting to each other to slow down, rather than by the officers (though conceivably it could have been officers doing the shouting).

The contrast between such running charges and the slow advance of the Spartans is illustrated at Mantinea (418):

> 'After this they joined battle, the Argives and their allies advancing with haste and fury, the Lacedaemonians slowly and to the music of many flute-players – a standing institution in their army, that has nothing to do with religion, but is meant to make them advance evenly, stepping in time, without breaking their order, as large armies are apt to do in the moment of engaging.' (Thucydides 5.70.1)

This is good (though not conclusive) evidence that Spartans did march in step, and Thucydides provides the thinking behind the steady advance – that it maintained order in the phalanx, avoiding the likely ruptures, gaps and general loosening of the formation that would inevitably arise with a headlong charge. A slow, silent advance could also frighten the enemy, by demonstrating the confidence and discipline of an army; the royal Persian army at Cunaxa was said by Xenophon to have advanced, 'not with shouting, but in the utmost silence and quietness, with equal step and slowly' (Xen., *Anab.* 1.8.11), unlike the Greeks. Whether a noisy, rapid advance or a silent, controlled one was more frightening was clearly still a debatable point, and other tactical considerations (such as the presence of archers) might also help determine the speed required.

In fighting outside Syracuse (414), a Syracusan force had to make an especially long forced march:

> 'Diomilus with his six hundred and the rest advanced as quickly as they could, but they had nearly three miles to go from the meadow before reaching them. Attacking in this way in considerable disorder, the Syracusans were defeated in battle at Epipolae and retired to the town, with a loss of about three hundred killed, and Diomilus among the number.' (Thucydides 6.97.3–4)

The need for haste in this situation resulted in dangerous levels of disorder and hence in defeat, but rapid advances were still used successfully over shorter distances, as at Coronea (394):

> 'Now as the opposing armies were coming together, there was deep silence for a time in both lines; but when they were distant from one another about a stadium [180 metres], the Thebans raised the war-cry and rushed to close quarters on the run. When, however, the distance between the armies was still about three *plethra* [about 100 metres], the troops whom Herippidas commanded, and with them the Ionians, Aeolians, and Hellespontines, ran forth in their turn from the phalanx of Agesilaus, and the whole mass joined in the charge.' (Xenophon, *Hellenica* 4.3.17)

We might guess that these sorts of distances – 100–200 metres – were the typical limit for a running charge if it was not to cause harmful confusion and disorder.

A rapid advance, as well as its psychological value, could also have specific tactical uses, particularly taking advantage of a sudden and perhaps fleeting opportunity. Such was the case in Plutarch's version of the Battle of Leuctra (371), where the Spartan army was attempting to outflank the Theban column:

> 'But at this point [the Theban general] Pelopidas darted forth from his position, and with his band of three hundred on the run, came up before Cleombrotus had either extended his wing or brought it back again into its old position and closed up his line of battle, so that the Lacedaemonians were not standing in array, but moving confusedly about among each other when his onset reached them.' (Plutarch, *Pelopidas* 23.2)

Here, a rapid advance (over what distance is not specified) allowed the Thebans to close with a temporarily disordered enemy, although the Spartans were disciplined enough to stand and fight as they were. Compare this with the attack of Brasidas on a similarly disordered Athenian phalanx, caught in the middle of a withdrawal manoeuvre:

'[Brasidas] ran at the top of his speed along the straight road, where the trophy now stands as you go by the steepest part of the hill, and fell upon and routed the centre of the Athenians, panic-stricken by their own disorder and astounded at his audacity.' (Thucydides 5.10.6)

One thing that we can be sure of is that the purpose of the running charge was not to build up physical momentum in order to slam into the opposing line on contact, in the manner of the popular conception of a charging knight on horseback. A human can reach top running speed from a standing start in a few metres, so there would be no benefit in running over a longer distance, and it would not justify the loss of cohesion. There is also little reason to suppose that hoplites ever body-slammed their opponents in this way, as we will see in the following chapter.

Tactical manoeuvres

In Chapter 3 we looked at the various tactical manoeuvres and special formations that were available to a hoplite army, at least to one with sufficient drill and discipline to carry them out. We now need only review a few examples of these manoeuvres in practice. The earliest and simplest hoplite deployment was of course the simple phalanx, at a depth determined by the need to balance solidity with width (to avoid being outflanked) and, of course, by tradition. It may be that all battles of the Archaic period and from before the fifth century, before we have any detailed accounts of hoplite battles, involved the simple face-to-face clash of phalanxes of equal length and parallel alignment, and the usual assumption is that tactical sophistication increased in the course of the fifth and fourth centuries, with more complex manoeuvres being attempted. This is a reasonable assumption, but is supported largely by absence of evidence. We start to see more complex manoeuvres right from the start of the Peloponnesian War, when we also start to get detailed battle accounts, so we cannot be sure that this is because they were invented then for the first time. However, it does seem likely that new tactical innovations were introduced in the fourth century at least, as we will see below. So the basic tactical deployment was two parallel phalanxes of equal length, and with depth varying, if necessary, in order to achieve this equality of length. This is the deployment we see at Marathon, where the Athenians were formed so as to match the width of the Persian army, although this meant forming the centre at reduced depth (assuming this is the correct interpretation of Hdt. 6.11.3).

The first clear example of outflanking manoeuvres is the well-known Spartan manoeuvre at Mantinea (418), and this was (in Thucydides' account) another accidental situation, arising from the phenomenon that two equally matched, parallel-deployed phalanxes tended to outflank each other naturally and accidentally through the 'rightward drift' of the phalanx as each man sought to keep his own unshielded right side away from the enemy:

> 'All armies are alike in this: on going into action they get forced out rather on their right wing, and one and the other overlap with this their adversary's left.' (Thucydides 5.71.1)

On this occasion, the frontage of the armies was not perfectly matched so there was already a natural overlap arising from the difference in numbers:

> 'On the present occasion the Mantineans reached with their wing far beyond the Sciritae, and the Lacedaemonians and Tegeans still farther beyond the Athenians, as their army was the largest.' (Thucydides 5.71.2)

Agis conceived the idea, at the last minute, of extending his left to prevent its being outflanked, while using extra units transferred from his right to plug the resulting gap in the centre (Thuc. 5.71.3). Recalcitrant Spartan officers refusing to move as ordered, this resulted in the Spartan left having an exposed inner flank, its right, with a gap in the line, and so the Spartan left was rapidly defeated. Meanwhile, the Spartan right outflanked the allied (Athenian) left – 'the Lacedaemonian and Tegean right simultaneously closing round [*ekuklouto*] the Athenians with the troops that outflanked them' (Thuc. 5.73.1) – this being a 'natural' outflanking caused by the greater Spartan numbers combined with the rightward drift common to all armies. Thucydides (5.72.2) attributed the victory to Spartan prowess as much as to the overlap. So there were outflanking manoeuvres at Mantinea, but they were largely accidental and involuntary, and the only real tactical manoeuvre (ultimately unsuccessful) was Agis' attempt to prevent the outflanking of his left wing. Precisely how the Lacedaemonian and Tegean right 'closed round' the Athenians is not recorded – we might envisage this (as discussed in Chapter 3) as a wheel by the outflanking force, by an envelopment in column, or as a simple lapping around, on a small-unit level, of the exposed Athenian flank.

At any rate, the result was something that we will encounter again: the right (most prestigious) wing of each army was victorious. On this occasion:

> 'Agis also on perceiving the distress of his left opposed to the Mantineans and the thousand Argives, ordered all the army to advance to the support of the defeated wing ... Meanwhile the Mantineans and their allies and the picked body of the Argives ceased to press the enemy, and seeing their friends defeated and the Lacedaemonians in full advance upon them, took to flight.' (Thucydides 5.73.2–3)

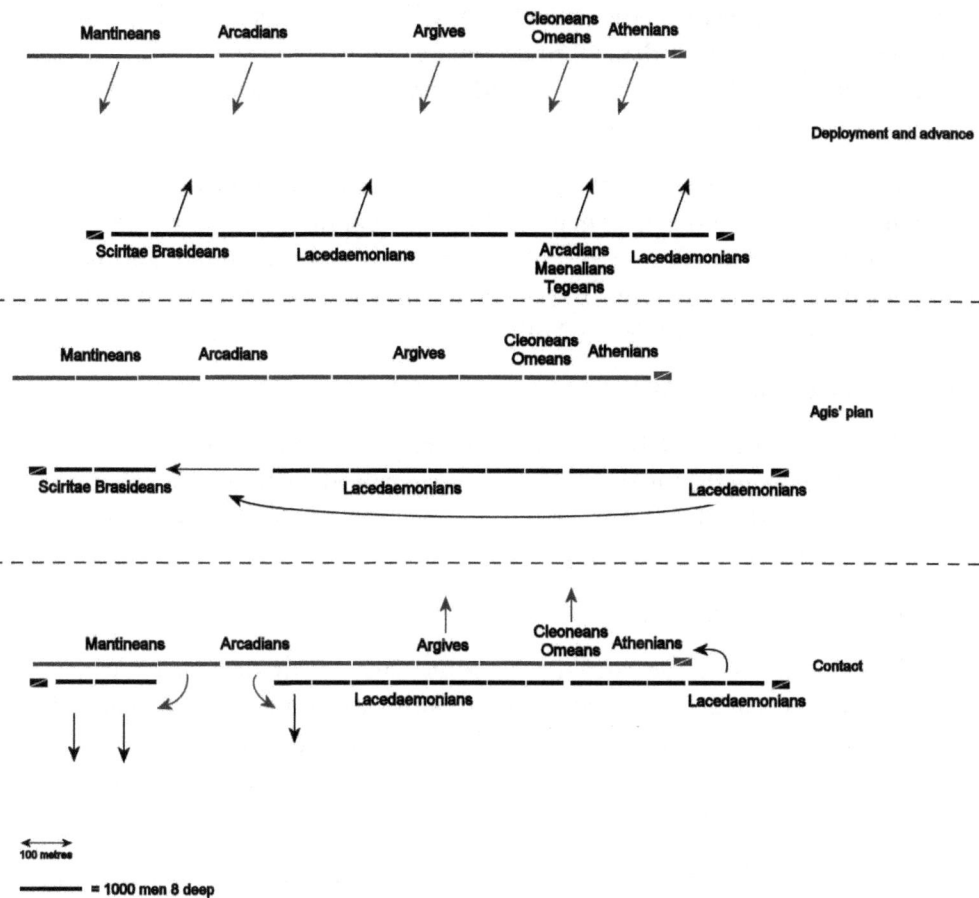

Scale diagram of manoeuvres at First Mantinea (418).

So the Spartan right turned (with a wheel in line or by individual contingents – Thucydides does not specify) to attack the victorious allied right, which fled rather than staying to fight.

This feature of hoplite battles – a result of the combination of the 'rightward drift' that led armies to outflank their opponents on the right, and the most prestigious position being on the right and so being occupied by the highest-quality hoplites (these two factors being no doubt historically related) – has aptly been called the 'revolving door'. Each right wing would defeat the enemy left wing opposed to it, and then might, if it was well disciplined and capable of the required drill, wheel to face the enemy's right wing, as happened at Mantinea. Alternatively – if it was ill-disciplined or drilled – it might simply pursue the beaten enemy contingent into the distance, perhaps returning to the battlefield after a while, or it might be content with the success achieved and attempt nothing further, especially if the enemy right also declined a rematch. If the victorious wings did not engage each other, then the overall outcome (determined by which side had control of the battlefield) could remain in doubt.[19]

'Revolving door' battles of this sort, and variants on them, are quite common. Examples include Potidaea (432):

> 'The wing of Aristeus, with the Corinthians and other picked troops round him, routed the wing opposed to it, and followed for a considerable distance in pursuit. But the rest of the army of the Potidaeans and of the Peloponnesians was defeated by the Athenians, and took refuge within the fortifications.' (Thucydides 1.62.6)

Similar is Laodocium (423/422):

> '[T]he Mantineans and Tegeans, and their respective allies, fought a battle at Laodocium, in the Oresthid. The victory remained doubtful, as each side routed one of the wings opposed to them, and both set up trophies and sent spoils to Delphi.' (Thucydides 4.134.1)

So also Miletus (412):

> '[T]he Athenians first defeated the Peloponnesians, and driving before them the barbarians and the ruck of the army, without engaging the Milesians, who after the rout of the Argives retreated into the town upon seeing their comrades worsted, crowned their victory by grounding their arms under the very walls of Miletus.' (Thucydides 8.25.4)

An example of a battle where the victorious wings did turn and engage each other is Coronea (394), which Xenophon says 'proved to be like no other of the battles of our time' (Xen., *Hell.* 4.3.16), perhaps because the two victorious wings did seek a rematch and engaged each other frontally. Agesilaus and the Spartan right wing were, as usual, victorious over the forces opposed to them:

> 'Thereupon some of the mercenaries were already garlanding Agesilaus, when a man brought him word that the Thebans had cut their way through the Orchomenians and were in among the baggage train. And he immediately wheeled his phalanx and led the advance against them; but the Thebans on their side, when they saw that their allies had taken refuge at Mount Helicon, wishing to break through to join their own friends, massed themselves together and came on stoutly.' (Xenophon, *Hellenica* 4.3.18)

Note the means of gaining essential battlefield intelligence was a man bringing word of what had happened. Xenophon characterizes this decision to seek a frontal confrontation as brave but rash; his preferred solution would have been to allow the Thebans to march across the Spartan front then attack them in flank and rear. This indeed was the ideal outcome of a 'revolving door' battle – for the victorious wing, having routed its opposite number, to turn inwards and roll up the enemy line, marching across the battlefield and defeating each enemy contingent in turn. This is what the Spartan army achieved at the Nemea earlier in the same year, a battle which demonstrates some tactical innovation as well as a perfect execution of the victorious wing scenario. This was the battle in which the allies (Thebans, Athenians and others) had agreed on a sixteen-rank depth for their armies:

> 'And in the first place, disregarding the sixteen-rank formation, they [the Thebans] made their phalanx exceedingly deep, and, besides, they also veered to the right in leading the advance, in order to outflank the enemy with their wing; and the Athenians, in order not to be detached from the rest of the line, followed them towards the right, although they knew that there was danger of their being surrounded.' (Xenophon, *Hellenica* 4.2.18)

This is not the involuntary rightward drift we saw at Mantinea, but a deliberate march to the right, specifically intended to outflank the enemy. The Athenians evidently did not expect this manoeuvre and were forced to follow it to avoid a large internal gap appearing in the line (compare with Mantinea). The precise nature of this manoeuvre is also unclear – the translation 'veered to the right in leading the advance' implies an oblique advance, a deliberate rightwards drift while moving forward, crabwise. The Greek does not particularly support this idea – more literally, 'they led to the right'. This might suggest a march in column toward the right rather than an oblique advance, aiming to get into an outflanking position before starting the advance. However, the allied army did need to close with the Spartans (which they did under cover of the trees covering their approach), so an oblique advance might be meant in this case. It may also be that the advance was at an oblique rather than a right angle. Unfortunately, certainty about what happened on this occasion is impossible.

However, the Spartans also made a deliberate outflanking manoeuvre:

> 'And when they had been drawn up together in the positions which the Lacedaemonian leaders of the allies assigned to the several divisions, they passed the word along to follow the van. Now the Lacedaemonians also veered to the right in leading the advance, and extended their wing so far beyond that of the enemy that only six tribes of the Athenians found themselves opposite the Lacedaemonians, the other four being opposite the Tegeans. And when the armies were now not so much as a stadium

346 The Greek Hoplite Phalanx

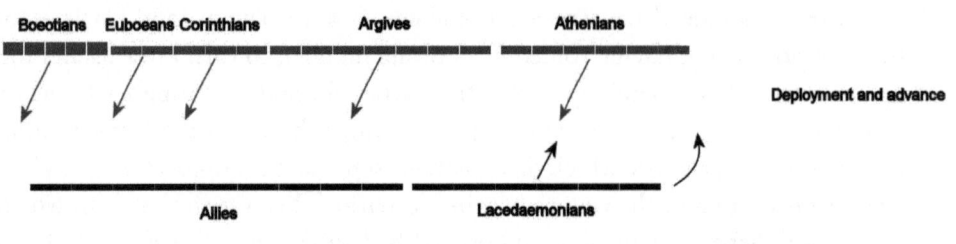

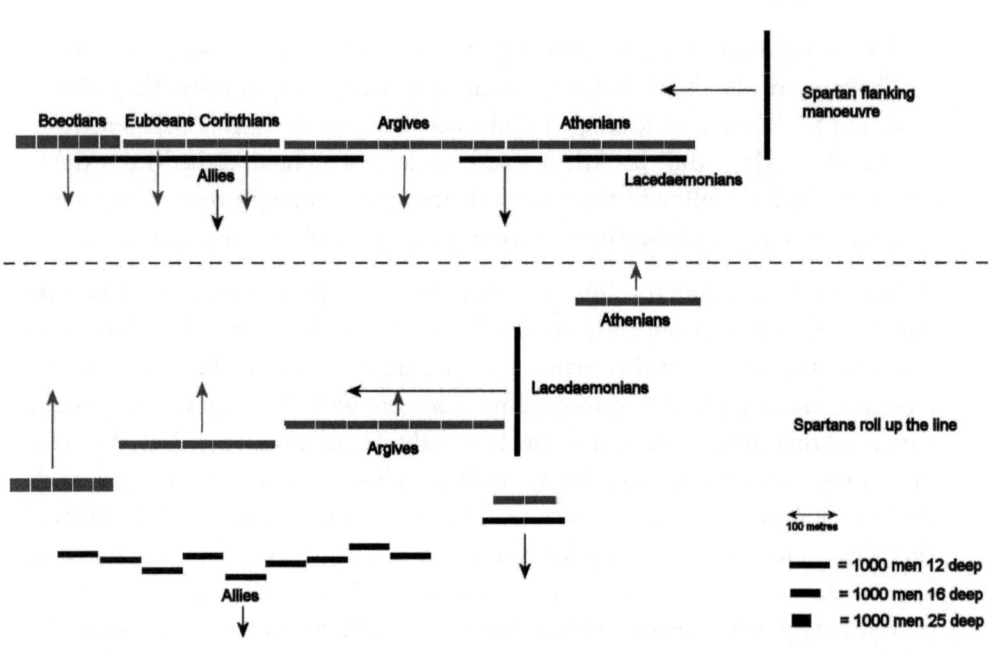

Spartan *kyklosis* manoeuvre at Nemea (394).

apart, the Lacedaemonians sacrificed the goat to Artemis Agrotera, as is their custom, and led the charge upon their adversaries, wheeling round their overlapping wing in order to surround them.' (Xenophon, *Hellenica* 4.2.19–20)

Again, the translation sounds more like an oblique advance than Xenophon's words warrant ('the Lacedaemonians too led to the right'). This manoeuvre, as discussed in Chapter 3, is probably a march in column to the right (hence the order to 'follow the van' or leader), followed by a wheel to the left and a turn to face inward, toward the Athenian flank, forming a new line at right angles to the original.

The Spartans quickly routed those Athenians they faced, then marched across the battlefield, defeating allied contingents as they met them one after another, for the Thebans on the allied right had also been victorious. The first group of Athenians were still pursuing their opponents:

> '[B]ut the Lacedaemonians did come upon the Argives as they were returning from the pursuit, and when the first *polemarchos* was about to attack them in front, it is said that someone shouted out to let their front ranks pass by. When this had been done, they struck them on their unprotected sides as they ran past, and killed many of them. The Lacedaemonians also attacked the Corinthians as they were returning. And, furthermore, they likewise came upon some of the Thebans returning from the pursuit, and killed a large number of them.' (Xenophon, *Hellenica* 4.2.22)

Note the contrast with Coronea, here an unnamed 'someone' shouting from the ranks not to seek a frontal collision with the victorious enemy wing, which Xenophon thought so risky.[20]

So by the early fourth century we see armies deliberately creating overlaps on one wing in order to outflank their opponents, and in the case of the Thebans using especially deep formations both to maximize their chances of breaking through and perhaps to make an oblique advance – if there was an oblique advance – easier to execute (as the formation is deeper and narrower, closer to a manoeuvre column). These two features were to be brought to perfection by Epaminondas and Pelopidas at Thebes, along with another seemingly obvious innovation, that of stationing their best units on the left of the line rather than the right, so that they could meet and defeat the strongest part of the enemy and there would be no need for a rematch. Theban tactics also involved a more active role for the cavalry, which I will consider further below.

Unfortunately, our knowledge and understanding of the tactical innovations of Epaminondas at his first great victory, Leuctra (371), are obscured by the poor state of the accounts of the battle that have come down to us. In one sense we are spoiled, since there are three reasonably full accounts of the battle: those of Xenophon, Plutarch and Diodorus. But unfortunately, Xenophon, who was in a position to know and understand exactly what happened and why, was clearly in no mood to dwell on the defeat of his Spartan heroes, while Plutarch was a biographer in search of a lively story rather than a sober military historian, and Diodorus embellishes all his battle accounts with stock phrases or his own imagination, which tends to obscure the details. Leuctra, like Marathon, is another classic example of the 'inverted pyramid'. Nevertheless, a general view of what took place can be extracted from the three sources.[21]

The first element of Theban tactics was to mass their best forces (the Thebans and the elite Sacred Band) in depth on their left opposite the Spartan contingent, led by the king (Cleombrotus) himself:

> 'The Thebans, however, were massed not less than fifty shields deep, calculating that if they conquered that part of the army which was around the king, all the rest of it would be easy to overcome.' (Xenophon, *Hellenica* 6.4.12)

The deep formation followed a tradition already seen at Delium and Nemea, while forming on the left was an attempt to 'crush the head of the snake' by defeating the enemy's best forces (the snake metaphor comes via Polyaenus, *Stratagems* 2.3.15).

The second element was, as at the Nemea, to lead out toward the flank, as described by Plutarch:

> 'In the battle, while Epaminondas was drawing his phalanx obliquely towards the left, in order that the right wing of the Spartans might be separated as far as possible from the rest of the Greeks, and that he might thrust back Cleombrotus by a fierce charge in column with all his hoplites…' (Plutarch, *Pelopidas* 23.1)

The Spartans attempted to counter this movement by extending their own phalanx to the right:

> '[T]hey were opening up their right wing and making an encircling movement, in order to surround Epaminondas and envelop him with their numbers.' (Plutarch, *Pelopidas* 23.2)

It was at this point that Pelopidas and the Sacred Band made the sudden attack that caught the Spartans mid-manoeuvre, as we saw above, though precisely where the Sacred Band was deployed relative to the rest of the Theban column (in front, on the flank or in reserve behind) remains one of the great unknowns of the battle.

The final component of Theban tactics comes from Diodorus:

> '[Epaminondas] selected from the entire army the bravest men and stationed them on one wing, intending to give battle to the finish with them himself. The weakest he placed on the other wing and instructed them to avoid battle and withdraw gradually during the enemy's attack. So then, by arranging his phalanx in oblique formation, he planned to decide the issue of the battle by means of the wing in which were the élite.' (Diodorus 15.55.2)

The Hoplite Battle 349

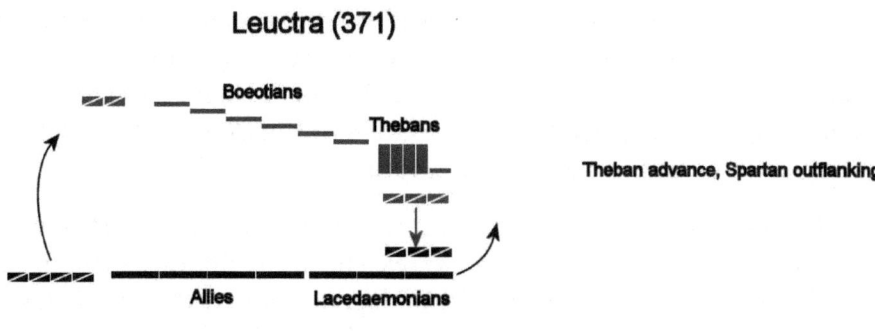

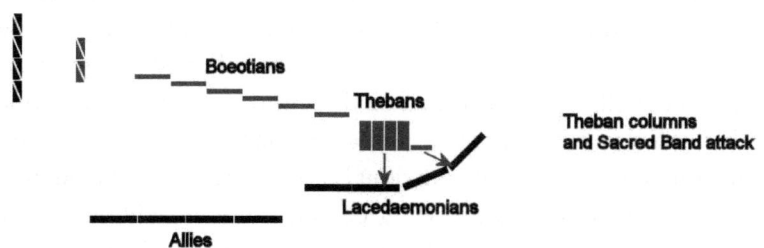

Theban tactics at Leuctra (371).

Diodorus confuses matters by claiming that the Spartans attacked in crescent formation on both wings, though the key point is that 'the Boeotians retreated on one wing, but on the other engaged the enemy in double-quick time' (Diod. 15.55.3). Both Plutarch and Diodorus mention this 'oblique' (*loxe*) formation, and while Xenophon does not, he does hint at the outcome:

> '[T]he [Spartans on the right wing] fell back under the pressure of the Theban mass while those who were on the left wing of the Lacedaemonians, when they saw that the right wing was being pushed back, gave way.' (Xenophon, *Hellenica* 6.4.14)

The Boeotian army was able to avoid a 'revolving door' battle and win a decisive victory by defeating the Spartan contingent while avoiding contact between its weaker allied wing and the Spartan allies, who fled upon the defeat of the Spartans. The Spartan defeat was itself brought about by a combination of the deeper Theban formation, the rapid charge of the Sacred Band, the bungled Spartan attempt to redeploy to counter the threat, the victory of the Theban cavalry over their opposite numbers and (in Xenophon's version) Spartan drinking on the morning of the battle (Xen., *Hell.* 6.4.8). Boeotian tradition also attributed the victory to superior Theban fitness, training and courage, something that tends to get passed over in modern accounts.[22]

The Battle of Second Mantinea (362) was to see a repeat of several of these factors in a second Boeotian victory over the Spartans. Epaminondas on this occasion first advanced his army in battle formation toward the left and made as if to set up camp on the hills on the edge of the plain (Xen., *Hell.* 7.5.21), causing a relaxation of readiness in the Spartan army, as we saw above:

> 'It was not until he had moved along [*paragagon*] successive *lochoi* to the wing where he was stationed, and had wheeled them into line thus strengthening the mass formation [*embolon*] of this wing, that he gave the order to take up arms and led the advance.' (Xenophon, *Hellenica* 7.5.22)

The nature of the *paragoge* has been examined in Chapter 3 – marching files or units (*lochoi* in this case) up alongside others, usually as a component of deployment into line – while the translation of 'wheeled' is inaccurate; the reinforcing units will have turned to face the front. The *embolon* described here is not to be understood as a wedge, as in some modern accounts, but as a deep formation created not by forming individual *lochoi* at greater depth, but by placing *lochoi* brought over from the right wing behind those already on the left. The result was, as at Leuctra, an especially deep phalanx on the Boeotian left.

> 'Meanwhile Epaminondas led forward his army prow on, like a trireme, believing that if he could strike and cut through anywhere, he would destroy the entire army of his adversaries. For he was preparing to make the contest with the strongest part of his force, and the weakest part he had stationed far back, knowing that if defeated it would cause discouragement to the troops who were with him and give courage to the enemy.' (Xenophon, *Hellenica* 7.5.23)

These are exactly the same tactics as at Leuctra, advancing in strength on the left while 'refusing' the weaker right, in order to win a decisive victory on the favoured wing:

> 'Thus, then, he made his attack, and he was not disappointed of his hope; for by gaining the mastery at the point where he struck, he caused the entire army of his adversaries to flee.' (Xenophon, *Hellenica* 7.5.24)

As at Leuctra, there was also an important role for the cavalry, elucidated in greater detail by Diodorus (whose account of the infantry fighting is so confused and formulaic as to be useless). On this occasion, Epaminondas was himself mortally wounded in the fighting and the accordingly demoralized Boeotian army failed to follow up its initial victory with a vigorous pursuit.[23]

Another tactical innovation (if we are correct to see an innovation rather than a survival of more evidence) is the use of reserves. On the whole, Greek armies did not use reserves, certainly not in the large scale and systematic way in which later Roman armies did, where the infantry were formed in three lines, the second line being used to support any threatened sections of the first line, and the third line providing a last-ditch reserve on which the others could fall back. The Greek phalanx was usually formed in a single line that was committed in one attack, and any break in the line led, at least locally, to inevitable defeat. There are examples, as we have seen, of ambushes, detached forces or reinforcements, expected or unexpected (Thuc. 3.108.1; 4.43.4; 4.96.5), but these are not true battlefield reserves – which the Roman-era military writer Onasander defines as:

> '[A] picked corps, stationed apart from the phalanx as military reserves [*ephedrous*], that he [the general] may have them ready to give assistance to those detachments of his force that are exhausted.' (Onasander, *The General* 22.1)

Onasander distinguishes these from detached forces in ambush or making flank or rear attacks (22.3–3) as in the examples above. The late Roman historian Vegetius provides a similar account of how reserves should be used:

> 'Some should be posted in rear of the wings and some near the centre, to be ready to fly immediately to the assistance of any part of the line which is hard pressed, to prevent its being pierced, to supply the vacancies made therein during the action and thereby to keep up the courage of their fellow soldiers and check the impetuosity of the enemy. This was an invention of the Lacedaemonians, in which they were imitated by the Carthaginians. The Romans have since observed it, and indeed no better disposition can be found.' (Vegetius, *De Re Militari* 3.17)

If the Spartans did invent the use of reserves, we see no trace of this in the literary battle accounts that have survived, nor does Xenophon mention them in his *Constitution of the Lacedaemonians*. At Amphipolis (422), the Spartan Brasidas ordered a follow-up attack by a second force 'as a fresh assailant has always more terrors for an enemy than the one he is immediately engaged with' (Thuc. 5.9.8), but this was an attack from inside the walls of a town rather than a pitched battle by a deployed phalanx. The earliest reference I am aware of to true battlefield reserves in Vegetius' and Onasander's sense occurs in the Athenian deployment at Syracuse in 415:

> 'Half their army was drawn up eight deep in advance, half close to their tents in a hollow square, formed also eight deep, which had orders to look out and be ready to go to the support of the troops hardest pressed. The camp followers were placed inside this reserve.' (Thucydides 6.67.1)

The Greek word translated as 'reserve' here, *epitaktos*, is very little used elsewhere, and the fact that this force was intended to protect the baggage suggests it had at least a dual function (camp guard as well as reserve). However, the important point is their purpose, to 'be ready to go to the support of the troops hardest pressed', which identifies them as a tactical reserve in the sense understood by the Romans. Thucydides makes no further comment on this purpose, nor does he record them actually lending assistance in the battle (which the Athenians seem to have won without needing to call on their aid).

Tactical reserves of this sort do, however, appear in Xenophon. While advancing against an enemy force in Bithynia, Xenophon proposed the following order of battle:

> "It seems to me, fellow generals, that we should station reserve [*phulakas*] *lochoi* behind our phalanx, so that we may have men to come to the aid of the phalanx if aid is needed at any point, and that the enemy, after they have fallen into disorder, may come upon troops that are in good order and fresh." All shared this opinion.' (Xenophon, *Anabasis* 6.5.9)

This was done (Xenophon stationing reserve *taxeis*, a nice switch in terminology as we saw in Chapter 3, and note the use of a different word again for 'reserve' – literally 'guards'). In the event, these too seem to have seen no action as the engagement that followed was disrupted by a ravine and the rapid flight of the enemy forces. Nevertheless, the principle and purpose of the tactical reserve is clear. Xenophon also proposed a somewhat similar role for detached forces of Persian cavalry and infantry in his account of Thymbrara, although in this case they had the specific purpose of attacking the Lydian outflanking columns, so are not quite reserves in the same sense.[24]

Why such reserves do not appear more often in Greek battles is an open question. Evidently, the theory behind them, and their usefulness, was well understood, by Nicias, Brasidas and Xenophon at any rate; yet in most pitched battles there is no trace of similar reserve units being posted. It is likely that the reasons are mostly cultural; with position in the line of battle being a matter of prestige, and with right wing and left wing placements in particular being hotly contested, there may have been a perceived loss of status in being stationed out of, and indeed behind, the main battle line. It may also be related to the fact that hoplite combat tended to be quite quickly decided, with decisive breakthroughs leading to the loss of whole sections of the line (see the following chapter), so that as a matter of practicality, in hoplite versus hoplite battles, reserves were found to be less useful. Also, with deep formations and outflanking both being favoured ways to gain an advantage in battle, committing some proportion of the hoplites to a supporting role may have been felt to be a waste of manpower. Note also that in later military theory, the value of reserves was that the commanding general would commit them to the battle at the crucial moment, but Greek generals would be fighting near the front of the main line and so were in no position to judge when reserves were needed nor to send them where they were required. The commitment of any such reserves as did exist was therefore up to the initiative of their own commanders. No doubt a combination of such factors tended to reduce the value of reserves, but whatever the reasons, although the concept of reserves was clearly already well understood by the later fifth century (and I have no doubt, much earlier), they saw little use.

Finally, there was a possible manoeuvre that poses some difficulties of interpretation – the deliberate opening of gaps in a phalanx to allow an enemy to pass through. Xenophon recounts the effect of the royal Persian chariots at Cunaxa (401):

> 'As for the enemy's chariots, some of them plunged through the lines of their own troops, others, however, through the Greek lines, but without charioteers. And whenever the Greeks saw them coming, they would open a gap for their passage; one fellow, to be sure, was caught, like a befuddled man on a race-course, yet it was said that even he was not hurt in the least, nor, for that matter, did any other single man among the Greeks get any hurt whatever in this battle, save that some one on the left wing was reported to have been hit by an arrow.' (Xenophon, *Anabasis* 1.8.20)

It is not certain that the hoplites met the chariots here. Xenophon also records that Tissaphernes, at the head of Persian cavalry in this battle,

> 'charged along the river through the Greek peltasts; he did not kill anyone in his passage, but the Greeks, after opening a gap for his men, proceeded to deal blows and throw javelins upon them as they went through.' (Xenophon, *Anabasis* 1.10.7)

It is easier to envisage peltasts, in loose formation and open order, making such gaps for cavalry, than it is to see how hoplites could open passages wide enough for chariots to pass through. We also do not know how the hoplites would have opened such gaps, whether by controlled drill manoeuvres or by a more ad hoc process of those in front of the chariots quickly getting out of the way. A similar manoeuvre was performed by the Macedonian phalanx 170 years later at Gaugamela (Arr., *Anab.* 3.13.6), although it is equally uncertain precisely what form the manoeuvre took. Obviously, in normal circumstances a phalanx would try to avoid gaps developing, but driverless chariots changed the situation – the horses, presented with the opportunity to avoid a collision, would gladly do so.

At the Battle of Tegyra (375), however, Spartan hoplites are said to have tried to open a passage for Theban hoplites:

> '[W]hen those about them [the Spartan *polemarchoi*] were being wounded and slain, their whole army was seized with fear and opened up a lane for the Thebans, imagining that they wished to force their way through to the opposite side and get away. But Pelopidas used the path thus opened to lead his men against those of the enemy who still held together, and slew them as he went along, so that finally all turned and fled.' (Plutarch, *Pelopidas* 17.4)

The result here, of opening a gap in the phalanx, was more what we would expect, turning out badly for the Spartans, and it is not clear anyway how a phalanx engaged in combat could have opened such a passage for an enemy. Perhaps a section of the phalanx simply gave way or fled, and this was rationalized as a controlled attempt to open a passage. At Coronea (394), Xenophon remarks of Agesilaus that he did not have to fight the Thebans face to face:

> 'For while he might have let the men pass by who were trying to break through and then have followed them and overcome those in the rear, he did not do this, but crashed against the Thebans front to front.' (Xenophon, *Hellenica* 4.3.19)

Whether this means let them pass by across the front of his force or make a passage for the 'break through' is not clear. At any rate, it seems to have been possible for hoplites, perhaps only if well-disciplined and drilled, to open such passages, but against any opponent other than driverless chariots it would have been very risky to do so. It is thus not surprising that we do not see this done often.

Hoplites versus Persians

We have already seen hoplites fighting Persians in the three great battles of the Persian Wars – Marathon, Thermopylae and Plataea – to which should be added Mycale, fought in Asia Minor on the same day as Plataea (according to tradition). Greek hoplite armies also fought against the Persians in Asia Minor in the late fifth and early fourth centuries, particularly in the march of the Ten Thousand and the campaigns of Agesilaus, as well as both for and against Alexander the Great in Asia later in the fourth century. This encounter of hoplites and Persian infantry (leaving aside, for the moment, the role of the Persian cavalry) is seen as one of the great confrontations of different weapons systems in military history, and was characterized by the Greeks as a contest between themselves – steadfast, heavily equipped, fighting face to face, and also manly, courageous and free – and the Persians (and their numerous subject peoples), characterized as lightly equipped, dependent on distant shooting with missiles and also (at least on occasion) womanly, cowardly and slavish. Whether this is an accurate characterization of the Persian army is a very large subject in itself (and clearly it is not accurate in important respects), and the number of modern studies of the Persian army is tiny in comparison with the mountain of books on Greek hoplites. This present addition to that book mountain, being a general study, cannot hope to do justice to this subject, so I will limit discussion to the classic conception of Greeks versus Persians as presented in our main (Greek) sources.[25]

The Persian army was a large imperial army consisting of a central core of Persian and Median infantry and cavalry, to which was added, as occasion demanded, smaller or larger contingents of subject or allied peoples – including of course many Greeks – who fought with their own traditional equipment and fighting styles. Our sources tend to refer to a particular contest, between spear-armed Greeks and largely bow-armed Persians, but given the wide variety of infantry types in Persian armies (elucidated in Herodotus' famous catalogue), this is itself a gross simplification, and actual battles must have been far more varied and complex affairs.

The classic confrontation is with the type of Persian infantry now generally referred to as *sparabara* (which is the word given by one Greek source as the Persian name for *gerrophoroi*, 'wicker-shield-bearers'). These were a form of heavy infantry, though less heavily equipped than the Greek hoplite: they had quilted armour of some sort (perhaps similar to the Greek linen or leather armour discussed in Chapter 2), and spears perhaps similar in length to (though thought by the Greeks to be shorter than) Greek spears (Hdt. 5.49.3; 7.211.2). They carried large rectangular wicker shields, called *gerrai* in Greek, almost as tall as the man, but of lighter construction than the hoplite *aspis* (though not necessarily much lighter, as wicker can make a very strong shield, and Greek shields were probably neither as thick nor as heavy as is often supposed). But *sparabara* is also used to refer to a composite unit made up of a front rank of these shield-bearers, backed up by several, perhaps many, ranks of more lightly equipped archers. Persian tactics appear to have been to advance within bow range of the enemy, then halt while the front rank propped their shields up with a stick or support to make a continuous, if perhaps rather flimsy, shield wall. The archers would then shoot from behind this wall, while the front-rankers were ready to repel with their spears any enemy that ventured too close. Persian tactics, it is usually supposed, were to bring down on any opponents a barrage of arrows hopefully sufficient to cause them to break and run, or at least to keep them at bay until the Persian cavalry could intervene against their flanks. In symmetric warfare against similarly or more lightly equipped opponents, these tactics evidently worked

Hoplite v. *sparabara*; note the propped-up Persian shield, leaving the Persian shieldless at close quarters, although he is armoured (presumably in textile armour) (Baseggio Cup, after Sekunda (1992))

well, allowing the Persians to carve out their extensive empire, but against well-shielded and armoured Greeks prepared to close immediately to hand-to-hand combat, they were to prove less effective.[26]

It is also worth touching briefly on the matter of Persian numbers, traditionally estimated by Greeks to range from tens of thousands (the basic Persian military unit being the 'myriad', or unit of 10,000 men) to millions. This subject has also generated much modern scholarship, aimed at replacing the wilder estimates of the Greeks (Herodotus estimates Xerxes' invasion force at two-and-a-half million men, plus as many servants and camp followers) with more realistic figures. I certainly do not think we can take the Greek estimates of Persian numbers at face value, but I also fear that it is hopeless trying to replace them with anything more specific when there is so little hard evidence to work from. Broadly speaking, armies in antiquity tended to have similar numerical strengths, although there are cases of larger but poorly trained, equipped or motivated armies being defeated by small, highly trained and disciplined and well-led ones. Establishing the relative values of quality and quantity is a difficult task, and I believe we are left with no choice but to conclude that while Persian armies were large – certainly larger than the Greek armies they faced – there is no way now to tell just how large.[27]

The Persians had a formidable reputation in their early encounters with the Greeks. Their efficient suppression of the Ionian Revolt, prelude to the Persian Wars, seems to have left the Greeks somewhat in awe of them, as Herodotus (6.112.3) remarks. As late as Plataea, there was still a fear of the Persians, as seen in the passage of Herodotus quoted above where the Spartans tried to get out of having to fight them (Hdt. 9.46.2). Herodotus preserves no details of the nature of the fighting at Marathon (490) – indeed it may surprise the modern reader that the entirety of our knowledge of the course of the actual fighting is seventy-six (translated) words of Herodotus:

> 'They fought a long time at Marathon. In the centre of the line the foreigners prevailed, where the Persians and Sacae were arrayed. The foreigners prevailed there and broke through in pursuit inland, but on each wing the Athenians and Plataeans prevailed. In victory they let the routed foreigners flee, and brought the wings together to fight those who had broken through the centre. The Athenians prevailed, then followed the fleeing Persians and struck them down.' (Herodotus 6.113.1–2)

We do know, though, that the Athenians, as discussed above, advanced at the run, perhaps for the first time, and it is supposed that this was partly in order to avoid long exposure to Persian archery. The normal Greek tactics of closing immediately to hand-to-hand combat where their close formation and heavier equipment gave them an advantage worked well here against a more lightly equipped enemy, denied by the running charge of much effect from its archery. But the fighting was far from one-sided, as the defeat of the Athenian centre demonstrates. Crucially, there was no Persian cavalry at Marathon, or none that had any recorded impact on the battle, for reasons that are obscure and the subject of much modern speculation.

At Thermopylae (480), an outnumbered Greek army attempted to hold the pass against a much larger Persian force that was prevented by the narrow ground from using its size to its advantage. At first, a force of Medes (actually Medes on this occasion) was tasked with dislodging the Greek defenders. Herodotus' brief account tells us little about the fighting and more about Greek feelings of superiority:

> 'The Medes bore down upon the Hellenes and attacked. Many fell, but others attacked in turn, and they made it clear to everyone, especially to the king himself, that among so many people there were few real men. The battle lasted all day.' (Herodotus 7.210.2)

The Medes were evidently attacking hand-to-hand, not just standing off and shooting, despite the well-known Persian threat before the battle that their arrows would blot out the sun (Hdt. 7.226.1). Next, the Persian king sent in the Persian Immortals, a myriad of elite Persian infantry:

> 'When they joined battle with the Hellenes, they fared neither better nor worse than the Median army, since they used shorter spears than the Hellenes and could not use their numbers fighting in a narrow space. The Lacedaemonians fought memorably, showing themselves skilled fighters amidst unskilled on many occasions, as when they would turn their backs and feign flight. The barbarians would see them fleeing and give chase with shouting and noise, but when the Lacedaemonians were overtaken, they would turn to face the barbarians and overthrow innumerable Persians. A few of the Spartans themselves were also slain. When the Persians could gain no inch of the pass, attacking by companies and in every other fashion, they withdrew.' (Herodotus 7.211.1–3)

The surprising suggestion that the Spartans fought in this way, with feigned retreats rather than a steadfast, close-ordered phalanx fighting face-to-face and hand-to-hand, is supported by other sources and bears comparison with similar events at Plataea (see below). Onasander, writing in the first century AD but drawing on earlier material, includes such feigned retreats in his list of advice for generals:

> 'It is sometimes a useful stratagem for an army facing the enemy to retire gradually, as if struck by fear, or to about-face and make a retreat similar to a flight but in order, and then, suddenly turning, to attack their pursuers.' (Onasander, *The General* 21.9)

If the story told of Thermopylae is accurate, it does suggest a lack of discipline on the Persian side and a very high level of individual skill as well as overall discipline by the Spartans.[28]

Further Persian attacks were similarly unsuccessful, until Xerxes was informed of a way through the mountains that would allow him to turn the Greek position. Learning of this, Leonidas, the Spartan king in overall command, sent away the rest of his army to safety and made a last stand with his famous picked force of 300 men, along with Thespian and Theban allies, advancing out of the narrows to face the Persian army to his front:

> 'Now, however, they joined battle outside the narrows and many of the barbarians fell, for the leaders of the companies beat everyone with whips from behind, urging them ever forward. Many of them were pushed into

the sea and drowned; far more were trampled alive by each other, with no regard for who perished.' (Herodotus 7.223.3)

The notion that Persian armies would fight (or work, or march) only under the compulsion of the whip is a common theme of Herodotus, and also crops up in Xenophon. This may indeed be historical, since less-disciplined subject peoples might indeed have had officers behind the line equipped with whips, performing a similar role to the file closers in the Greek phalanx whose job was to prevent anyone from fleeing and keep the phalanx moving forward. To the Greeks, however, the use of the whip was a particular sign of Persian servility, and contrasted with their own status as free, politically independent citizens.[29]

That Persian spears were shorter than Greek ones may be true, though in normal spear fighting small differences of spear length would not offer any particular advantage or disadvantage, since it was a simple matter to 'add a step' to the reach of the spear by stepping forward to strike and parry the enemy spearpoint (this was vastly less easy against the multiple rigid pike points of the Macedonian phalanx), and the reach of a spear was determined as much by its point of balance as by its length. Indeed, once a fighter was inside the minimum reach of his opponent's spear, a longer spear would be a positive disadvantage, and in vase paintings of combats, spears tend to be held in such a way that the spear point is aligned more or less with the reach of the shield, there being no obvious benefit in a longer reach. The exception of course is when there is an obstacle of some sort between the combatants so that they cannot get closer, and in the case of Greeks facing Persians, such an obstacle was provided by the Persian shield wall. The Persians behind their propped shields (which were probably intended for protection from missiles) would have been unable to close with the Greeks, who, if they had slightly longer spears, could have stood just outside Persian spear range and stabbed over the shields. The Persians' only recourse would be to back away from the shields, which would have given the Greeks an opportunity to knock down the shield wall. This may be why Herodotus draws particular attention to short Persian spears in these fights, when we hear little of spear length in any other type of fighting.[30]

Nevertheless, the Greek resistance at Thermopylae was courageous but ultimately doomed. First Leonidas was killed:

> 'There was a great struggle [*othismos*] between the Persians and Lacedaemonians over Leonidas' body, until the Hellenes by their courageous prowess dragged it away and routed their enemies four times.' (Herodotus 7.225.1)

Then the survivors were surrounded on a low hillock and shot down with missiles (Hdt. 7.225.2–3), the diminishing Greek numbers meaning that they were unable to march out against the Persians, so the Persian tactics of standing off and shooting were able to come into play.

At Plataea the following year (479), a large Greek army headed by Spartan and Athenian contingents faced a Persian army (traditionally vast, but probably similar in size to the Greek army) left in Greece when Xerxes returned home after the naval defeat at Salamis, along with a large force of Medizing Greeks. A long stand-off before the battle saw Persian cavalry and light forces constantly harassing the Greeks, forcing them eventually to attempt a night-time withdrawal from the plain into the nearby hills. When dawn found this retreat not completed and the Greek contingents divided and scattered across the plain, the Persian infantry and cavalry under Mardonius attacked in haste. Despite their earlier efforts to swap places with the Athenians, the Spartans found themselves facing Persian infantry, and as we saw above, initially on the defensive enduring Persian archery:

> 'They could get no favourable omen from their sacrifices, and in the meanwhile many of them were killed and by far more wounded (for the Persians set up their shields for a fence, and shot showers of arrows).' (Herodotus 9.61.3)

Here we see the Persian use of the propped up *gerrai*, behind which the archers could safely engage the Spartans, who lacked any missile support of their own. Eventually the sacrifices came good and the Spartans attacked:

> 'Now they too charged the Persians, and the Persians met them, throwing away their bows. First they fought by the fence of shields, and when that was down, there was a fierce and long fight around the temple of Demeter itself, until they came to blows at close quarters [*othismos*]. For the barbarians laid hold of the spears and broke them short. Now the Persians were neither less valorous nor weaker, but they had no armour; moreover, since they were unskilled and no match for their adversaries in craft, they would rush out singly and in tens or in groups great or small, hurling themselves on the Spartans and so perishing.' (Herodotus 9.62.1–3)

There are intriguing, if puzzling, details of the nature of the fighting in this account. Why did the Persians 'throw away their bows' (or put them aside) when there were already spearmen among them? It may be that all Persian infantry were equipped with shield, spear and bow, and would shoot until the enemy came to close quarters, then take to the spear (previously stuck in the ground, perhaps). How did the three types of fighting – 'by the fence of shields', 'a fierce and long fight', '*othismos*' – differ from one another? I will consider this particularly knotty question in the next chapter. Here, the Persians, rather than being short-speared women, are valorous and strong, but ill-equipped (*anoplia*, 'unarmed' or 'ill-equipped', translated as 'had no armour', although we know at least some Persian infantry did wear armour of some sort). Perhaps above all, how in these phases of close fighting were the Persians able to rush out singly or in tens? Was this similar to the 'feigned retreats' we saw at Thermopylae, above, which also saw small groups of Persians overwhelmed by the Spartans? Plato's anecdote about this battle would suggest so:

> 'For they say that at Plataea, when the Spartans came up to the men with wicker shields, they were not willing to stand and fight against these, but fled; when, however, the Persian ranks were broken, the Spartans kept turning round and fighting like cavalry, and so won that great battle.' (Plato, *Laches* 191c)

'Fighting like cavalry' means alternating attacks and retreats, as cavalry customarily did. Evidently, various traditions arose among Greek accounts of this battle, not necessarily mutually compatible, and it is probably a futile task trying to establish precise tactical details from such anecdotal reports. At any rate, better Greek equipment and superior Spartan discipline and training were enough to overcome the Persian shield wall, archers and spearmen. After a hard fight, Mardonius himself was killed, this finally breaking Persian resistance, although Herodotus again cites the inadequacy of Persian equipment:

> 'For what harmed them the most was the fact that they wore no armour [*hopla*] over their clothes and fought, as it were, naked against men fully armed [literally 'as *gymnetes* against hoplites'].' (Herodotus 9.63.2)

I think that *anoplia*, 'unarmed', above could be understood to mean 'unshielded' (explained by this observation that they fought as if *gymnetes* against *hoplitai*, the key lack of *gymnetes* being shields), which would be the case if the shield-bearing ranks set up their shields in a fence and the archer component did not have shields. Once the shield fence was down, the Persians behind would indeed be unshielded.[31]

The same year, and according to tradition on the same day, an Athenian, Spartan and allied expeditionary force fought the Persians at Mycale in Asia Minor. Here too, the Persians 'set their shields close to make a barricade' (Hdt. 9.99.3) to fight the Greeks:

> 'As long as the Persians' shields stood upright, they defended themselves and held their own in the battle, but when the Athenians and their neighbours in the line passed the word and went more zealously to work, that they and not the Lacedaemonians might win the victory, immediately the face of the fight changed. Breaking down the shields they charged all together into

the midst of the Persians, who received the onset and stood their ground for a long time, but at last fled within their wall [of their camp].' (Herodotus 9.102.2–3)

The battle proceeded just like Plataea, though with Athenians bearing the brunt of the fighting (the Spartans were making a flanking move and were delayed by difficult ground). The shield wall held up the Athenians for a while, but, knocking it down, they got in among the Persians and overwhelmed them.

These successes of hoplite armies against the Persians were to cause Greeks to be in high demand as mercenaries in Persian armies – those of Persian satraps (regional governors), of rebel Persian claimants to the throne (Cyrus, who hired the Ten Thousand) and ultimately of the King of Kings himself (Darius III hiring large Greek contingents to face the invasion of Alexander the Great). We have already seen the success of the Ten Thousand against Persian arms at Cunaxa (401), where the charge alone was sufficient for local victory (although the battle was lost), and there was no need for long and hard fights around walls of shields. The retreat of the Ten Thousand back through Persian territory saw numerous encounters between Greek hoplites and light infantry and various light and irregular native defenders, though these generally took the form of irregular actions and skirmishes rather than pitched battles.

Light infantry and cavalry

In Chapter 6 we saw how vulnerable hoplites could be to light infantry in certain circumstances – specifically, where there was space and time enough for the light infantry to fall back when attacked and avoid contact, returning to shower the hoplites with missiles when they fell back, or where difficult ground made it difficult for the hoplites to attack the light infantry without losing formation. This applied particularly where forces of light infantry were attacking hoplites on the march or were able to harry them during a long advance.

Here we are concerned more with the role of light infantry in pitched battles, where hoplites were present on both sides and where a decisive outcome was being sought on the field of battle rather than through prolonged skirmishing. We have seen that light infantry tend to be deployed (where we hear anything at all about how they are deployed) either between the lines to oppose enemy light infantry, or else on the flanks, perhaps accompanying cavalry. As such, hoplites would rarely be opposed by light infantry in the course of a pitched battle – they would be more likely to come up against their own kind, other hoplites. Partly this may have been due to the traditional nature of hoplite battle, matching the citizen soldiers of each side against each other, but as usual there are also practical reasons. However long it might take for hoplites to defeat hoplites, light infantry harrying hoplites would take even longer, and there was no point in opposing an enemy indecisively on one wing with light infantry while on the other one's own heavy infantry was quickly defeated. Large, formal battles therefore mostly saw hoplites employed against hoplites.

Nevertheless, there were occasions where light infantry engaged hoplites, particularly in smaller or irregular encounters, starting with some of the earliest accounts of hoplite battle we have. As early as the mid-fifth century, at Megara (458), after a hoplite battle between Athenians and Corinthians – the result of which was (as so often) disputed – the Athenians cut off a part of the Corinthian army which was trying to erect a surreptitious trophy (see further below on trophies):

> 'In the retreat of the vanquished army, a considerable division, pressed by the pursuers and mistaking the road, dashed into a field on some private property, with a deep trench all round it, and no way out. Being acquainted with the place, the Athenians hemmed their front with heavy infantry, and placing the light troops [*psiloi*] round in a circle, stoned all who had gone in. Corinth here suffered a severe blow.' (Thucydides 1.106.1–2)

In this case, the presence of the trench preventing the Corinthians from counterattacking and of Athenian hoplites who could have repelled them if they did, made things easy for the *psiloi*. We may also note the lack of any gentlemen's agreement for light infantry not to kill hoplites, or indeed of any reluctance to slaughter trapped and helpless men.

At Potidaea (432), a Corinthian and Peloponnesian force fought a classic 'revolving door' battle with the Athenians, the stronger wing of each army defeating and pursuing those opposed to them:

> 'Returning from the pursuit, Aristeus [with the Corinthians] perceived the defeat of the rest of the army. Being at a loss which of the two risks to choose, whether to go to Olynthus or to Potidaea, he at last determined to draw his men into as small a space as possible, and force his way with a run into Potidaea. Not without difficulty, through a storm of missiles, he passed along by the breakwater through the sea, and brought off most of his men safe, though a few were lost.' (Thucydides 1.63.1)

Here, the usual tactics of hoplites to draw together into the most compact formation possible and to move at the run were sufficient to rescue the force safely, since they had a definite nearby objective (the city of Potidaea) in which they could take refuge. It is not clear what the Athenian hoplites were doing at this point.

Spartolus (429) provides an occasion where a hoplite force did have to face light infantry and cavalry on a regular battlefield. Here, an Athenian hoplite force engaged and defeated Chalcidian hoplites outside their town of Spartolus, but their light forces were defeated by their opposite numbers. The Chalcidians then renewed the attack on the Athenian hoplites.

> 'Whenever the Athenians advanced, their adversary gave way, pressing them with missiles the instant they began to retire. The Chalcidian horse also, riding up and charging them just as they pleased, at last caused a panic amongst them and routed and pursued them to a great distance.' (Thucydides 2.79.6–7)

The most well-known examples of hoplite forces cut off and defeated by light infantry – Aegitium (426, Thuc. 3.97–98), Sphacteria (425, Thuc. 4.29–37) and Lechaeum (390, Xen., *Hell.* 4.5.11–17) – have been considered above in Chapter 6. These are sometimes regarded as new developments in the practice of warfare, marking a break from the more formal hoplite versus hoplite battles of earlier times, but they should surely be seen as part of an ongoing pattern of light infantry posing a serious threat to hoplites, and being used against hoplites, when conditions allowed. What was most remarkable about these three occasions is the scale of the defeat inflicted, and in the latter two cases that it was Spartan hoplites defeated, fleeing or surrendering, rather than that light infantry should defeat hoplites at all.

We have also seen already above, and in Chapter 6, that cavalry could be actively employed against hoplites, from the earliest detailed battle accounts we have onwards. Because closely formed, disciplined and confident hoplites were safe from the charge of any sort of cavalry used at the time (whether javelin-throwing Greek cavalry or heavier, armoured Persian cavalry), formed hoplites in formation had little to fear. Cavalry were therefore used to skirmish and harry enemy light infantry and cavalry, but also to work around the flanks or rear of enemy hoplites, and particularly to restrict the mobility of enemy hoplites by forcing them to remain in close formation – this latter case was most useful in preventing victorious hoplites from carrying out an aggressive pursuit of their defeated opposite numbers.

Of the four major battles of the Persian War, despite their assumed reliance on cavalry, the Persians used cavalry actively only at Plataea. They were mysteriously absent from Marathon, unsuitable to the task of frontal assault required at Thermopylae and also absent from Mycale. At Plataea, they successfully harassed the Greek positions and operated against their supply lines, forcing the retreat that brought on the final battle, but are absent from the main fighting, aside from the initial encounter in which Masistius was killed (Hdt. 9.22.1–3). But the

Persian cavalry (and Greek cavalry fighting with them) performed their classic role of delaying pursuit by threatening any pursuers who broke ranks:

> '[The Medizing Greeks] accordingly all fled, save the cavalry, Boeotian and other; this helped the fleeing men in so far as it remained between them and their enemies and shielded its friends from the Greeks in their flight.' (Herodotus 9.68.1)

There was one significant disaster for the Greek forces:

> 'During this steadily growing rout there came a message to the rest of the Greeks, who were by the temple of Hera and had stayed out of the fighting, that there had been a battle and that Pausanias' men were victorious. When they heard this, they set forth in no ordered array, those who were with the Corinthians keeping to the spurs of the mountain and the hill country, by the road that led upward straight to the temple of Demeter, and those who were with the Megarians and Philasians taking the most level route over the plain. However, when the Megarians and Philasians had come near the enemy, the Theban horsemen (whose captain was Asopodorus son of Timander) caught sight of them approaching in haste and disorder, and rode at them; in this attack they trampled six hundred of them, and pursued and drove the rest to Cithaeron.' (Herodotus 9.69.1–2)

It is in this role of screening a retreat, and making opportunistic attacks on hoplites out of formation, that we mostly see Greek cavalry for the next century-and-a-half. For example, they covered the Athenian retreat after the defeat at Mantinea (418):

> 'Indeed they [the Athenians] would have suffered more severely than any other part of the army, but for the services of the cavalry which they had with them.' (Thucydides 5.73.1)

Cavalry performed a similar role in Syracuse:

> 'The Athenians did not pursue far, being held in check by the numerous and undefeated Syracusan horse, who attacked and drove back any of their heavy infantry whom they saw pursuing in advance of the rest; in spite of which the victors followed so far as was safe in a body, and then went back and set up a trophy.' (Thucydides 6.70.3)

Yet there are examples too of cavalry directly defeating infantry, sometimes hoplites. We need not imagine that this means that the cavalry charged frontally against a wall of spears and shields, rather that they were able to threaten exposed flanks or exploit gaps that opened due to ill-discipline, rough ground or missile casualties, as at Solygeia (425):

> 'After holding on for a long while without either giving way, the Athenians aided by their horse, of which the enemy had none, at length routed the Corinthians, who retired to the hill and halting remained quiet there, without coming down again.' (Thucydides 4.44.1)

The Syracusan cavalry, not successfully countered by the Athenians, was particularly effective:

> 'A battle now ensued, in which the Athenians were victorious, the right wing of the Syracusans flying to the town and the left to the river. The three hundred picked Athenians, wishing to cut off their passage, pressed on at a run to the bridge, when the alarmed Syracusans, who had with them most of their cavalry, closed and routed them, hurling them back upon the Athenian right wing, the first tribe of which was thrown into a panic by the shock.' (Thucydides 6.101.4–5)

On a later occasion:

> 'During the engagement the cavalry attacked and routed the left wing of the Athenians, which was opposed to them; and the rest of the Athenian army was in consequence defeated by the Syracusans and driven headlong within their lines.' (Thucydides 7.6.3)

Cavalry could be particularly effective against the rear of a hoplite phalanx. Before Delium (424), the Athenians hoped to take advantage of this:

> 'Meanwhile Hippocrates at Delium, informed of the approach of the Boeotians, sent orders to his troops to throw themselves into line, and himself joined them not long afterwards, leaving about three hundred horse behind him at Delium, at once to guard the place in case of attack, and to watch their opportunity and fall upon the Boeotians during the battle.' (Thucydides 4.93.2)

But in the event it was Boeotian cavalry threatening the rear of the Athenians that decided the battle (even if by mistaken identity):

> 'It so happened also that Pagondas, seeing the distress of his left, had sent two squadrons of horse, where they could not be seen, round the hill, and their sudden appearance struck a panic into the victorious wing of the Athenians, who thought that it was another army coming against them.' (Thucydides 4.96.5)

Presumably, the small cavalry force was intended simply to cover the Boeotian retreat, but it is notable how easily the Athenians panicked upon sighting such a small (and in theory harmless) force, a reminder of how hoplite armies were constantly teetering on the edge of panic, as we will see in the next chapter. Cavalry could be said to have won this battle, though not by fighting.

Cavalry's mobility and ability to work around the flank or rear of an infantry formation could make up, to a large extent, for its inability to threaten formed hoplites with a frontal charge. In the fourth century, the Thebans Epaminondas and Pelopidas were to make particularly effective use of cavalry, though generally in a supporting role and by opposing and neutralizing enemy cavalry, rather than as a main striking force. The value of cavalry was established at Tegyra (375), where Pelopidas was surprised by a Spartan force:

> 'Then he at once ordered all his horsemen to ride up from the rear in order to charge, while he himself put his hoplites, three hundred in number, into close array, expecting that wherever they charged he would be most likely to cut his way through the enemy, who outnumbered him.' (Plutarch, *Pelopidas* 17.2)

Pelopidas' hopes were fulfilled, the three hundred men of the Sacred Band cutting through and routing the Spartan contingent. It is not clear what actual role the cavalry played, but they may have softened up the Spartan line or, by threatening their flank, prevented them from closing around the outnumbered Thebans.

Boeotian cavalry played a greater part in the two notable victories over the Spartans, at Leuctra (371) and Mantinea (362):

> '[S]ince the space between the armies [at Leuctra] was a plain, the Lacedaemonians posted their horsemen in front of their phalanx, and the Thebans in like manner posted theirs over against them.' (Xenophon, *Hell*enica 6.4.10)

It is odd to post cavalry in front of a phalanx rather than on the flanks, and it is quite likely that they were in their capacity as advanced scouts, not expecting a full battle to take place, rather than this being a planned battle array. Evidently there was much confusion on the Spartan side at least, as Xenophon's vague account indicates (Xen., *Hell*. 6.4.13), as we saw above. The Spartan cavalry were driven back onto the phalanx, itself taken by surprise by the rapid Theban advance. It feels rather as if Xenophon is making excuses, but it is perfectly possible that the Spartans were disorganized by their own cavalry retreating past or through them. Yet the fighting that followed was all about the hoplites, and the Theban cavalry seem not to have followed up their initial success.

Boeotian cavalry also played a major role at Cynoscephalae (364), intervening decisively against a combined-arms Thessalian army, although it is is difficult to extract precise tactical details from Plutarch's confused account (Plut., *Pelop.* 32.2–7). At the following Theban victory over the Spartans at Mantinea (362), cavalry again played a relatively major part, according to Xenophon:

> 'Again, while the enemy had formed their horsemen like a phalanx of hoplites – six deep and without intermingled foot soldiers – Epaminondas on the other hand had made a strong column of his cavalry, also, and had mingled foot soldiers among them, believing that when he cut through the enemy's cavalry, he would have defeated the entire opposing army; for it is very hard to find men who will stand firm when they see any of their own side in flight.' (Xenophon, *Hellenica* 7.5.23–24)

That Epaminondas made his phalanx deep at this battle in order to 'crush the head of the snake' is well known, but it is interesting that he did the same with his cavalry, with the same intent, and that Xenophon should suggest that the defeat of the Spartan and allied cavalry would serve to demoralize their whole army. Boeotian cavalry was also employed in their other important role of dissuading infantry forces from operating freely on the battlefield, rather than attacking them frontally:

> 'And in order to prevent the Athenians on the left wing from coming to the aid of those who were posted next to them, he stationed both horsemen and hoplites upon some hills over against them, desiring to create in them the fear that if they proceeded to give aid, these troops would fall upon them from behind.' (Xenophon, *Hellenica* 7.5.24)

According to Xenophon's apologetic and inadequate account, the cavalry achieved little in the battle, defeating their opposite numbers but falling back anyway, dismayed by the death of Epaminondas, although Diodorus (Diod. 15.85.7--8) has Theban cavalry trying to outflank the Athenian phalanx on the allied left, and only being prevented from doing so by the intervention of an Eleian cavalry reserve.

So cavalry were certainly not the tactical afterthought that they are sometimes portrayed as being, and did play a significant role against hoplites in a number of battles. It was, however, to be the Macedonians who would elevate cavalry to a true battle-winning force.[32]

Terrain

According to the stereotypical view of hoplite combat, hoplite battles were fought on flat featureless plains in which terrain played no role. However, we have already seen a number of cases where terrain was significant, and it is in fact quite difficult to find a hoplite battle in which terrain played no part at all. We might divide the different terrains into a number of types: firstly relief, that is hills, mountains and high ground generally; then linear features such as rivers, streams and ravines; and finally ground cover, the vegetation, trees or crops that covered the land and which played an underrated part in many battles.

In terms of relief, it is true that hoplites – and in fact armies generally – might prefer to fight on level ground and to avoid hilly terrain. However, Greece being a mountainous country, it was difficult to find any large extent of ground that was completely flat. This has led to the often repeated claim that hoplite battles, and the Greek way of war generally, 'defy geomorphic logic' for not adopting, as 'common sense dictates', arms, armour and tactics suitable for mountain warfare. However, there are two obvious counters to this argument. One is that, while Greece is mountainous, the plains are (on the scale of the armies deployed on them) of sufficient size and numerous, and of course form the most important component of the territory of the cities, and are therefore perfectly suitable for the operation of the armies of the time. The second is that, as argued in the previous chapter, Greek cities fighting similarly equipped and constituted armies from other Greek cities had every reason to engage in a mutually convenient form of warfare that emphasized the role of the heavy infantry.[33]

The part hills and high ground played in battles and in campaigns generally was to provide a refuge for infantry against cavalry, to provide a strong position in which a weaker force could avoid an attack from a stronger, and to provide a rallying point for beaten forces which, while not in full flight, did not wish to re-engage the victorious enemy. An army formed on a hill was presumed to be at a great advantage, and so long as they maintained their position, the enemy might well have been unwilling to offer them battle – which, if battle was the preferred outcome of the campaign, would be counterproductive to both sides, which is why armies which were not hopelessly outmatched would usually come down from the hills into the plains. Before Mantinea (418), the Spartan army first advanced towards the city:

> 'Here they were seen by the Argives and their allies, who immediately took up a strong and difficult position, and formed in order of battle. The Lacedaemonians at once advanced against them, and came on within a stone's throw or javelin's cast, when one of the older men, seeing the enemy's position to be a strong one, hallooed to Agis that he was minded to cure one evil with another; meaning that he wished to make amends for his retreat, which had been so much blamed, from Argos, by his present untimely precipitation.' (Thucydides 5.65.1–2)

The nature of the strong position Thucydides does not spell out here, though later he says that Agis hoped to 'make the Argives and their allies come down from the hill' (Thuc. 5.65.5), and it is generally supposed that the allies were formed on the hills just to the east of the city, and so had the advantage of being upslope. At any rate, Agis was persuaded and withdrew, for the moment.

The idea that an army on a hill was at a great advantage explains the 'fact', related by Mardonius, that Greeks 'come down to the fairest and most level ground that they can find and fight there' (Hdt. 7.9b.1). A pitched battle is, by definition, fought by mutual consent, and to remain in an advantageous position on a hill (or later in the Hellenistic and Roman periods to remain in a fortified camp) was to declare that one did not consent to battle at this time. A stronger army would not risk throwing away its advantage by attacking a defended position, while a weaker army would not risk defeat by leaving superior ground. Only if both sides felt themselves in with a fair chance would they leave any such position and engage in battle.

In the speech that Xenophon gives to Thrasybulus at Munychia (403), the benefits of high ground are spelled out:

> '[The enemy] because they are marching up hill, cannot throw either spears or javelins over the heads of those in front of them, while we, throwing both spears and javelins and stones down hill, shall reach them and strike down many. And though one would have supposed that we should have to fight with their front ranks at least on even terms, yet in fact, if you let fly your missiles with a will, as you should, no one will miss his man when the road is full of them, and they in their efforts to protect themselves will be continually skulking under their shields. You will therefore be able, just as if they were blind men, to strike them wherever you please and then leap upon them and overthrow them.' (Xenophon, *Hellenica* 2.4.15–16)

Note that the advantage here is in use of missiles (even though these were mostly hoplites fighting), a reminder that hoplites too could resort to missiles where necessary (and a suggestion that use of missiles by the phalanx may have been more common than we think). A similar case, and also an example of a defeated enemy taking refuge on a hill, is Haliartus (395):

> 'Now when Lysander had been killed and his troops were fleeing to the mountain, the Thebans pursued stoutly. But when they had reached the heights in their pursuit and came upon rough country and narrow ways, the hoplites of the enemy turned about and threw javelins and other missiles upon them. And when two or three of them who were in the van had been struck down, and the enemy began to roll stones down the hill upon the rest

and to attack them with great spirit, the Thebans were driven in flight from the slope, and more than two hundred of them were killed.' (Xenophon, *Hellenica* 3.5.19–21)

Lyncestis (423) was a cavalry and hoplite battle in which the hills served as a refuge for the defeated after they had voluntarily fought on the plain (Thuc. 4.124.3), as we saw earlier. At Amphipolis (422) too, high ground provided a refuge for defeated hoplites, though not impunity from missiles:

'The Athenian right made a better stand, and though Cleon, who from the first had no thought of fighting, at once fled and was overtaken and slain by a Myrcinian peltast, his infantry forming in close order upon the hill twice or thrice repulsed the attacks of Clearidas, and did not finally give way until they were surrounded and routed by the missiles of the Myrcinian and Chalcidian horse and the peltasts.' (Thucydides 5.10.9)

We have already seen a similar example at Solygeia (425):

'[The Athenians] at length routed the Corinthians, who retired to the hill and halting remained quiet there, without coming down again ... The rest of the army, broken and put to flight in this way without being seriously pursued or hurried, retired to the high ground and there took up its position. The Athenians, finding that the enemy no longer offered to engage them, stripped his dead and took up their own and immediately set up a trophy.' (Thucydides 4.44.1–3)

Note the defeated forces retiring to high ground, and the victors unwilling to follow them there, relying instead on their offering battle by descending to the plain – which in this case they declined to do. At Mantinea, when the Spartan army returned from its initial abortive advance, the allies did offer battle by advancing from the hill on which they were deployed into the plain (Thuc. 5.66.1). After the defeat at Leuctra, the Spartans retreated to their camp:

'The camp, to be sure, was not on ground which was altogether level, but rather on the slope of a hill.' (Xenophon, *Hellenica* 6.4.14)

Here, the Spartans debated what to do next, and decided, rather than leaving the hill and re-engaging, to admit defeat.

High ground could also of course constrain the battlefield by protecting one or both flanks of an army, preventing it from being outflanked by a larger or more mobile enemy. Passes could be defended in this way – as most notably at Thermopylae – but as we saw in Chapter 6, an army would rarely prefer to defend a pass rather than accept battle on more equitable ground because it was usually a simple matter to turn a defensive position in a pass, and either come round behind the defending army or bypass them completely, leaving the defenders' agricultural land at the mercy of an attacker.

Hoplite armies repeatedly showed that they were very willing to deploy on hills, and actively sought them out for defensive purposes, to protect against cavalry or a victorious enemy. However, because this offered a great advantage to the uphill side, it was unlikely that the enemy would take the risk of forcing the issue by advancing and attacking uphill, and if they did so they often came off badly. The ideal way to dislodge hoplites from a hill was to use missiles, as happened at Lechaeum when the Spartans tried to take refuge from peltasts on high ground:

> 'Therefore in desperation they gathered together on a small hill, distant from the sea about two stadia [360 metres], and from Lechaeum about sixteen or seventeen stadia [about 3km]. Then the troops, being now desperate, because they were suffering and being slain, while unable to inflict any harm themselves, and, besides this, seeing the Athenian hoplites also coming against them, took to flight.' (Xenophon, *Hellenica* 4.5.17)

As well as providing refuges, hills could also block line of sight and so cover the manoeuvres of an army. Delium provides an example of this, being fought over a hill which may have been lower and with gentler slopes than some of the hills we have encountered, so that the two armies were willing to deploy and engage across its slopes:

> 'Pagondas persuaded the Boeotians to attack the Athenians, and quickly breaking up his camp led his army forward, it being now late in the day. On nearing the enemy, he halted in a position where a hill intervening prevented the two armies from seeing each other, and then formed and prepared for action ... when everything was arranged to their satisfaction [they] appeared over the hill, and halted in the order which they had determined on.' (Thucydides 4.94.1–3)

The Athenians accepted battle in this position, advancing to meet the Boeotians coming down the hill. The hill also provided cover for the flanking move by the Boeotian cavalry we saw above (Thuc. 4.96.5). The element of surprise provided by the limited visibility allowed a very modest force of cavalry to turn the battle against the Athenians.

Greece not being a country with many large rivers, such features do not feature prominently in hoplite battles (unlike the battles of Alexander the Great in Asia, where three of the four major battles were fought across rivers, indicating both the greater prevalence of large rivers and the more defensive mindset of Persian commanders, at least in the view of their Greek enemies). At Plataea, the armies faced each other across a river (the Asopus), but the Greeks made no attempt to prevent the Persians crossing the watercourse, which was probably shallow, before the battle. A river played a greater role at the Battle of the Crimisus River (339), fought in Sicily between a Syracusan, allied and mercenary force under the Corinthian general Timoleon, and a Carthaginian army. Both sides fielded a hoplite phalanx (in the loose sense – it is not certain how Greek in style Carthaginian equipment was at this date), the Carthaginians being noted (Plut., *Timol.* 28.1) as being especially heavily armoured and with large shields. On this occasion, Timoleon's army surprised, under cover of a morning mist, the Carthaginian army while it was partway through crossing the river:

> '[As the mist cleared,] the Crimisus came into view, and the enemy were seen crossing it, in the van their four-horse chariots formidably arrayed for battle, and behind these ten thousand hoplites with white shields. These the Corinthians conjectured to be Carthaginians, from the splendour of their armour and the slowness and good order of their march. After these the other nations streamed on and were making the crossing in tumultuous confusion ['with *othismos* and disorder']. Then Timoleon, noticing that the river was putting it in their power to cut off and engage with whatever numbers of the enemy they themselves desired, and bidding his soldiers observe that the phalanx of the enemy was sundered by the river, since some of them had already crossed, while others were about to do so [ordered the attack].' (Plutarch, *Timoleon* 27.2–4)

The river thus served to divide and disorder the larger Carthaginian army, and Timoleon's men, aided by a fortuitous downpour and consequent swelling of the river, routed the Carthaginians with heavy loss. Battles in Greece tended not to be fought over larger watercourses, nor in such inclement weather, which was rather a feature of battles in Sicily (compare Thuc. 6.70.1).

A ravine and hills on the Greek right at Mycale delayed the arrival of the Spartan contingent, which was marching inland (Hdt. 9.102.1), leaving the Athenians to win the battle alone. Terrain played a similar delaying role at Delium, preventing some part of each army from coming into action:

> 'The extreme wing of neither army came into action, one like the other being stopped by the water-courses in the way; the rest engaged with the utmost obstinacy, shield against shield ["with *othismos* of shields"].' (Thucydides 4.96.2)

Linear features – man-made as well as natural – could be of most value in protecting the flanks of an army, especially one outnumbered or concerned by the presence of enemy light infantry and cavalry, like the Athenian army at Syracuse:

'In the meantime, as the march before the Syracusans was a long one, the Athenians quietly sat down their army in a convenient position, where they could begin an engagement when they pleased, and where the Syracusan cavalry would have least opportunity of annoying them, either before or during the action, being fenced off on one side by walls, houses, trees, and by a marsh, and on the other by cliffs.' (Thucydides 6.66.1)

Where natural or pre-existing features did not suffice, Greek armies could supplement them with fieldworks (as they frequently did about their camps), as at Corinth in 392 where a Cointhian force sheltered within their long walls:

'The walls, however, are a long distance from each other; [the Corinthian] troops, in consequence, when they formed in line for battle, thought themselves to be few in number, and therefore made a stockade and as good a trench as they could in front of them, to protect them until their allies should come to their aid.' (Xenophon, *Hellenica* 4.4.9)

Similar field works were constructed by the Athenians at Pylos, attempting to cut off the Spartan garrison. Demosthenes, the Athenian general, initially tried to persuade his officers of the need to fortify the place, but in the end, Thucydides tells us 'the soldiers themselves wanting occupation were seized with a sudden impulse to go round and fortify the place':

'Accordingly they set to work in earnest, and having no iron tools, picked up stones, and put them together as they happened to fit, and where mortar was needed, carried it on their backs for want of hods, stooping down to make it stay on, and clasping their hands together behind to prevent it falling off; sparing no effort to be able to complete the most vulnerable points before the arrival of the Lacedaemonians, most of the place being sufficiently strong by nature without further fortification.' (Thucydides 4.4.2)

In both these cases, however, the objective was to avoid an open field battle rather than to influence the course or outcome of one.[34]

Hills, and to some extent rivers, are long-lasting terrain features, and are likely to exist on the ground today much as they did in antiquity, meaning that the modern desire to understand the course of a battle by studying the ground over which it was fought (where this can be identified) makes a good deal of sense. However, another vital feature of terrain, and one that may have changed utterly in the intervening centuries, is the nature of the ground cover – both natural vegetation and man-made features such as crops, olive groves, farm walls and buildings. For all that the impression often gained from ancient battle accounts is that, where not fought over explicitly mentioned hills or linear features, they were fought on entirely featureless plains, there must in reality have been a range of different types of ground cover. It must, on the heavily agriculturally developed plains surrounding Greek cities, have been exceedingly rare to find any extensive area of completely featureless ground. Outside the plains, natural vegetation would presumably have been an even bigger factor; much of modern Greece, where it is not agriculturally developed, is covered in thick *maquis* or *phrygana*, a mixed vegetation of numerous shrubs which can grow very dense indeed. Looking at the vegetation cover of many of the battlefields of Greece today, particularly on the hills, it is hard to see how a hoplite phalanx could ever have manoeuvred through it (dense *maquis* is difficult enough just to walk through), and we can only guess that the uplands were more extensively cleared for grazing than they are today. Some areas of Greece were also more heavily wooded than today, and woodland features in a number of battles, or at least in their preliminaries. Perhaps surprisingly, the chief role attributed to such vegetation is in blocking visibility rather than slowing or disordering formed bodies of men, though they must have done this too.

Woodland could play as simple a role as covering the Persian flanking movement at Thermopylae, for example:

> 'The Phocians learned in the following way that the Persians had climbed up: they had ascended without the Phocians' notice because the mountain was entirely covered with oak trees.' (Herodotus 7.218.1)

The Spartan army at Mantinea, crossing the open plain before the city, was surprised by the sudden appearance of the allied army in front of them (Thuc. 5.66.1–2), presumably because the Spartans had been advancing through the woods believed at that time to have covered part of the plain (and, characteristically, had not bothered to send out any scouts). Nemea (394) saw another Spartan army taken by surprise by an enemy advance:

> 'Now for a time the Lacedaemonians did not perceive that the enemy were advancing; for the place was thickly overgrown; but when the latter struck up the paean, then at length they knew, and immediately gave orders in their turn that all should make ready for battle.' (Xenophon, *Hellenica* 4.2.19)

Woods, and lack of visibility, could have a direct impact on the course of a battle. The Athenian general Demosthenes learned the danger of woods in the disaster at Aegitium (Thuc. 3.97–98), and this made him apprehensive of attacking the Spartans on Sphacteria (425):

> 'He had been at first afraid, because the island having never been inhabited was almost entirely covered with wood and without paths, thinking this to

be in the enemy's favour, as he might land with a large force, and yet might suffer loss by an attack from an unseen position. The mistakes and forces of the enemy the wood would in a great measure conceal from him, while every blunder of his own troops would be at once detected, and they would be thus able to fall upon him unexpectedly just where they pleased, the attack being always in their power.' (Thucydides 4.29.3–4)

As Thucydides remarks, 'the Aetolian disaster [Aegitium], which had been mainly caused by the wood, had not a little to do with these reflections' (Thuc. 4.30.1). In the event, of course, the terrain worked in the Athenians' favour, especially once an accidental forest fire reduced much of it to ash.

On the whole, though, while undergrowth and woodland sometimes hid an advance, the actual fighting was done in more open ground; but we cannot be certain precisely how open such ground was, and it certainly cannot have been totally devoid of features. This should affect the image we sometimes have of a large phalanx as a single monolithic block, in which the appearance of a single gap would be fatal. Any phalanx must have possessed flexibility enough to cope with common features like trees, walls, ditches, outbuildings, crops, patches of scrub or piles of stones, and the division of the phalanx into smaller subunits would have been intended, in part, to allow it to open out to pass around such obstacles, closing together again on the far side.[35]

Fighting

I do not intend to say much at this point about what happened when two phalanxes (despite all the impediments of weather, terrain, light infantry, cavalry, incompetent generals and religious scruples) actually came into contact and fought each other, since that will form the entire subject of the following chapter. For now I will just point out that detailed descriptions of the nature of hoplite battle are lacking, and our sources instead use a variety of expressions to convey what happened. Thucydides, for example, will sometimes call such fighting *en chersi*, literally 'in hand', or hand-to-hand, but a variety of words and expressions are used to convey the scene, as we would expect from such a stylish writer. A good example is Solygeia:

> 'The Corinthians first attacked the right wing of the Athenians, which had just landed in front of Chersonese, and afterwards the rest of the army. The battle was an obstinate one, and fought throughout hand to hand [*en chersi*]. The right wing of the Athenians and Carystians, who had been placed at the end of the line, received and with some difficulty repulsed [*eosanto*] the Corinthians, who thereupon retreated to a wall upon the rising ground

behind, and throwing down the stones upon them, came on again singing the paean, and being received by the Athenians, were again engaged at close quarters [*en chersin*].' (Thucydides 4.43.2–3)

We also hear (particularly in Xenophon) of phalanxes coming *epi* or *eis doru*, 'to spear', as for example at Coronea (394):

'When, however, the distance between the armies was still about three *plethra*, the troops whom Herippidas commanded, and with them the Ionians, Aeolians, and Hellespontines, ran forth in their turn from the phalanx of Agesilaus, and the whole mass joined in the charge and, when they came within spear thrust [*eis doru*], put to flight the force in their front.' (Xenophon, *Hellenica* 4.3.17)

Other words frequently employed to describe such fighting are the verb *otheo*, 'to force', 'push' or 'drive', and compounds or derivatives of this. There has been a tendency among modern scholars to identify these varied terms with specific and different types of fighting, and in particular to suppose that the use of 'push' verbs (or the related noun) indicate a unique hoplite way of fighting that involved literally pushing in the manner of a rugby scrum. I am certain that this tendency is mistaken, and that in fact such words all refer to much the same thing, namely two phalanxes closing within weapons reach of each other and fighting with their weapons – that is (until many of them broke), with their spears. We are no more obliged to assume that the use of *otheo* implies literal pushing than we are to believe that *en chersi* implies literally fighting with one's hands. I will explore this question at length in the next chapter.[36]

One other point to note is that the result of a charge – or indeed of a slow march into contact – would sometimes be not hand-to-hand fighting but the flight of the enemy before a blow was even struck. This was indeed the ideal outcome (since it entailed no risk to one's own side), and was part of the reason for a frightening, fast and loud – or disconcertingly slow and quiet – attack, as we saw above. Faced with such an attack, a phalanx lacking sufficient training, order, discipline or motivation might simply revert back to the mob that dwells inside every army, and run away. Such a bloodless rout happened at Mantinea:

'But the Lacedaemonians ... fell on the older men of the Argives and the five companies so named, and on the Cleonaeans, the Orneans, and the Athenians next them, and instantly routed them; the greater number not even waiting to strike a blow, but giving way the moment that they came on, some even being trodden under foot, in their fear of being overtaken by their assailants.' (Thucydides 5.72.3–4)

The Spartan left – the Sciritae and Brasideans – had themselves put up the briefest resistance, fleeing 'as soon as they came to came to close quarters [*en chersin*]' (Thuc. 5.72.3).

A similar rout took place at Amphipolis, as we saw above, where Brasidas' men 'fell upon and routed the centre of the Athenians, panic-stricken by their own disorder and astounded at his audacity', while his flanking force had similar success:

> 'The result was that the Athenians, suddenly and unexpectedly attacked on both sides, fell into confusion; and their left towards Eion, which had already got on some distance, at once broke and fled.' (Thucydides 5.10.7–8).

The Spartans afterwards returned the Athenian dead:

> 'About six hundred of the latter had fallen and only seven of the enemy, owing to there having been no regular engagement, but the affair of accident and panic that I have described.' (Thucydides 5.11.2)

Clearly, that battle was not really a matter of fighting at all, and the Athenian dead will mostly have been killed in the pursuit. Confident armies could expect to win effortless victories of this sort, though sometimes this confidence led to their own undoing, as at Miletus (412):

> '[T]he Argives rushed forward on their own wing with the careless disdain of men advancing against Ionians who would never stand their charge, and were defeated by the Milesians with a loss little short of three hundred men.' (Thucydides 8.25.3)

At Coronea, we have already seen the flight of one contingent as soon as they came within spear thrust. In other parts of the line at this battle, victory was even more rapid:

> 'As for the Argives, they did not await the attack of the forces of Agesilaus, but fled to Mount Helicon.' (Xenophon, *Hellenica* 4.3.17)

The most notable example was the so-called 'tearless battle' (Plut., *Ages*. 33.3), where Agesilaus' son, Archidamus, led a Spartan army against the Arcadians:

> 'And when Archidamus led the advance, only a few of the enemy waited till his men came within spear-thrust [*eis doru*]; these were killed, and the rest were cut down as they fled, many by the horsemen and many by the Celts.' (Xenophon, *Hellenica* 7.1.31).

As a result 'not so much as one of the Lacedaemonians had been slain, while vast numbers of the enemy had fallen' (Xen., *Hell*. 7.1.32). The Spartans at home wept tears of joy when they received news of the outcome, according to Xenophon, so it was not such a tearless battle after all. Nevertheless, rapid, bloodless (at least bloodless to one side) collapses of a phalanx in combat are a not infrequent occurrence.

Victory and defeat

In the following chapter I will look further at the factors leading to victory or defeat. Here, it is necessary only to point out that victory would often not be decided by technical questions of arms and armour, nor by niceties of formations and tactics, nor even by individual fighting skill. Often, fighting was not even involved, and where it was it might be of only short duration, or be ended by a sudden panic that led to the collapse of one side, perhaps for seemingly trivial causes. Ancient Greek writers generally are far more honest about the presence of panic on the ancient battlefield than authors of other periods tended to be. Greek hoplites are not, on the whole, depicted either as mindless automata obeying the will of their generals, nor as superhuman heroes glorying in battle, and there is a high degree of psychological insight in many accounts of Greek battle (at least among the better historians).[37]

Panic and fear (*phobos*) are indeed ever-present on the Greek battlefield, from the fear felt by Greeks at the sight of Median clothes (Hdt. 6.112.3) onward. At Spartolus (429), when the Chalcidian cavalry routed the Athenians, it was by utilizing *phobos*:

'The Chalcidian horse also, riding up and charging them just as they pleased, at last caused a panic [*phobeo*] amongst them and routed and pursued them to a great distance.' (Thucydides 2.79.6)

The defenders of Stratus laid ambushes against their Chaonian attackers (429):

'[A]s soon as they approached [the defenders] engaged them at close quarters from the city and the ambuscades. A panic seizing the Chaonians, great numbers of them were slain.' (Thucydides 2.81.5-6)

Strictly speaking, the Chaonians were 'barbarians', not Hellenes, which demonstrates that in this regard at least, barbarians and Greeks were not thought to be greatly different. When Demosthenes sprang his ambush at Olpae (426/425), the result was similar:

'The Peloponnesians were now well engaged and with their outflanking wing were upon the point of turning their enemy's right; when the Acarnanians from the ambuscade set upon them from behind, and broke them at the first attack, without their staying to resist; while the panic into which they fell caused the flight of most of their army, terrified beyond measure at seeing the division of Eurylochus and their best troops cut to pieces.' (Thucydides 3.108.1)

At Delium, as we have already seen, the appearance of a mere two squadrons of cavalry was enough to undo all the success in combat of the Athenians:

'[T]heir sudden appearance struck a panic into the victorious wing of the Athenians, who thought that it was another army coming against them. At length in both parts of the field, disturbed by this panic, and with their line broken by the advancing Thebans, the whole Athenian army took to flight.' (Thucydides 4.96.5–6)

Brasidas pointed out the effectiveness of this sort of stratagem in his speech at Amphipolis:

'That is our best chance of establishing a panic among them, as a fresh assailant has always more terrors for an enemy than the one he is immediately engaged with.' (Thucydides 5.9.8)

Note that the objective of the attack is simply to panic the enemy, not to gain an advantageous position for subsequent fighting. Clearly, for Thucydides the role of *phobos* is paramount in hoplite battle; he gives fear as the reason for the rightward drift he attributes to all armies (Thuc. 5.71.1). We do not see quite the same emphasis in Herodotus or Xenophon, but it is still clear that fear and panic are the most important factors in battle, and the way to defeat an enemy army is to induce panic in its men. Of course, this could be achieved in a number of ways, not least through the high fighting skill, or high reputation, of one's own side.

Flight and pursuit

Once an enemy army was defeated, it would flee in rout from the field of battle. 'Rout' is one of those words commonly used in military history, the actual meaning of which is rather poorly defined. Greek authors generally used the word *trepo* to refer to the phenomenon, literally 'turn around' (from which the Greeks and we derive the word 'trophy', see below), as descriptive a term as any since it would involve the men within a phalanx (or any formation) literally turning around and running away – thereby presenting their backs to the enemy. This was of course an extremely dangerous undertaking for those fleeing, since with their backs turned and rapidly losing all order and cohesion, they would be easy prey to the enemy, who would now 'pursue', that is chase after and kill as many of the fleeing foe as they could (often, we imagine, fuelled by the released pent-up tension of the preceding fighting, which made the victors particularly enthusiastic to kill). Statistics for battle casualties, and specific examples, suggest that losses on both sides might have been quite even during the actual fighting, with the losing side suffering most of its losses – a disproportionately high number – in the rout and pursuit.[38]

This means of course that rout was a largely irrational act, since it exposed the defeated to the greatest risk of death, which is why panic was such an important factor. It is also the case that the rear ranks, further from the enemy, with a head start and with several ranks of their own men between themselves and the enemy, had the greatest chance of getting away safely, which is why routs tended to start at the rear of formations not the front, and why the second most important position in the phalanx, after the file leader who did most of the fighting, was the file closer standing at the rear. There will be more on this topic in the next chapter.

Naturally enough, cavalry and light infantry were particularly suited to pursuit of heavily encumbered hoplites, even if the hoplites (as they routinely did) threw away their heavy shields so as to run faster. After their defeat at Delium, the Athenian army scattered:

> 'Some made for Delium and the sea, some for Oropus, others for Mount Parnes, or wherever they had hopes of safety, pursued and cut down by the Boeotians, and in particular by the cavalry, composed partly of Boeotians and partly of Locrians, who had come up just as the rout began. Night however coming on to interrupt the pursuit, the mass of the fugitives escaped more easily than they would otherwise have done.' (Thucydides 4.96.7–8)

Pursuit by cavalry lasting until nightfall is a not uncommon feature, as in the case of Abydus (409/408), where a Persian force was defeated by Alcibiades:

> 'And Alcibiades pursued him with the Athenian cavalry and one hundred and twenty of the hoplites, under the command of Menander, until darkness covered the retreat.' (Xenophon, *Hellenica* 1.2.1)

Pursuit was not always pressed closely or for long, however. This was not likely through a desire to ameliorate the horrors of war or to avoid killing hoplites of one's own class, since armies showed little compunction about killing when they had the opportunity. When a Spartan army trapped a fleeing Corinthian army between the walls of Corinth, far from showing any reluctance to kill, they embraced the opportunity with enthusiasm:

'And the Lacedaemonians were in no uncertainty about whom they should kill; for then at least heaven granted them an achievement such as they could never even have prayed for. For to have a crowd of enemies delivered into their hands, frightened, panic-stricken, presenting their unprotected sides, no one rallying to his own defence, but all rendering all possible assistance toward their own destruction – how could one help regarding this as a gift from heaven? On that day, at all events, so many fell within a short time that men accustomed to see heaps of corn, wood, or stones, beheld then heaps of dead bodies.' (Xenophon, *Hellenica* 4.4.12)

Yet Spartan armies (which were traditionally weak in cavalry) were noted for not pressing a pursuit, according to Thucydides' account of Mantinea:

'Many of the Mantineans perished; but the bulk of the picked body of the Argives made good their escape. The flight and retreat, however, were neither hurried nor long; the Lacedaemonians fighting long and stubbornly until the rout of their enemy, but that once effected, pursuing for a short time and not far.' (Thucydides 5.73.4)

The reluctance of Spartan armies to pursue is also mentioned by Pausanias (assigning this practice to a much earlier date):

'Also it was an ancient practice with them not to carry out a pursuit too quickly, as they were more careful about maintaining their formation than about slaying the flying.' (Pausanias 4.8.11)

The reasons for this are probably entirely practical, since the pursuers were not themselves out of danger. We have already seen how cavalry on the defeated side could intervene to protect their own fleeing infantry, and the disaster that befell the Megarians and Phliasians at Plataea, caught in disorder by cavalry while pursuing the fleeing Persians. Hoplites in pursuit would lose cohesion almost as badly as those fleeing, which could be fatal if they encountered a fresh enemy. For this reason, the victorious Greeks at Cunaxa (where large numbers of Persian cavalry were present) were especially cautious:

'[T]he Greeks pursued with all their might, but shouted meanwhile to one another not to run at a headlong pace, but to keep their ranks in the pursuit.' (Xenophon, *Anabasis* 1.8.19)

At Tegyra (375), the Thebans achieved their first victory over the Spartans:

'The pursuit, however, was carried but a little way, for the Thebans feared the Orchomenians, who were near, and the relief force from Sparta.' (Plutarch, *Pelopidas* 17.4)

The dangers of pursuit were well demonstrated at Olynthus (381):

> 'Now when the Olynthians saw the peltasts sallying forth, they turned about, retired quietly, and crossed the river again. The peltasts, on the other hand, followed very rashly and, with the thought that the enemy were in flight, pushed into the river after them to pursue them. Thereupon the Olynthian horsemen, at the moment when they thought that those who had crossed the river were still easy to handle, turned about and dashed upon them, and they not only killed Tlemonidas himself, but more than one hundred of the others. But Teleutias, filled with anger when he saw what was going on, snatched up his arms and led the hoplites swiftly forward, while he ordered the peltasts and the horsemen to pursue and not stop pursuing. Now in many other instances those who have pressed a pursuit too close to a city's wall have come off badly in their retreat, and in this case also, when the men were showered with missiles from the towers, they were forced to retire in disorder and to guard themselves against the missiles. At this moment the Olynthians sent out their horsemen to the attack, and the peltasts also came to their support; finally, their hoplites likewise rushed out, and fell upon the Lacedaemonian phalanx when it was already in confusion.' (Xenophon, *Hellenica* 5.3.4–6)

A short pursuit was also normal where a single contingent was routed, rather than a whole army. The battle of Nemea, for example, saw the allied contingents that had routed their opponents pursuing but then coming back to the battlefield, presumably in some disorder, in time to be caught in the flank by the victorious Spartan wing as they marched across the field.

So a vigorous pursuit was the way in which a victory could be made decisive, by killing large numbers of the enemy, but in order to conduct such a pursuit, a number of factors had to come into play. The victors ideally needed a stronger force of cavalry or light infantry to press the pursuit, there needed to be no nearby safe haven for the fugitives such as a camp, friendly city or even convenient hill, and ideally the whole of the enemy army had to have fled, rather than just isolated contingents. Doubtless it is because these factors did not always apply that we do not see so many vigorous pursuits, nor such high casualties, in hoplite battles as are sometimes seen in other periods of warfare. It is doubtful that there was any deliberate effort to avoid bloodshed – as we have seen, killing the enemy (including, perhaps specifically, enemy hoplites), was not something that any Greek army shied away from, and it would embrace the opportunity to do so enthusiastically when it had the chance. Evidently, Greek armies would kill when they could, but killing was not the primary objective, merely the means to the end of demonstrating and asserting dominance over the defeated side.

Wholesale slaughter of the enemy hoplites might help to reinforce this message and to make future confrontations easier, but if circumstances, or fear of enemy cavalry, made that difficult, a victory was still a victory.[39]

Aftermath

Following a battle, events followed a well established and quite formal course. The defeated side would send a herald to the victors, requesting the return of their dead and thereby acknowledging defeat – a request that was not normally refused. The victors, who would remain on the field of battle to indicate their status, would erect a trophy – a collection of captured armour attached to a tree or post – at the point at which the rout (*trope*) of the enemy began, and would deal with their own dead.[40]

This pattern is repeated with clockwork regularity in the battle accounts of Thucydides. To give some examples:

> Cotyrta (424): 'A single garrison which ventured to resist, near Cotyrta and Aphrodisia, struck terror by its charge into the scattered mob of light troops, but retreated, upon being received by the heavy infantry, with the loss of a few men and some arms, for which the Athenians set up a trophy, and then sailed off to Cythera.' (Thucydides 4.56.1)

> Nisaea (424, a cavalry action): 'The Athenians killed and stripped the leader of the Boeotian horse and some few of his comrades who had charged right up to Nisaea, and remaining masters of the bodies gave them back under truce, and set up a trophy; but regarding the action as a whole the forces separated without either side having gained a decisive advantage.' (Thucydides 4.72.4)

> Amphipolis (422): 'The men who had taken up and rescued Brasidas, brought him into the town with the breath still in him: he lived to hear of the victory of his troops, and not long after expired. The rest of the army returning with Clearidas from the pursuit stripped the dead and set up a trophy.' (Thucydides 5.10.11)

Winged *Nike* (Victory) pins a helmet to a trophy.

The pattern continues in the pages of Xenophon:

> Pygela (409): 'Thereupon the peltasts and two *lochoi* of the hoplites came to the aid of their light troops and killed all but a few of the men from Miletus; they also captured about two hundred shields and set up a trophy.' (Xenophon, *Hellenica* 1.2.3)

> Ephesus (409): 'All these contingents directed their first attack upon the hoplites at Coressus; and after routing them, killing about a hundred of them, and pursuing the rest down to the shore, they turned their attention to those by the marsh; and there also the Athenians were put to flight, and about three hundred of them were killed. So the Ephesians set up a trophy there and a second at Coressus.' (Xenophon, *Hellenica* 1.2.9–10)

> Coronea (394): 'And in the morning [after the battle] Agesilaus gave orders that Gylis, the *polemarchos*, should draw up the army in line of battle and set up a trophy, that all should deck themselves with garlands in honour of the god, and that all the flute-players should play. And they did these things. The Thebans, however, sent heralds asking to bury their dead under a truce.' (Xenophon, *Hellenica* 4.3.21)

Matters could get complicated when the outcome of the battle was disputed, or there was some other complicating factor such as the victors failing to control the battlefield or the territory in which it lay. We have already seen above the wrangles over the Athenian dead after Delium, and there are other examples of disputed, shared or otherwise contentious erection of trophies.

> Megara (458): 'After a drawn battle with the Corinthians, the rival hosts parted, each with the impression that they had gained the victory. The Athenians, however, if anything, had rather the advantage, and on the departure of the Corinthians set up a trophy. Urged by the taunts of the elders in their city, the Corinthians made their preparations, and about twelve days afterwards came and set up their trophy as victors.' (Thucydides 1.105.5–6)

> Mytilene (428): 'The Mytilenians made a sortie with all their forces against the Athenian camp; and a battle ensued, in which they gained some slight advantage, but retired notwithstanding, not feeling sufficient confidence in themselves to spend the night upon the field.' (Thucydides 3.5.2)

> Laodocium (423/422): 'After heavy loss on both sides the battle was undecided, and night interrupted the action; yet the Tegeans passed the night on the field and set up a trophy at once, while the Mantineans withdrew to Bucolion and set up theirs afterwards.' (Thucydides 4.134.1–2)

Panormus (412): 'The same summer the Athenians in the twenty ships at Lade blockading Miletus, made a descent at Panormus in the Milesian territory, and killed Chalcideus the Lacedaemonian commander, who had come with a few men against them, and the third day after sailed over and set up a trophy, which, as they were not masters of the country, was however pulled down by the Milesians.' (Thucydides 8.24.1)

Following their first great defeat at Leuctra (371), some Spartans were inclined not to accept the outcome:

'After the disaster some of the Lacedaemonians, thinking it unendurable, said that they ought to prevent the enemy from setting up their trophy and to try to recover the bodies of the dead, not by means of a truce, but by fighting.' (Xenophon, *Hellenica* 6.4.14)

They were dissuaded, however:

'And as all thought it best to recover the bodies of the dead by a truce, they finally sent a herald to ask for a truce. After this, then, the Thebans set up a trophy and gave back the bodies under a truce.' (Xenophon, *Hellenica* 6.4.15)

The result of Mantinea (362) was disputed following the death of Epaminondas and the Boeotians' failure to follow up their victory:

'[T]he deity so ordered it that both parties set up a trophy as though victorious and neither tried to hinder those who set them up, that both gave back the dead under a truce as though victorious, and both received back their dead under a truce as though defeated, and that while each party claimed to be victorious, neither was found to be any better off, as regards either additional territory, or city, or sway, than before the battle took place.' (Xenophon, *Hellenica* 7.5.26–27)

Note that the defeated side's dead were stripped of their armour (in part for it to be added to the trophy, in part as plunder), and perhaps also of their clothes, although sometimes this step could be skipped as a mark of respect, as in the Athenian versus Athenian battle at Munychia (403):

'And the victors took possession of their arms, but they did not strip off the tunic of any citizen. When this had been done and while they were giving back the bodies of the dead, many on either side mingled and talked with one another.' (Xenophon, *Hellenica* 2.4)

Greeks valued the recovery of their own dead extremely highly, as some of the anecdotes above indicate, and sometimes we see a fixation on the recovery of the dead reminiscent of some of the 'no man left behind' policies of modern armies. After a raid on Solygeia (425), for example, reserves drove off the Athenian attackers:

> 'The Athenians seeing them all coming against them, and thinking that they were reinforcements from the neighbouring Peloponnesians, withdrew in haste to their ships with their spoils and their own dead, except two that they left behind, not being able to find them, and going on board crossed over to the islands opposite, and from thence sent a herald, and took up under truce the bodies which they had left behind.' (Thucydides 4.44.5–6)

The significance of this was that the Athenians, by asking for a truce to retrieve the two bodies, were formally admitting defeat in a battle they could claim to have won. The seriousness with which the Athenians took this matter is shown by the fact that the commanders of the Athenian fleet after the naval victory at Arginusae (406) were executed on their return to Athens because they had failed to recover the Athenian shipwrecked in the battle (Xen., *Hell.* 1.6.34–1.7.35).[41]

Recovering bodies after a truce was not the only option – the victor would not have to ask permission. Following a disastrous Spartan attack on Haliartus (395) in which Lysander was killed after approaching too close to the enemy city, Pausanias arrived with reinforcements:

> 'Accordingly Pausanias and the other Lacedaemonians who were in authority, considering that Lysander was dead and that the army under his command had been defeated and was gone, while the Corinthians had altogether refused to accompany them and those who had come were not serving with any spirit; considering also the matter of horsemen, that the enemy's were numerous while their own were few, and, most important of all, that the bodies lay close up to the wall, so that even in case of victory it would not be easy to recover them on account of the men upon the towers – for all these reasons they decided that it was best to recover the bodies under a truce.' (Xenophon, *Hellenica* 3.5.23)

On his return to Sparta, however, Pausanias was charged 'with having recovered the bodies of the dead by a truce instead of trying to recover them by battle', and condemned to death (Xen., *Hell.* 3.5.25) – compare this with the events at Leuctra.

Depending on the location of the battle, the bodies would not necessarily be returned to the home city (which would pose obvious practical difficulties), but might be buried or cremated, the victors perhaps on the battlefield, the defeated side at some convenient location. After Ephesus (409):

'As for the Athenians, after obtaining a truce and so recovering the bodies of their dead, they sailed back to Notium, buried the dead there, and sailed on towards Lesbos and the Hellespont.' (Xenophon, *Hellenica* 1.2.11)

The Athenians were perhaps unusual in their practice (developed maybe in the early to mid-fifth century) of returning the bones of their dead to Athens, where possible, for special burial and commemoration:

'In the same winter the Athenians gave a funeral at the public cost to those who had first fallen in this war. It was a custom of their ancestors, and the manner of it is as follows. Three days before the ceremony, the bones of the dead are laid out in a tent which has been erected; and their friends bring to their relatives such offerings as they please. In the funeral procession cypress coffins are borne in carts, one for each tribe; the bones of the deceased being placed in the coffin of their tribe. Among these is carried one empty bier decked for the missing, that is, for those whose bodies could not be recovered. Any citizen or stranger who pleases, joins in the procession: and the female relatives are there to wail at the burial. The dead are laid in the public sepulchre in the most beautiful suburb of the city, in which those who fall in war are always buried; with the exception of those slain at Marathon, who for their singular and extraordinary valour were interred on the spot where they fell. After the bodies have been laid in the earth, a man chosen by the state, of approved wisdom and eminent reputation, pronounces over them an appropriate panegyric; after which all retire. Such is the manner of the burying; and throughout the whole of the war, whenever the occasion arose, the established custom was observed.' (Thucydides 2.34.1–7)

On this occasion, Pericles provided the panegyric, the well-known Funeral Oration. The names of the dead were also inscribed on monuments in this burial area, equivalent to the lists of names on war memorials so familiar from the wars of the first half of the twentieth century AD. Like modern war memorials, these monuments can be a graphic illustration both of the individual lives lost (listed individually, by name) and of the expanse of territory over which wars were waged. A surviving Athenian example, the casualties of the Erechtheid tribe in 460 or 459, lists those who 'died in the war in Cyprus, in Egypt, in Phoenicia, in Hales, at Aegina and at Megara in the same year' (*GHI²* no.33).

Spartan practice was different:

'It was Spartan custom, when men of ordinary rank died in a foreign country, to give their bodies funeral rites and burial there, but to carry the bodies of their kings home.' (Plutarch, *Agesilaus* 40.3)

Spartans seem to have avoided public commemoration of their dead, but the names of the fallen were collected if only so that next of kin could be notified (Xen., *Hell.* 6.4.16). Whatever specific practices were in place in other cities, a hoplite would know that his death would be marked in some way and he would not be forgotten or become a faceless statistic.[42]

We do not hear much about the wounded on either side, although these would presumably, comparison with other eras suggests, have been more numerous than the dead. The wounded among the victors would have had to make their way back home, assisted by their comrades and their slaves or attendants. There is no record of any centralized hospital train or of treatment for the wounded, although armies would certainly take their wounded with them as they marched on from the battlefield. Xenophon records the Ten Thousand slowed by the need to carry along their wounded:

> '[N]ecessity taught them to encamp in the first village they caught sight of, and not to continue the plan of marching and fighting at the same time; for a large number of the Greeks were *hors de combat*, not only the wounded, but also those who were carrying them and the men who took in charge the arms of these carriers.' (Xenophon, *Anabasis* 3.4.32)

Not only battle wounded needed to be carried in this way – Xenophon also records (Xen., *Anab.* 4.5) the need to help, cajole or carry those afflicted by frostbite and snowblindness on a mountainous section of their march (see also Xen., *Anab.* 5.8.8).

The wounded on the defeated side, if they were not able to make their own way off the battlefield, were probably either left to die or were finished off by looters and scavengers (most likely human). Thucydides provides a melancholy account of the Athenian retreat from Syracuse:

> 'The dead lay unburied, and each man as he recognized a friend among them shuddered with grief and horror; while the living whom they were leaving behind, wounded or sick, were to the living far more shocking than the dead, and more to be pitied than those who had perished. These fell to entreating and bewailing until their friends knew not what to do, begging them to take them and loudly calling to each individual comrade or relative whom they could see, hanging upon the necks of their tent-fellows in the act of departure, and following as far as they could, and when their bodily strength failed them, calling again and again upon heaven and shrieking aloud as they were left behind.' (Thucydides 7.75.3–4)

In the disorganized rout that most defeated armies experienced in pitched battle, such scenes would have been avoided, but the final outcome, for the wounded, would have been much the same.[43]

Another post-battle activity that may have occurred frequently but is less often mentioned is the awarding of prizes. Herodotus in particular is keen to list the most courageous units and individuals on both sides – victorious and defeated – and there are mentions of a similar procedure in other battles. For example, after the victory at Ephesus (409) over the Athenians, the Ephesians first set up a trophy:

> 'They also gave to the Syracusans and Selinuntines, who had especially distinguished themselves, the prizes for valour, not only general prizes, but many to particular individuals among them, while upon any one of them who at any time might desire it they conferred the privilege of dwelling in Ephesus tax free; and to the Selinuntines, after Selinus had been destroyed, they gave the rights of Ephesian citizenship as well.' (Xenophon, *Hellenica* 1.2.10)

These features of Greek warfare were all to be continued and if anything elaborated by the Macedonians who were to conquer Greece, and then Persia, in the fourth century. At Chaeronea (338), Philip II of Macedon defeated the Athenian and Boeotian armies:

> 'After the battle Philip raised a trophy of victory, yielded the dead for burial, gave sacrifices to the gods for victory, and rewarded according to their deserts those of his men who had distinguished themselves.' (Diodorus 16.86)

The demise of the Classical Greek hoplite and his eventual replacement by the Macedonian phalangite will be the subject of the final chapter, but first, as promised, we must turn to the nature of the fighting that took place when hoplite met hoplite.

Chapter 8

Fighting in the Phalanx

The meaning of *othismos*

The question of how hoplites actually fought when formed up in the phalanx is one with a relatively brief but interesting history. So far as we can tell, historians of the nineteenth century (AD) and earlier saw nothing unusual about the way hoplites did battle, and assumed that they would have fought in much the same way as all infantry fought (or are believed to have fought) prior to the widespread adoption of firearms. Exactly what way this was is never clearly set out, as like historians of the ancient world, modern historians have either assumed a certain degree of familiarity with the nature of combat, or else have not taken any particular interest in the subject. Of the historians contemporary with the Greek phalanx – Herodotus, Thucydides and Xenophon – the last two at least certainly had first-hand knowledge of hoplite fighting, but never felt the need to spell out its nature explicitly, probably assuming that it would be familiar to all their readers (though even so, Thucydides in particular does explain some aspects, like the 'rightward drift' at Mantinea, as if he was talking to those unfamiliar with the phenomenon). Later historians such as Diodorus or Plutarch do not display any great familiarity with hoplite battle, but were probably familiar, if not at first hand, with the combat of their own day (the Roman Republic or Empire), which would likely have been broadly similar. Historians in more recent periods also wrote about ancient battles in much the same terms as historians of all eras have written about battle, which is to say with an emphasis on high-level tactical manoeuvres and the role and decisions of the commanding general. The weaknesses of this approach to the historiography of battle were pointed out in the 1970s in the seminal *The Face of Battle*, which attempted to look beyond these high-level accounts in which armies are faceless masses performing the will of their commanders, and instead to see armies as made up of individuals with their own motivations and experiences, and to examine the low-level 'nuts and bolts' of how men fought on the battlefield. This 'face of battle' approach has been widely adopted in the study of ancient battles and armies, and has taken on two slightly divergent forms: the study of the experience of battle as an end in itself, emphasizing the hardships and horrors involved in ancient warfare; and the study of the mechanics of battle, concentrating on the way in which men with different weapons and styles of fighting interacted and fought on the battlefield.[1]

In the case of the Greek hoplite, this 'mechanics of battle' approach has a longer history. Historians in the early twentieth century (AD) began to take an interest in the question of how hoplites fought, and to devise a model of hoplite combat that set them apart from all other infantry before or since (or contemporary), with this becoming a central aspect of the hoplite orthodoxy, and a form of 'hoplite exceptionalism' which we have encountered several times already. This model was based on the occurrence in Ancient Greek historians of words derived from the Greek for 'to push' (in its verb form, *othein*, ὠθεῖν; or the derived noun, *othismos*, ὠθισμός, 'pushing'; or in compound verbs derived from this, such as *exothein*, ἐξωθεῖν, 'to push out') in descriptions of phalanx fighting. According to the theory that developed and was widely adopted by the middle of the twentieth century, these words, where they occur in the context of hoplite battle (and only there), must be taken absolutely literally; they mean that hoplites did not fight each other with weapons as had always been assumed (or at least did not only fight with weapons), but physically shoved each other, in an effort to force their opponents off the field of battle like a reverse tug of war. Furthermore, the attested or assumed close-order nature of the hoplite phalanx with its deep formations and narrow file intervals was taken to mean that hoplites did not push as individuals, but rather that the phalanx closed right up into a single mass, and pushed as one, so as to force the entire opposing phalanx from the battlefield. The natural metaphor for such coordinated mass pushing was the rugby scrum (taking 'scrum' here to mean the formal encounter in rugby where forwards lock arms with each other and brace their shoulders against their opponents to push as one, not the more informal use of 'scrum' to mean a confused melee), and so for several decades of the twentieth century – and in some quarters still to this day – the predominant model of hoplite combat is that it took, uniquely, the form of a giant rugby scrum. I will refer to this idea below as 'the hoplite scrum', though it usually goes by the name *othismos*, this Greek word having been adopted (perhaps surprisingly, as we will see) by modern historians to describe the phenomenon.[2]

There have been variations on this model. Some see the scrum as a second stage of combat after an initial bout of fighting with weapons, others have it as occurring immediately upon contact with hoplites running into and colliding with each other, while yet others see the pushing stage as something that arises involuntarily in the later stages of combat as the lines close up, forming a situation analogous to a crowd crush.

In the later part of the twentieth century, challenges to this orthodox view began to emerge, following two threads. As we saw in Chapter 1, the necessity for hoplites to fight in tight, close-order formations was challenged, with the suggestion that they could and did fight in looser formations in which there was plentiful scope for individual skill at arms, and seeing a continuous, gradual evolution from the open-order duelling apparently described by Homer, to the fully developed classical phalanx of the Peloponnesian War (or later). Accompanying this new perspective was the suggestion that hoplite battles were not giant rugby scrums, but remained battles between individual fighters, albeit formed up in large continuous formations. According to this view, the use of 'push' verbs in the sources should not be taken literally – a phalanx could be 'pushed back' in Greek just as it can in English (where the expression is used frequently in this figurative or metaphorical sense), without there being any requirement for actual physical contact, still less literal pushing.[3]

Although various other arguments and analogies have been brought to bear on this question, the fundamental underlying disagreement is this: should 'push' words in ancient battle accounts be taken literally, in every case that involves hoplites (and only in those cases), to mean physical mass pushing, or should they be taken figuratively (in most, but not necessarily all, cases) in the same way such words are used in English? Proponents of the literal interpretation (that 'push' always means a literal mass push, when applied to hoplites) will sometimes call this the 'natural' reading of the texts, to which their less-literal opponents will respond that it is a more literal reading, but in no way more natural. Little progress has been made on this question, with scholars often dividing themselves into two camps matching those formed for the wider 'hoplite debate' (the literal or 'orthodox' view, and the figurative or 'heretical' view), with little meeting of minds.

A brief digression is in order on the use of metaphor (note that in this discussion I will term all the figurative uses encountered 'metaphors', though it may be more accurate to describe the Greek combat usage as a metonym or possibly synecdoche, as we will see; I hope that referring to metaphor throughout will simplify the discussion without obscuring the meaning). English, of course, makes very free use of metaphor, and is in a state of constant change and development in the metaphors it uses. In recent years, 'push back' has been adopted in certain (chiefly corporate and office) circles to mean, broadly, 'disagree with' (and also in a single-word noun form, 'pushback', meaning 'opposition'). Such metaphors tend to be very noticeable when they are first adopted, but pass unnoticed once they are firmly embedded in the language (as 'dead metaphors'). It is becoming common now to use 'reach out to' to mean 'speak to', 'get in touch with' or 'contact', and note also how these last two are themselves metaphors of physical contact; we might usually 'get in touch with' someone on the phone or 'contact' them by email. In military language, we are so used to metaphors such as an army having 'wings' ('horns', κέρατα, in Greek) that we do not even notice their use, and understand with no hesitation that the usage is not literal. Similarly, modern military expressions such as 'heavy fire' or 'the big push' cause no confusion among readers (these may be examples of 'conceptual metaphors'). Context is everything in deciding whether a usage is literal or metaphorical, but in the case of ancient Greek combat, that context is often now obscure to us, and the literal underpinnings, if any, of an adopted metaphorical usage – such as the Greek conception of the 'weight' and 'heaviness' of an infantry formation – may not be well understood. Greek also was in a state of constant change, like any language, and usage inevitably varied through time. While later authors often made a conscious effort to emulate the language of earlier times, they may themselves not always have clearly understood metaphorical language or have used the words in question in quite the same way.[4]

Because close-quarters hand-to-hand combat has been unknown in the modern world for at least a century, because of the lack of clear literary, artistic or archaeological evidence on the nature of hoplite fighting (or any sort of fighting in the ancient world), and because of the obvious limitations of any sort of simulation or re-enactment, it is difficult for either side of the scrum debate to formulate a truly decisive argument. I believe, however, that the starting point should be a more careful examination of the language, vocabulary and usage of our key ancient authors to establish whether it is really the case that 'the natural reading' of our sources favours the literal interpretation. From there I will go on to examine the other literary evidence for the nature of ancient infantry combat with specific reference to hoplites, as well as some comparative material.

So the best starting point for an examination of this issue would seem to be the range of meanings that could be applied to the 'push' words used by the ancient sources – the noun *othismos* (ὠθισμὸς, 'the push', or 'pushing') and the verb *othein*, in the first person *otheo* (ὠθέω), 'push', along with various compounds such as *exotheo*, 'push out'.

The argument about whether or not there was literal pushing in hoplite battles has become known as 'the *othismos* debate', but what is perhaps surprising is how rarely the word *othismos* is actually used by ancient authors contemporary with the period of hoplite warfare. In fact, the word is used just once (by Thucydides) to describe a battle of hoplite against hoplite – the oft-quoted '*othismos aspidon*', 'push of shields', that occurred at Delium. Yet this phrase, unique though it is, is often adopted by modern writers as shorthand for hoplite battle generally. Consider this quote from a recent popular history:

> 'Xenophon, in his eyewitness account of the battle of Second Koroneia, laconically recalls: "[Agesilaus' phalanx] crashed into the Thebans front to front. So with shield pressed against shield (*othismos aspidon*) they

struggled, killed and were killed." (Xenophon *Hellenika* 4.3.19). At the battle of Delion (424 BC) according to Thucydides, the Thebans "got the better of the Athenians, pushing them back (*othismos aspidon*) step by step at first and keeping up their pressure" (4.96.4).'

In fact, Xenophon (Xen., *Hell.* 4.3.19) does not use the word *othismos* or the phrase *othismos aspidon* to describe the fighting at Coronea at all (he does use the verb *otheo*, and described the armies *symbalontes tas aspidas* – 'striking together their shields' – as we shall see below). Thucydides does use *othismos aspidon* at Delium, but at 4.96.2, not in the passage quoted, where again the verb *otheo* is used (*osamenoi kata brachu*, 'they pushed by a bit'). The usage by Thucydides at 4.96.2 is the only occurrence in Classical literature of the phrase *othismos aspidon* (it occurs once more in a Byzantine author nearly 1,000 years later, as we will see). Yet the idea of *othismos* as the defining feature of hoplite combat is pervasive. Even one of the main modern opponents of the hoplite orthodoxy begins his analysis of hoplite combat with the observation that battles 'were won or lost in a single great "push" (*ôthismos*)'.[5]

So let us first examine the uses of the word *othismos* in the three historians contemporary with Classical hoplite battle (Herodotus, Thucydides and Xenophon). In the quoted passages that follow, I will leave the word *othismos* untranslated so as not to pre-judge the issue of what is the correct meaning. I have also sometimes slightly modified the translations (which are otherwise those in the Loeb translations or other editions as noted in the bibliography) in order to remove any paraphrasing around the critical words. Note also that while it is sometimes claimed that the verb *otheo* and the noun *othismos* are used interchangeably, my view is that they are (or may be – the point must be established) different words in terms of usage, although obviously related in terms of grammar, and that the framing of the discussion of hoplite fighting techniques as 'the *othismos* debate' implies that this word, specifically, was used by ancient writers to describe something particular (the hoplite scrum). It is therefore important to establish the actual meaning and usage of this word, specifically, as a starting point. I will move on to the verbal forms (*otheo* and related words) later.[6]

Of the three contemporary historians, the word *othismos* occurs four times in Herodotus, once in Thucydides, and once in Xenophon, in the following passages:

> 'There was much *othismos* between the Persians and Lacedaemonians over Leonidas' body, until the Hellenes by their courageous prowess dragged it away and routed their enemies four times.' (Herodotus 7.225; Spartans fighting Persians, Thermopylae)

'Among the generals at Salamis there was much *othismos* of words.' (Herodotus 7.78; debate before Salamis)

'During the drawing up of battle formation there arose much *othismos* of words between the Tegeans and the Athenians.' (Herodotus 9.26; debate before Plataea)

'First they fought by the fence of shields, and when that was down, there was a fierce and long fight around the temple of Demeter itself, until they came to *othismos*. For the barbarians laid hold of the spears and broke them short. Now the Persians were neither less valorous nor weaker, but they had no armour; moreover, since they were unskilled and no match for their adversaries in craft, they would rush out singly and in tens or in groups great or small, hurling themselves on the Spartans and so perishing.' (Herodotus 9.62; Spartans fighting Persians, Plataea)

'The Athenians hastened forward, and the two armies met at a run. The extreme right and left of either army never engaged, for the same reason; they were both prevented by water-courses. But the rest closed, and there was much *othismos* of shields.' (Thucydides 4.96.2; Delium)

'After no long interval a shout arose within and men came pouring forth in flight, some carrying with them what they had seized, then soon a number of men that were wounded; and there was much *othismos* about the gates.' (Xenophon, *Anabasis* 5.2.17; plundering a town)

We thus have a total of three uses in the context of battle: two in Herodotus, where Spartans fight Persians, and one in Thucydides, where Spartans fight Athenians (which is the only hoplite versus hoplite use of the word in Classical literature). Other uses are for arguments (*othismos* of words, twice), and a crowd exiting a gate (with no hostile forces involved, and no fighting).[7]

Herodotus' usage at Plataea is interesting, since the reasons he gives for the Persian defeat (that they were unarmoured and that they engaged in small groups) does not fit with the idea of a mass shove; in such a shove, the presence or absence of armour would be irrelevant, while by definition the whole force would have to be involved. It would also cast into doubt the idea that *othismos* was a style of fighting exclusive to hoplites – in order for a hoplite scrum to develop, both sides must be pushing or there is nothing to push against, and so the Persians must also have been pushing *en masse* if *othismos* is taken to mean a mass shove in this case. It would seem either that *othismos* can be applied to hand-to-hand combat generally, not just to hoplites, or if it is something special (a mass shove), then it was not something practised exclusively by Greeks.[8]

Let us now extend the search to include later authors who, though not eyewitnesses of hoplite warfare, wrote about it and may preserve earlier usage. Plutarch uses the word *othismos* seven times in historical contexts. The most significant usage for our purposes is:

> 'Xenophon says that this battle was unlike any ever fought, and he was present himself and fought on the side of Agesilaus, having crossed over with him from Asia. The first impact, it is true, did not meet with much resistance, nor was there much *othismos*, but the Thebans speedily routed the Orchomenians, as Agesilaus did the Argives.' (Plutarch, *Agesilaus* 18; Coronea)

This is an interesting passage, since Plutarch is using Xenophon as a source, but we know that Xenophon does not use the word *othismos* himself about this battle (Xenophon instead says at this phase of this battle that at the 'first impact' to which Plutarch refers, the Spartans 'when they came within spear thrust, put to flight the force in their front' (Xen., *Hell*. 4.3.17), while in the rematch, as we shall see, Spartans and Thebans 'clashed shields, fought and pushed [*otheo*]' (Xen., *Hell*. 4.3.19, and see below for discussion of the verb *otheo*). In other words, the use of *othismos* here is not Xenophon's usage, but Plutarch's own.

The other passages in Plutarch refer variously to crowds, generally around doors in Rome (Plut., *Brutus* 18; Plut., *Caesar* 64; Plut., *Crassus* 31), to ships (Plut., *Aristides* 9), to disorderly Carthaginians crossing the River Crimisus (Plut., *Timoleon* 27) or to Romans and elephants in battle (Plut., *Marcellus* 26).

There is, however, another interesting use by Plutarch, in his discussion of boxing, wrestling and running as athletic contests:

> 'The first task of fighters is to strike out and to defend themselves. And their next task, when they are now met in hand-to-hand conflict, is to attack with *othismoi*, and overthrow each other. By this especially, it is reported, the Spartans at Leuctra were overpowered by our men [Boeotians] who were practised wrestlers ... And finally the soldier's third task is to run away when beaten and to pursue when winning. It is reasonable therefore for boxing to lead off the list, for wrestling to have second place, and for racing the last, because boxing mimics attack and defence, wrestling the twisting and *othismos* of close quarter combat, and in the foot race one practises the art of fleeing the battlefield and of pursuing those who do so.' (Plutarch, *Moralia* 639f = *Quaestiones Convivales* 2.5)

Note that Plutarch is using wrestling as a simile for hand-to-hand combat, as he is using boxing as one for spear-fighting, and we should not conclude that the skills and movements involved were actually identical.

So it appears that to Plutarch, *othismos* (in the singular) is used with a range of meanings encompassing 'struggling', 'crowding' or what we might generally call 'pushing and shoving', and could be applied to all sorts of agents (hoplites, Romans, crowds, ships, elephants), while *othismoi* (in the plural) refers to the individual struggling of wrestlers or of soldiers in hand-to-hand combat.

Polybius (a contemporary witness of Hellenistic warfare, though not of the Classical era of hoplites) uses the word once, in reference to routing men fleeing through a gate (much like Xen., *Anab.* 5.2.17 above):

> 'Alexander himself fell fighting in the actual battle; but Archidamus was killed in the *othismos* and crush at the gates.' (Polybius 4.58.9)

Arrian's *Anabasis of Alexander* (second century AD) contains three uses (we cannot be certain if these are his own usage, or if he found them in his sources, which were chiefly Alexander the Great's contemporaries Ptolemy and Aristobulus. At any rate, though Arrian was an admirer of Xenophon, he will not have taken his use of the word from Xenophon, who uses it only once, and not in a military context, as we saw):

> 'Then ensued an *othismos* of horses, on the one side to emerge from the river, and on the other to prevent the landing.' (Arrian, *Anabasis* 1.15.2; Granicus, cavalry versus cavalry)

> 'For a short time there ensued a hand-to-hand fight; but when the Macedonian cavalry, commanded by Alexander himself, attacked with strong *othismoi*, and striking the Persians' faces with their spears...' (Arrian, *Anabasis* 3.14.3; Gaugamela, cavalry versus cavalry)

> 'The elephants being now cooped up into a narrow space, their friends were no less injured by them than their foes, being trampled down in their turning and *othismoi*.' (Arrian, *Anabasis* 5.17.5; Hydaspes, elephants versus infantry)

The word also occurs in Arrian's *Tactica*, describing the Macedonian phalanx:

> 'The man next in file to the file leader must be second to the latter in courage. For his spear reaches all the way to the enemy and he supports the *othismoi* of the man deployed in front of him.' (Arrian, *Tactica* 12)

So Arrian uses the singular for the pushing or struggle of horses or elephants, and the plural seemingly carries a meaning something like 'thrusts' (of weapons, or possibly of the bodies of horses).

Amongst other Roman-era authors (Cassius Dio, Josephus, Dionysius of Halicarnassus, Pausanias, Appian), there are a number of further uses, none to do with hoplite battles, but variously Roman legionaries, crowds, rioters or ships.[9]

An exception is a use by Pausanias (second century AD), describing the possibly mythical battle between Spartans and Messenians that we encountered in Chapter 1:

'And now with their taunts they come to deeds, *othismos* of crowd against crowd, especially on the Lacedaemonian side, and man attacking man.' (Pausanias 4.8)

There are a further handful of uses in minor authors and non-military contexts, which I will not enumerate. However, one sixth-century AD Byzantine (Late Roman) use is rather more significant for our purposes, occurring in the tactical handbook attributed to the Emperor Maurice (Maurikios), and writing about cavalry:

'As far as the depth of the line is concerned, the ancient authorities wrote that it had formerly been regarded as sufficient to form the ranks four deep in each *tagma*, greater depth being viewed as useless and serving no purpose. For there can be no *othismos* from the rear up through the ranks, as happens with an infantry formation, which may force the men in front to push forward against their will. Horses cannot use their heads to push people in front of them evenly, as can infantry.' (Maurice, *Strategikon* 2.6; concerning cavalry)

Note that Maurice is writing in a long tradition of tactical manuals going back to Hellenistic times (the 'ancient authorities' he refers to), and similar passages about the ineffectiveness of deep formations for cavalry because of their inability to push occur in the tactical manuals of Asclepiodotus (7.4), Aelian (*Tactics* 18) and Arrian (*Tactics* 16.10–14), as well as the anonymous Byzantine *On Strategy* (17), though none of them use the word *othismos*, and they vary in the extent to which they explicitly refer to pushing in infantry formations. Asclepiodotus, for example, the earliest, writes that depth 'does not have the same importance as in the infantry, rather it may work more havoc than the enemy themselves, for when the riders run afoul of one another they frighten the horses' (Asclep. 7.4).[10]

Finally, there is the remarkable total of sixteen uses in the sixth-century AD Byzantine historian Procopius' *De Bellis*, almost outnumbering all other authors put together. It seems unnecessary to quote every one of these instances. Almost all are in a military context, and describe hand-to-hand fights between Byzantine, Persian and Gothic infantry and cavalry and the crowding of forces (for example on siege ramps or embankments), with several times 'much *othismos*', once an '*othismos* of ships' and at 8.29.18 an '*othismos aspidon*' as Byzantine infantry repel Gothic cavalry – the only other occurrence of this phrase in the whole of ancient literature, some thousand years after its first use by Thucydides. Procopius was undoubtedly deliberately imitating Thucydides' style in his narrative, and was likely in this case also imitating his vocabulary.[11]

I think that given the scarcity of uses of the word *othismos* to describe battles between hoplites, or even battles involving hoplites, it is very hard to maintain the position that *othismos* is a characteristic and diagnostic feature of hoplite warfare, and that the word is used in a specific technical sense (of a coordinated mass shove) when applied (and only when applied) to such battles. We must be cautious of course, since as we have seen (in Chapter 1), the word 'phalanx' is never used by Herodotus or Thucydides in a military context, and becomes common only with Xenophon, although we can be tolerably certain that a phalanx (in some form) did exist before Xenophon's time. But as I argued before, the idea that the phalanx of Xenophon is something unique and specific to the Greek hoplite is likely incorrect, and 'phalanx' just meant 'line of battle' in a more general sense. Similarly, it seems most likely that *othismos* is used as a general word for (according to context) 'pushing and shoving', 'struggle', 'crowding', 'jostling', 'argument', 'melee' or 'hand-to-hand fighting', or in the plural, 'thrusts', and that the occasional use of the word applied to hoplites cannot be taken as an indication of a special hoplite tactic or of a type of fighting unique to the hoplite.

The two cases which seem to lend some support to the scrum theory are Pausanias and Maurice (both writing centuries after the demise of the Greek hoplite phalanx, of course). Pausanias' '*othismos* of crowd against crowd, and man attacking man', however, is better interpreted as in Chapter 1, as two types of combat, a more fluid and open-order engagement between individual fighters (in the 'traditional' style), as opposed to the mass formation closing to engage as one, in the more disciplined way pioneered (perhaps) by the Spartans. Alternatively, Pausanias may be describing the same combat on two different scales – the big picture being the struggle of masses, the detail being men fighting men. Nevertheless, there is no reason to suppose that mass shoving is involved. However, Maurice uses the word *othismos* to refer quite specifically to pushing by the rear ranks of an infantry formation, and earlier writers refer to the same phenomenon though without using the same word. I will return to this question below.

First, however, we should consider the fact that while *othismos*, the noun, is not much used by ancient authors, *otheo*, the verb, occurs much more frequently, along with various compounds ('push back', 'push out'; for the purposes of this chapter I will look particularly at *exotheo*, 'push out'). Is there good reason to suppose that ancient authors used the verb in a literal sense, always or sometimes, or when applied only to hoplite battles? In the passages quoted below, I will render *otheo* simply and literally as 'push' (in italics, and in appropriate English tense and case), remembering that the meaning of 'push' is not a given, and will consider only those uses in a military context (the word 'push', naturally enough, is often used with no military implication).

The usages for our key authors (Herodotus, Thucydides and Xenophon) are as follows, starting with Herodotus:

> 'After Miltiades had *pushed* the Apsinthians by walling off the neck of the Chersonese…' (Herodotus 6.37.1; strategic)

> '[F]or when they [the Athenians] had *pushed* the Persians and the war was no longer for their territory but for his…' (Herodotus 8.3.2; strategic)

> 'So the barbarians honoured Masistius' death in their customary way, but the Greeks were greatly encouraged that they withstood and *pushed* the charging horsemen.' (Herodotus 9.25.1; hoplites and archers versus Persian cavalry)

Thucydides uses *otheo* six times:

> 'The Athenians had thus to defend themselves on both sides, from the land and from the sea; the enemy rowing up in small detachments [of ships], the one relieving the other – it being impossible for many to bring to at once – and showing great ardour and cheering each other on, in the endeavour to *push* a passage and to take the fortification.' (Thucydides 4.11.3; ships attacking the shore at Pylos)

> 'The Athenians pursuing, unable to surround and hem them in, owing to the strength of the ground, attacked them in front and tried to *push* the position.' (Thucydides 4.35.3; peltasts attacking fortified Spartan hoplites on Sphacteria)

> 'The right wing of the Athenians and Carystians, who had been placed at the end of the line, received and with some difficulty *pushed* the Corinthians, who thereupon retreated to a wall upon the rising ground behind.' (Thucydides 4.43.3; hoplites versus hoplites)

> 'In this part of the field the Boeotians were beaten, and retreated upon the troops still fighting; but the right, where the Thebans were, got the better of the Athenians and *pushed* them bit by bit, though gradually at first.' (Thucydides 4.96.4; hoplites versus hoplites, Delium)

> 'At last the Argives *pushed* the Syracusan left, and after them the Athenians routed the troops opposed to them, and the Syracusan army was thus cut in two and took to flight.' (Thucydides 6.70.2; hoplites versus hoplites, Syracuse)

'[T]he Athenians first defeated the Peloponnesians, and *pushed* the barbarians and the ruck of the army, without engaging the Milesians.' (Thucydides 8.25.4; hoplites versus mixed forces)

Xenophon has several uses:

'At this juncture one may say without fear of contradiction that Agesilaus showed courage; but the course that he adopted was not the safest. For he might have allowed the men who were trying to break through to pass, and then have followed them and annihilated those in the rear. Instead of doing that he made a furious frontal attack on the Thebans. Striking together their shields, they *pushed* and fought and killed and were killed.' (Xenophon, *Hellenica* 4.3.19; hoplites versus hoplites, Coronea)

'Thereupon some of them, climbing up by the steps to the top of the wall, jumped down on the other side and were killed, others perished around the steps, being *pushed* and struck by the enemy, and still others were trodden under foot by one another and suffocated.' (Xenophon, *Hellenica* 4.4.11; hoplites versus hoplites)

'But when Deinon, the polemarch, Sphodrias, one of the king's tent-companions, and Cleonymus, the son of Sphodrias, had been killed, then the royal bodyguard, the so-called aides of the polemarch, and the others fell back *pushed* by the Theban mass, while those who were on the left wing of the Lacedaemonians, when they saw that the right wing was being *pushed*, gave way.' (Xenophon, *Hellenica* 6.4.14; hoplites versus hoplites, Leuctra)

'As a result, therefore, of all these things, it is reported that the soldiers were inspired with so much strength and courage that it was a task for their leaders to restrain them as they *pushed* to the front. And when Archidamus led the advance, only a few of the enemy waited till his men came within spear-thrust.' (Xenophon, *Hellenica* 7.1.31; hoplites)

'When, however, the citizens gained possession of some of the towers on this side and on that, they closed in desperate battle with those who had mounted upon their walls. And the enemy, as they were *pushed* by them – by their courage as well as by their fighting – were being crowded together into an ever smaller space.' (Xenophon, *Hellenica* 7.2.8; fighting on the walls of a town)

'When, however, they had pursued the enemy to the space between the senate house and the temple of Hestia and the theatre which adjoins these buildings, although they fought no less stoutly and kept *pushing* the enemy towards the altar...' (Xenophon, *Hellenica* 7.4.31; hoplite street fighting)

There is a single use in Xenophon's *Anabasis* (Xen., *Anab.* 3.4.48), where Xenophon physically pushes a complaining man out of the line of march.

Then we have the *Cyropaedia*, Xenophon's fictional account of the early years of Cyrus, which he used to expound some of his own ideas on leadership and tactics. Here we have a flurry of uses of *otheo*, which need to be considered in some detail:

> 'You, Medes, march on our left; and you, Armenians, half keep to our right and half lead on in front; while you, cavalrymen, shall follow behind, to encourage and *push* us on upward; and if anyone is inclined to show weakness, do not allow it.' (Xenophon, *Cyropaedia* 3.2.5)

> 'And the Persians on their part, following them [Assyrians, including chariots] up to the gates, mowed many of them down as they were *pushing* one another; and upon some who fell into the ditches they leaped down and slew them, both men and horses.' (Xenophon, *Cyropaedia* 3.3.64)

> 'The infantry that you will fight against, you have fought before – all but the Egyptians; and they are armed and drawn up alike badly; for with those big shields which they have they cannot do anything or see anything; and drawn up a hundred deep, it is clear that they will hinder one another from fighting – all except a few. But if they believe that by *pushing* they will *push us off* [*exotheo*] the field, they will first have to sustain the charge of horses and of steel driven upon them by the force of horses.' (Xenophon, *Cyropaedia* 6.4.17 f.; Thymbrara)

> 'Here, then, was a dreadful conflict with spears and lances and swords. The Egyptians, however, had the advantage both in numbers and in weapons; for the spears that they use even unto this day are long and powerful, and their shields cover their bodies much more effectually than corselets and targets, and as they rest against the shoulder they are a help in *pushing*. So, locking their shields together, they advanced and *pushed*. And because the Persians had to hold out their little shields [*gerrai*] clutched in their hands, they were unable to hold the line, but were forced back foot by foot, giving and taking blows, until they came up under cover of the moving towers. When they reached that point, the Egyptians in turn received a volley from the towers; and the forces in the extreme rear would not allow any retreat on the part of either archers or javelinmen, but with drawn swords they compelled them to shoot and hurl.

'...

'At this juncture Cyrus came up in pursuit of the part that had been opposed to him; and when he saw that the Persians had been *pushed* from their position, he was grieved; but as he realized that he could in no way check the enemy's progress more quickly than by marching around behind them, he ordered his men to follow him and rode around to the rear. There he fell upon the enemy as they faced the other way and smote them and slew many of them. And when the Egyptians became aware of their position they shouted out that the enemy was in their rear, and amidst the blows they faced about. And then they fought promiscuously both foot and horse; and a certain man, who had fallen under Cyrus's horse and was under the animal's heels, struck the horse in the belly with his sword. And the horse thus wounded plunged convulsively and threw Cyrus off. Then one might have realized how much it is worth to an officer to be loved by his men; for they all at once cried out and leaping forward they fought, *pushed* and were *pushed*, struck and were struck. And one of his aides-de-camp [*hyperetes*] leaped down from his own horse and helped him mount upon it.' (Xenophon, *Cyropaedia* 7.1.33–38; Thymbrara)

In these examples, I think we see the importance of the scale at which the events described by *otheo* take place, in understanding the meaning of the word. Two out of Herodotus' three examples are at the strategic level (the Aspinthians pushed out of the Chersonese, the Persians pushed out of Greece). Nobody would insist that any literal pushing is involved here. Xenophon, particularly in the *Cyropaedia*, often uses the term at the individual level: the Egyptian shield is suitable for pushing, and the Egyptians advance and push; the opposing hoplites at Coronea advance and push, fight, kill and are killed; Cyrus' cavalry fight Egyptian infantry and push and are pushed – note incidentally that the construction here ('they fought, pushed and were pushed, struck and were struck') is very similar to that used by Xenophon when describing Coronea ('they pushed and fought and killed and were killed', Xen., *Hell.* 4.3.19), and he uses a similar construction to describe the fighting of missile-armed light infantry ('they javelined and threw and shot and slung', Xen., *Hell.* 2.4.33). I think it is uncontroversial to suppose that this pushing at the individual level is physical pushing – usually with the shield, though also physical jostling with horses in the case of the cavalry of Cyrus, similar to the examples of *othismos* in Arrian quoted earlier.

What is harder to determine is the intermediate, tactical level. When Athenian hoplites, as a body, are '*pushed*' by Spartan hoplites, is this a result of physical mass pushing by the whole body? According to the literal hoplite scrum view, it is, but I hope that it is clear from the above quotes that this is by no means certain from the language alone. Herodotus uses *otheo* at this tactical level once, and it is in reference to Persian cavalry being 'pushed back' by Greek infantry – which is highly unlikely to have involved a mass push and certainly is not a scrum as normally envisaged, since horses cannot push this way, as Maurice observes, and men on foot cannot push cavalry. Thucydides has one case of fortifications being *pushed* and one of ships attempting to *push* a landing – neither of which can have involved any physical pushing – and four cases of hoplites against hoplites. Xenophon's *Hellenica* has two cases of hoplite formations being *pushed* (6.4.14 and 7.4.31), three (4.3.19, 4.4.11 and 7.2.8) where a case could be made either way as to whether the usage is individual or tactical, and one (7.1.31) where the pushing is in the sense of 'pushing forward' (without an enemy in contact). One passage quoted above ('the enemy, as they were *pushed* by them – by their courage as well as by their fighting', Xen., *Hell.* 7.2.8) seems to me especially informative, as it shows infantry being 'pushed back' explicitly not by physical pushing. The forces here are engaging in combat (with weapons, presumably), and the superior skill and courage of one 'pushes' the other. Compare this with Diodorus' description of Leuctra, where the Spartans 'were unable to endure the weight of the courageous fighting of the elite corps' of Thebans (15.55.4), showing that a 'weight' metaphor, like a 'pushing' metaphor, could apply to courage and skill at arms. The advancing and falling back are real and physical enough, but hoplites pushed back 'by courage' are not being shoved back by a mass scrum.

In the passages from the *Cyropaedia*, we have two uses (cavalry 'pushing on' their own forces, and men struggling about a gate) which are certainly not evidence of a mass push, and while physical pushing is likely involved in the latter, it is unlikely to be in the former (the men were most likely not physically shoved by the cavalry horses, and the translation 'driving on' seems to give the sense best). But at the fictional Battle of Thymbrara, the Egyptians clearly push physically with their shields, and the Persians are said to have been *pushed* from their position. But Xenophon is explicit that the Egyptians did not take part in a mass shove by the whole formation – 'drawn up a hundred deep, it is clear that they will hinder one another from fighting – all except a few' – while the Persians, in a much shallower line, are forced back by the Egyptians because their light hand-held shields were not suitable for pushing, not because the pressure of 100 men overwhelmed their resistance, as it surely must have done, if mass pushing were involved. Xenophon has Cyrus make the same point earlier:

> "'And do you think, Cyrus,' said one of the generals, 'that drawn up with lines so shallow we shall be a match for so deep a phalanx?'
>
> "'When phalanxes are too deep to reach the enemy with weapons,' answered Cyrus, 'how do you think they can either hurt their enemy or help their friends?'" (Xenophon, *Cyropaedia* 6.3.22)

This is surely a clear case of only the front ranks taking part in the pushing, at an individual level, and of the Persian formation being, as a whole, 'pushed back', in the sense of being forced to retreat (we might wonder in this case why the Egyptians would have formed up 100 deep in the first place, a question to which I will return below). I think Coronea is a similar example – individual hoplites undoubtedly pushed, as well as fighting, killing and being killed, but this does not mean the whole formation pushed *en masse* (and indeed if it had, there would have been no scope for fighting). The distinction is again made in Xen., *Cyr.* 6.4.18, 'But if they believe that by *pushing* they will *push us off* the field', which could best be translated as 'if they think that by pushing they will drive us from the field', the simple verb and the compound representing action at the individual (literal, physical) and tactical (more figurative) levels respectively in this case, since as a matter of style two different words were needed. Needless to say, Xenophon is also not talking about Greek hoplites, but (semi-fictional) Persian and Egyptian infantry and cavalry, so if these passages were to be taken as evidence for the literal shove, then it could not have been a tactic exclusive to Greek hoplites.[12]

There is, however, one more case, a passage which we have already encountered in these pages, which I will mention here and then return to later:

> '"Your analogy is perfect, Socrates," said the youth; "for in war one must put the best men in the van and the rear, and the worst in the centre, that they may be led by the one and *pushed* by the other."' (Xenophon, *Memorabilia* 3.1.7–8)

This does appear to be evidence for the role of the rear ranks (or at least, the rearmost rank) in physical pushing. I will return to this passage below.

We should also consider (if more briefly) usage of *otheo* in Homer, and also in later authors. The noun *othismos* does not occur in Homer at all, but the verb *otheo* appears in the *Iliad* an impressive forty-six times, almost always in a military context. It is not necessary to list every occurrence, so instead I will present a selection, again translating as *push*:

> 'Then once again the Olympian aroused might in the hearts of the Trojans; and they *pushed* the Achaeans straight toward the deep ditch; and amid the foremost went Hector exulting in his might.' (Homer, *Iliad* 8.336)

> 'Then Hector rushed forth to tear from the head of great-hearted Amphimachus the helm that was fitted to his temples, but Aias lunged with his bright spear at Hector as he rushed, yet in no wise reached he his flesh, for he was all clad in dread bronze; but he smote the boss of his shield, and *pushed* him back with mighty strength, so that he gave ground backward from the two corpses, and the Achaeans drew them off.' (Homer, *Iliad* 13.193)

'Hector of the flashing helm spake no word in answer, but hastened by, eager with all speed to *push* the Argives and take the lives of many.' (Homer, *Iliad* 5.691)

'For neither could the mighty Lycians break the wall of the Danaans, and make a path to the ships, nor ever could the Danaan spearmen *push* the Lycians from the wall.' (Homer, *Iliad* 12.420)

'And as he pondered, this thing seemed to him the better, that the valiant squire of Achilles, Peleus' son, should again *push* toward the city the Trojans and Hector, harnessed in bronze, and take the lives of many.' (Homer, *Iliad* 16.655)

As we have seen, there is much discussion over the nature of the fighting in Homer, with its seeming mix of massed infantry formations and individual chariot-borne warriors. But few would insist that Homeric infantry performed a mass shove like that of the hoplite scrum attributed to Classical hoplites, and in a number of cases (Lycians being 'pushed' from the wall, Hector 'pushing' the Argive army, Achilles 'pushing' the Trojan army) it is perfectly clear that a literal push is not meant. It seems reasonable, therefore, to conclude that, like the Classical authors, Homer too uses *otheo* in a non-literal sense when applied to massed bodies of infantry, and that the correct translation is something like 'drive' or 'drive off', even though in the case of individuals (such as Hector 'pushed back' by Aias), literal pushing could well have been (and probably was) involved.[13]

It might be objected that what was true for Homer was not necessarily true for Thucydides and Xenophon, but in fact, aside from the fact that Homer was traditionally considered the first military writer and the model for all later Greek historians, there are parallels between Xenophon's language at least and Homer's. Take for example Xenophon's description of the fighting at Coronea (quoted above), particularly in the slightly modified version he presented in his *Agesilaus*:

'Clashing their shields together, they *pushed*, they fought, they killed, they died. There was no screaming, nor was there silence, but the noise that anger and battle together will produce … When the fighting ended, one could see, where they met one another, the ground stained with blood.' (Xenophon, *Agesilaus*, 2.12–14)

This has parallels with a well-known passage from Homer:

'Now as these advancing came to one place and encountered, they clashed their shields together and their spears, and the strength of armoured men in bronze, and the shields massive in the middle clashed against each other, and the sound grew huge of the fighting. There the wails of despair and the cries of triumph rose up together of men killing and men killed, and the ground ran with blood.' (Homer, *Iliad* 4.446f.)

'Clashing/clashed their shields' uses the same verb, *symbalein*, in Xenophon and Homer, though the use of this verb here by Xenophon is often taken by hoplite scrum advocates as evidence of a shield-against-shield scrum. The phrase *symbalontes tes aspidas* also appears in poetry – for example, the well-known single combat between Eteocles and Polyneices:

> 'Then clutching their sword-hilts they closed, and round and round, with shields clashing, they fought a wild battle.' (Euripides, *Phoenissae* 1405)

Compare also Aristophanes, *Peace* 1275, 'The fight begins, the hollow shields clash against each other.' There is no reason to suppose this indicates any sort of mass push (clearly not in a single combat), but rather is a typical part of hand-to-hand fighting, with individual warriors striking their shield against their opponent's shield in an attempt to unbalance or knock them back (as is frequently depicted in vase paintings). So there is no reason to suppose that *otheo* was used by Thucydides and Xenophon in a different way from how it was earlier used by Homer.[14]

As for later authors, Polybius is a very sparing user of *otheo* with just two uses – one non-military (8.15.13) and one definitely physical and individual:

> 'The way in which elephants fight is this: they get their tusks entangled and jammed, and then *push* against one another with all their might, trying to make each other yield ground until one of them proving superior in strength has pushed aside [*parotheo*] the other's trunk; and when once he

Symbalontes tes aspidas; hoplites clash shields in combat on the Nereid Monument. (*British Museum*)

can get a side blow at his enemy, he pierces him with his tusks as a bull would with his horns.' (Polybius 5.84.3)

If only some ancient source had given such a clear description of the way hoplites fought, although according to hoplite scrum theory, it was pretty much the same, without the tusks and trunks.

Polybius is also the source (along with the Hellenistic tacticians, as we will see below) of the account much quoted by hoplite scrum advocates of the role of the rear ranks in a Macedonian phalanx:

> 'These rear ranks, however, during an advance, press forward those in front by the weight of their bodies; and thus make the charge very forcible, and at the same time render it impossible for the front ranks to face about.' (Polybius 18.30.4)

I will return to this passage below, and here will just note that the verb Polybius uses is *piezo*, 'press', a word used very commonly by Polybius (no less than fifty-five uses) with the same sort of range of meanings as *otheo*, for example:

> '[I]f, in the course of the battle, the men had been *pressed* ever so little from their ground...' (Polybius 2.33.8; Roman legions, in battle with the Gallic Insubres)

> 'For he left himself no place of retreat, and by allowing the enemy to reach his position, unharmed and in unbroken order, he was placed at the disadvantage of having to give them battle on the very summit of the hill; and so, as soon as he was *pressed* by the weight of their heavy armour and their close order to give any ground, it was immediately occupied by the Illyrians.' (Polybius 2.68.9; Spartans and allies versus Illyrians, Sellasia)

There are many other similar uses of *piezo* in Polybius for all sorts of military forces (never Classical hoplites) being 'pressed', 'pushed' or 'forced' in battle. It should also be noted that with forty uses in Thucydides and twelve in Xenophon's *Hellenica*, *piezo* is more common than *otheo* in these authors, and is generally translated as 'hard pressed', or 'press hard' in the active. Often, in Thucydides and Xenophon, the word is used in a strategic sense ('hard pressed by their enemies') or referring to shortages of food ('hard pressed by famine'), but it also has tactical uses, such as this passage which we will visit again below:

> 'There, as it chanced, the whole body of the light troops and likewise the hoplites of the men in Piraeus were arming themselves. And the light troops, rushing forth at once, set to throwing javelins, hurling stones, shooting arrows, and discharging slings; then the Lacedaemonians, since many of

them were being wounded and they were *hard pressed*, gave ground, though still facing the enemy; and at this the latter attacked much more vigorously ... Now Thrasybulus and the rest of his troops – that is, the hoplites – when they saw the situation, came running to lend aid, and quickly formed in line, eight deep, in front of their comrades. And Pausanias, being *hard pressed* and retreating about four or five stadia to a hill, sent orders to the Lacedaemonians and to the allies to join him. There he formed an extremely deep phalanx and led the charge against the Athenians. The Athenians did indeed accept battle at close quarters; but in the end some of them were *pushed* [*exeosthesan*] into the mire of the marsh of Halae and others gave way; and about one hundred and fifty of them were slain.' (Xenophon, *Hellenica* 2.4.33 f.)

It may seem tempting to see a distinction between being 'pressed', using *piezo*, by light infantry or 'pushed', using *otheo*, by hoplites, but other uses of the verb do not support this distinction. Hoplites are as likely to 'press' hoplites as are other forces, and the use of the different verbs seems more a matter of style than of different meaning. In neither case is it necessary to assume literal pushing (and it is clearly out of the question in the case of missile-users and light infantry).

Polyaenus, writing in the second century AD, provides an interesting example. He relates that Iphicrates, in some unknown battle, gave a signal to his men upon which they all took one step forward:

'[H]e gave the signal; the army responded with a shout, after which they advanced a pace and *pushing back* [*osamenoi*] the enemy put them to flight.' (Polyaenus 3.9.27)

Polyaenus also tells the same story of the Theban general Epaminondas, specifying that this took place at the Battle of Leuctra:

'The battle remained finely balanced for a long time, until Epaminondas called on his troops to give him one step more, and he would ensure the victory. They did as he asked; and they gained the victory.' (Polyaenus 2.3.2)

In this case, Polyaenus does not use *otheo* (saying simply that 'they obeyed, they won').[15]

Plutarch uses *otheo* fairly frequently in military and non-military contexts, and in reference to battles among both Greeks and Romans. A sample will suffice, this time using the translations given in the Loeb editions to highlight (in italics) the many different ways in which *otheo* can be translated:

'Agesilaus was carried away by passion and the ardour of battle and advanced directly upon them, wishing to *bear them down* by sheer force ... But since it proved too hard a task *to break* the Theban front...' (Plutarch, *Agesilaus* 18.2, 4; Coronea, hoplites versus hoplites)

'[T]he Romans, having no opportunity for sidelong shifts and counter-movements, as on the previous day, were obliged to engage on level ground and front to front; and being anxious to *repulse* the enemy's men-at-arms ['hoplites'] before their elephants came up, they fought fiercely with their swords against the Macedonian spears.' (Plutarch, *Pyrrhus* 21.6; Battle of Asculum, Romans versus Macedonian phalanx)

'Accordingly, he [Pyrrhus] led on faster, *pushing* along the horsemen in front of him.' (Plutarch, *Pyrrhus* 32.3)

'[A] soldier, while Caesar in person was watching the battle, *dashed* into the midst of the fight, displayed many conspicuous deeds of daring, and rescued the centurions.' (Plutarch, *Caesar* 16.3; Romans versus Britons)

'With difficulty and after much strenuous effort he *repulsed* the enemy and slew over thirty thousand of them.' (Plutarch, *Caesar* 56.2; Caesar versus Pompey, Munda)

Diodorus Siculus (who does not use *othismos*) has a couple of Hellenistic usages of *otheo*:

'Eumenes, although he and a few troopers were left unsupported at the extremity of the wing, regarded it as shameful to yield to fortune and flee ... he *pushed* his way toward Antigonus himself. A fierce cavalry battle ensued.' (Diodorus 19.42.5; cavalry versus cavalry)

'[T]hey manned the three staunchest ships with picked men, whom they instructed to try to sink with their rams the ships that carried the engines of the enemy. These men, accordingly, *pushed* forward although missiles in large numbers were speeding against them.' (Diodorus 20.88.5; ships)

Arrian has some interesting uses – again a sample will suffice, giving the Loeb translations:

'[W]hen the [Macedonian] cavalry, no longer shooting, but actually *thrusting* them [the Triballians] with their horses, fell on them here, there and everywhere...' (Arrian, *Anabasis* 1.2.6)

Compare with Arrian's two uses of *othismos* for cavalry, above. Arrian clearly is a firm believer in the ability of horses to push physically, something that may have coloured, or been coloured by, his experience commanding a Roman army against the nomadic, cavalry-using Alans.

> 'The Thebans were *pushed* inside the gates [by the Macedonian phalanx]; their flight became a panic, so that while being *thrust* through the gates into the city they could not shut them in time.' (Arrian, *Anabasis* 1.8.5)

> 'He did not expect that any of the enemy would dare *force* a way through the gaps between the elephants on horseback.' (Arrian, *Anabasis* 5.15.6; Porus' deployment at Hydaspes)

Dionysius of Halicarnassus is positively Homeric in his use of *otheo*, with many uses, in a variety of Roman contexts (swords being thrust, crowds going here and there, armies repulsing their enemies). A couple of examples will suffice:

> 'For the Bovillani not only *repulsed* the assailants from the walls, but even threw open their gates, and sallying out in a body, forcibly *thrust back* down hill those who opposed them.' (Dionysius 8.20.2; Romans versus Volscii)

> 'For the enemy, having *thrust* forward against them and cleared palisades of those who defended them, mounted the ramparts.' (Dionysius 11.23.5; Romans versus Sabines)

Again, there is no evidence that there is anything unique or specific to hoplites about using *otheo* to describe fighting. The verb is used by later authors in much the same contexts as it was used by Homer, Herodotus, Thucydides and Xenophon – to describe being 'pushed back' (repulsed, defeated) in various ways by various agents, with a variety of possible English translations according to context, some of which certainly could involve physical pushing (crowds in gates, horses jostling, individual men with shields, elephants) and some of which do not (ships, pushing through missiles or from walls or palisades).

Various compounds of *otheo* are also used in a similar way. I will briefly examine the usage of one common example (already encountered a couple of times above), *exotheo*, 'push out'. *Exotheo* is used nine times by Thucydides. Sometimes it is in a similar context to *otheo*, for example:

> '[T]he Mantinean right broke the Sciritae and Brasideans, and bursting in with their allies and the thousand picked Argives into the unclosed breach in their line cut up and surrounded the Lacedaemonians, and *pushed them out* to the wagons, slaying some of the older men on guard there.' (Thucydides 5.72; Mantinea)

'Drove them' seems the best translation in this case. The majority of uses in Thucydides refer to ships being 'driven ashore', usually by enemy action (Thuc. 2.90.5; 7.36.5; 7.52.2; 7.63; 8.104.4; 8.105.1). Physical pushing is of course not involved in these cases.

One particularly interesting passage is this, already familiar from earlier discussions:

> 'All armies are alike in this: on going into action they are *pushed out* on their right wing, and one and the other overlap with this their adversary's left ... The man primarily responsible for this is the first upon the right wing, who is always striving to withdraw from the enemy his unarmed side; and the same apprehension makes the rest follow him.' (Thucydides 5.71.1)

Again this is not physical pushing; the line is 'pushed out' to the right by the movement of the man on the extreme right, and of those who follow him. The movement at the individual level (which does not involve pushing) results in the formation as a whole being 'pushed out'.

Xenophon uses *exotheo* three times in *Hellenica* – twice for ships, as in Thucydides' examples (4.3.12, twice), and once for hoplites (as we have already seen above):

> 'The Athenians did indeed accept battle at close quarters; but in the end some of them were *pushed out* into the marsh of Halae and others gave way.' (Xenophon, *Hellenica* 2.4.34)

Hoplites being driven into a marsh seems to be a similar phenomenon to triremes being driven onto the shore, and contains no necessary implication of a mass shove. The example of *exotheo* from Xenophon's *Cyropaedia* I have considered above.

Herodotus never uses *exotheo* in a military context, but only for being driven by winds or for being driven out of a city or country.

The word is much used by later authors. Polybius amasses seventeen uses, largely military, of which a few examples will give an impression:

> 'The maniples in front were thrown into utter confusion by the crushing weight of the animals: *pushed out* and trampled upon by them they perished in heaps upon the field; yet owing to its great depth the main body remained for a time unbroken.' (Polybius 1.34.5; defeat of Regulus, elephants versus Romans)

> 'A single charge, however, of the Illyrians, whose numbers and close order gave them irresistible weight, served to *push out* the light-armed troops, and forced the cavalry who were on the ground with them to retire to the hoplites.' (Polybius 2.3.5; Illyrian infantry versus Aetolian light troops and cavalry)

Compare the above example with the use of *piezo* at Sellasia quoted earlier.

'At the same time the Greek mercenaries stationed near the phalanx, and behind the elephants, charged Ptolemy's peltasts and *pushed them out*, the elephants having already thrown their ranks also into confusion.' (Polybius 5.84.9; Raphia)

As with other authors, in some cases (elephants trampling Romans) literal pushing of some sort is implied, but in others it is not (and these cases do not speak of Classical hoplites or even heavy infantry anyway) or is clearly impossible.

It would be possible to give more examples of the usage of these terms, but I think that the pattern has been established – *exotheo* is used much like *otheo*, as well as particularly in the sense of 'driving out' or 'driving away' (or ashore), for a wide range of forces across a range of periods.

Similarly, I think it is clear, given the way *otheo* is used of a variety of forces (hoplites, peltasts, cavalry, ships, elephants) and situations (pitched battles, struggles on walls or in towns, attacks on fortified positions), that the burden of proof is firmly on those who insist that *otheo*, when and only when used to describe battles of hoplite against hoplite in the open field, should be taken to mean a specific sort of massed shove by the whole formation, rather than being used in the same ways and contexts as the English word 'push' would be used. It seems apparent that *otheo* can variously be translated as 'push', 'force', 'push back', 'drive', 'urge' or a range of similar words and expressions.[16]

We can see that, as far as the linguistic evidence is concerned, there is no reason to suppose that any of these words – *othismos*, *otheo*, *piezo*, *exotheo* and others – carried a different meaning when applied to hoplites fighting hoplites than they did when applied (as they frequently were) to all sorts of other combats involving cavalry, light infantry, ships, Romans, walls and palisades, elephants and so on, in many of which physical or mass pushing quite clearly cannot be the meaning, for all that there are cases concerning individual men (Greek, Egyptian and Roman), horses or elephants where literal, physical pushing probably or certainly is involved. It is not good enough to claim that the 'natural reading' of the texts supports the idea of a specific tactic or battlefield occurrence – the mass shove or hoplite scrum – since the linguistic evidence alone does no such thing. However, this does not mean that there is no other evidence for pushing in ancient infantry formations, some of which we have already encountered above, and it is to this that I will now turn.

'Shoved by the rear ranks'

So the linguistic evidence alone does not offer any *a priori* reason why 'push' words should be read literally in the context of hoplite battles. But there does exist direct evidence for a pushing role for rear ranks in at least some circumstances. For the Classical period, this is provided by Xenophon:

> 'In war one must put the best men in the van and the rear, and the worst in the centre, that they may be led by the one and *driven on* by the other.' (Xenophon, *Memorabilia* 3.1.8)

This sounds like individual (and therefore literal) pushing of the rest of the formation by the rear rank. The similarity to Polybius' comment on the role of those ranks behind the first in the Macedonian phalanx is obvious – 'These rear ranks ... *press forward* (*piezo*) those in front by the weight of their bodies' (Pol. 18.30.4). The evidence for the Macedonan phalanx is more extensive and will be considered below, but first we need to consider the possibility that even this case does not describe physical pushing.

Xenophon also describes Cyrus giving an encouraging, exhortatory role to his cavalry, using the verb *otheo* – 'you, cavalrymen, shall follow behind, to encourage and *push* us on upward' (Xen., *Cyrop.* 3.2.5) – and compare also this passage where the same phenomenon is described, this time without a 'push' verb:

> '[Cyrus] called in the officers of the rear-guard [*ouragoi*] and gave them the following instructions: ... "For as you are behind, you can observe those who are valiant and by exhorting them make them still more valiant; and if anyone should be inclined to hang back and you should see it, you would not permit it ... And if those in front call to you and bid you follow, obey them and see that you be not outdone by them even in this respect but give them a counter cheer to lead on faster against the enemy."' (Xenophon, *Cyropaedia* 3.3.40)

So it appears that 'pushing' can take the form of encouragement, those 'pushing' include cavalry, and the objects of the pushing include light troops – the sense is clearly of 'keeping them facing and moving forward' rather than 'shoving them in the back', and this can be done by 'encouragement' as much as by 'pushing'.

A small digression on pushing among Roman infantry may be illustrative here. A modern supporter of the hoplite scrum theory quotes Livy's account of the Battle of Zama (202) as evidence of the hoplite practice of a mass shove continuing into Roman times: 'Livy said of the famous battle at Zama that the Romans pushed their own men ahead by thrusting at their backs with the centres of their shields.'[17]

What Livy actually says is:

> '[The Romans] beating them [the Carthaginians] back with their shoulders and the bosses of their shields, being now in close contact with men forced

from their position, they made considerable progress, as no one offered any resistance, while as soon as they saw that the enemy's line had given way, even the rear line pressed upon the first, a circumstance which of itself gave them great force in repulsing the enemy.' (Livy 30.34)

So it is the enemy Carthaginians that the Romans pushed with their shields, not the backs of their own men. Nevertheless, this passage does seem to suggest that the rear ranks at least 'pressed upon' the first. But this translation is from 1949, and if we look back to that from 1850 (before the idea of ancient battles as shoving matches had become fashionable), we find a slightly different rendition of the crucial clause:

'those who formed the rear urging forward those in front when they perceived the line of the enemy giving way; which circumstance itself gave great additional force in repelling them.'

'Urging forward' is rather different from 'pressed upon', and even further from 'thrusting at their backs'. The original Latin is this:

'urgentibus et novissimis primos, ut semel motam aciem sensere, quod ipsum vim magnam ad pellendum hostem addebat.'

The key word here is *urgentibus*. *Urgeo* has numerous meanings, including 'to press, push, force, drive, impel, urge'; we might opt for (from the *Latin Dictionary of Lewis and Short*) 'To follow up, keep to, stick to, ply hard, push forward, urge on'. Incidentally, *pellendum* (*pello*) is the Latin equivalent to the Greek *otheo*, with meanings including 'to beat, strike, knock, push, drive, hurl, impel, propel' and the specific meaning (from Lewis and Short): 'In milit. lang., to rout, put to flight, discomfit.' Latin, like Greek, can use a verb of physical pushing as a general metaphor for defeating an enemy. A literal translation of this sentence could therefore be:

'the rearmost also urging on those in front, once they perceived the movement of the battleline, which itself added great strength to the defeat of the enemy.'

It might be hard to decide which is the better of the possible translations ('press' or 'urge'), but in this case we need not be in any doubt as to Livy's meaning, as we have the benefit of Polybius' Greek account of this battle, which Livy used as his source. Here is the Loeb edition translation of Polybius:

'The rear ranks of the Romans followed close on their comrades, cheering them on.' (Polybius 15.13.2)

The crucial words 'cheering on' here being a translation of παρακαλούντων, *parakaleo*, with meanings like 'exhort, encourage' (from Liddell and Scott's lexicon). So the 1850 translation of Livy is the correct one – the Roman rear ranks did not push their comrades in the back with their shields, they simply cheered them on.[18]

The purpose of this digression is not to establish any incontrovertible facts about Roman combat at Zama, but to point out the ease with which meanings can be modified to match the preconceptions of the reader or translator. Anyone who is certain that *urgeo* means 'press upon' may well also be certain that *otheo* means 'shove', but this says more about their preconceptions than about the original author's intent or the physical reality on the battlefield.[19]

Consequently, I think that at the least, caution is necessary before deciding that Xenophon's reference to the rear rank 'pushing' the other ranks is evidence of a literal shove – he may instead mean that the rear rank encouraged, urged or drove on the rest (after all, we have no difficulty understanding that the 'leading' done by the first rank is not physical and literal). In practice, the 'encouragement' may well have taken a very physical form. We know from much later, better-attested periods, such as the linear infantry formations of the eighteenth century AD, that officers would be stationed behind a formation to force the men to keep their ranks and prevent them from running away, using whatever verbal or physical means were at their disposal, sometimes including a spear (the 'spontoon') which could be held horizontally and used to push the men in the back, or hitting them with the flats of their swords. This, note, was in infantry formations which engaged the enemy with firearms and where there is absolutely no question of a scrum or mass shove.[20]

Yet there exists later evidence, from the Hellenistic period and after, which is much clearer as to there being a pushing role for the rear ranks of the Macedonian phalanx, the successor to and replacement of the Classical hoplite phalanx. We have already seen Polybius' account of how the rear ranks of the Macedonian phalanx 'press forward with the weight of their bodies' (Pol. 18.30.4), and the Hellenistic tacticians, which possibly trace their origins back to the writings of Polybius, contain similar words ('bear forward with their bodies', Asclep. 5.2; 'pressing forward with the weight of their bodies', Ael., *Tact*. 14; 'press on with the weight of their bodies', Arr., *Tact* 12.10).

There is a related thread of evidence, through the Hellenistic tacticians to their Byzantine imitators, which assigns a pushing role to the rear rank of file closers (*ouragoi*) and contrasts the inability of cavalry to push their own men with that of infantry to do so (this is the chain of evidence that includes the passage from Maurice quoted above).

The role of the file closers is defined as being:

'bringing back to position any who may leave their places through fear, and forcing them to close up in case they lock shields.' (Asclepiodotus 3.6)

'[to] see to it that every man in his line holds their position in both rank and file, and [to] compel any man who is quitting his post, either from cowardice or on any other account, to resume it again.' (Aelian, *Tac*tics 14)

'[to] control the ranks' advance and not allow deserters to run away from the formation.' (Arrian, *Tac*tics 12.11)

'[to] keep the men ahead of them in close order.' (Anonymous, *On Strategy* 15)

'In combat, also, they should push forward the men in front of them, so that none of the soldiers will become hesitant and hold back.' (Maurice, *Strategikon* 12 B 16)

The contrast of the pushing ability of infantry and that of cavalry is given in these passages:

'[T]he depth of the cavalry unit, provided it is enough to hold the squadron firm and in line, does not have the same importance as in the infantry … for when the riders run afoul of one another they frighten the horses.' (Asclepiodotus 7.4)

'The rear ranks do not contribute to how well an enemy charge is resisted, nor do they increase the momentum of those before them, nor close up with them, nor, holding on to each other, make a solid mass. If the leading ranks are pressed forward from the rear, the horses become annoyed, create disorder and are more likely to do harm to themselves than to the enemy.' (Aelian, *Tactics* 18)

'For horsemen neither push those in front of them, not being able to press horse upon horse, in the way foot soldiers push with chests and shoulders. Nor, becoming continuous with those deployed in front of them, do they achieve the single weight of the entire throng. Rather, if they were to press together and densify, they more likely will upset the horses.' (Arrian, *Tactics* 16.10–14)

'The cavalry phalanx, however, does differ from the infantry one. The latter is closed up very tightly, which gives it an irresistible weight as the men crowd together and push one another forward upon the enemy.' (Anonymous, *On Strategy* 17)

'For there can be no pressure [*othismos*] from the rear up through the ranks, as happens with an infantry formation, which may force the men in front to push forward against their will. Horses cannot use their heads to push people in front of them evenly, as can infantry.' (Maurice, *Strategikon* 2.6)

There can be no doubt from these passages that in the Macedonian phalanx, and presumably in Byantine infantry formations (it is unclear how much of the Byzantine tactical tradition reflects actual Byzantine practice, and how much it just quotes the ancient authorities), the rear ranks did have a pushing role of some sort. The file closers were there to push the rest of the formation into place and prevent any hanging back or desertion, which could well be similar to the 'encouraging' role Xenophon assigns to the file closers in the Classical phalanx. But the rear ranks generally were able, in some way, to close up into a single mass to increase the weight and momentum of the formation, in a way that sounds very much like the traditional conception of the hoplite scrum.[21]

But great caution is necessary before concluding that this provides evidence for the way Classical hoplites fought. Polybius, the Hellenistic tacticians and the Byzantine authors writing in the same tradition are all describing the Macedonian phalanx (or formations derived from it), and this was a different beast from the earlier hoplite phalanx. The Macedonian phalanx was specifically designed to fight in closer order than the hoplite phalanx, and replaced the individual actions of men with shield and spear with the mass effect of multiple rows of pike points projecting ahead of the formation and held rigidly advanced. Unfortunately, some modern authors will still quote the Hellenistic tacticians as if they provide evidence for the Classical hoplite phalanx, but this is invalid as (despite the continuing use of the word 'hoplite' to describe the men in the Macedonian phalanx – used in its generic sense of 'heavy infantryman', not in any specific technical sense) we simply cannot project backwards the formation and tactics of the Macedonian phalanx on to its Classical predecessor.

Furthermore, while this is not the place to examine in detail the pushing role of the rear ranks of the Macedonian phalanx, there are serious problems with interpreting the evidence given above to mean that the Macedonians fought in the style of a scrum, whether or not Classical Greeks did so. For one thing, Polybius and the tacticians assign (in normal formation) a rank spacing of 3ft to the Macedonian phalanx, and with spacing this wide, there could be no question of physically shoving in the back the man in front (the pressures exerted by a full scrum would close men up to half this distance or less; see below), and the number of pike points projecting beyond the front rank, so carefully enumerated by Polybius and the tacticians, would be completely changed if in action the phalanx closed up much tighter. More seriously, the hoplite scrum theory requires that hoplites press their shields or bodies up against their enemies' shields in order to shove them back (which is how 'clashing together their shields' has sometimes been interpreted), but men armed with 5-metre pikes cannot close up to this sort of close contact – any pushing would have been transmitted down the length of the shaft of the pike, not shield against shield, and this would surely just snap the pike, or at best impale all the front-rankers, long before it pushed back the enemy formation.

For these reasons, I do not think that the Hellenistic and Byzantine evidence can be used to argue for a Macedonian or Byzantine scrum. The Macedonian phalanx undoubtedly could close up into a very tight formation, but the difficulties involved in interpreting this to mean that it fought using a mass shove in the way envisaged for the hoplite scrum are to my mind insurmountable. Rather, we must be seeing something else at work, and something that I believe does shed light on the way hoplites fought. The tightening of the Macedonian formation, and the men closing up, if necessary, into actual bodily contact, evidently gave an advantage in combat, but this advantage was not to do with a scrum, pushing against an enemy formation. One benefit is clearly stated by Polybius – that the tight formation made it impossible for the front ranks to run away – and the tacticians repeat this idea in various words, as we can see from the excerpts above. So at least a part of the purpose of the rear ranks pushing forward is to keep those in front of them moving forward, to 'force the men in front to push forward against their will', as Maurice expresses it. This would fit well with the testimony of Xenophon for the Classical phalanx, with the similar examples in the *Cyropaedia* and with what we know of infantry combat in other periods. The aim is to keep up the momentum of the advance into contact by keeping the whole formation moving forward, not to physically drive back the enemy once in contact.

Another purpose of the rear ranks pushing would be similar, but subtly different: to provide 'solidity' to the formation so as to prevent it being broken through. A breakthrough of a phalanx – a gap appearing from front to back through which the enemy could penetrate – would be the greatest danger a phalanx faced, since it would cause panic to be transmitted rapidly along the line, often causing the entire formation to collapse. I will look at the mechanics of this below, but here it will suffice to say that one way to prevent such a breakthrough was to have a deeper and more densely packed formation (even if this meant having a narrower one). Xenophon makes this point in the *Anabasis*, on the occasion when the Ten Thousand had to advance against a Colchian force stationed on the crest of a hill, in a passage that we have already encountered:

> 'Xenophon accordingly said that in his opinion they should give up the phalanx and form the *lochoi* in column. "For the phalanx," he continued, "will be broken up at once; for we shall find the mountain hard to traverse

at some points and easy at others; and the immediate result will be discouragement, when men who are formed in phalanx see the line broken up. Furthermore, if we advance upon them formed in a line many ranks deep, the enemy will outflank us, and will use their outflanking wing for whatever purpose they please; on the other hand, if we are formed in a line a few ranks deep, it would be nothing surprising if our line should be cut through by a multitude both of missiles and men falling upon us in a mass; and if this happens at any point, it will be bad for the whole phalanx.'" (Xenophon, *Anabasis* 4.8.10–11)

Note the fear that a shallow phalanx might be 'cut through' (*diakopto*) by missiles and a mass of men (who are not themselves equipped as or fighting as hoplites). Being 'cut through' in this way was a mortal danger for any line of battle. Compare Epaminondas' tactics at Leuctra, where 'Epaminondas led forward his army prow on, like a trireme, believing that if he could strike and cut through anywhere, he would destroy the entire army of his adversaries' (Xen., *Hell*. 7.5.23), or the intended role of the scythed chariots at Cunaxa, to 'drive into the ranks of the Greeks and cut through them' (Xen., *Anab*. 1.8.10). One way to reduce the risk was to make the formation deeper (for reasons to be examined below). Another, closely related, way to prevent such a breakthrough would be to increase the solidity of the line – to prevent the backward movement of the ranks that could lead to a breakthrough, and thus to make the line more resistant to enemy attack. The rear ranks in other words were not pushing the enemy (through the bodies of their colleagues in the front ranks), rather they were bracing the men in the front ranks to prevent them giving ground. The aim was to produce a line with greater solidity, and this was necessary in cavalry as well as in infantry formations – as Asclepiodotus (7.4), 'provided it [the depth] is enough to hold the squadron firm and in line' makes clear – where there is explicitly no physical contact between ranks at all.[22]

Finally, this closing up would have given the formation greater 'weight'; but we need not interpret 'weight' to mean 'shoving power in a scrum', since the Greek metaphorical use of 'weight' when applied to infantry is much more subtle and varied than that. 'Weight' was derived from the density of a formation (the narrowness of the intervals in rank and file) and from its depth in ranks, but also from the armour and the weapons of the men – heavy armour produced greater weight, which could 'push' or 'press' an enemy – or from their courage and fighting ability, as we saw above in Xenophon and in Diodorus. It would be incorrect to state that the use of the word is always purely metaphorical, since the depth and density of the formation did give a literal weight (or mass) of a kind, but it would also be wrong to claim that it must be strictly literal and physical, since non-physical factors contributed to it. The weight of a formation was the combination of formation, equipment, human factors and fighting style, and was not simply a physical mass to be fed into a scrum. It might be argued that a weight and pushing metaphor developed because there was an underlying reality of mass and pushing. This argument might have some weight were it not for the fact that the same pushing metaphor was applied also to light infantry, cavalry and ships, where there is no question of literal pushing, and were we not so familiar in English with similar military metaphors where there is no underlying literal physical reality.

We must also remember that even though Macedonian formations and later formations derived from them could close right up into bodily contact, this was not the normal combat formation. The Hellenistic tacticians are clear that the normal fighting formation was that with two cubit (1 metre) spacing, in which there would be no bodily contact between men. If the Macedonian phalanx did literally push their own men, this would only have been in the tightest one cubit spacing (*synaspismos*), which the tacticians tell us was used as a static defensive formation ('when the enemy are marching upon us', Asclep. 4.3). So either the literal '*othismos* up through the ranks' of Maurice was something only adopted when the tightest formation was adopted, and then only on the defensive, or else we should not imagine that 'bearing forward with the weight of their bodies' involves literally shoving the man in front in the back and pressing as tightly as possibly up against him as in the scrum models. It may be that the required bracing and closeness could be achieved while still around 1 metre apart and therefore not in the most compressed bodily contact, and that physical proximity, rather than actual physical shoving, was all that was required to achieve the desired effect. This is made especially likely given what we know of equivalent infantry formations in later periods. The pike formations of the Renaissance and early modern periods used the same file intervals as the Macedonian phalanx (being based on the same tactical manuals), and also had a concept of pushing, but despite some strange modern reconstructions in which men do shove each other in the back, it is clear that these pike formations did not use a literal scrum either. It may well also be that Late Roman and Byzantine infantry formations habitually used a tighter formation than the Macedonian phalanx, not least because they were often facing forces of heavy cavalry which required an especially rigid formation to oppose (as shown by Arrian's second century AD recommendations in *Battleline against the Alans*), and the '*othismos* up through the ranks' of Maurice reflects this tradition, rather than going back to Hellenistic, still less Classical, formations. We must also remember that a Macedonian phalanx probably advanced to the attack at a walk, while hoplites at least sometimes ran into combat, and there can be no question of men pushing those ahead of them in the back while running, an idea which is (I hope) self-evidently absurd. Therefore, the effect of the rear ranks in keeping the whole formation moving forward must be something that could be achieved without physical shoving.[23]

For all these reasons, I think that the literary evidence gives us no reason to suppose that the Classical (and Archaic) phalanx ever used a hoplite scrum (at least not deliberately). As argued earlier, it is likely that most Greek formations did not have rigidly defined file intervals, but the normal fighting spacing may well have been about 1 metre (given the size of the shield), and a similar spacing was probably adopted between ranks, as was the normal practice in the Hellenistic period. This means that hoplites fought as individuals; they did not close up into a single mass and push as one. But even while fighting as individuals, the depth and 'weight' of the formation could result in an effect for which an appropriate metaphor was 'pushing' – just like the Medieval 'press' or the 'push of pike' of Early Modern armies – and the depth and 'weight' of a formation could indeed be a crucial factor in victory and defeat, for reasons I will consider below.

Other arguments for and against the scrum

Before abandoning the idea of the hoplite scrum completely and moving on, we should first recognize that arguments are sometimes proposed other than the literal interpretation of particular words in literary sources, for supposing that there was a scrum. One such is that *othismos* and the *otheo* verbs describe not so much a deliberate tactic as a particular phase of hoplite combat, in which fighting at the front of the formation has proven indecisive, but the back ranks continue to press forward in an attempt to keep the formation moving forward (as they would have done during the advance), resulting in the formation compressing from front to back as it meets equal and equivalent pressure from the other side. Both sides would then compress into a single mass in a situation closely analogous to a crowd crush, with tremendous pressure generated by the rear ranks simply leaning into those in front of them – pressure that builds so long as there is an equal and opposite resistance from the enemy. If the enemy gave way, then pressure was released and the formation would expand out again.[24]

It must be said that this model ('crowd crush *othismos*') is not significantly different from the normal hoplite scrum model of combat, in that the crucial component is the crush or scrum, but this version does eliminate some of the conceptual confusion that has dogged other models of the scrum. In particular, the assertion is sometimes made that while engaged in the scrum, hoplites could maintain a side-on or three-quarters stance and continue to wield spear and perhaps shield. This is impossible, since the pressure from rear ranks shoving would inevitably compress the formation into the smallest possible space, in which men would be pressed into a chest-forward position, and arms and shields would be jammed up against and between the bodies of the hoplites (though it would have been possible to keep the right arm free above the press to wield a spear or sword in the high hold). If there were a scrum at all, then it would have taken the form of a crush like this, otherwise what is being described is more like the more tentative bracing pressure I argued for above, and this could not have had the effect of literally pushing back the enemy, since no significant forward pressure could have been generated. Pushing pressure from the rear great enough to force back an enemy formation must, of necessity, also have been enough to force all the men into a front-on stance, compressed to the minimum possible space. If the men were able to maintain a different fighting posture, then pressure was by definition not being transmitted through the formation to the enemy.[25]

Other versions of the scrum suggest that combat began with a run into direct contact, in which the two sides would crash into each other at the run. This idea is inherently highly improbable, and again is conceptually confused since in this model the charge and collision is then sometimes followed by a period of fencing with spears, so the hoplites (or the survivors of the collision) would have had to separate out again after impact – the rear ranks presumably backing off to make room – in order to fight. The 'crowd crush' model has the benefit of doing away with this difficulty, by having the crush develop in the course of fighting under steadily mounting pressure from the rear, without any initial collision.[26]

Aside from the usual literal interpretation of the sources, a supporting argument for the 'crowd crush' is a 'form follows function' one based on the shape of the hoplite shield. The argument goes that the hoplite shield is unusually over-engineered for the normal function of a shield in combat (protecting the bearer from weapon blows), and that much of the weight and strength of the shield comes from its back-curved edge section, which creates the characteristic hollow bowl. Since this bowl shape and the strength of the sides of the shield do not have an obvious protective function, the argument goes, their

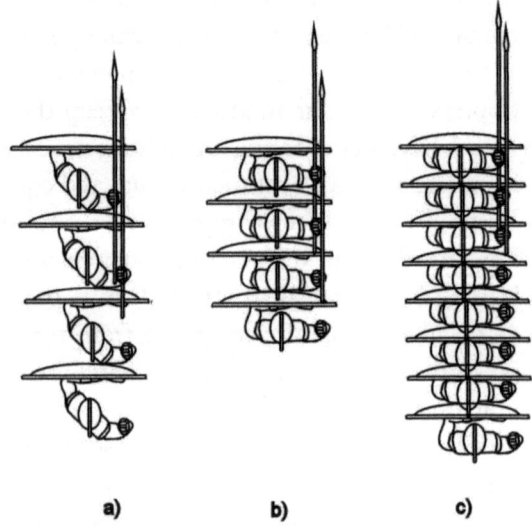

Combat stances: a) three-quarters stance, no pressure could be applied to the enemy formation without causing the men to close up to face-on stance as b); continuous pressure from the rear would then compress the formation to c).

role was to prevent the hoplite being asphyxiated in the crush (as the pressures generated would have been high enough to do), by providing a breathing space for the chest, with the rim braced against shoulders and thighs. I think that this argument is also highly improbable – we have no *a priori* reason to suppose there was a crush at all, so the argument that the shape of the shield (designed to survive a crush) is evidence for the existence of a crush is circular, unless we can establish that there is no other possible reason why the shield was made this way. As we have seen, the Greeks probably adopted their shield from the East, and shields of varying and different shapes were used by various earlier and contemporary peoples, including heavily dished shapes. The shape that Xenophon says was 'suitable for pushing' – the Egyptian shield – seems to have been a tall shield, not a bowl-shaped one. It could well be that the strengthened rim of the Greek shield was considered important to provide strength against spear or sword blows, and we do not understand enough about the relative benefits of different shield shapes and constructions to be able to state that the only explanation of the Greek shield shape is to survive a crush. Also, other formations that adopted close-order and bodily contact – the Macedonian, Roman and Byzantine formations considered above – did not also adopt the hoplite shield. It therefore seems unlikely that the shape of the shield is evidence of the existence of a crush, given the absence of other evidence in favour of the idea.[27]

A further objection is that the pressure of such a scrum would have killed or incapacitated those in the front ranks, though recent experiments have shown that the pressures exerted, though considerable, would not have caused death (even without the life-preserving shield shape, it seems). Even so, though this means that the scrum is physically possible, it would still be the case that men so crushed together could not have fought in any meaningful way with spears or swords; provided their arms were held up above the mass, all they could have done is strike at the head or face of an enemy several ranks away, given the closeness of the crush. With the men packed immobile together, it would also have been impossible to take any defensive action against such a thrust (such as dodging, or parrying with the shield), with the only chance of survival being to parry with the spear or sword and hope that the head of the man in front provided some protection – a possible manoeuvre for sure, but in general such a mutually lethal form of combat, especially for the supposed best men in the front rank, seems unlikely to have been favoured.[28]

Hoplites could also have dropped their spears (which may have been broken at this stage) and taken to their swords or daggers, though these could still only have been wielded above the scrum (and could not be drawn once the scrum developed). It may indeed have been a prerequisite of a crush or scrum forming that the combatants took to their swords and so closed to a range (face-to-face contact, almost literally) at which spears would no longer have been of value. The problem here is that, though swords certainly were used in close combat, and spears did break, there are sufficient depictions in art both of single combats with the spear and of spears being wielded in mass formations to suppose that spear fighting must still have been the major form of hoplite combat. If spear fighting was only a precursor to the truly decisive element of combat, a scrum with swords, then it becomes difficult to see the point or purpose of spear fighting. Xenophon's description of the aftermath of fighting at Coronea – 'shields smashed to pieces, spears snapped in two, daggers bared of their sheaths, some on the ground, some embedded in the bodies, some yet gripped by the hand' (Xen., *Ages.* 2.14) – shows that swords or daggers were a feature of such combat. But that swords were used when spears broke does not suggest that the sword was the primary weapon for the decisive stage of combat, and the idea that hoplite phalanxes routinely broke all their spears in an initial round of combat and so were forced to close to face-to-face contact to use their swords remains at best highly speculative.[29]

Evidently, hoplite fighting was varied and violent in nature, with some serious forces applied (enough to smash shields), and swords were used at close quarters (or after spears broke). Yet there is also evidence that the fighting was more open than would have been possible in a scrum – in particular the ability of men to fall to the ground and either stand up again or be pulled out of the action by their comrades. For example, Xenophon records the Spartan Archidamus at Leuctra 'fell three times and was the first of the citizens to lose his life in the midst of the enemy' (Xen. *Hell.* 5.4.33); he must therefore have been able to get up again after the first two falls. The fate of the Spartan king Cleombrotus at Leuctra is similar:

> 'Nevertheless, the fact that Cleombrotus and his men were at first prevailing in the battle may be known from this clear indication: they would not have been able to take him up and carry him off still living, had not those who were fighting in front of him been holding the advantage at that time.' (Xenophon, *Hellenica* 6.4.13)

Men cannot fall to the ground, be passed over by the fighting, or be extracted from the fighting and taken to the rear, in a scrum in which all men are crushed up against each other at maximum pressure and minimum distance. In such a scrum, the dead and wounded would necessarily not have fallen at all, but would have been pinioned in place by the pressure, upright but helpless and immobile. This leads to another objection to the idea of such a scrum – that such pressures and forces would be highly unstable, as shown by the various re-enactment groups which have attempted to recreate the scrum, which often rapidly collapses into a struggling heap of men on the ground, or spins and fragments as men pop out of the sides of the press. Such scrums appear to be inherently unstable, and it is difficult to see how these fights could ever have continued for 'a long time' as they are said to have done (although for how long is another matter, to which I will return below), or been carried on continuously along a front of hundreds of metres, without the dynamics involved causing the crush to spin, collapse or break up very rapidly.

Consequently, I am profoundly unconvinced by the idea of the hoplite scrum. There is no evidence in the literary sources that would lead us to suppose that Greek hoplites fought in a fundamentally different way than other types of infantry or formations. While there is evidence that rear ranks had a pushing role in later formations which may go back to the Classical phalanx, this pushing is more to do with maintaining a steady advance and retaining the solidity of the formation than with pushing the enemy off the battlefield. There are also practical objections to the possibility of fighting in this way, when compared with the plentiful evidence for Greeks fighting in a more conventional manner, using their weapons in the usual way. The notion of the hoplite scrum is a part of the general trend of 'hoplite exceptionalism', seeing hoplites as inherently different from other types of infantry before, contemporary or since. The Greeks themselves were certainly open to the idea that they were better than everyone else, but even they never stated that they fought in a way that was fundamentally different from all other cultures. I believe, therefore, that the reality of hoplite combat must be sought in a type of fighting that was basically similar to the way such close-order infantry formations have fought throughout history, at close quarters certainly, at times very violently, and sometimes involving an element of pushing and shoving, but still as individuals, with weapons. The rest of this chapter will consist of an attempt to provide a model of such fighting.

Charge and contact

We saw in Chapter 7 that hoplites might either run or walk to cover the last hundred metres or so before contact. As we saw above, what precisely happened upon contact (and note that 'contact' here does not mean literal physical contact between men, but in the military sense of contact between two formations, when they are close enough to engage with the weapons they carry) is one of the great unknowns of combat in all ages, not just for the hoplite phalanx, since nobody alive today or for many decades has any experience of the contact of close-order formations of infantry or cavalry. According to some versions of the scrum theory, the formations would literally collide with each other at the run, while other interpretations have them then separate again to fight for a while with weapons. However, it is unlikely that the purpose of the charge was to run into direct bodily contact (colliding together), except perhaps in exceptional circumstances.

This is an issue that remains unclear and has not been definitively settled for any other period of military history. It has most often been studied from the point of view of cavalry charging infantry, where the question is whether horses would or could actually collide with infantry (and knock them over, unless they were braced by the men behind them). The question remains open, with many modern authors of the opinion that such charges into contact were impossible for cavalry, though it seems likely that this is overstating the difficulties involved, and that on occasion cavalry could indeed charge into full bodily contact with infantry. Yet the result would often be equally bad for both sides, with injury to all concerned, and men and horses falling. It seems that the purpose of a charge was rather to break up a defending infantry formation through the fear of full contact, so that when the cavalry arrived at the infantry formation, the infantry were already starting to scatter and lose their cohesion. In these circumstances, with the infantry no longer presenting a united front, there would be gaps into which the cavalry could enter, with at worst sidelong rather than full frontal collisions, which would be survivable for the cavalry (and psychologically achievable for the horses) and would tend to further open up and disrupt the infantry formation. A cavalry charge against infantry was therefore very much a battle of wills, or several battles of wills. The cavalrymen had to persuade their horses to run at what appeared to be a solid obstacle (a close-order line of men), something to which horses are naturally averse. The infantry, meanwhile, had to persuade the cavalry that they would stand firm in the face of such a charge, and not break order and lose their formation, presenting gaps into which the cavalry could penetrate. Furthermore, the cavalry had to persuade the infantry that their charge was serious and that they would indeed charge home, and that the infantry were well advised to get out of their way. In this battle of wills, training and experience (and in many gunpowder-era armies, iron discipline) were essential. The infantry had to know that, provided they stood firm, they would be safe against the cavalry and could drive them off – *otheo* them, as Herodotus has it (9.25.1). The cavalry had to press their charge with enough determination to leave the infantry in doubt as to whether standing in the face of such an attack was really a wise course of action. The natural response of both sides was to avoid a collision; the infantry by fleeing, the cavalry by pulling up or swerving away, either of which could be fatal if the other side held their nerve. This made a cavalry charge like a game of chicken (using the game theory application of the idea) in which the outcome if both sides did hold their nerve would be difficult to foretell, but probably violent and damaging to both. Yet in the vast majority of cases, both sides did not hold their nerve, one or the other flinching, which in turn would give confidence to their opponents to press on or to stand firm.[30]

We have seen in Chapter 1 how the origins of the hoplite phalanx may be seen as a way for a middling class to assert its dominance over an aristocratic cavalry, in battlefield terms by creating a formation of close-order heavy infantry with a strong ethos of standing their ground and sticking to their comrades no matter what – exactly what is required for infantry to face down cavalry. Other contemporary civilizations, particularly the Persians, with a monarchical government and dominated by a still horse-borne aristocracy, had no incentive to specialize in close-order heavy infantry to this extent. Their infantry formations, though highly capable, emphasize instead shooting and defence against shooting, and a more mobile style of fighting. Once the Greek cities had established battle-dominating forces of heavy infantry, their main opponents would of course have been each other, and so heavy infantry phalanx came to fight heavy infantry phalanx.

As we saw in the previous chapter, there were two schools of thought as to how best to initiate such combat, by either running into contact or by advancing slowly and in step, with advantages and disadvantages for each. Running minimized exposure to missiles (where this was a factor, as at Marathon), but could cause the formation to lose order and cohesion (Thuc. 5.70.1), exactly what a phalanx aimed to avoid. But chiefly, as with a cavalry charge, it might be hoped that the enemy would be frightened into flight. Advancing in step, meanwhile, kept the formation tight and could be intimidating to the enemy in itself, though it required greater self-confidence and discipline to carry out. We can be certain that no part of the purpose of a running charge was to build up momentum for a collision. A human can reach top running speed in a very few metres, and no extra momentum can be gained after this, so any longer charge would disrupt the formation for no gain; besides which, if this was essential to hoplite fighting, we would expect the Spartans, masters of the art, to utilize it. So we must seek some other purpose for a running charge, and this is not hard to do, since surely the intent was much the same as with cavalry – to encourage the attacking side with a burst of adrenaline and to minimize the waiting time before contact, while at the same time forcing the enemy to lose their nerve and turn to flight even before contact occurred, or if not to turn to flee at least to waver, back off, create gaps and openings, and cause some to face about and begin to break away from the back of the formation. All of this would have further undermined the confidence of the defending force while encouraging the attackers, and would have made gaps and openings into which the charging force could penetrate to gain local advantages – two on one, or blows from the side – against those individuals who still stood firm. An infantry charge, like a cavalry charge, was an attempt to win the combat before it had even begun, by denting the confidence and breaking up the formation of the enemy while boosting the morale of one's own side. That it was sometimes successful in this is shown by those occasions where an enemy ran away before contact, or as soon as contact was made. Such was the case at Cunaxa (against Egyptian and Persian infantry), where 'before an arrow reached them, the barbarians broke and fled' (Xen., *Anab.* 1.8.17–19), and in the first phase of the fighting at Coronea, where both sides experienced such flights: 'when they came within spear thrust, [they] put to flight the force in their front. As for the Argives, they did not await the attack of the forces of Agesilaus, but fled to Mount Helicon' (Xen., *Hell.* 4.3.17). The Spartan army of Archidamus had similar success in the 'tearless battle' in 368:

> '[I]t is reported that the soldiers were inspired with so much strength and courage [by Archidamus' speech and a series of good omens] that it was a task for their leaders to restrain them as they *pushed* forward to the front. And when Archidamus led the advance, only a few of the enemy waited till his men came within spear-thrust; these were killed, and the rest were cut down as they fled, many by the horsemen and many by the Celts.' (Xenophon, *Hellenica* 7.5.14)

If a charge ended in contact between well-formed formations, then it had already failed in its primary purpose.

What happened when both formations did hold their nerve is never clearly spelled out, but we have no reason to suppose it was a mass collision, still less a collision in which the rear ranks piled into the backs of the ranks in front of them. The one example that is given as evidence of such a collision, the second stage of fighting at Coronea, is not good evidence for this at all:

> '[Agesilaus] immediately wheeled his phalanx and led the advance against them; but the Thebans on their side ... wishing to break through to join their own friends, massed themselves together and came on stoutly ... [Agesilaus' phalanx] crashed against the Thebans front to front; and setting shields against shields they shoved, fought, killed, and were killed.' (Xenophon, *Hellenica* 4.3.18–19)

The translation 'crashed against' of course gives an impression of a mass collision, but the verb used – *sunarrasso* – does not require such an interpretation, being used for example for Athenian cavalry attacking Theban and Thessalian cavalry (Xen., *Hell.* 7.5.16), where there can have been no collision (unless concussing the horses was a part of Greek cavalry tactics). A less dramatic translation such as 'come together' would be more accurate, though Xenophon no doubt wished to emphasize the violence of the encounter. Similarly, 'setting shields against shields' gives an impression of mass pushing, but as we have already seen above, it is better translated as 'clashing together their shields' as individuals fence, strike and parry.

Looking also at the idea of the back ranks pushing those in front of them in an advance – 'press forward with the weight of their bodies' (Pol. 18.30.4) – we can be certain that such pressing forward could not have taken place during a charge at the run. Men cannot run while being pushed in the back, still less pushed in the back by a whole file of men behind them, and doing so would of course achieve nothing, except to push the first few ranks over, as no pressure could be applied by them, there being no enemy to push against. As argued above, this is another reason to suppose that such pushing cannot be ascribed to hoplite formations which habitually did attack at the run, and instead was a feature of the Macedonian phalanx which advanced in closer order and at a steady pace, and which needed to maintain a tight formation throughout the advance.[31]

This need to maintain momentum (the reason that Xenophon recommends having the file closers 'drive on' the men in front) comes back to the psychological importance of the charge or approach to contact, both to maintain the confidence of one's own side and to harm that of the enemy. Julius Caesar, who understood the psychology of men in battle, describes this well in the context of Roman legionary combat at the Battle of Pharsalus (48), in terms that must also have some applicability to hoplites:

> 'Between the two lines there was only as much space left as was necessary for the charge of each army. But Pompey had previously ordered his men to await Caesar's attack without moving from their position, and to allow his line to fall into disorder. He is said to have done this on the advice of G. Triarius, in order that the first charge and impetus of the troops might be broken and their line spread out, and that so the Pompeians marshalled in their proper ranks might attack a scattered foe … Now this seems to us to have been an irrational act on the part of Pompey, because there is a certain keenness of spirit and impetuosity implanted by nature in all men which is kindled by the ardour of battle. This feeling it is the duty of commanders not to repress but to foster, nor was it without good reason that the custom was instituted of old that signals should sound in every direction and the whole body of men raise a shout, by which means they thought that the enemy were terrified and their own men stimulated.' (Caesar, *Civil War* 3.93)

On this occasion Caesar's disciplined and experienced army were able to halt in the middle of the neutral ground, reform their ranks and then resume their advance.[32]

But advancing into contact with an enemy slowly while maintaining order required even greater discipline and experience, which is why the Spartan and Macedonian armies with their full-time professional soldiers were able to attack this way. Less-experienced Greek city militias would have been more dependent on 'keenness of spirit and impetuosity' being kindled by a rapid, determined, aggressive charge, for all the loss of order it entailed. So the purpose of a running charge was not just to terrify the enemy but to embolden one's own side, as Caesar says. If a force of doubtful experience or enthusiasm lost forward momentum it might not be able to regain it, and would be in danger of losing heart and running before contact against a bolder enemy. This need for confidence in an advance, on both sides, leads to some interesting effects – such as the ploy of the Athenian commander Chabrias when his Theban and allied force was attacked by Agesilaus' Spartans:

> 'Chabrias the Athenian, however, leading his mercenary troops, ordered his men to receive the enemy with a show of contempt, maintaining all the while their battle lines, and, leaning their shields against their knees, to wait with upraised spear. Since they did what they were ordered as at a single word of command, Agesilaus, marvelling at the fine discipline of the enemy and their posture of contempt, judged it inadvisable to force a way against the higher ground and compel his opponents to show their valour in a hand-to-hand contest, and, having learned by trial that they would dare, if forced, to dispute the victory [he withdrew his forces].' (Diodorus 15.32.5–6)

Having the mercenary force show their discipline and confidence by standing in an 'at ease' position was enough to convince Ageislaus not to try his luck against them. Such displays of confidence were an essential part of the approach to combat; the Spartans showed their confidence by a steady advance with pipers and by stopping in full view to perform a religious ceremony. A running charge was a way to imbue confidence in what otherwise might have been a reluctant army.[33]

I have no doubt that on some occasions, or with some particularly aggressive or deranged individuals, attackers would charge into and physically collide with their opponents. It was a commonplace among Greek and Roman writers that 'barbarians' – meaning particularly the peoples of central and northwestern Europe, rather than Persians or other Asians – fought in this way, with great ferocity but little steadfastness. Furthermore, Romans, who also prided themselves on their *furor*, might on occasion have body-slammed their enemies, bracing themselves behind their large shields. We do not need to believe that such things never happened, and they may perhaps be reflected by 'clashing their shields' at Coronea, where both sides were particularly determined, had exceptionally high self-confidence and motivation, and met in a tactical situation in which defeating the enemy directly to the front was the only path to safety. Yet the fact that Xenophon singles this out as exceptional – 'it proved to be like no other of the battles of our time' (Xen. *Hell.* 4.3.16), assuming it is this aspect of the battle he is referring to – shows that this cannot be taken as typical of hoplite fighting, and what usually happened at the moment of contact must have been less violent and more tentative. The charging side (both sides, if both were charging) must have slowed down to a walk as contact approached (the Spartans, experts at such fighting, would always have been at a walk), and then stopped when in weapon range (*eis doru*, 'within spear-thrust') to fence with spears.[34]

Fighting with weapons

So the charge, or steady intimidating advance to contact, having failed to break up or chase away the enemy, the two forces came within spear-thrust to fight with their weapons. We need not try to assign any fixed distance to spear-thrust: the actual maximum thrust range of a hand-held spear is a matter of some academic interest perhaps, but would never have been a particular fixed distance in antiquity, since we can be sure that different hoplites had spears – and arms – of different lengths, and fought using slightly different techniques and with greatly varying degrees of enthusiasm and aggression, so that the practical, battlefield reach of a thrust spear would have varied considerably. Nevertheless, we can perhaps assume something from 1–3 metres as a reasonable estimate. This would have put the hoplites in the front rank a step or two apart, though they could (like Eteocles and Polyneices) have stepped forward to closer range in order to clash together their shields in an effort to unbalance or temporarily unshield an opponent. The spear being a long weapon held around its point of balance, as well as a maximum it would have had a minimum range within which it could not strike a target, without the wielder stepping back to reopen the separation between the antagonists. Where both sides had spears, we must imagine a series of these lunges and withdrawals, clashes of shield and thrusts of spear against any exposed part of the enemy's body, in an effort to wound or kill the immediate opponent.[35]

Such individual weapons skills, as we saw in Chapter 4, were assumed to come naturally to a man and did not require special training; yet as we also saw, trainers were available who could, for a fee, teach particular tricks and skills. Plato likened the movements of wrestling and of dancing to those required in combat, in particular the Pyrrhic dance:

'[I]t represents modes of eluding all kinds of blows and shots by swervings and duckings and side-leaps upward or crouching; and also the opposite kinds of motion, which lead to active postures of offence, when it strives to represent the movements involved in shooting with bows or darts, and blows of every description.' (Plato, *Laws* 7.815a)

Of course we need not suppose that Pyrrhic dancing actually used the same movements as spear-fighting (any more than wrestling did), but that it was considered good training for it means that there must have been some commonality in the skills required. Compare also with the comparison that Plutarch makes between boxing, wrestling and fighting, as we saw above. Plato (in the mouth of Laches) does say of weapons training that:

'[T]his accomplishment will be of some benefit also in actual battle, when it comes to fighting in line with a number of other men; but its greatest advantage will be felt when the ranks are broken, and you find you must fight man to man.' (Plato, *Laches* 182a)

So individual weapons skill was still of use in the battle line (*taxis*), even though it was more useful still in single combat where the range of movement available would obviously have been much greater, as there was more space in which to fight. Phalanx fighting was certainly physically constrained by the presence of men to left and right and behind, resulting in a smaller space in which to fight and no opportunity to 'wheel round and round' or 'fight a wild battle', like Eteocles and Polyneices or any of the vast number of single combats depicted in art. Indeed, single combat, or at least irregular combat in broken ranks, must have been the main experience of fighting for most hoplites, as Plato points out. Precisely how constrained the space was, whether 1 or 2 metres square, or some indeterminate and variable distance in between, is largely beside the point, and no doubt varied from case to case. Hoplites in the phalanx had less room in which to fight than they would in single combat or open fighting, but not no room at all, and if they could not 'wheel round and round' they could still have used ploys like the 'crafty Thessalian trick' carried out by Eteocles against Polyneices:

'Disengaging himself from the immediate contest, he drew back his left foot but kept his eye closely on the pit of the other's stomach from a distance; then advancing his right foot he plunged the weapon through his navel and fixed it in his spine.' (Euripides, *Phoenican Women* 1410–1414)

Such movements required a certain amount of backward and forward freedom of movement, and at least a little sideways movement, in order to dodge or parry with the shield and aim spear blows. The Athenian Sophanes, the 'best

Athenian fighter' at Plataea, either cast an anchor attached to his belt so that the enemy, 'as they left their ranks (*taxeis*)', would not be able to shift him, or had an anchor on his shield 'which he constantly whirled round and never held still' (Hdt 9.74.1–2). Herodotus is non-committal as to which story he believes, but either would require Sophanes to have a certain amount of space and freedom of movement (within the limits of the anchor rope, in the first case). At the same time, the amount of space needed to fight effectively is sometimes exaggerated. Clearly there must be a difference between single combat and fighting in the phalanx for Plato's comments to have any meaning, so space in the phalanx was undoubtedly more constrained than in single combat. But there is no indication in any of our sources that hoplites would not need or be able to fight with spears – weapons handling and fighting are important, whether that fighting is analogous to boxing, to wrestling or to dancing.

We may also imagine that in most cases, spear fighting was much more tentative than either the popular conception of a swirling, confused melee or the scrum theory of heaving masses would suggest. The hoplite ethos required each man to hold his place and not to abandon his comrades, which suggests a relatively stationary and static style of fighting, without a lot of dynamic movement. The picture of Sophanes planted by his anchor is probably an exaggerated version of reality, in which hoplites would stand their ground rather than aggressively attempting to penetrate the enemy formation (of course, there will have been exceptions). Recent models of Roman-era infantry combat have suggested that rather than toe-to-toe duelling with swords, the default position was a 'dynamic stand-off' in which opposing sides would stand in close proximity but not in (military) contact, and that fighting would take place only in flurries as small groups, led by dynamic individuals such as the Roman centurions, built up enough aggression and enthusiasm to charge into contact for a brief exchange of blows before dropping back again to the default stand-off and to rest and allow their aggression to build up again. This may well have been the default form of combat in antiquity, and the form out of which the hoplite phalanx developed. We have seen such a stand-off in confrontations between hoplites and cavalry, though I am not yet convinced it applies to hoplites fighting hoplites (at least from the fifth century on), where the sources, unclear as they are, at any rate do not clearly suggest any degree of separation but give an impression of continuous contact (where there is contact at all) for spear fighting. It may well be the case, however, that such contact involved only low-intensity fighting, with tentative stabs with the spear and more time spent sheltering behind shields and dodging blows than in actively trying to kill the enemy. Self-preservation rates highly in the objectives of all soldiers, and while this can sometimes be overcome by training or fanaticism, citizen militias are not the usual place to look for such extremes. Many hoplites would undoubtedly have been happy to show their courage by standing firm in the line, and were less concerned about killing the enemy or performing conspicuous deeds of individual courage or recklessness than they were about returning home alive (at least until the enemy turned their backs in flight and changed from predators into prey).[36]

There are also some indications that hoplite fighting, at least against Persians, could be more mobile than the usual picture of massed phalanxes in close contact would suggest. We have already seen Herodotus' account of the difference between Spartans and Persians at Plataea (Hdt. 9.62.3, after the fight 'had come to *othismos*', 9.62.2). Compare with Sophanes above, holding off the enemy 'as they left their ranks (*taxeis*)'. At Thermopylae, Herodotus (7.211.3) records that the Spartans used feigned flights or withdrawals, and this tradition of Spartans fighting 'like cavalry' (by alternating attacking and retiring in the characteristic way described by Xenophon) also appears elsewhere, as we saw in Chapter 7. It is difficult to reconcile such feigned flight with the traditional image of hoplite fighting, and we do not see similar tactics in later (better-documented) battles.

This may be evidence that the phalanx was a looser formation at this period than it became by the later fifth century, or it may just show that the disciplined Spartans had more strings to their bow than other Greek armies. At any rate, it indicates that it was normal, among Persians at least, for individuals or small groups to charge out of the main body to make separate, localized attacks, as predicted by the dynamic stand-off model, and this may have been how at least some combat between hoplites also transpired if neither side was determined or aggressive enough to press forward into closer contact all along the line. If this style of fighting – a stand-off with flurries of combat – was the normal default in antiquity, those Greek (and other) armies that closed to full contact in a single mass may have been the exception.

As to how the spear was wielded, it seems likely as discussed in Chapter 2 that it was held in the high or underhand position, with the arm above the head and thrusts made forward and angled downward. This is the most common position shown in art, especially where massed formations are shown. It is true that this pose sometimes depicts, or may depict, spears being readied to throw (in the Archaic period at least), but this way of wielding the spear was probably common at the time when spears could be either thrown or thrust, and was carried over to the pure thrusting spear of the Classical phalanx. The advantage of such a hold is that it can be used in a crowded phalanx without fouling or accidentally stabbing one's own side, which would have been a danger with the low hold, with the spear held at waist height. The high hold also offered the chance to thrust over the shield of the enemy at his head, face or neck, while an underarm hold would be more likely to encounter the enemy's shield (though it might pass below to hit abdomen, groin or thighs). The underarm hold is also commonly depicted in art in single combats, so we need not doubt that the spear could be held either way, especially in combat outside the phalanx, and it was probably a matter of personal preference which style was used, or indeed a hoplite could switch from one to the other in the course of fighting. However, in the phalanx, it is most likely that the high hold was the standard.[37]

The front rank could naturally reach the enemy to strike with spears, and the rank behind could probably do so also, depending on precisely how close the front ranks were, which would not have been a constant either within or between combats. As Xenophon remarked with some rhetorical exaggeration, 'when phalanxes are too deep to reach the enemy with weapons' (Xen., *Cyrop.* 6.3.22), the extra ranks are of no use – how many ranks could reach with weapons he does not define, but it would not be more than two or three (his 'ideal' Persians are perhaps two ranks deep, or alternatively twelve ranks). Men not in reach to stab an opponent could still have used their spears to help parry blows aimed at the men in their own front ranks. Unlike the Macedonian phalanx, where multiple rows of pike points projected in front of the formation and the pikes were held rigidly in two hands, hoplite spears held in one hand could not produce an impenetrable barrier of spear points (hoplites probably stood further apart, their shorter spears projected less far and spears held in one hand could more easily be knocked or parried aside), so there is no question with a hoplite phalanx of being unable to close to the desired fighting distance due to a fence of spear points. Rather, hoplites would prefer to maintain some separation so as to wield their spears more effectively, unless they closed to clash shields as discussed above.[38]

This type of fighting with spears is sometimes termed, in modern authors, *doratismos* (spear-play), though as is the case with *othismos*, this is not a word used by ancient authors to describe such combat. I am aware of just two occurrences of the word in ancient literature, both in Plutarch, with one in the account of fighting between Timoleon's mercenaries and Carthaginians at the Crimisus (which is clearly imagined as a hoplite versus hoplite battle):

> '[Timoleon] made his vanguard [*promachoi*] lock their shields in close array [*puknosas toi sunaspismoi*], ordered the trumpet to sound the charge, and fell upon the Carthaginians. But these withstood his first onset sturdily, and owing to the iron breastplates and bronze helmets with which their persons were protected, and the great shields which they held in front of them, repelled the spear thrusts [*ton doratismon*]. But when the struggle came to swords and the work required skill no less than strength, suddenly, from the hills, fearful peals of thunder crashed down, and vivid flashes of lightning darted forth with them.' (Plutarch, *Timoleon* 27.6–28.1)

Note here the initial spear fighting, followed by sword fighting (the spears presumably having broken or the range being closed). The only other use of the word is to describe a Hellenistic single combat between Pyrrhus and Pantauchus, translated as: 'At first they hurled their spears, then, coming to close quarters [*en chersin*], they plied their swords with might and skill' (Plut., *Pyrrh.* 7.5), though the similarity with the previous passage suggests we need not imagine the spears being 'hurled', but rather thrust. That the word *doratismos* was not a usual Greek term for spear fighting does not of course mean that there was no spear fighting; clearly there was, and it might be an important and long-lasting (how long is a matter we will return to below) period of fighting.[39]

However, there were two alternative forms of fighting to such spear-fencing. One we have already encountered of course – the use of the shield for pushing. Xenophon's Egyptians at Thymbrara, as we saw, used 'spears that are long and powerful', but also their shields, which as they 'rest against the shoulder ... are a help in pushing. So, locking their shields together [*sugkleisantes tas aspidas*], they advanced and pushed' (Xen., *Cyrop.* 7.1.33). It would be odd if the pushing with shields was instead of using the spears, since the spears are worthy of special mention, so somehow the two things (wielding spears and pushing with shields) were done together. As we have seen, this is not evidence for a mass scrum by the Egyptians (since the ranks behind the first were 'no help' in the fighting), but it is further evidence, to set alongside the *othismos aspidon* at Delium (Thuc. 4.96.2), and 'they fought, they pushed' (Xen., *Hell.* 4.3.19) at Coronea, that pushing with shields could be an integral part of hoplite (or hoplite-style) fighting with spears. As argued above, this should be seen as individual pushing, punching or swinging of the shields of the front-rankers against their immediate opponents, something that is attested also in other armies and eras. The Romans in particular, with their tall, hand-held shields, are noted on a number of occasions as swinging or punching with the shields, particularly with the central iron boss, as part of their fighting method. The most-often quoted example is Zama, as we have seen, but use of the boss by Romans is referred to on a number of occasions. Romans of course did not take part in a mass scrum any more than Greeks did. In the Roman case, punches with the shield could be alternated with stabs with the sword, and hoplites could well have fought in a similar way, blows of the shield being used to knock aside the opponent's shield or knock him back off-balance, so creating an opening for a quick stab with the spear. The spear-fencing phase of fighting would thus have involved, for the more aggressive or confident hoplites willing to get so close to the enemy, a mixture of spear stabbing and shield punching, pushing and clashing. This is how we should imagine the 'clashing together their shields' of Xenophon, as well as the *othismos aspidon* of Thucydides – two different ways of describing the same phenomenon.[40]

But spear fighting would not last indefinitely. Spears were brittle, and the force of blows against shields or bodies was often enough to snap them. Additionally, the enemy might deliberately seek to break the spears (as Herodotus says the Persians did at Plataea, Hdt. 9.62.2). This was not a peculiarity of hoplite combat, of course. Xenophon describes an encounter of Greek and Persian cavalry:

> 'When they came to a hand-to-hand encounter [*eis cheiras*], all of the Greeks who struck anyone broke their spears, while the barbarians, being armed with javelins [*palta*] of cornel-wood, speedily killed twelve men and two horses.' (Xenophon, *Hellenica* 3.4.14)

The implication is that the Greek spears broke without doing any other damage. It is surprising perhaps to find Greek spears breaking so easily, and we might imagine that on at least some occasions the prevalence of broken spears is exaggerated for rhetorical effect. Nevertheless, spears clearly did break, or after a strike they might become lodged in an opponent's shield or body and were impossible to withdraw, or else the lines might simply have closed up to a shorter range in the course of the fighting. A spear has a minimum range at which it can be wielded effectively, and once an opponent was inside this range, the spearman had two options: to step back and open up the range again, giving him space again to thrust his spear (difficult when there is a file of men standing behind), or to drop his spear and take to his sword. This is one reason why the length or reach of spears, where it varied only by a modest amount (discounting the exceptionally long Macedonian pike), was probably not of great importance in combat. A man with a shorter spear could, if he was able to parry or knock aside his opponent's spear point, take a step closer to a position where he was inside his opponent's minimum range, and the longer spear would then be a disadvantage. We are reminded of one of the sayings of Spartan women recorded by Plutarch:

> 'Another, in answer to her son who said that the sword which he carried was short, said, "Add a step to it."' (Plutarch, *Sayings of Spartan Women* 18; *Moralia* 241F)

So in the course of combat there would have been a natural tendency, if both sides were equally aggressive and maintained the fighting at close quarters, for lines to close up with each other. This might not have applied in a dynamic stand-off, where fighting was more tentative and punctuated only by brief flurries of close combat. But if the lines did come to close quarters, and remained close, then two factors would come into play: as spears broke, hoplites would take to their swords and therefore would have to move even closer to the enemy in order to continue to strike (the maximum and minimum range of a sword both being much shorter than those of a spear); and as fighting continued, the formations would naturally compress somewhat due to the presence of the rear ranks (more on this below). If one or other side, rather than standing its ground, fell back as a result of this natural compression, in order to maintain separation for spear fighting or, lacking confidence and aggression, in order to open up a safe distance, then the result would be that one side moved backwards, and their opponents, following up in order to re-establish and maintain close contact, and gaining confidence from their success, would press forward. The fact that maintaining this sort of close contact was not a given (as it would have been if the front ranks were shoved together by those in the rear ranks) is shown by the stories of Spartans with their short swords or with pictures of flies on their shields, who made a point of honour of how close they intended to get to the enemy. This, I believe, is how we are to understand a phalanx being 'pushed back'. Is such 'pushing back' literal or metaphorical? It is a false dichotomy, since there is real physical pushing and pressure involved (clashing shields and stabbing with swords or spears, rear ranks bracing and holding front ranks in place), but there is no scrum or crowd crush.[41]

This phase of fighting would tend then to be at swordpoint and at particularly close range. Talk of phases should not fool us into believing that hoplite battles took place in a sequence of strictly delineated and sequential stages. It may be that in some cases, fighting went straight to such close quarters without an initial more tentative period of spear fencing (as perhaps at Coronea), or that it did not go to close quarters at all, depending on the confidence, aggression and experience of the two sides and a wide range of other factors. There is also, as usual, no particular Greek expression for such close fighting – *en chersi*, 'in hand' or 'hand-to-hand', is sometimes offered as a possible candidate (also *eis cheiras*, with the same meaning), but it seems apparent that this is just a general expression for all forms of hand-to-hand fighting and not a particular word for one type as opposed to another. Be that as it may, this distinction between an initial more open period of fighting with spears and a later, closer period, largely with swords, is reflected in, for example, the distinction Plutarch makes between boxing, which 'mimics attack and defence', and wrestling, which mimics 'the twisting and pushing of close quarters combat' (Plut., *Mor.* 639f). In this type of fighting, the opposing front ranks would close right up and engage, no doubt, with swords, shields or, in desperation, hands (Hdt. 7.225.3; and *in extremis*, even teeth, according to Herodotus), using unconventional moves that may well have had a similarity to wrestling (but Plutarch's analogy is just an analogy, and should not be taken to mean that hoplites actually wrestled). I believe that, so far as ancient authors regarded the noun *othismos* as denoting a distinct type of fighting at all, it was this type, the very close, very confused and hotly contested fighting that occurred between two equally determined phalanxes where both were willing to close, and neither was willing to give ground. This usage is analogous to the Medieval and Early Modern English use of 'the press' or 'push of pike'.

So we should not envisage hoplites shoving each other *en masse* and barely, if at all, using their weapons, nor must we imagine that there were distinct and separate phases of hoplite combat. Rather, hoplite combat (and in this it was probably very similar to all heavy infantry combat before the advent of gunpowder) could take a number of forms and go through a number of stages. An initial charge was intended to buoy the spirits of one's own side and frighten the enemy. If the charge failed, combat would ensue at the length of the spear, then as spears broke, lines closed up and the fighting became particularly close and intense at the shortest possible range. If one side then gave ground, it would be 'pushed back' by the enemy (or perhaps before that, break; see below). If both sides were equally determined, a period of intense close-quarters combat would follow for a while.

Now we must return to the question that has been hanging over this discussion: if only the first few ranks could use weapons, as Xenophon asserts, what was the purpose of those ranks behind the first two or three, and why did Greeks – and many others – habitually form up in such deep formations?

The purpose of depth

The first point that must be stressed is that Greeks were of course not alone in forming deep formations. We do not have reliable figures for the depths used by any other armies in antiquity, not even (rather astonishingly) Romans – the depth usually adopted by Roman legions is never specified, despite all the vast and intimate details we have about other aspects of legionary organization. Depths of six ranks are often assumed, and may be about right, though in reality – as with Greek armies – depths could be varied to suit the tactical requirements. We know that Persians sometimes formed very deep – Xenophon credits them with formations 100 men deep (and forming into blocks 100 men wide by 100 deep, the 'myriad' or body of 10,000, of which Greeks believed the Persians deployed so many). So far as we know, the eight-rank (or so) depths typical of hoplite formations may have been on the shallow side in comparison with other ancient armies, and certainly do not make the Greek phalanx a remarkably or unusually deep formation. The twenty-five- or fifty-deep Theban phalanxes were more unusually deep, but not disproportionate to other armies, if Greek descriptions of Eastern armies are at all accurate. In more recent periods, firm figures for Medieval armies are again mostly lacking, and it is only for the 'pike and shot' armies of the Early Modern period that we again have firm figures, and as they used drill manuals based on the Hellenistic tacticians, depths of around eight men are again typical, though as firearms became more widespread, depths were reduced to allow more men to shoot. We should also note that cavalry (which as we have seen were explicitly said to be unable to close up into a single mass) also formed multiple ranks deep. Polybius tells us that 'to be really useful cavalry should not be drawn up more than eight deep' (Pol. 12.18.3). While Byzantine cavalry, for example, formed more shallowly – perhaps only four deep, according to their drill manuals – these are still far more ranks than could have reached the enemy with weapons, and they cannot, being on horses, have contributed to any pushing by the whole formation, whether in the advance or in combat.[42]

This being so, it is strange that some modern authors talk about the depth of Greek phalanxes as if it is a unique phenomenon that requires some special explanation, and even lapse into sarcasm when discussing the apparent pointlessness of having rear rankers if they were not there to push. Even one of the chief opponents of the literal scrum suggests that 'if these [pushing] expressions were meant figuratively, the other ranks might simply have stood by, waiting their turn'. Yet surely three-quarters or more of the strength of any army would not have been employed just to stand by (and 'waiting their turn' is not a likely role for them, as we will see). We must seek another, more active task for them to perform. We must keep in mind at all times that these arguments will apply to all infantry and cavalry who formed in deep formations, though here I will consider the matter with particular reference to hoplites, for obvious reasons.[43]

A number of reasons for infantry to form in deep formations have been proposed. The simplest is that suggested by the quote above, that they served as reserves or replacements for those in the front ranks (waiting their turn to fight). When front-rankers were killed or wounded, the man behind would be expected to step forwards to take their place, continuing to fight and keeping the front rank continuous. Certainly this will explain the existence of a second rank, and is indeed expressly given as the task of the second rank in the Hellenistic tacticians (for example, Asclep. 3.6). A third rank would also be necessary for the same reason, to replace a fallen second-ranker. However, the figures for casualties in hoplite battles suggest that it would be very unusual to ever need more than three ranks to provide replacements – losses are in the region of 5 per cent for the winning side and 14 per cent for the losers, and would mean that the winner in the fight might expect to lose no more than half of their front rank, and the loser only about one-and-a-half ranks (and many of these losses would be in the flight and pursuit). These figures are of course averaged across the whole army, while in reality there would have been local successes and failures leading to locally higher or lower losses, but even so, it is hard to see how more than four ranks at the absolute outside would have been needed to provide casualty replacements.[44]

A related possibility is that front-rankers, when tired out but not killed or incapacitated, could drop back through the ranks and be replaced by men behind, in a manner analagous to, but on a smaller scale than, the Roman manipular line replacement tactic. There are several problems with this idea. One is that the file was structured with the best men at front and back, so replacing the file leader with less capable men would mean weakening the phalanx, just when greater resilience was needed in a hard fight. Another problem is that assuming the phalanx formation was fairly close with narrow file intervals, it is hard to see how there would have been room for men to filter forward and backward between the files. This would apply particularly in the Macedonian phalanx, where file intervals could be as narrow as half a metre, men were armed with 5-metre-long pikes and the formation was typically sixteen ranks deep. It is highly improbable that tired front-rankers could retire to the rear in such a formation, yet depth (sometimes up to thirty-two ranks) was still considered valuable. A third (and to my mind clinching) problem is that there is no reference to such manoeuvres in any ancient sources – not just for hoplites, but for infantry of any type or period. So while casualty replacements might account for four ranks or so, deeper formations require some other explanation.

The next most obvious explanation is the one discussed above, that the rear ranks were intended to drive on and perhaps sometimes push forward those ahead when advancing. As we saw, this clearly was at least one of the roles of the rear ranks in a Macedonian phalanx, and in those formations inheriting from the Macedonian phalanx, but there are good reasons to suppose that it was not the main role of rear ranks in looser formations such as Roman legions (not least Polybius' explicit statement, 18.30.10), which still formed up multiple ranks deep. The Macedonian phalanx depended on the mass effect of its rigidly held projecting spears, adopted a particularly close formation, and advanced to the attack (in all probability) at a slow, regular march. As such, the rear ranks certainly had the tasks of pushing forward to maintain the forward momentum of the advance, denying the front-rankers any opportunity to avoid contact and keeping the formation close and compact. Such roles might also have applied to some extent to the rear ranks of a hoplite phalanx, and as we saw, Xenophon does suggest that the last rank at least was expected to perform these tasks, and indeed rear ranks of Roman infantry, of cavalry and of other forces could have a similar role, provided we do not insist on physical shoving. There can be little doubt that these are important functions of rear ranks in deep formations throughout history, right into the era of firearms when infantry might still advance to the attack in very deep formations, much deeper than would have been adopted if use of weapons was the only consideration. But hoplites were different in important respects from Macedonian phalangites, relying on individual fighting with spear and shield, not on a mass advance with levelled pike. Furthermore, as argued above, this does not mean that the rear ranks were expected to keep pushing once contact was made, thus (inevitably) forming a crowd crush or scrum. The Macedonian phalanx could not have fought this way because of the 5-metre pikes, and hoplites would not have because all the evidence suggests they continued to fight with their weapons. So we can certainly accept that rear ranks were intended, if not for 'shoving', then for maintaining forward momentum by their presence during the advance, and this could well explain the preference for deep formations up to the typical eight ranks or so; but it is hard to see how any additional advantage could be gained by any deeper formation, least of all twenty-five, fifty or 100 ranks of depth. Such extra ranks can have added little to the forward momentum or compactness of the formation, nor made it any harder for the front-rankers to avoid contact, so this explanation cannot be the whole story.[45]

Another explanation arises from the fact the best men were to form the front and rear ranks, with, presumably, less brave, fit or capable men in between. Assuming 'the best' made up some limited proportion of the whole, then the number of 'best' available would have limited the frontage, and therefore determined the depth, of the formation. An eight-deep formation with two ranks of 'best men' (front and rear) would have 25 per cent of its strength assigned to such superior status. If this proportion was typical, then formed four deep, of necessity half of the front and rear rank (or all of one or the other) would be made up of 'lesser' men. The deeper the formation, the fewer brave, experienced, fit, well-trained or well-equipped men would be required to achieve the same overall level of combat efficiency. This might well explain why infantry considered (by the Greeks) of lower quality adopted deeper formations. It also probably reflects early Archaic or Homeric warfare, where a front line of well-equipped *promachoi* was backed up by a mass of retainers and followers. It does not explain why the Thebans, for example, who especially in the fourth century were considered (or considered themselves) to be excellent hoplites, adopted particularly deep formations, although the early role of the Sacred Band in providing the front rank would be relevant, by fixing the maximum width of the formation; if the front rank was 300 men strong, then all other, less capable, men would form up behind them to a depth determined by their overall numbers.

We have noted the role of rear ranks in forcing the front ranks to continue into contact, and how this added to the effect of a confident advance. By making clear a formation's willingness to advance to full contact, the hope was that an enemy might in turn lose confidence and so lose formation, or even break and run, before contact was made. This is, I am sure, the main significance of Polybius and the tacticians' observation that the rear ranks denied the front any possibility of flight – in the game of chicken that was the advance to combat, a very deep formation was, by virtue of its depth, almost certain not to flinch. But what was true in the advance would also be true in combat, as a deep formation behind the front ranks while they were fighting would limit the possibility of their losing heart and turning to flight, both by its physical presence as an unmovable mass behind the front-rankers, and also by the psychological benefit of having a close, steady body of friends to one's rear while fighting. Some modern historians express scepticism of this idea, characterizing the rear ranks as 'standing idly around'. This makes a rhetorical point by deliberately obscuring the actual role of the rear ranks, which is to encourage those in front by word if not by deed, as we have seen in examples above, and by their comforting physical presence. The rear ranks, as well as preventing the front ranks from fleeing or flinching, and as well as providing the psychological comfort of a dense mass of friends behind the fighters, could have verbally encouraged the front-rankers – the role assigned to rear-rankers and to cavalry behind a formation in Xenophon's *Cyropaedia*, and observed also with the Romans at Zama. Motivational speeches from the rear ranks were not required, but encouraging shouts, warnings, battle cries and exhortations might all have been of value in bracing the front ranks' courage to face combat. (We should note though that Hellenistic and Byzantine manuals suggest the men should remain silent, perhaps to allow orders to be heard, at least during the advance.) The psychological and morale benefit of such presence and encouragement was undoubtedly of great importance, and I will return to this idea below. However, it is also difficult to see how very deep formations provided a greater benefit than merely sufficiently deep ones.[46]

Another possible purpose of deep formations reminds us of Xenophon's observations in the fight against the Carduchians (Xen., *Anab.* 4.8.11), that a phalanx advancing over broken ground was likely to develop many gaps, which would be damaging to the men's morale, and that columns would avoid this danger by making it easier to go around obstacles, so replacing unintended, morale-damaging gaps with intended and harmless spaces between columns. This reflects, as has often been observed, the similar division of the Roman Army into distinct bodies (maniples in the Roman case) with gaps between them, which was probably adopted in part to facilitate battle on uneven ground and which famously contributed to Roman victories over the Macedonian phalanx. The problem with this as an explanation for hoplite depth is that most of the time, hoplite armies did not form separate columns, they formed a phalanx, a more or less continuous linear formation, whatever its depth. The ability to go around obstacles would therefore not have applied most of the time. It is true though that a deeper line might be less seriously disrupted by local obstacles than a shallow line, as the hoplites could flow around the obstacle and close up again on the other side, without at any time leaving a gap all the way through the formation from front to back. This may well have been a useful feature of a deep formation, along with the related way that deep formations allowed a given number of men to fight on a narrower frontage, thus requiring a smaller amount of flat open ground, and also reducing the command and control difficulties inherent in very wide formations. A deep formation could also survive a greater loss of cohesion in the advance – due to men running at different rates or with different degrees of enthusiasm – without being completely broken up, for similar reasons: the scattering of multiple ranks would still provide a dense crowd of men, while a shallow formation which became scattered this way would naturally arrive at the enemy with much-reduced strength and more dangerous gaps right through the formation. Such explanations carry some force, but cannot be the whole reason why deep formations were used, given the benefits available from forming shallow and wide and so outflanking the enemy.[47]

We should also return to the idea of pushing, though in the sense here of 'bracing' and 'holding firm', rather than shoving front-rankers into the enemy or building up a crowd crush. What happened when infantry equipped for pushing with shields (individually) met men not so equipped is described by Xenophon at Thymbrara, as we have seen when the Egyptians 'advanced and pushed':

> '[B]ecause the Persians had to hold out their little *gerrai* clutched in their hands, they were unable to hold the line, but were forced back foot by foot, giving and taking blows, until they came up under cover of the moving towers.' (Xenophon, *Cyropaedia* 7.1.33–34)

At this point, a volley from the towers checked the Egyptians while the Persian rearguard compelled the front-rank men ('with drawn swords' – another form of encouragement) to stand their ground. Being 'forced back foot by foot' (*epi poda anechazonto*) is surely the same phenomenon as the Athenians being 'pushed back bit by bit' at Delium (*osamenoi kata brachu*, Thuc. 4.96.4), where both sides were similarly equipped but one proved superior to the other in the fighting. In Xenophon's mind, such a retreat could be halted by the encouragement of a rearguard and by missile-equipped towers, while at Delium there was no such rearguard and so the Athenians continued to be 'pushed back'. Here we come to the essence of the way close-order infantry fighting worked. Front-rankers might fight with sword or spear, or clash shields, or advance with shield held up and braced by the arm or shoulder, or with spear rigidly advanced (a variation which was perfected in the Macedonian phalanx). In each case, the opponent could stand his ground to fight back and parry with shield, spear or sword. But a man who felt himself overmatched by his opponent, or by his opponent's equipment, or who felt his own courage or willingness to die for his city failing, would naturally tend to step back, to avoid immediate death and danger and to restore the safety distance – just a metre or two for spear- or sword-equipped men – between the lines. Such backward movement was itself an admission of inferiority, and would tend to encourage the enemy while further discouraging those falling back; we still today are familiar with the idea of being 'on the back foot', which would have applied very literally in close-quarters infantry combat. In addition, once one man began to fall back, his neighbours to either side, seeing or sensing his absence, would start to feel themselves exposed and would be afraid of unseen attackers striking them from the side (the whole point of a close-order formation with a continuous front rank was to deny the enemy the chance for such unanswered, unblocked attacks from the sides). In these circumstances, these men too would step back for their own safety, and so a backward movement could be transmitted from man to man along the line. In a very shallow formation, just two ranks deep say, the second-rankers, themselves already close to the point of danger, might also take a step or two back ('forced back foot by foot'), to keep themselves out of reach of the enemy or because they began to feel their side overmatched and the fight turning against them. Even if they preferred to stand their ground, if the man ahead actually stepped back into them (or even beside them in the gaps between the files), it would take great resolve to hold firm, with the evidence of the fight going badly and the immediate danger of the closing enemy in front. The fewer the ranks, the more likely it was that any such retrograde movement at the front would be transmitted, involuntarily, back through the ranks, with each rank on the back foot, stepping back, and the enemy following up. A shallow formation would soon find itself stepping backward not just once or twice but continuously, with the retrograde movement being transmitted down the line side-to-side as well as front-to-back, as men felt themselves exposed and sought to distance themselves from the enemy – without turning to flight – and seek the safety of the mass behind them.

A deeper formation would resist this backward movement in two ways. Firstly, the very depth of the formation would tend to insulate the men at the back from the sense of danger and of impending defeat that might infect the front ranks. With several ranks of friends between them and the enemy, and with the backward movement of the men in front absorbed by the rank spacings of the formation ahead, men eight ranks back – or twelve, or twenty – would be less immediately aware of the danger to their front, less jostled by the backward movement of the men ahead and less inclined to feel afraid of threats from their left and right. A deep formation therefore acquired an inherent solidity and resistance to the tendency to retreat; a deeper formation was harder to 'push back' because of its very depth. But in addition, as men tried to step backward from the front, we can well imagine that those behind, 'encouraged' by ranks further back and by their officers to stand firm, could brace themselves and their shields and present a solid front which would tend to physically prevent the front ranks from retreating, just as the rearguard compelled the front ranks of Xenophon's Persians to fight or the front ranks of the Macedonian phalanx were denied any possibility of retreat by the ranks behind. Physical shoving and jostling might well enter the picture here; not a scrum or crush by the whole formation, but bracing, standing firm, pushing back on their own men, not the enemy, to give the formation solidity and prevent it being forced ever further onto the back foot.[48]

If a phalanx did find itself edged backwards, then the enemy, according to the dynamic stand-off model of combat, might not follow up, but might instead allow a safe distance to open up between the two sides. There would follow a period of threatening and the exchange of insults until bolder spirits took over and, perhaps in localized flurries, the lines would close again to combat. However, a more confident, aggressive and determined enemy might not allow this lapse into low-intensity combat, but would instead immediately follow up and continue the fight at high intensity. This is where the depth of the superior, winning formation would come into play. In a shallow formation, it would be very much up to the personal preference of the front-rankers as to whether combat was resumed, and the ranks immediately behind, themselves close to the point of danger, might be no more inclined to maintain combat than the more experienced front-rankers. But if the formation was deep, then it would be much easier for the rear ranks to continue stepping forward, encouraged by the file closers at the rear and being not yet themselves in any immediate danger, while seeing or sensing from the movement of the lines the discomfiture of the enemy ahead. The front ranks need not have been shoved forward by the men behind with a shield in the back, as in the usual scrum model, to feel the forward momentum of a mass of men behind, which would be akin to being at the front of a slow-moving crowd rather than being involved in a crowd crush. I am sure many of us have personal familiarity with being in a moving crowd of this sort, and of how there is a definite sense of forward momentum even without actual bodily contact. This, I believe, is what the Hellenistic tacticians meant by the ranks behind 'bearing forward with their bodies', and the same phenomenon could have applied to hoplites. Sensing the fighting at the front going well and the enemy falling back, the rear ranks could continue stepping forward, so that the front ranks – whether they wished to or not, and irrespective of their own levels of aggression – would also step forward to continue the fight and prevent a stand-off developing. This way the forward and backward motions of the two phalanxes would be continued, and the psychological pressure on the losing side would be maintained and intensified, with no opportunity to step back and draw breath. The attacking formation would 'push forward' in a way that is not entirely

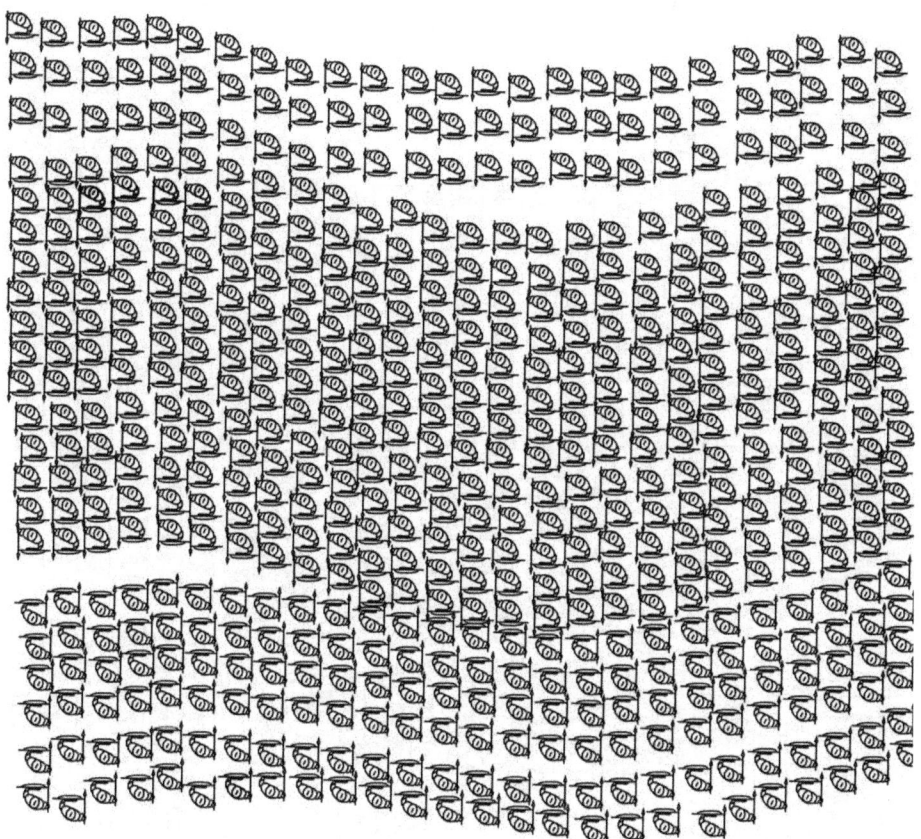

A deep formation (top, sixteen ranks) pushes back a shallow formation (bottom, six ranks)

figurative or metaphorical. Once the front ranks were fighting in place again, the forward movement would temporarily ease and the rear ranks would no longer surge forward, though they would be ready to do so again as soon as they sensed the enemy giving way.

We must also understand the way in which a phalanx, or any close-order infantry formation, would be defeated. It is unlikely that the fighting would go uniformly in favour of one side or the other along the whole front – there must in a fairly equal fight have been points where one side was getting the upper hand, and points where it was being defeated. At the points of defeat, a backward surge might tend to cause the phalanx to billow backwards, a movement which would be transmitted along the line to left and right but not, at first, to the whole formation. Eventually, a critical point would come, a collective loss of nerve perhaps just at one point in the line. So long as there were rear ranks standing firm and resisting, perhaps by physical bracing, any backward movement of the front ranks, the line would maintain its cohesion. If an extended section of the

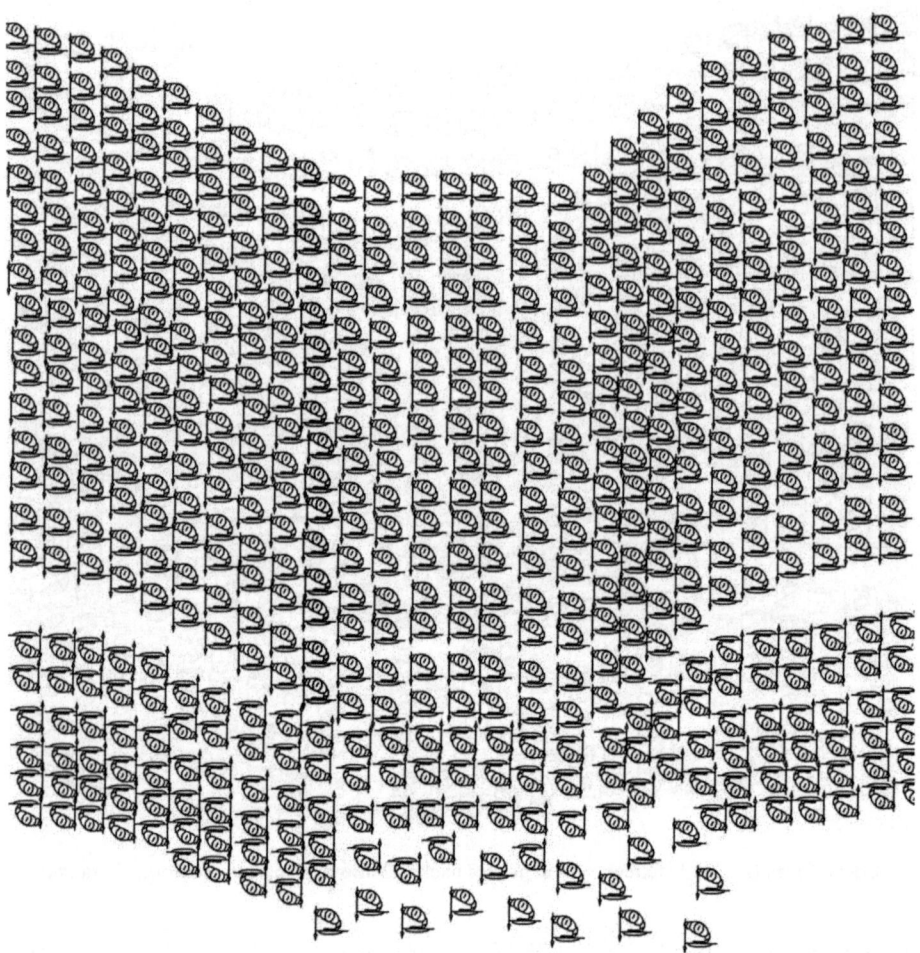

Trope: the shallow formation gives way from the rear at the point of greatest psychological pressure

line was forced back, it might retreat for some distance while remaining basically cohesive, and without losing formation entirely or ceasing to offer resistance. But eventually, the stresses and tension would become too great, and at some point along the line, backward steps and the billowing back of the phalanx would turn into flight, some men turning their backs, the collective effort collapsing into every man for himself. It has often been observed that formations rout not from the front but from the back, and we can see why this would be so when the rear ranks were what held the front ranks in place, which was why Greeks recommended placing some of the best men at the rear, to hold the formation together.[49]

As men at the front began to panic and struggled harder to get away, and as men at the back, instead of resisting this flight, joined it, the ability of the formation to maintain cohesion would be lost and the result would be the 'turn' and 'breakthrough' to which Xenophon and other authors refer. We can readily imagine that once this happened at one point, panic and flight would rapidly be communicated to left and right, as the victorious enemy poured into the opening gap, and as men to each side of the gap felt their flanks exposed and themselves in immediate danger. Panic is infectious, and men who have lost the support of those to left, right or behind will all the more readily turn to flight. Only the bravest or most reckless could keep their heads while all around were losing theirs. A breakthrough at one point would thus rapidly spread up and down the line, leading to the collapse of a whole contingent or even a whole army.[50]

This is why a deep formation, by delaying the point in time at which even a losing fight became a breakthrough, could offer great advantages in combat. When the Athenians and Syracusans fought outside Syracuse in 415, Thucydides records that:

> 'The armies now came to close quarters, and for a long while fought without either giving ground. Meanwhile there occurred some claps of thunder with lightning and heavy rain, which did not fail to add to the fears of the party fighting for the first time, and very little acquainted with war; while to their more experienced adversaries these phenomena appeared to be produced by the time of year, and much more alarm was felt at the continued resistance of the enemy.' (Thucydides 6.70.1)

Leaving aside the inclement weather, we can well understand the dynamics here. The Athenians, better trained and more experienced, would hope to break through the Syracusan line at some point, hopefully quickly; but every minute that passed in which this had not happened increased the danger that in some place the Athenians' own line might be broken through (in this case, the Argive contingent of the Athenian army eventually 'pushed back' those opposite them while the Athenians 'broke' their own opponents, Thuc. 6.70.2). Men in the front ranks concentrating on fighting those in front of them, men in the centre ranks with a very limited view of what was going on around them, or men in the rear rank concentrating on holding the phalanx firm, could not know exactly what was happening to their left and right more than a few metres away. Nevertheless, shouts and screams, and the sensation of movement as the phalanx to each side surged forward or back, would offer the constant fear that their own side might collapse somewhere along the line. Collapse and breakthrough would rapidly turn to rout, and it was in the rout – with formations broken, mutual support lost and backs turned to the enemy – that most men were killed. Hoplites must have known that their chances of living through a face-to-face battle were good, but their chances of surviving a rout and pursuit (especially against an enemy strong in cavalry or light infantry) were far worse. There would have been a constant fear that the line would break at some point the longer a fight went on, with disastrous consequences. Small wonder then that it was the point at which the enemy phalanx first 'turned' and was broken that was marked by the most consistent and ancient of Greek battle rituals, the setting up of a trophy.

This, I believe, is the main reason why deep formations were always used, and sometimes formations even deeper than the norm. But I do not believe that any single factor was decisive in offering deep formations an advantage, as all the factors considered above will have been relevant in making deeper formations more effective. Rather than seeking a single reason for deep formations, we should accept that all of the above reasons could apply, to a lesser or greater extent depending on circumstances. This leads to another important point to keep in mind: that it was not certain that deep formations did offer a decisive advantage, particularly given the concomitant risk of being outflanked, and there was doubtless an ongoing debate among Greek military thinkers as to whether deeper formations were more effective, and if so how deep was the ideal. We see this in the debates before battle as to how deep to form the phalanx. Deeper formations offered some safety against breakthrough, but they also meant a shorter front and so a danger of outflanking, and it was an open question as to whether the benefit of depth outweighed the risk to the flanks (which might matter less if the flanks were occupied by allied or inferior forces, not one's own citizens). Xenophon's *Cyropaedia* in particular can be seen as an ode to the shallow formation, exaggerated to such rhetorical effect that Xenophon almost grants no value at all to deep formations. Xenophon was no lover of the Thebans with their longstanding preference for deep formations, and his resentment of the Theban victory at Leuctra against his favoured Spartans is obvious, a victory he attributes to 'the Theban mass' (Xen., *Hell.* 6.4.14), using a pejorative word, *ochlos*, often applied to crowds or mobs. Epaminondas for his part found a way – with the oblique line and by attacking the enemy right – in which his deep formation could be victorious before the enemy had a chance to work round against his flanks.

This ongoing debate – which remained unsettled across four centuries – is analogous to the debate over linear or columnar infantry tactics in late eighteenth- and early nineteenth-century AD Europe, where the successes of French columnar tactics in the Revolutionary and Napoleonic wars were apparent. The case as to which was better, column or line, remained very much open, with British linear tactics in particular winning a string of victories over French armies. A similar debate can be seen in earlier centuries, for example in the writings of Roger Boyle, Earl of Orrery, in the seventeenth century. Boyle particularly admired the military expertise of the Greeks and Romans, and recognized the continuity between ancient times and his own:

'And though they Fought with their Files exceeding deep, which we with much Reason have Alter'd; yet as to the Main, we owe to them most of our Knowledge: And the Difference seems Little more, than between Old-fashion'd Plate, hammer'd into New; where though the Form is chang'd, yet the Substance remains.'

He continued that the reason for the deeper formations of the ancients was that 'use of all Fire-Arms, and of Cannon, were Intirely unknown to them, which has much alter'd the manner of making War'. Boyle went on to argue against even the practice of his own day, of 'the drawing up our Shot, and Pike, six deep; and our Horse, three deep', and he 'judged it best for me to Fight my Foot four deep, and my Horse two deep', thereby allowing them to extend their frontage:

'For I was fully satisfied, that it was likelier I should be worsted by the Enemy, if he fell into my Flanks and Rear, holding me also to equal Play in the Front, than if four Ranks of my Foot should be broken, or two Ranks of my Horse, that the third Rank of the Horse, and the fifth and sixth Ranks of my Foot, should recover all again.'

Boyle could make his personal recommendation of fighting in four ranks for infantry ('after having as thoroughly weighed all the Arguments for and against it, as my weak judgment could suggest to me'), but he also observed that:

'I must confess, that he who makes such an alteration in Military Discipline, (unless he be a Sovereign Prince, or have sufficient Orders to do it) ought to resolve, his success only must Apologize for it; that is, to be victorious, or be kill'd.'[51]

This is always the difficulty with experiments in this field – the only way to test an innovation is in combat, where a failed experiment could mean disaster for the city and likely death for the commanding general and many of his fellow citizens. This is not a situation which favours innovation or 'blue sky thinking', and conservatism in military matters is very much to be expected, especially in a setting such as ancient Greece where most cities lacked a strong central military authority or tradition of military professionalism, where all cities fought with similar armies on similar terms, and were broadly satisfied with the social and political status quo that went with this. A depth of around eight ranks having been settled on in the early days of the phalanx as a good compromise between solidity and width, we can well imagine that there was no great incentive to experiment with other possibilities. The Thebans tried deeper formations for the best part of a century before eventually achieving victory under Epaminondas, yet it cannot be claimed that the deep formation was the sole cause of their victory even then.

454   The Greek Hoplite Phalanx

So I think it is apparent that there were many reasons why deep formations might be favoured that have nothing to do with a mass shove or hoplite scrum; but also that it remained an open question (and one not liable to a definitive answer) as to what depth was best.

Duration

One final aspect of the nature of hoplite fighting remains to be considered, that of how long such combat might continue and how long a hoplite battle could take. The question of the length of hoplite battles (and indeed of all battles in the period before accurate timekeeping and detailed written records, that is broadly before the eighteenth century AD) is dogged by some basic methodological uncertainties. Ancient sources do sometimes contain references to the length of battles, but only in the vaguest terms – often that a battle continued 'for a long time' (with no indication of how long that might be) or that it carried on 'until evening' (with no mention of when it began). The ancients lacked portable devices for telling the time and could only judge time (in the field) by the movement of the sun, so the smallest increment of time available is an hour or so, one-twelfth of the available daylight on any given day. Even where there are indications of elapsed time, we cannot tell when a battle was thought to begin – whether when armies marched out of camp, when they completed deployment and began to advance or when the first blow was struck – nor when they were thought to end – with the first breakthrough, when the whole line was in flight or after a possibly lengthy pursuit.[52]

As an example of the difficulties involved, Thucydides records that at the Battle of Olpae, the Ambraciots defeated the force opposite them, pursued them to the nearby town of Argos, returned to the battlefield, found their main body defeated and made their own retreat to Olpae (about 5km from Argos), then remarks that the 'battle did not end until late in the day' (Thuc. 3.108.3). Presumably, all these pursuits, returns and flights are included in 'the battle'. A similar example of such a prolonged battle, including all the preliminaries and aftermath, is Cunaxa (401). Xenophon describes events in detail, from the first reports in the late morning that the enemy were advancing (Xen., *Anab.* 1.8.1), to the afternoon when they first came in sight (1.8.8), through the deployment and preparation of Cyrus' army until 'at length' (1.8.17) the armies were a few hundred metres apart, when the Greeks advanced, drove off their enemies without striking a blow and pursued for thirty stadia (about 5.4km), while the royal army defeated and killed Cyrus and plundered the camp. Both armies then reformed on the original battlefield, marched against each other, the royal army withdrew again to a nearby hill and the Greeks advanced again, driving them off. 'At about this time the sun set', Xenophon tells us (Xen., *Anab*. 1.10). So this was a long battle, but one in which (if Xenophon is to be believed) the duration of actual combat involving the Greek phalanx was precisely zero minutes.

There are also indications of duration for different types of fighting. Low-intensity combat, especially involving light infantry, might last for a long time because casualties were low, fighting produced no decisive result, and the physical and psychological pressure on men able to drop back out of contact to rest and recover was less than that on men in static close-order combat. Even where hoplites were involved, not every encounter was a pitched battle between phalanxes: of the Spartan attempt to force a landing on Pylos, Thucydides (4.13.1) says that 'their attacks [continued] during that day and most of the next', but these are attempts to make a seaborne landing, so how much of that time was spent actually fighting is unknown. In the later battle on the island of Sphacteria, the fighting went on 'for a long time, indeed for most of the day' (Thuc. 4.35.4), but this was not a regular hoplite battle either, rather an attempt by light infantry backed by hoplites to force a fortified position and to grind down with missiles the beleaguered Spartan force. Some ancient writers (Diodorus, to name names) were also less than careful about their temporal designations, employing stock phrases such as 'for a long time' to describe numerous battles, and often doing so to indicate the battle's importance and hard-fought nature rather than as an accurate indication of its length. As a result of all these factors, although careful collections have been made of all the indications of battle duration in ancient accounts, they do not produce definitive answers and sometimes rather carelessly fail to distinguish between different types of combat in wildly varying circumstances that are not applicable to hoplite combat specifically.[53]

An alternative approach is to take comparative material from other eras of combat to draw general conclusions about how long men could endure in close-quarters combat. The problem here is that the period for which we start to have more precise and accurate records of duration is also the period in which the widespread adoption of firearms completely changed the nature of battle. In particular, firearm-equipped armies from the eighteenth century AD onward deployed in more dispersed order (with larger gaps between units, although individual units themselves were still in close order) and in multiple lines, with large proportions of an army held in reserve, none of which was normal practice in antiquity, and especially not in hoplite battles, as we have seen. It is therefore difficult to draw any conclusions from the duration of more recent battles. Arguments from the physical capabilities of combatants are also indecisive. Modern re-enactors tend to find that a few minutes of dynamic close-quarters combat is exhausting, which might suggest a very low upper limit for the length of a hoplite combat, tens of minutes at the absolute most and certainly not hours.

But it may be that real combat, especially where the life of the combatants was at stake, was far more tentative than modern reconstructions would suggest, and therefore could be sustained for longer. We must also be careful about assigning physical limitations to people in the past based on what we know of modern physiology, when the endurance and strength of ancient people who were used to lives of physical exercise and hardship from birth may have been much greater than we would suppose.[54]

In addition, if the dynamic stand-off model is correct, men in contact would mostly not be in combat, but just outside weapons range and not exerting themselves at all (other than to think of new insults to shout at the enemy), a situation which could be sustained for a long time; indeed, the dynamic stand-off model was developed in part to account for combats apparently lasting for several hours. That they may have lasted so long is suggested by a 'battlefield clock' analysis of ancient battles, that is by comparing known grand tactical movements of forces with combats in another part of the line. If, for example, two opposing wings engage, while on the other wing one side flees without contact, and if the victorious wing is then able to march across the battlefield to take the other wing in the flank while it was still engaged, then we can make (very approximate) estimates of the times such movements must have taken, and can infer that the fighting on the contested wing must have taken an equal amount of time. Using this method, combat durations of hours, not minutes, have been suggested for Punic War battles between Roman and Carthaginian armies. A related consideration is the lethality of combat. If combat casualties were as low (to the winning side, and to both sides before flight and pursuit) as we believe they were (around 5 per cent at most), then either combat must have had very low lethality or very short duration. The dynamic stand-off model attempts to square the long duration suggested by the battlefield clock with the low lethality suggested by casualty figures, by assuming that most of the time opposing lines were not actually fighting each other. It is indeed difficult to see how the traditional picture of hoplite combat, in which phalanxes are face-to-face and constantly stabbing with spear or sword – still less the scrum or crowd crush models, in which phalanxes are at maximum physical effort, within biting distance of each other, pinioned immobile and unable to use their shields or take any evasive action – can be reconciled with the idea that hoplite combat might last for hours while producing low casualties.[55]

As far as actual indications of the duration of battles go, the most often quoted case is that of the Battle of Pydna (168), which Plutarch (Plut., *Aem.* 22.1) tells us was unusually short in taking one hour. All the usual problems apply to this duration – we do not know at what point in proceedings the battle is reckoned to have begun (starting the stopwatch) or when it ended (stopping it). The hour may include potentially large amounts of marching before the main fighting, and flight and pursuit after it, so this tells us very little about the time spent actually fighting. Plutarch was also most familiar with Roman battles, which were more similar to those of the age of firearms in that Romans habitually fought in three or more lines with large parts of their force held in reserve, and could replace or reinforce a wavering line, avoiding the more sudden collapses that were such a feature of hoplite battles. It may therefore be that Plutarch was making a comparison with Roman battle, which is not really helpful. The same applies to the assertion by the late fourth-century AD Roman writer Vegetius that battles were 'commonly decided in two or three hours' (Veg., *Mil.* 3.9.2) – we do not know at what point a battle was considered to have begun, when it was decided or how much of this time was actually spent fighting. We might conclude that Roman battles – reckoning their start to be after deployment was complete but before a blow was struck, and their conclusion when one side is definitively beaten and in rout all across the field – might take from one to three hours. However, this time could include preliminary skirmishing, the engagement and possible reinforcement and replacement of multiple lines, and grand tactical redeployment such as of one wing to another, and as a result it tells us little about the duration of actual hand-to-hand combat.

There are some indications of the duration of hoplite combat, such as at Solygeia, where Thucydides records the two armies engaging, and 'after holding on for a long while [*chronon polun*] without either giving way, the Athenians ... at length routed the Corinthians' (Thuc. 4.44.1). This indication of duration must apply specifically to the actual fighting, but how long is 'a long while'? We might reasonably suppose that 'a long while' for two phalanxes to be engaged in close combat is not the same thing as 'a long time' for a succession of battlefield manoeuvres, flights and pursuits. The same applies to the fighting at Syracuse in 415, where Thucydides tells us that after preliminaries including indecisive skirmishing, sacrifices and advances, the two armies 'now came to close quarters [*en chersi*], and for a long while [*epi polu*] fought without either giving ground' (Thuc. 6.70.1). In this instance, clearly it was specifically the fighting that lasted 'a long while', but how long 'a long while' might be in the circumstances is not stated. As we saw above, on this occasion the Athenians were increasingly alarmed by the continuing Syracusan resistance, so a few minutes might have felt like an age.

There are also occasions where one side runs away before contact is even made, so that there is no fighting at all. Cunaxa is one example as we saw above, as is the victory of Archidamus' Spartans over an allied army in 368, where 'only a few of the enemy waited till his men came within spear-thrust; these were killed, and the rest were cut down as they fled' (Xen., *Hell.* 7.1.31), the Spartans taking no losses, or Mantinea (418), where another Spartan king charged another allied force and 'instantly routed them; the greater number not even waiting to strike a blow, but giving way the moment that they came on' (Thuc. 5.72.4). These occasions where forces flee without striking a blow, combined in the same battle with occasions where forces do fight, provide an opportunity to apply the 'battlefield clock' to hoplite battles, though it is difficult to draw many firm conclusions. A good example is the Battle of the Nemea (394), as we saw in the previous chapter. This was a large battle by the standards of the time – Xenophon (Xen., *Hell.* 4.2.16) gives the Spartans and their allies 13,500 hoplites plus some cavalry and light infantry, and the Athenians, Boeotians, Argives and allies 24,000 hoplites plus cavalry and light infantry, although the precise numbers are not of great importance. The battle proper began with the Spartan goat-killing and charge at one stadium (180 metres) apart (Xen., *Hell.* 4.2.20). When the armies came into contact, all the Spartan allies except the Pellenians fled, while the Spartans outflanked part of the Athenians and defeated them almost without loss. The Spartans then marched across the back of the pursuing Athenians, and also caught each of the Argives, Corinthians and Thebans in the flank as they returned from their pursuit. We do not know how far each of these contingents pressed their pursuit, nor can we tell how quickly the Spartans overcame the Athenians, but given that they took them in the flank and themselves suffered very few losses, this must surely have been another very swift combat. The total duration of the battle might have been fairly long – the Spartans effectively marched across a battlefield which might have been (taking 24,000 men in 1 metre intervals and an average of ten deep, the Thebans being deeper than usual) up to 2.4km across, which in itself could have taken an hour. But the battle surely makes little sense if the combats themselves were prolonged; the Spartans outflanking and routing the Athenian left took the same time as the Athenians defeating and pursuing off the battlefield the men opposed to them. Combat durations of hours are certainly out of the question, and they probably are better measured in minutes.

The Battle of Mantinea (418) offers a similar example. The Spartan left was defeated by the Mantineans 'as soon as they came to close quarters [*en chersin*]' (Thuc. 5.72.3) and was pursued to their wagons, while the Spartan centre defeated those opposite them without striking a blow (5.72.4). The Athenians on the allied left were thus outflanked on left and right and fled, while the Spartans turned on the victorious Mantineans. There are no long and stubborn fights here, and indeed it is hard to discern any fights at all, as opposed to pursuits of broken or fleeing enemies. Each of the engagements was over in minutes or seconds, although the totality of the movements on the field will have taken longer. Of course this was a small battle (fewer than 10,000 combatants on each side), equivalent to only a small component of one of the great battles of the Hellenistic period or the Punic Wars. We should not be surprised then to find the total duration of the battle extending only to tens of minutes at most, and the combats themselves, such as they were, measured in minutes or seconds.

There is a modern tendency to believe that major battles, given their importance, must have lasted a long time, a tendency also common in the ancient world among some historians, with talk of battles lasting all day. But we can see that many hoplite battles must have been over very swiftly, and many hoplite combats were over even quicker than that, before they began in fact. It is doubtful whether even two equally determined opponents could actually engage in combat for periods above ten or twenty minutes at the outside without exhaustion or mutual annihilation rendering the fighting moot. There is very little evidence available that would allow firm conclusions to be drawn, but it is my belief that many hoplite combats lasted only a few minutes, that the most stubborn and drawn-out combats may have lasted only ten or fifteen minutes at most (which would have felt like an age), and that whole battles, including marching to and fro across the battlefield in some cases, would still clock in at well under an hour. Roman experience (even where it can be clearly discerned, which mostly it cannot) is not strictly relevant because of the 'sudden death' nature of hoplite combat, fought in single lines without reserves and ended by a decisive rout with no opportunity to fall back and rally.

The nature of hoplite combat

It only remains to pull together the threads of this chapter and the last to offer a model of hoplite combat. Such a model must respect the evidence of the sources that we have, but without reading particular words or passages in forced or overly literal ways that go against what we know about infantry combat in other periods. It must also avoid the trap of seeing hoplites as something fundamentally different from other heavy infantry at the same time or since; hoplites certainly differed in some ways from their contemporaries, and were particularly effective at close-quarters fighting, but we have no reason to suppose that their combat abilities were different in kind rather than in degree from that of other heavy infantry in antiquity.

We have seen that hoplite phalanxes formed in linear formations typically eight deep, or whatever depth was agreed with allies or determined by the officers. They advanced towards contact, whether at the run or at a walk, and with varying degrees of loss of cohesion in the advance, according to training, enthusiasm and terrain. If one side greatly intimidated the other at this stage, the weaker side might break and run before contact, but the file closers would be doing their best to keep the phalanx together and moving forwards, and any tendency to slow to a halt from the front ranks could result in the ranks behind, urged on by the file closers, forcing the men in front forwards. If the efforts of the file closers were successful, and the file leaders were determined enough, the phalanxes would continue to close. Depending on the enthusiasm of both sides, three things might happen at this point. Less well-motivated or determined men, or those without a strong cultural tradition for 'getting stuck in', might halt outside spear range, close enough to exchange insults, and missiles if they had them, and for bold individuals and those around them to dash forward for brief flurries of fighting before falling back on the main body (in the manner of Homeric warfare, or according to Herodotus, the Persians) – but the Greek phalanx developed specifically to avoid this outcome. Alternatively, and ideally, they might come to a halt within spear range, a couple of metres apart perhaps, and begin fencing with their spears. In some cases, where both sides were fully motivated and determined, they could get even closer, the front ranks using their shields to try to force aside the shields of their opponents or knock them back off balance, and closing up to an intense struggle of shield bashing and close-range stabbing. If they fought first at spear range, then eventually spears would get broken and hoplites would take to their swords, and the ebb and flow of the fighting, in a space restricted by the presence of the rear ranks which would prevent any attempt by the front ranks to step back and increase the range, would also tend to make the combat increasingly close, likely resulting over time in a similar state of very close combat with shields, swords or spears. All such combat – spear fighting, shield bashing, sword play – went by the general name of *en chersi*, 'hand-to-hand' (or similar terms), as distinguished from the long-range missile fighting of skirmishers, and was occasionally called *othismos*, though this is not a specific word for a particular type of fighting, just a general word for what has in other ages been called 'the press', 'push of pike' or simply 'melee'. This sort of fighting would go on for a variable amount of time, depending on the skill and enthusiasm of the combatants and other circumstances, but probably only ever measured in minutes; fights said to have lasted 'a long time' might thus have lasted around ten minutes at the outside.

After some time spent fighting like this, two things might happen: one side might collapse as, with a general failure of nerve, the phalanx disintegrated from the back – this was known by various terms, most often *trope*, 'rout'; or a phalanx might retain cohesion but with the front ranks, feeling themselves outmatched, backing off, and the rear ranks, instead of standing firm, backing off also, so that the whole battle line flexed backwards, the opponents surging forwards in turn. The latter could be called 'pushing back' or 'being pushed back' (using an *otheo* or *piezo* verb), though such terms were also used for being repulsed or driven off generally, and for light infantry, cavalry and ships as well as hoplites, so there is no simple correlation between the word used and the type of combat – further qualification would be needed (such as 'pushed back slowly at first' or 'step by step') to identify this as a slow withdrawal. Different levels of success would also be met with in different parts of the field, and would not be uniform all along the line.

A failure of nerve at any one point would lead to a breaking of the line, and almost certainly the rapid rout of those hoplites stationed alongside the break, at least as far as the dividing line between this city contingent and the next (where a small gap between units might have the effect of a fire break, preventing the rout being transmitted further). A push back would also usually lead quickly to a rout, though in rare cases a line might have been pushed back some distance while still retaining order and facing the front. More usually, once a phalanx found itself on the back foot, without any reserves or rear support to halt the retreat, it was only a matter of time before a push back lead to a break at one or more points along the line, men losing their nerve and turning to run, and this loss of nerve would rapidly be communicated along the rest of the line. A complete rout might or might not have been followed by a pursuit. Often, a victorious wing would need to avoid pursuit in order to retain formation and wheel to fight other enemy contingents, provided discipline was great enough to prevent the men just running after their enemies to relieve pent-up emotions in a burst of killing, at least until the fleeing side got out of range.

I hope that this provides a plausible picture of the way hoplite combat usually worked, although of course any degree of certainty on this matter is impossible, and we can never know for sure whether our models of combat are totally accurate. This model, however, fits with the attested evidence and, importantly, is not very different from the nature of close-quarters combat that we see in other eras, requiring of the hoplite neither superhuman fortitude nor historically unique fighting techniques. It is a type of fighting that would be familiar to Egyptian, Assyrian, Persian, Carthaginian, Etruscan, Macedonian and Roman infantry contemporary with the hoplite and in the centuries shortly before and after. This is not to say that there were no differences between Greek hoplites and these other types of infantry, as clearly there were many differences of detail – in equipment, technique and above all in culture and mindset. Greeks considered themselves to be – and judging by their success in battle against Persians, actually were – particularly effective at this type of fighting, as shown also by the demand for Greeks as mercenaries. Etruscans and Romans, who used essentially the same equipment and the same style of fighting as the Greeks, also had a high level of success and a high reputation, although the Romans soon altered their fighting methods for reasons of their own.[56]

Two main reasons can be identified for Greek hoplites' high reputation. One is their equipment, which was heavier than that of much Eastern infantry at least. The hoplite shield was strongly built and would have offered good protection in a close-quarters fight, as well as being effective for the shield clashing and pushing that such fighting often called for, and being attached to the forearm it could be braced more strongly than a hand-held shield. Greeks also, especially in the earlier period, wore more bronze armour than their contemporaries, with bronze helmets covering much of the head and face, and at least partial bronze armour for the body, all of which would have offered good protection against wounds, as well as increasing the confidence of the wearer and their willingness to enter into such a dangerous form of fighting in the first place (this may indeed be more important than the actual protection offered). The second difference was mental and cultural. The relatively egalitarian lifestyle of Greek citizens and the sense of shared purpose among a socially and economically relatively homogenous group, whose political involvement in affairs of state gave them a personal commitment to the success of the state's endeavours, made them well suited to a style of fighting in which keeping formation (so as to protect one's neighbour) and standing one's ground (so as to protect the formation) were both essential features. Greeks depended on, and (at least sometimes) felt they could rely on, the shared willingness of their fellow citizens to stand and fight alongside them, not to seek temporary personal safety in retreat or flight at the expense of their neighbours. This combined with small advantages of equipment gave Greek hoplites tremendous self confidence, in a virtuous spiral of positive reinforcement, especially once they had won a few victories against the initially feared Persians. Greeks did not invent a new style of fighting, but they became experts at a particular existing type, and this expertise saw them dominate eastern Mediterranean battlefields (chiefly, it must be said, facing each other) for several centuries.

Eventually, however, Greek hoplites did meet their match, both at the hands of their old enemies, light infantry and cavalry – now wielded by a power that did not have the shared motivation to minimize the role of these arms – and also by a variant on the hoplite phalanx that was more revolutionary than anything that had gone before. The eventual decline of the hoplite phalanx will form the subject of the final chapter.

Chapter 9

Decline of the Hoplite

The Macedonian phalanx

The Greek hoplite had been the dominant force on the battlefields of Greece from his first emergence, some time in the Archaic period, well into the fourth century, and Greeks – as colonists or mercenaries – had settled much of the Mediterrranean coastline as far west as modern France and had played an important role in the wars of the Persian Empire. Men armed in the Greek fashion also ruled the battlefields of Italy for several centuries, until supplanted by a new and distinctive Roman method of fighting.

However, in the mid-fourth century, a new force arose in northern Greece, as the kingdom of Macedon – under its dynamic king Philip II and his son, Alexander the Great – began its meteoric rise to prominence, powered in large part by a new way of making war and a new form of heavy infantry. The details of the armament and organization of these new infantry have been dealt with at length elsewhere, so here it is necessary only to provide a brief outline. These new infantry were technically speaking also hoplites, the word for the Classical Greek infantryman being simply applied to them also (as armoured heavy infantry, the term was still fully applicable), and they also fought in a phalanx, but a phalanx with subtle but important differences from the familiar Greek formation.[1]

The Macedonian phalangite (a word used mainly by modern authors to distinguish such men from the earlier hoplites) was, like his Greek counterpart, a heavy infantryman, but his main offensive weapon, rather than a spear around 2 metres long held in one hand, was a pike (*sarisa* in Greek, usually spelled sarissa in English) up to 5 metres long and held in both hands. This pike gave the phalangite a much greater reach in combat and also allowed formations to be packed together much more tightly, since room was no longer required for individual spear play; instead, the Macedonian phalanx could rely on the mass effect of a dense, deep formation with numerous serried rows of pikes projecting from its front. This in turn required a much higher level of training, discipline and drill to prevent the dense formation with its long pikes falling into hopeless disarray, so the Macedonian phalangite became a professional soldier. This does not necessarily mean he served for a long term, since in peacetime at least most Macedonian soldiers made a living as farmers and were only conscripted into the army at need. However, he was highly trained in centrally organized programmes of drill and instruction unknown to the Greeks, outside of Sparta and the elite units of picked men. This training and professionalism allowed the Macedonian phalanx to perform manoeuvres on the battlefield that most non-Spartan Greek armies could not have contemplated. The Macedonian phalanx was to a large extent a refined and perfected form of the Greek phalanx, bringing together offensive weaponry, formation and drill to produce the ultimate 'heavy' infantry formation, one not seriously threatened by anything on the battlefield except another similar phalanx, although with the flaw of great reliance on its good order and cohesion, the loss of which, due to difficult terrain or enemy tactics, could prove fatal.

The new Macedonian phalanx was also just one part of a more integrated military machine. Heavy cavalry were refined into a true shock arm, capable of delivering a decisive charge against most opponents (though probably not against a steady, formed phalanx, Greek or Macedonian). The Macedonian army also used large numbers of light infantry, drawn from allied or subject peoples and from mercenaries, which could perform all those tasks for which the phalanx itself was ill-suited. Furthermore, the Macedonians refined the art of siegecraft, employing specialist engineers and enthusiastically adopting the technical advances in siege machine technology that were taking place in the fourth century, especially in outlying regions of the Greek world such as Sicily. Macedonian armies were also paid to serve for the duration of a possibly lengthy campaign, and did not expect or require to be sent home at the end of every campaigning season. This allowed them to operate for longer and (because enemy cities could routinely be captured by siege) in more decisive campaigns.

All such factors lie somewhat outside the scope of the central confrontation between Greek hoplite and Macedonian phalangite. Unfortunately, if we know little of how hoplite fought hoplite, we know even less of how hoplite fought phalangite, as the crucial battles of the later fourth century, culminating in the great Macedonian victory at Chaeronea (338) over Athens and Thebes, are very poorly documented, a second-rate account in Diodorus and a few scattered anecdotes in Polyaenus and Plutarch being all we have to go on. This has not prevented Chaeronea becoming another (if lesser) example of the 'inverted pyramid', of an edifice of modern speculation about a battle being constructed on the foundation of brief and wholly inadequate ancient accounts. These accounts are brief enough to quote in full:

> 'The armies deployed at dawn, and the king [Philip] stationed his son Alexander, young in age but noted for his valour and swiftness of action, on one wing, placing beside him his most seasoned generals, while he himself

at the head of picked men exercised the command over the other; individual units were stationed where the occasion required. On the other side, dividing the line according to nationality, the Athenians assigned one wing to the Boeotians and kept command of the other themselves. Once joined, the battle was hotly contested for a long time and many fell on both sides, so that for a while the struggle permitted hopes of victory to both. Then Alexander, his heart set on showing his father his prowess and yielding to none in will to win, ably seconded by his men, first succeeded in rupturing the solid front of the enemy line and striking down many he bore heavily on the troops opposite him. As the same success was won by his companions, gaps in the front were constantly opened. Corpses piled up, until finally Alexander forced his way through the line and put his opponents to flight. Then the king also in person advanced, well in front and not conceding credit for the victory even to Alexander; he first forced back the troops stationed before him and then by compelling them to flee became the man responsible for the victory. More than a thousand Athenians fell in the battle and no less than two thousand were captured. Likewise, many of the Boeotians were killed and not a few taken prisoners.' (Diodorus 16.86.1–6)

'Engaging the Athenians at Chaeronea, Philip made a sham retreat: and Stratocles, the Athenian general, ordered his men to push forwards, crying out, "We will pursue them to the heart of Macedonia." Philip observed, "The Athenians know not how to conquer": and ordered his phalanx to keep close and firm, and to retreat slowly, covering themselves with their shields from the attacks of the enemy. As soon as he had by the manoeuvre drawn them from their advantageous ground, and gained an eminence, he halted; and encouraging his troops to a vigorous assault, he attacked the Athenians and won a brilliant victory.' (Polyaenus 4.2.2)

'Philip, at Chaeronea, knowing the Athenians were impetuous and inexperienced, and the Macedonians inured to fatigues and exercise, contrived to prolong the action: and reserving his principal attack to the latter end of the engagement, the enemy weak and exhausted were unable to sustain the charge.' (Polyaenus 4.2.7)

'He [Alexander] was also present at Chaeronea and took part in the battle against the Greeks, and he is said to have been the first to break the ranks of the Sacred Band of the Thebans.' (Plutarch, *Alexander* 9.2)

'It is said, moreover, that the [Theban Sacred Band] was never beaten, until the battle of Chaeronea; and when, after the battle, Philip was surveying the dead, and stopped at the place where the three hundred were lying, all where they had faced the long spears of his phalanx, with their armour, and mingled one with another, he was amazed, and on learning that this was the band of lovers and beloved, burst into tears and said: "Perish miserably they who think that these men did or suffered aught disgraceful."' (Plutarch, *Pelopidas* 18.5)

Note that in this last passage, the phrase 'the long spears of his phalanx' is a slightly free translation, since the Greek only mentions 'long spears' (*sarisais*, sarissas), and it is a moot point whether these were indeed carried by the Macedonian phalanx rather than cavalry for example. None of the accounts make clear which forces Alexander was commanding, and while in later battles he always fought at the head of his Companion cavalry, that was as king and commander-in-chief, and as heir and second-in-command he might not necessarily have done so. The 'companions' he is said to have broken the line with in Diodorus' account could be those senior officers placed around him already referred to, rather than the Companion cavalry (companions and Companions both being *hetairoi*, it is not possible to be certain which is meant). As a result, we cannot be sure which forces – Macedonian infantry or Macedonian cavalry – Alexander used to break the line and defeat the Thebans, although Plutarch's anecdote about the Sacred Band offers some support for the view that it was infantry.

Be that as it may, Philip was in command of his 'picked men', *epilektoi*, which in this context probably means the royal guard, probably known at this time as Foot Companions, and shortly afterwards to be renamed Hypaspists by Alexander. These probably were equipped and fought in the same way as the rest of the Macedonian phalanx, though they may have had slightly lighter shields and armour. Whether Polyaenus' story of a feigned withdrawal by this unit is to be believed is a matter of dispute. If it is, then it is perhaps reminiscent of the Spartan feigned retreats at Thermopylae and/or Plataea discussed previously. The two stories in Polyaenus, however, are not entirely consistent, nor is there any mention of this manoeuvre in Diodorus. So all we can say is that Alexander probably first broke the Theban wing, though with what forces and in what circumstances we can only guess, while Philip, later in the action (which need not mean after any considerable lapse of time), broke the Athenians.[2]

Further tactical niceties remain elusive. What we are left with is the clear outcome: the Athenians and Thebans were both broken in pitched battle and routed with heavy loss, demonstrating, seemingly, the superiority of the Macedonian system of arming and organizing a phalanx over the Greek. However, we must remember that other factors than arms and armour were at play. Polyaenus' mention of the Macedonians being 'inured to fatigues and exercise' reminds us of the intensive training regime through which Philip put his army, and that the

Athenians, still at this late date, were no more than an enthusiastic militia. It may be that the Macedonians' discipline and professionalism were as important as Macedonian equipment in securing this victory.

Hoplite reforms

The Macedonians were not the first to reform or alter the equipment of the phalanx, at least according to one story. Diodorus, supported by a similar account with minor variations of detail by the Roman historian Cornelius Nepos, relates that the early fourth-century Athenian general Iphicrates also devised a reform of hoplite equipment, which we have already encountered in Chapter 1:

> 'Hence we are told, after he had acquired his long experience of military operations in the Persian war [not the great Persian War of the early fifth century], he devised many improvements in the tools of war, devoting himself especially to the matter of arms. For instance, the Greeks were using shields which were large and consequently difficult to handle; these he discarded and made suitably sized light shields [*peltas summetrous*], thus successfully achieving both objects, to furnish the body with adequate cover and to enable the user of the *pelte*, on account of its lightness, to be completely free in his movements. After a trial of the new shield its easy manipulation secured its adoption, and the infantry who had formerly been called "hoplites" because of their heavy shield, then had their name changed to "peltasts" from the light shield [*pelte*] they carried. As regards spear and sword, he made changes in the contrary direction: namely, he increased the length of the spears by half, and made the swords almost twice as long. The actual use of these arms confirmed the initial test and from the success of the experiment won great fame for the inventive genius of the general.' (Diodorus 15.44.1–4)

This reform has attracted much discussion and poses many questions. Diodorus' account read literally would have us suppose that hoplites in general all adopted this new equipment, and henceforth became known as peltasts. There are many problems with this idea. For instance, peltasts had already been known throughout the fourth and fifth centuries as a type of light infantry, so Diodorus' suggestion that the name was simply adopted by hoplites is not plausible. There is also no trace of this reform in other writers (aside from Cornelius Nepos), and we see hoplites, under the name of hoplites – and as far as we can tell with normal hoplite equipment – continuing in use long after Iphicrates' time. It is difficult to be absolutely certain on this latter point, since with figurative vase painting falling out of fashion in the fourth century, the largest source of artistic depictions of infantry equipment largely dries up. A few tombstones and depictions of Macedonians mostly on funerary monuments are the only evidence for infantry equipment from the later fourth century. Even so, the traditional Greek hoplite *aspis* continues to be depicted in art, carried by Greeks and Macedonians (as well as mythological figures) well into the Hellenistic period and beyond, and it appears that when Greeks did adopt Macedonian equipment, in the late third century (see below), it was, at least on occasion, a direct transition from Classical hoplite equipment. There are also plentiful references to hoplites throughout the fourth century and Hellenistic period, and while the exact details of their equipment are rarely specified and could certainly have varied (Macedonian phalangites, for example, are also usually called hoplites), the impression gained is that they continued to be armed in the traditional style of the Classical hoplite.

There have been various attempts to interpret Diodorus' testimony. One is that he was describing a reform not of hoplite but of peltast equipment, an attempt not to make a lighter, more active hoplite, but a heavier, more-steady peltast. This is then seen as an attempt to create a cheaper version of the hoplite from the material of light infantry, that was able to stand in battle against traditionally armed hoplites. The problem with this interpretation is that it is the exact opposite of what Diodorus says – in his account, it is clearly hoplites whose equipment is lightened. So it may be that Iphicrates invented a more battle-capable peltast who could stand in the line of battle – in the phalanx – rather than just being a skirmisher, and it may also be that some or all of the peltasts we encounter in the later fourth century and into the Hellenistic period were equipped in this way, a light-heavy infantry hybrid capable of fighting at close quarters as well as of resisting cavalry, while retaining some of the speed and manoeuvrability of light infantry. But this is not what Diodorus tells us, and there is little supporting evidence for this theory, the equipment of peltasts from the time of Philip and Alexander onwards being obscure. Indeed, the very word 'peltasts' seems to have fallen out of use in this period, with light infantry being referred to by more generic terms or in terms of their specific armament – javelinmen, archers, slingers – and going into the Hellenistic period we see new types of infantry lighter than the hoplite emerge, using new terms such as *thureophoroi* ('long-shield-carriers'), *thorakitai* ('armour-wearers') and *euzonoi* ('well-equipped'), on all of which see more below. Peltasts reappear in the Hellenistic period as a lighter version of the Macedonian phalangite, particularly the royal guard units of the Hellenistic kingdoms.[3]

Another possibility is suggested by the apparent similarity between the Iphicratean equipment described by Diodorus and that of the Macedonian phalanx devised by Philip II. A longer spear and a smaller, lighter shield called a *pelte* or *pelta* match what we know of the early Macedonian phalanx equipment, where Philip's crucial reform was to introduce the sarissa or pike and to lighten the rest of the equipment, in particular replacing the traditional Greek *aspis* with a lighter Macedonian shield, a shield which because it was lighter and did not have the traditional *aspis* rim, could be referred to as a *pelte*. According to this theory, rather than Philip II inventing the equipment of the Macedonian phalanx (the evidence for which is another passage of Diodorus), he simply adopted the reforms already trialled by Iphicrates. This is an appealing theory in some ways, though there is still the problem that there is no trace of Iphicrates' reforms having been enacted anywhere before Philip's supposed adoption of them, and it would be necessary to reject Diodorus' identification of Philip as the inventor of the Macedonian phalanx. There is also a mismatch of tactical intent between Iphicrates' reforms, intended to make hoplites into a lighter, more mobile form of infantry, and Philip's invention, intended to create a heavier, specifically closer-order form of heavy infantry.[4]

Another final possibility – more likely to my mind – is that Iphicrates did indeed introduce a reform of hoplite equipment but that its scope and adoption was much more limited than Diodorus' account suggested, perhaps being restricted to a particular campaign or even a specific body of hoplites – a possibility being that it applies specifically to mercenary hoplites fighting in Iphicrates' campaigns in Egypt. The reform might then have been real and was 'widely adopted' within one particular army or campaign, but was not taken up generally in Greece. It might still have provided some inspiration for the later reforms of Philip, who probably had many influences (Iphicrates, Epaminondas and Pelopidas, Homer) on his tactical thinking.

The likelihood is then that it was hoplites armed in the traditional, Classical way that continued to form the backbone of Greek city and mercenary armies well into the Hellenistic period. The Greek mercenary infantry fighting with the Persians against Alexander, the Corinthian League Greek contingents fighting alongside Alexander and the various Greek contingents who fought against Alexander's Successors in the Lamian War and other uprisings against Macedonian domination probably all continued to be armed and equipped much like hoplites of the early fourth century and before, for all that there was probably a general trend to lighten equipment.

Unfortunately, details of these Greek contingents are extremely sketchy. The allied hoplites fighting for Alexander are passed over in almost complete silence in the ancient accounts of Alexander's battles, and, perhaps for political reasons, they seem to have been relegated to at most a supporting role in his campaigns, used largely as garrisons and as (often unwilling) settlers in his new city foundations. Of the Greeks fighting against Alexander, we have accounts of a large contingent being massacred in the closing stages of the Granicus (334) and fighting in the Persian front line at Issus (333). They were also present at Gaugamela (331), though we hear little of their actions there. The fighting at Issus provides the best evidence for hoplite capabilities at this time.[5]

The Persian king Darius had, according to Arrian, 30,000 Greek mercenaries (the number is almost certainly exaggerated) in his army at Issus:

> 'He placed the Greek mercenaries, about 30,000, foremost of his hoplites, facing the Macedonian phalanx; next, on either side, 60,000 of the so-called Cardaces, who were also hoplites.' (Arrian, *Anabasis* 2.8.6)

The identity of the Cardaces is much disputed, but they were probably native infantry equipped entirely for close combat, rather than the mixed archer-shieldbearer units typical of earlier Persian armies. Note that Arrian is certainly using 'hoplites' in a generic sense to mean 'heavy infantry' rather than with any specific equipment in mind, but even so it is most likely that these Greek mercenaries were equipped in the traditional manner. Note again the absence of specific technical terminology – both Greeks and Cardaces are 'hoplites'. Arrian goes on to say that:

> 'The general mass of his [Darius'] *psiloi* and hoplites, arranged by their nations in such depth that they were useless, was behind the Greek mercenaries and the barbarian force drawn up in phalanx formation.' (Arrian, *Anabasis* 2.8.8)

Hoplites, phalanxes and deep formations are thus a feature just as much of the Persian forces as of the Greek. These Greek mercenaries were to acquit themselves well in the battle:

> 'But Darius' Greek mercenaries attacked the Macedonian phalanx, where a gap appeared as it broke formation on the right ... the Macedonian centre did not set to with equal impetus, and finding the river banks precipitous in many places, were unable to maintain their front in unbroken line; and the Greeks attacked where they saw that the phalanx had been particularly torn apart. There the action was severe, the Greeks tried to push off [*apotheo*] the Macedonians into the river and to restore victory to their own side who were already in flight, while the Macedonians sought to rival the success of Alexander, which was already apparent, and to preserve the reputation of the phalanx, whose sheer invincibility had hitherto been on everyone's lips. There was also some emulation between antagonists of the Greek and Macedonian races.' (Arrian, *Anabasis* 2.10.5–7)

The references to 'push off' should not provoke too much excitement, in light of the discussion in the previous chapter, *apotheo* having a similar range of uses and meanings as *exotheo*. Arrian also has Alexander's cavalry 'come hand-to-hand [*en chersi*] with the Persians and push them back [*exothei*]' (Arr., *Anab.* 2.10.5), vocabulary obviously inspired by the usage of Thucydides and Xenophon and, like theirs, not carrying a literal meaning.

The performance of the Greeks at Issus indicates that the advantage of the Macedonian phalanx over the Greek was not decisive, and that in the right circumstances (in this case, the broken ground of the river bank), the Greek phalanx could still do well, although in this case Alexander and the Macedonian right flank were able to outflank the Greeks and drive them back, saving the Macedonian centre from possible defeat. Again, the discipline and high motivation of the Macedonians in not breaking into flight when things went against them, and the ability of the right flank not to press on in pursuit of the fleeing Persians opposite them but to come to the aid of their embattled centre, are what turned the battle in the Macedonians' favour, rather than the technical differences of formation and armament of their respective phalanxes. In many Classical battles, the Macedonian centre might simply have collapsed when attacked in this way, while the right would not have been under sufficient control to perform its flanking movement.

The third century

That the superiority of Macedonian equipment was marked but not decisive perhaps explains (alongside the usual Greek military conservatism) why, despite the apparent invincibility of the Macedonian phalanx to which Arrian refers, Greeks did not themselves adopt the Macedonian formation and equipment for at least another century. Unfortunately, we have no clear picture of the armament of most Greeks in the closing years of the fourth century. Alexander's Successors, battling each other for control of Asia, all used phalanxes 'armed in the Macedonian fashion', most of which were actually Macedonians, though there were also 'mixed' forces of Greek mercenaries or Asians who may have been armed as Macedonians, and Alexander had himself armed a large force of Persians in the Macedonian fashion, as well as experimenting with introducing them, with their native equipment, into the ranks of his phalanx. These experiments with the use of native forces do not seem, however, to have been continued by the Successors or in the Hellenistic states which they established. Successor armies were all built around a core phalanx of ethnic Macedonians, and possession of Macedonians, along with land and treasure, became one of the prizes over which the Successors fought. Meanwhile, in Greece, the city armies which fought to shake off (or support) Macedonian control were, so far as we know, still equipped as hoplites, though few details are preserved about either their equipment or the battles in which they fought, and the terminology of later authors – to whom all heavy infantry are simply 'hoplites', and all battle lines a 'phalanx' – does not make for greater clarity.

With the establishment of the major Hellenistic kingdoms – the Seleucids in Syria and the former Persian lands, the Ptolemies in Egypt, the Antigonids (eventually) in Macedon and various other kingdoms such as Lysimachid Thrace and the city states of Rhodes and Pergamum – each had to establish its own source of Greco-Macedonian manpower. Macedonians, along with Greek mercenaries, serving in the Successor armies were granted plots of land in the new kingdoms to farm (or to sub-let) and were encouraged to settle in new city foundations within the kingdoms (Macedon of course simply continued its existing methods of conscription). These men – a mix of Macedonians and Greeks, with a preponderance of Macedonians – then formed the phalanx of each kingdom. Hellenistic kings were slow and reluctant to enrol native peoples from their kingdoms into the phalanx for internal political reasons.[6]

In Greece, things were thrown into turmoil by a Gallic invasion in the early third century that overran Macedon and penetrated as far as Delphi. The invaders were eventually expelled, but left a lasting legacy in their types of shields: the long rectangular or oval shield familiar from its use in Italy by the Romans as well as by many Celtic peoples, and known to the Greeks as the *thureos* or 'door'. Evidently, this shield was found to be a convenient shape, size, weight and design, and was adopted by many Greek infantry. Unfortunately, as with Iphicrates' reforms, it is not clear exactly which infantry adopted it. A common assumption is that various forms of light infantry upgraded their equipment to *thureos* and long spear, forming a sort of hybrid light-heavy infantry with greater mobility than traditional hoplites, but still able to fight in the battle line and stand up to cavalry (just like the 'Iphicratean peltast'). An alternative interpretation is that hoplites themselves lightened their equipment, adopting the *thureos* to form a lighter, more mobile heavy infantry. There is conflicting evidence to support both interpretations. In the Hellenistic kingdoms, *thureophoroi*, as such infantry were called ('*thureos*-carriers'), appear to be a sort of heavy infantry outside of the Macedonian-armed phalanx, still forming part of the battle line in pitched battle but also called on to perform rapid marches or assaults over difficult ground for which the heavily equipped phalangites would have been unsuitable. In Greece, the evidence for the later adoption of Macedonian equipment by the Achaean League (formed from many of the cities of the northern Peloponnese) suggests that the whole of their infantry were equipped as *thureophoroi*. It may be that different regions adopted *thureophoros* equipment to different extents and for different purposes; it appears, for example, that Spartans at least retained traditional hoplite equipment, while the Boeotians may have armed themselves as peltasts, in the Hellenistic sense.⁷

For alongside these *thureophoroi* were other, even more obscure, forms of infantry. I have already mentioned *thorakitai* and *euzonoi*, and alongside these were *peltophoroi* ('*pelta*-carriers') as well as 'peltasts'. This latter word seems to be applied not to a type of light infantry, as in Classical times, but to men armed in the Macedonian fashion, in the phalanx, but with lighter equipment (smaller shield, shorter spear, probably less armour) than the main phalanx. In the Hellenistic kingdoms, such peltasts formed the guard infantry, drawn from the younger age classes and perhaps kept on permanent, paid active duty. In pitched battles, they formed part of the phalanx (usually the right wing), while they could perform special duties requiring speed and mobility in other operations, much in the same style as Alexander's Hypaspists (who may themselves have been the original peltasts, in this new sense). Whether *peltophoroi*, who are known particularly from Boeotia, were armed and organized in a similar fashion is an open question. Similarly, little is known of the equipment or battlefield roles of the *thorakitai* and *euzonoi*, who remain little more than names, though they too seem to have been a more mobile form of infantry intermediate between the traditional hoplite phalanx and the old-syle skirmishing light infantry. All of these new types of infantry appear to be an attempt to create a force capable of rapid, irregular operations on difficult terrain, things for which the traditional Classical phalanx was capable but for which it was very far from optimized. The more specialized Macedonian phalanx was even less suited to such operations, hence the need for a more mobile form of infantry in Hellenistic armies, while in Greece it may be that the experience of the Gallic invasions, which would have involved fighting fast-moving yet powerful infantry equipped with *thureos* and sword, suggested the need, in some cities at least, for adopting lighter equipment. Once the enemy of Greek hoplites was no longer other Greek hoplites, the tactical needs which infantry had to meet obviously changed. The overall picture remains confused, but it appears that Greece in the third century saw varied use of traditional hoplites and these lighter variants in different proportions at different times and circumstances.⁸

Adoption of Macedonian equipment

Only towards the end of the third century do we see Greek cities beginning to adopt Macedonian equipment, as they engaged in a series of wars in support of or opposition to Macedonian domination, and a resurgent Sparta, despite its long demographic decline from the fourth century on, sought to reassert itself in opposition to both the Macedonians and the Achaeans. The Spartan army was the first that we hear of explicitly to have adopted Macedonian equipment, directed by their king Cleomenes, in preparation for the campaign that was to culminate in their defeat by the Macedonians at Sellasia (222). Cleomenes enrolled a body of helots and *perioikoi*, in the time-honoured manner of the Spartans when faced with self-inflicted manpower shortages:

> 'Then he filled up the body of citizens with the most promising of the *perioikoi*, and thus raised a body of four thousand hoplites, whom he taught to use a sarissa, held in both hands, instead of a spear, and to carry their shields by a handle instead of by an armband.' (Plutarch, *Cleomenes* 11.2)

> 'Cleomenes ... set free those of the Helots who could pay down five Attic minas (thereby raising a sum of five hundred talents), armed two thousand of them in Macedonian fashion as an offset to the White Shields of Antigonus.' (Plutarch, *Cleomenes* 23.1)

The 'White Shields' were (probably) one of the main constituent units of the Macedonian phalanx, though their identity is not absolutely certain. These passages provide useful evidence for the equipment of the Macedonian phalanx – the important point is that the shield used is called an *aspis*, and the suggestion is that the carrying arrangements of the shield were altered (to free up the left hand to carry the sarissa), not that new types of shields were issued. This may be evidence that the Spartan army was still armed in Classical hoplite fashion, and that the reform was to replace spear with sarissa and change the way the shield was carried, rather than anything more fundamental. In the ensuing battle at Sellasia, the Spartan phalanx acquitted itself well against the Macedonians:

> 'Each side now recalled by bugle their light-armed troops from the space between them, and shouting their war-cry and lowering their sarissas, the two phalanxes met. A stubborn struggle followed. At one time the Macedonians gradually fell back facing the enemy, giving way for a long distance before the courage of the Lacedaemonians, at another the latter were pushed from their ground by the weight of the Macedonian phalanx, until, on Antigonus ordering the Macedonians to close up in the peculiar formation of the double phalanx with its serried line of sarissas, they delivered an attack which finally forced the Lacedaemonians from their stronghold.' (Polybius 2.69)

The falling back on each side seems here not to be a 'feigned retreat' like that at Chaeronea or Thermopylae, but a gradual falling back (or being pushed back) in combat, as discussed in the previous chapter, while the 'long distance' involved is probably only long relatively speaking. The Spartans had evidently rapidly picked up the basics of fighting in the Macedonian fashion, but the extra experience and discipline of the Macedonians was sufficient to win the day. It is not clear precisely what is meant by the 'double phalanx' in this case, but the Macedonians probably closed up into the one cubit, closest possible order, termed *synaspismos* by the Hellenistic tacticians, who use this word to refer to this specific drill, rather than just in the general sense of massing closely together. We also cannot be certain whether the existing Spartan citizens, rather than just these newly enrolled contingents, also adopted Macedonian equipment, or whether they continued to fight as Classical hoplites. It seems most likely, especially given the events at Sellasia, that the whole of the Spartan phalanx was rearmed.[9]

The Achaeans, allies of the Macedonians at Sellasia, though they alternated in support and hostility to Macedon, were next to take the leap, at the instigation of their general Philopoemen:

> 'In the first place, however, he changed the faulty practice of the Achaeans in drawing up and arming their soldiers. For they used *thureoi* which were easily carried because they were so light, and yet were too narrow to protect the body; and spears which were much shorter than the Macedonian sarissa. For this reason they were effective in fighting at a long distance, because they were so lightly armed, but when they came to close quarters with the enemy they were at a disadvantage. Moreover, a division of line and formation into *speirai* was not customary with them, and since they employed a solid phalanx without either levelled line of spears [*probole*] or wall of interlocking shields [*synaspismos*] such as the Macedonian phalanx presented, they were easily dislodged and scattered. Philopoemen showed them all this, and persuaded them to adopt sarissa and *aspis* instead of spear and *thureos*, to protect their bodies with helmets and breastplates and greaves, and to practise stationary and steadfast fighting instead of the nimble movements of light-armed troops [*peltastikes*].' (Plutarch, *Philopoemen* 9.1–2)

If Plutarch is right, then the Achaean army had already adopted *thureophoros* equipment in place of traditional hoplite equipment, and were fighting as a form of infantry perhaps intermediate between light and heavy, but with more in common with light. This has been doubted by modern historians, but we do not know enough about early third-century Greek infantry to rule out the possibility that the Achaeans, and perhaps others as well, had indeed re-equipped in this way. At any rate, Philopoemen rearmed them in the Macedonian fashion (Achaeans had a head start in this, as the Macedonian king Antigonus Doson had earlier loaned some of them Macedonian shields, at least), and also (just as importantly) reorganized them into formal sub-units (the *speirai* to which Plutarch refers, the *speira* or *syntagma* being the basic building block of the Macedonian phalanx, usually 256 men strong). Even more importantly, he also introduced formal and competitive drill and training:

> 'After he had thus arrayed and adorned the young men, Philopoemen exercised and drilled them, and they eagerly and emulously obeyed his instructions. For the new order of battle pleased them wonderfully, since it seemed to secure a close array that could not be broken; and the armour which they used became light and manageable for them, since they wore or grasped it with delight because of its beauty and splendour, and wished to get into action with it and fight a decisive battle with their enemies as soon as possible.' (Plutarch, *Philopoemen* 9.7–8)

These three elements – Macedonian equipment, formal drill and organization, and training and professionalism – were all interconnected, with each part essential to the success of the overall reform. Merely issuing sarissas to otherwise poorly trained and organized infantry would achieve little.[10]

Precisely how other Greek armies of this period were armed, and whether any or all of the others adopted Macedonian equipment, unfortunately remains shrouded in mystery. The traditional Greek *aspis* continues to be depicted in art into the second century, but there are few artistic depictions of Greek soldiers of the period, and no clear literary accounts of either army organization or battles. Roman armies rapidly came to dominate Greece, defeating the Antigonid Macedonian phalanx first at Cynoscephalae (197) then Pydna (168), and the Seleucid phalanx at Thermopylae (191) and Magnesia (189). There were Greek contingents on both sides at these battles, chiefly on the Roman side, though details of their armament are lacking. The Achaeans themselves were the last to fall to the Romans, an ill-advised revolt against Roman control of Greece ending in defeat in battle at Corinth (146). Assuming the Achaeans had continued to use Macedonian equipment, then there may not have been any Classically armed hoplites present at this latter battle. The end of the hoplite as a battlefield presence thus comes with more of a whimper than a bang, with evidence for their final appearance sadly lacking. It is most likely that the final victories of the Romans against Greek, Macedonian, Seleucid and, later, Pontic armies were victories over a Macedonian-armed phalanx, not a Greek-armed one. Indeed, in Greece we have no clear evidence that a hoplite phalanx ever fought the Romans, though men armed with hoplite equipment will certainly have done so in the earlier wars in Italy, and Carthaginians were probably equipped much like Classical hoplites, at least in the early stages of their wars with Rome.[11]

Part of the argument of this book is that there was nothing particularly unusual or different about Greek hoplite equipment or tactics. Heavily armed infantry had fought in close-order formations with spear and shield for centuries before the Classical Greeks did so, and continued to do so for centuries afterwards. The triumph of Roman arms in the second and first centuries meant that Roman equipment – which was, broadly speaking, the equipment of the *thureophoros*, a long shield, javelins and sword, used in a relatively open-order formation for individual duelling – came to dominate warfare for the next several centuries, as the Roman Empire was carved out around the shores of the Mediterranean and in north-west Europe. However, heavy close-order infantry with spear and shield were far from obsolete, and a number of enemies of the Romans continued to use them, though the particular combination of social and cultural circumstances that made the Greek hoplite precisely what he was did not arise again. The Romans themselves appear to have adopted something closer to hoplite equipment and tactics in the later Empire, perhaps finding them more effective than their traditional arms against the increasingly cavalry-based armies they faced. The Eastern Roman (Byzantine) Empire armed its infantry in a manner more reminiscent of the Classical hoplite, with spear and shield, and adopted drill and organization derived from the Hellenistic tactical manuals of the Macedonian phalanx, so can be seen as descendants, at one remove, of the hoplite phalanx.

Conclusions

The aim of this book has been, in part, to emphasize that the Greek hoplite, and the phalanx in which he fought, while certainly distinctive, was not unique, and that what I have called 'hoplite exceptionalism' – the tendency to view the hoplite and the Greek phalanx as something qualitatively different from other heavy infantry formations of the time – is largely misguided. The hoplite fits into a tradition of close-order heavy infantry that can be traced back through the previous centuries in the Near East and Mediterranean, rather than representing a break from this tradition. The factors that made the hoplite so successful are not to be found in technical matters of equipment and formation, so much as in the culture and psychology of the hoplites themselves. It is also to some extent to be found in the particular circumstances in which the hoplite operated, in which possibly technically superior formations and armaments – such as the Macedonian and the Roman – were not encountered (or did not exist) for several centuries, during which time hoplites mostly fought against each other, against poorly equipped and disorganized tribal forces on the edges of the Greek world, or against Persian armies which did not have such a strong tradition of native heavy infantry.

To recap some of the findings of this study:

- The origins of Greek hoplite equipment and tactics are obscured by time and lack of evidence, but it is likely that hoplite equipment came into widespread use during the seventh century and probable that hoplites were fighting in a phalanx formation of some sort at the same time as this equipment was adopted. However, this is not because such equipment could only be used in phalanx formation, nor because it was particularly specialized for phalanx use, but because a phalanx, a relatively close-order, multi-rank formation, was the natural formation for spear-armed heavy infantry, and had already been long used by other peoples and armies, if not under this name. We do not need to seek the birth of the phalanx in the Archaic period; rather we should see the slow adoption and (for some) perfection of existing organization and fighting styles, along with a military culture perfectly suited to this type of fighting.
- Greek equipment was adopted and adapted from neighbouring peoples – perhaps particularly from Carians and Egyptians – though with its own distinctive features. Greeks emphasized the use of bronze armour, especially in the Archaic period, when bronze body armour and enclosing bronze helmets were widely worn (at least by the wealthier infantry). Later, there were moves to make this equipment lighter and more convenient, even as the phalanx itself became more formalized, further evidence that heavy hoplite equipment did not require a close-order phalanx, even though in practice the two developed together. This lightening of equipment may have had social and economic causes – the extension of the hoplite franchise to poorer citizens, more widespread mercenary service by men who were not wealthy – as well as military, and the earlier emphasis on bronze may indicate only that earlier hoplites were a smaller, more closed caste of the wealthy.
- The hoplite phalanx was a close-order, deep formation of spearmen, functionally much like such formations in other armies of the same and earlier times. Early versions of the phalanx did not have all the elements of formal drill and organization that we see in the Spartan army, which itself was not typical of Greek armies, and although hoplites increasingly fought hand-to-hand, the early phalanx retained some more traditional elements including a somewhat less formal formation and a tendency to fight as a crowd of individuals rather than as a closely disciplined whole. The combination of heavy Greek armour and other factors made Greeks particularly effective at close-quarters fighting.
- These other factors include a cultural emphasis on close combat, on standing one's ground and on supporting the common good, as represented by the phalanx as the collective, armed citizenry of the home

city. This made the Greek phalanx particularly stubborn and determined, and increasingly willing to close *en masse* to hand-to-hand combat, unlike earlier aristocratic-dominated infantry formations in which individual duelling was more important than mass combat, and in which fighters, seeking individual glory rather than victory for the group, were more willing to advance or retire as convenient, rather than standing their ground.

- Most Greek hoplite armies were poorly trained and disciplined – deliberately so, since there was an ethos of amateurism that regarded military training as unnecessary, and a belief that hoplite skills should come naturally to men of the right sort. The exception is Sparta, whose peculiar and unique constitution meant that professionalism, training, drill and discipline were expected of all hoplites, making the Spartans the most effective and successful Greek army for several centuries.

- Hoplites were particularly effective in pitched battle on the open field, on good terrain and with a minimum of tactical finesse, where fighting skills and stubbornness were the most important factors. But they were also perfectly capable of fighting in other circumstances: in sieges and assaults on fortified positions, in more irregular encounters and above all from shipboard, with hoplite marines and seaborne forces of hoplites being effective and frequently used throughout the period. The idea that hoplites were capable of fighting only in a formed phalanx on flat terrain cannot be sustained.

- While masters of the battlefield, particularly when facing their own kind, hoplites were also vulnerable, like all such heavy infantry, to missile-armed light infantry and to cavalry, in the right circumstances. We only have detailed accounts of hoplite defeats to such forces from fairly late on, but given the tactical realities, and the fact that light infantry and cavalry forces had always been a part of Greek warfare, it is likely that hoplites always had these vulnerabilities and similar hoplite defeats to the famous ones of the later fifth and fourth centuries probably also occurred at earlier dates.

- However, given the relatively limited nature of Greek warfare in which cities usually fought neighbours in disputes over land, the opportunities for light forces and cavalry to harm a hoplite phalanx were fairly limited. In such symmetric warfare, the interests of the politically and socially dominant factions on both sides were best served by a simple battle between hoplites, minimizing the role of the socially inferior light infantry or (sometimes-resented) aristocratic cavalry.

- A number of customs and traditions developed around warfare of this sort, though these should not be confused with rules attempting to restrict or ameliorate the waging of war. Trickery and guile were always a part of Greek warfare, alongside and in uneasy competition with the preference for straightforward face-to-face fighting.
- Greek hoplites fought in phalanx formation using their weapons – spear and sword to strike with, shield to parry and protect. While there was an element of pushing and shoving in this sort of close-order fighting between deep formations, the idea of a unique hoplite mass scrum is fanciful and based on an over-literal reading of literary battle accounts.
- The Greek phalanx tradition did not die out with the dominance of Macedonian arms. Greeks continued to fight in the phalanx – alongside other forms of arming and organizing – into the Hellenistic period, and similar forces of close-order, spear-armed heavy infantry, functionally identical to the Greek phalanx, remained mainstays of warfare for centuries to come.

It may seem as if this study has sought to depose the hoplite from a special exalted position to which he has been elevated (a position to which some may feel he needs again to be 'reinstated'), and this has indeed been part of my intent. I think that 'hoplite exceptionalism' is largely misguided, and we understand the hoplite better by seeing him in the context of other similar infantry forces before, since and contemporary, forces from which the hoplite phalanx was not so very different. To a great extent, the modern historiography of the hoplite has been dominated by the over-strict and over-literal interpretation of a handful of words – particularly 'hoplite', 'phalanx' and *otheo/othismos* – and by the desire to see these words in ancient sources as specific, technical terms that mean something quite unique. I do not believe that Greek writers often used vocabulary in this specific, technical way, and we will gain a clearer picture of the Greek hoplite phalanx by taking these words to mean 'heavy infantry', 'battle line' and 'hand-to-hand combat', rather than assigning them more restricted meanings. In addition, I think there are no grounds for seeing the hoplite as an exemplar or even originator of a unique 'Western way of war', different in kind from the ways of war of peoples of other points of the compass, cultures or colours of skin.

But this is not to say that there were no differences at all between the Greek hoplite and his equivalent in other cultures. A particular combination of equipment, formation, culture and psychology served to make the Greek hoplite what he undoubtedly was, a master at his preferred type of warfare: the clash of heavy infantry forces on the regular battlefield. Greeks were in demand as mercenaries for a variety of reasons – not least the demographic dynamism that also led them to colonize much of the Mediterranean coast – but among these reasons was that the Greeks, even with their amateur ethos, were perceived as masters of close-quarters warfare. A large reason for this is the particular political and cultural circumstances of the Greek city states, in which numerous more or less egalitarian bodies of citizens (egalitarian within the citizen class, at any rate) were able to work towards a shared and common good while also being in a state of constant competition for status, honour and excellence, competition embodied by the Olympic Games, by competitive poetry and drama, and by hoplite warfare. Throwing off aristocratic rule by the horse-owning wealthy, and better organized and more disciplined than the light infantry of tribal peoples, Greeks were able to dominate their chosen style of warfare for several centuries. The Greek hoplite was not unique, but he did excel at what he did.

List of Battles

The following is a necessarily incomplete list of the battles fought by hoplite armies during the Classical period. Any such list is arbitrary, since many battles are known only by name if at all, and a decision must be made as to which to include and which to exclude. Similar lists in Schwartz (2013) and Krentz (2007) p.169 will fill some of the deficiencies in this list. I list only the major or most important source for each battle – the larger battles frequently have multiple accounts in later authors.

Major battles or those for which there is a reasonably detailed account are listed in bold capitals; battles with brief but useful descriptions are in bold lower case; battles known only from a single sentence or less, or only by name, are in lower-case plain text.

Persian Wars
MARATHON (490) – Athenians and Plataeans defeat Persians
Hdt. 6.111–117
THERMOPYLAE (480) – Spartans and other Greeks unsuccessfully attempt to hold pass against Persians
Hdt. 7.207–232
MYCALE (479) – Spartans and Athenians defeat Persians
Hdt. 9.97–105
PLATAEA (479) – Spartans, Athenians and other Greeks defeat Persians and Medizing Greeks (hoplites)
Hdt. 9.20–74

Fifth Century
Megara (458) – Athenians defeat Corinthians
Thuc. 1.105–106
Tanagra (457) – Spartans defeat Athenians and allies
Thuc. 1.108
Oinophyta (457) – Athenians defeat Boeotians
Thuc. 1.108; Diod. 11.83.1
Potidaea (432) – Athenians and allies defeat Potidaeans and allies
Thuc. 1.63

Peloponnesian War
Pheia (431) – Athenians defeat Elis
Thuc. 2.25.3
Alope (431) – Athenians defeat Locrians
Thuc. 2.26.2
Lycia (430/429) – Lycians defeat Athenians
Thuc. 2.69.2
Spartolus (429) – Chalcidians defeat Athenians
Thuc. 2.79
Stratus (429) – Stratians defeat Acarnanians and allies
Thuc. 2.81–82
Mytilene (428) – Mytilenians versus Athenians, indecisive
Thuc. 3.5.2
Nericus (428) – Leucadians defeat Athenians
Thuc. 3.7.4–5
Antissa (428) – Antissans defeat Methymnians
Thuc. 3.18
Sicily (426) – Athenians defeat Syracusans
Thuc. 3.90
Tanagra (426) – Athenians defeat Tanagrans and Thebans
Thuc. 3.91
Aegitium (426) – Aetolians defeat Athenians
Thuc. 3.97–98
Locris (426) – Athenians defeat Locrians
Thuc. 3.99
Locris (426/425) – Athenians defeat Locrians
Thuc. 3.103.3
Olpae (426/425) – Athenians defeat Peloponnesians and Ambraciots
Thuc. 3.107–108
Idomene (426/425) – Athenians defeat Ambraciots
Thuc. 3.112–113
Locris 426/425 – Locrians defeat Athenians
Thuc. 3.115.6
Naxos (425) – Naxians defeat Messinians
Thuc. 4.25.7–9
Messina (425) – Messinians defeat Leontians
Thuc. 4.25.10–11
Solygeia (425) – Athenians defeat Corinthians
Thuc. 4.42–44

PYLOS (425) – Athenians defeat Spartans
Thuc. 4.4–14
SPHACTERIA (425) – Athenians defeat Spartans
Thuc. 4.29–37
Cythera (424) – Athenians defeat Cytherans
Thuc. 4.54.1–2
Cotyrta (424) – Athenians defeat Spartans
Thuc. 4.56.1
Nisaea (424) – Athenians defeat Boeotians
Thuc. 4.72
Antandrus (424) – Athenians defeat Antandrians
Thuc. 4.75.1
DELIUM (424) – Boeotians defeat Athenians
Thuc. 4.91–98
Sicyon (424) – Sicyonians defeat Athenians
Thuc. 4.101.3–4
Lyncestis (423) – Peloponnesians defeat Macedonians
Thuc 4.124
Laodocium (423/422) – Mantineans versus Tegeans, indecisive
Thuc. 4.134
Amphipolis (422) – Spartans and allies defeat Athenians
Thuc. 5.7–11
Heracleia (420/419) – Athenians defeat Heracleots
Thuc. 5.51
MANTINEA (418) – Spartans and allies defeat Mantineans, Athenians and allies
Thuc. 5.66–74
Syracuse (415/414) – Athenians defeat Syracusans
Thuc. 6.66–71
Epipolae (414) – Athenians defeat Syracusans
Thuc. 6.97
Syracuse (414) – Athenians versus Syracusans, indecisive
Thuc. 6.101
Epipolae (414) – Athenians defeat Syracusans
Thuc. 7.5
Epipolae (414) – Syracusans defeat Athenians
Thuc. 7.6
Euesperitae (413) – Peloponnesians defeat Libyans
Thuc. 7.50.2
Panormus (412) – Athenians defeat Spartans and Milesians
Thuc 8.24.1

Cardamyle, Phanae, Leuconium (412) – Athenians defeat Chians
Thuc 8.24.3
Miletus (412) – Athenians and allies defeat Milesians and allies
Thuc. 8.25
Ta Kerata (409) – Athenians defeat Megarians
Diod. 13.65
Pygela (409) – Athenians defeat Milesians
Xen., *Hell.* 1.2.2–3
Ephesus (409) – Ephesians defeat Athenians
Xen., *Hell.* 1.2.7–11
Abydus (409/408) – Athenians defeat Persians
Xen., *Hell.* 1.2.16
Heraclea (409/408) – Oetaeans defeat Heracleots
Xen., *Hell.* 1.2.18
Calcedon (408) – Athenians defeat Spartans and Calcedonians
Xen., *Hell.* 1.3.5–6
Gaurium (407) – Athenians defeat Andrians and Laconians
Xen., *Hell.* 1.4.22–23
Phyle (403) – Athenian democrats defeat Athenian oligarchs
Xen., *Hell.* 2.4.4
Munychia (403) – Athenian democrats defeat Athenian oligarchs
Xen., *Hell.* 2.4.10–19

Fourth Century
CUNAXA (401) – Persians (Artaxerxes) defeat Cyrus and Ten Thousand
Xen., *Anab.* 1.8.1–28; 1.10.4–17
Haliartus (395) – Thebans and allies defeat Spartans and allies
Xen., *Hell.* 3.5.17–25
Naryx (395/394) – Pocians defeat Aenianians and Athamanians
Diod. 14.82
NEMEA (394) – Spartans and allies defeat Athenians, Thebans and allies
Xen., *Hell.* 4.2.14–23
CORONEA (394) – Spartans defeat Boeotians and allies
Xen., *Hell.* 4.3.15–23
Corinth (392) – Spartans and allies defeat Corinthians and Argives
Xen., *Hell.* 4.4.9–14
LECHAEUM (390) – Athenians and mercenaries defeat Spartans
Xen., *Hell.* 4.5.11–17
Olynthus (381) – Olynthians defeat Spartans
Xen., *Hell.* 5.3.3–7

Tegyra (375) – Thebans defeat Spartans
Plut., *Pel.* 17
LEUCTRA (371) – Thebans and allies defeat Spartans and allies
Xen., *Hell.* 6.4.8–15; Plut., *Pel.* 23; Diod. 15.55–56
Cromnus (365) – Arcadians defeat Spartans
Xen., *Hell.* 7.4.21.25
Cynoscephalae (364) – Thebans and allies defeat Thessalians
Plut., *Pel.* 32
MANTINEA (362) – Thebans and allies defeat Spartans and allies
Xen., *Hell.* 7.5.19–27; Diod. 15.85–87
Crimisus (339) – Syracusans defeat Carthaginians
Diod. 16.77–81; Plut., *Tim.* 27–29
CHAERONEA (338) – Macedonians defeat Athenians and Thebans
Diod. 16.86; Polyaenus 4.2.2; 4.2.7

Glossary

Agogē – a system of education, specifically used for the Spartan education and conditioning of boys aged from 7–20

Agonal – 'competitive', specifically used for the system of customs, rules or norms thought to have governed Archaic and Classical Greek warfare

Andreia – 'manliness', Greek word for courage

Aretē – 'excellence' or 'virtue'

Aspis – any type of shield, usually round, also specifically the larger types of round shield carried by heavy infantry; the classical Greek *aspis* had a bowl-like shape and prominent rim

Classical – the period of Greek history from around the sixth century BC to the time of Alexander the Great (late fourth century BC). Characterized by numerous small city states (*poleis*) throughout Greece and the Greek world (which included southern Italy and the coast of Asia Minor). Followed by the **Hellenistic** period

Column – a formation deeper than it is wide (with more **ranks** than **files**), usually used for marching. Compare with **phalanx**

Cubit – Greek measure of distance, approximately equivalent to 1½ft or 45cm (roughly 0.5 metres, for ease of calculations)

Cuirass – body armour, of unspecified construction but usually bronze or thickly padded linen

Dorians – one of the major ethnic groups in Greece, alongside **Ionians**; used particularly of inhabitants of the Peloponnese

Doru – the general Greek word for **spear**; can be used for both **spears** and **pikes**. Also transliterated as *dory*

Drachma – a coin or small measure of currency, of silver, and roughly equivalent to a day's pay. Consists of six **obols**

Ekklesia – assembly; the popular assembly of democratic states such as Athens, and by extension a meeting of any politically enfranchized group (such as soldiers)

Enomotia – 'sworn band', the smallest division of the Spartan phalanx, containing up to forty men

Ephebes (Greek *epheboi*) – 18–19-year-old youths undertaking a state-sponsored programme of military training

Ephors – the ruling magistrates of Sparta

Epibatēs – 'passenger', a hoplite serving as a marine aboard ship

Epilektoi – 'picked men', an elite unit in several armies

File – line of men in a formation standing one behind another, from front to back of a formation (such as the **phalanx**). Compare with **rank**

File leader – soldier at the front of a **file**, specially selected for courage and experience, and receiving higher pay. Commands the file, equivalent to a modern NCO (non-commissioned officer). The file leaders of a formation collectively form the formation's front **rank**

File closer – soldier at the rear of a **file**, selected for courage and experience after the **file leader**. The file closers of a formation collectively form the formation's rear **rank**

Greaves – shin protectors, armour for the lower leg

Gymnetēs – 'naked', unarmoured and probably unshielded light infantry

Hamippos – light infantryman running and fighting alongside cavalry

Hegemon – plural *hegemones*; Greek word for a leader or officer, which could be applied to men of various ranks. Sometimes used to refer to the **file leaders** collectively, or to more senior officers, but less senior than the *strategoi*

Hellenistic – the period of Greek history from the time of Alexander the Great and extending traditionally to the Battle of Actium (late fourth century to late first century BC). Followed the **Classical** period, and characterized by a small number of Greco-Macedonian kingdoms ruling much of the Near East

Helot – a Spartan serf

Hippeis – 'horsemen', cavalry, and also the elite 300-strong unit of the Spartan army

Hopla (hoplon) – 'tools', used for any or all of the equipment of a soldier

Hoplite – English name for the Classical Greek heavy infantryman, armed with *aspis* and *doru*. In Greek, *hoplitai* were any sort of heavy infantry, including **hoplites** and **phalangites**

Hoplomachos – a trainer in **hoplite** fighting (drill and weapons use)

Ionians – one of the major ethnic groups in Greece, alongside **Dorians**; used particularly of Athenians and islanders

Javelin – a **spear** specifically designed for throwing

Katalogos – 'list', specifically the list of men conscripted for a particular campaign

Keras – 'horn', the flank or wing of a formation, so (*epi keras*) a marching **column**

Kerux – 'herald', a public official used to communicate with foreign powers, and also sometimes to transmit orders in battle

Lacedaemonians – Spartans, including the *Spartiates* (citizens proper) and *perioikoi* (subordinate inhabitants)

Lance – a **spear** carried by cavalry

Lochagos – commander of a **lochos**

Lochos – general word for a military unit, sometimes used to refer to a **file**, usually in the **Classical** period to a larger unit

Medizing – term used for those Greeks who allied or collaborated with the Persian invasion of Greece

Metic (metoikos) – resident alien in a city, without citizen's rights

Misthos – pay, particularly military pay (whether paid to citizen soldiers, allies, subjects or mercenaries). Compare with *sitos*

Misthophoroi – receivers of pay, professionals; usually used to refer to mercenaries, non-citizen foreign soldiers, whether serving long term or for a specific contract. Compare with **xenoi**

Mora – the largest unit of the Spartan army, commanded by a *polemarchos*

Neodamodeis – 'new citizens', emancipated **helots** in Sparta

Obol – small bronze coin, one-sixth of a **drachma**

Othismos – 'thrust', 'push' or 'pushing', a word occasionally applied to battles when one side is 'pushed back' (defeated), and in the context of **hoplite** battles taken by some modern authors to refer to a particular tactic (a mass shove, scrum or crowd crush)

Ouragos – file closer, the man at the back of a **file**

Paean – religious song often sung preparatory to or during the advance to battle

Pelte or *pelta* – a type of shield, lacking a rim and either smaller in diameter, or lighter in construction, than the larger shield (**aspis**) of the heavy infantry. Usually round, though in earlier periods often crescent-shaped

Peltasts (Greek *peltastai*) – infantry carrying a *pelte*. In the fifth and fourth centuries applied to light infantry skirmishers typically armed with **javelins**

Pentekostys – unit of the Spartan army between an *enomotia* and a *lochos*

Perioikoi – inhabitants of a city or state (typically Sparta) with reduced citizen rights

Phalangite – English name for *sarissa*-carrying heavy infantry (in Greek, called *phalangitai* or *hoplitai*)

Phalanx – a close-order linear formation of heavy infantry. Also used in Greek to denote a line of battle that could be made up of several types of soldiers. Often used to refer specifically to the entire heavy infantry force, of **hoplites** or **phalangites**, in an army

Pike – a long **spear** held in both hands

Polemarchos – polemarch, 'war leader', senior commander in many armies, and commander of a **mora** in the Spartan army

Polis – plural *poleis* – Greek word for city, and usually used to mean an independent city state particularly of **Classical** Greece or Asia Minor. Citizens of a *polis* were required to fight in the **phalanx** as **hoplites**

Psiloi – lightest-equipped of the infantry, unarmoured and using missile weapons

Pteruges – 'feathers', flexible leather or linen strips attached to the bottom edges and arm holes of a **cuirass** to provide additional protection

Rank – line of men in a formation standing beside one another, from side to side of a formation (such as the **phalanx**). Compare with **file**

Rhipsaspia – 'shield flinging', the offence of throwing away one's shield when fleeing from battle

Salpinx – trumpet-like instrument used for military signals

Sarisa/sarissa – the Greek word for a pike, but perhaps also a dialect word for other types of spear

Sauroter – a butt spike (that is, a spike at the opposite end of the spear from the spearhead that was used in combat) of a **spear** (**doru** – it is not known for sure whether the **sarissa** also had a butt spike)

Sitos – 'food', particularly army rations. Compare *misthos*

Spear – a shafted weapon with a blade at one end and often, though not always, a butt spike at the other, wielded in one hand. A longer spear wielded in both hands is called a **pike**. A light spear for throwing is a **javelin**. A spear carried by cavalry is sometimes but not always called a **lance**

Sphagia – sacrifice, particularly by Spartans of a goat before battle

Stadium *(stadion)* – measure of distance, approximately 180 metres

Strategos – plural *strategoi* – Greek word for a general, often applied to more senior army commanders of various ranks

Tacticians – writers of organization and drill manuals describing the Macedonian **phalanx** (and associated arms). Three such manuals have survived to the modern day, written by Asclepiodotus, Aelian and Arrian. The names of several other authors are known but their works have not survived. The tradition of writing such manuals was continued under the Byzantines (Late Romans), using earlier manuals as a source for some of their material.

Talent – a measure of weight of about 26kg, commonly used to measure large sums of money (usually as talents of silver). Equivalent to 6,000 **drachmae** or 36,000 **obols**

Taxiarchos – commander of a *taxis*

Taxis – a general word for any organized body of soldiers (like 'unit' in English). Also the largest division of the Athenian army

Thetes – the lowest Athenian property class, often providing rowers for the navy

Thureophoroi – *thureos*-carriers – Greek infantry armed with the *thureos* and, probably, a long spear, and fighting as medium infantry, lighter than the **phalangites** or **hoplites**

Thureos – 'door', a rectangular or elongated oval type of shield, traditionally carried by Gallic infantry and widely adopted in the Hellenistic world in the third century.

Trophy (tropaion) – symbolic marker erected at the point where an enemy army 'turned' to rout in battle; indicated a claim of victory in the battle

Xenoi – 'foreigners', often used to refer to foreign mercenary soldiers, also used for Sparta's allies

Zeugitēs – the second-highest Athenian property class, broadly equivalent to those performing hoplite service

Notes

Chapter 1
1. Echeverría (2012) for usage of *hoplitēs* as a new, fifth-century coinage, with justification for what is admittedly an argument *ex silentio* (lack of examples of Archaic usage). Echeverría notes also that the modern usage of these terms (hoplite, phalanx) masks a lot of assumptions about organization and armament – see further below.
2. For the correct name of the shield, Lazenby and Whitehead (1996), with Echeverría (2012) for further clarification.
3. Wheeler (2007) p.192; Echeverría (2012) pp.299–303.
4. Wheeler (2007) p.192 on first uses of 'phalanx', with Echeverría (2012) pp.303–18 (312 f. for Homeric uses of *phalanges*). Wheeler's view is that 'not all linear formations of heavy infantry constitute a phalanx, which derives its character from the cohesion of the mass', but the question remains, did Greeks really, uniquely, fight as a 'mass'? If not, then what distinguishes a phalanx from other linear formations? Wheeler (p.193) also offers depth, but (see below) Greek formations were not uniquely or unusually deep. Brouwers (2013) pp.92–93 also makes the distinction between 'mass' and 'massed', with 'massed' meaning 'men organized in tight formations', as well as a formal command structure. I am not convinced that the tightness of the formation is of that much significance, and at p.165 Brouwers agrees that 'it is difficult to maintain' that 'the Greeks were the first to field armed men in tight formations'. Echeverría (2012) identifies Xenophon as the originator of the word 'phalanx' with a restricted technical meaning, although we cannot be sure of course that he really coined the usage, rather than that it developed in his lifetime. See further on these points below.
5. Krentz (2013) p.137 points out the the first occurrence of 'the phalanx of hoplites' (ἡ φάλαγξ τῶν ὁπλιτῶν) in literature is Xen., *Anab* 6.5.27 (the need to specify explained by 6.5.7, where the Persian commanders lead numerous cavalry 'in phalanx'). So also Echeverría (2012) p.308. The word 'phalanx' occurs multiple times in the *Anabasis* to refer to the hoplite phalanx where 'of hoplites' is to be understood (the first example being Xen., *Anab*. 1.2.17, the parade before the Cilician queen). Note that at Xen., *Anab*. 1.8.17, 'phalanx' refers to the two opposing lines, Greeks and Persians, at Cunaxa; at 1.10.10 to the Persian line specifically; at 2.1.6 to the Greek position as they provided themselves with food from the baggage animals; and so on – clearly the word is not being used in any strictly technical sense (see Echeverría (2012) for further examples). At Xen., *Anab*. 4.8.9–12 however, it is used to mean specifically a line, as opposed to column, formation. In Thucydides (Echeverría (2012) p.310) much of the same vocabulary is used to describe the infantry formation (in terms of its depth, density, width), but never the word 'phalanx'.
6. For a useful summary of the hoplite debate, on which this summary is based, see Kagan and Viggiano (2013). The modern conception of hoplite warfare has its origins, as well as in the nineteenth-century English tradition, in German scholarship of the late nineteenth and early twentieth centuries, for which see Konijnendijk (2018) ch.1 (the 'Prussian model').
7. Kagan and Viggiano (eds) (2013) pp.2–5, with Grote (1846) II pp.106–08. For the tyrants, Andrewes (1956), though military aspects of this phenomenon, such as reliance on mercenaries, are not so firmly believed as they once were.
8. Kagan and Viggiano (eds) (2013) pp.7–12; the quote is from Grundy (1911), p.268.
9. Nilsson – Kagan and Viggiano (eds) (2013) pp.12–14, with Nilsson (1929). Lorimer – Kagan and Viggiano (eds) (2013) pp.14–16, with Lorimer (1947) p.76. The history of the development of the linkage between hoplite shield, hoplite phalanx and hoplite revolution is

summarized by Krentz (2013) pp.137–40, tracing the development of the idea that the shield requires the phalanx (in its closest order form) to the nineteenth-century German historian Helbig.
10. Kagan and Viggiano (eds) (2013) pp.16–17, with Adcock (1957) and Andrewes (1956).
11. Kagan and Viggiano (eds) (2013) pp.21–35, with Hanson (1993), Hanson (1989) and Hanson (1995).
12. Hanson (1995). The difficult question of agricultural population and practice in Archaic Greece is addressed by Foxhall (1997) and (2013), a methodology that seems promising of decisive results though the findings are more negative than positive, but do suggest (1997 p.123) that 'there is no evidence for dramatic changes in cultivation practice' before the sixth century. See also for criticism both of Hanson's model and of the survey evidence, Forsdyke (2006). Van Wees (2013) for further criticisms of Hanson's model, particularly (p.224) that the evidence for this supposed Archaic transformation is all Classical. On the question of which came first, phalanx or equipment, see for example Viggiano and van Wees (2013) p.58. Schwartz (2013a) p.103 quotes with approval Hanson's (1991) p.77 n.28 analogy with First World War aerial tactics; Hanson compares the development of Greek equipment with Pentagon military procurement, notes that 'battle in the air antedated the appearance of true fighter aircraft' and that 'No one would suggest that air combat grew out of the discovery of novel aerially mounted automatic weapons.' The suggestion appears to be that phalanx tactics (close-order, hand-to-hand, pushing-based) came first and that hoplite equipment was developed to meet the needs of these tactics. But aside from the fact that the cultural evolution of weapons is as different from Pentagon weapons procurement as can be conceived, the comparison with aerial tactics is misguided; aerial tactics as we know them absolutely did grow out of the invention of the forward-firing synchronized machine gun, which predated and triggered the development of such tactics. The need for such a gun was of course demonstrated by earlier aerial combat, with inferior weapons and different, less sophisticated, tactics.
13. Kagan and Viggiano (eds) (2013) pp.35–38, with Snodgrass (1965).
14. Kagan and Viggiano (eds) (2013) pp.38–41 for other gradualists; Krentz and Cawkwell, pp.41–44, with Krentz (1985b) and Cawkwell (1989).
15. Kagan and Viggiano (eds) (2013) pp.46–49, with van Wees (1994) and (2004).
16. For the new approach, see for example Brouwers (2013) and (2014); Konijnendijk (2018); Brouwers and Konijnendijk (2020); Echeverría (2011) and (2012). Viggiano (2013) (amongst others referenced elsewhere in this chapter) offers a defence of the orthodox view, concluding (p.126) that 'A grand narrative involving the hoplite phalanx helps explain the rise of this unique phenomenon' (the Greek *polis* and its relatively egalitarian constitution).
17. Kagan and Viggiano (eds) (2013) pp.44–9. Brouwers (2013) pp.16–39 for Mycenaean warfare, pp.40–71 for the Dark Ages, and for thoughts on the applicability of Homer to any particular historic period, with further reading, pp.151–52. Widespread acceptance that Homer can tell us something about Archaic warfare is due largely to Latacz (1977), on which see also Snodgrass (2013); van Wees (1994). What it can tell us remains much disputed; I do not believe that Homer describes fully developed hoplite warfare identical to that of the Classical period, but the value of Latacz, as Snodgrass (2013) points out, is in doing away with the idea (from Lorimer (1947) and others) that Homeric warfare and hoplite warfare are of necessity utterly different. See also Schwartz (2013a) pp.105–15. The dates of Homer's composition, or commitment to writing, is a huge topic in itself – summarized in Nagy (1997).
18. For the supposed Homeric origin of the Macedonian phalanx, Taylor (2020) pp.82-5; Pol.18.28.6 and Diodorus 16.3.2. On Homer and the tacticians, Rance (2017). For the influence of Homer on Greek military thinking in general, Lendon (2005).
19. Arguing that Homer's picture of chariot use may reflect actual (Dark Age) practice, Brouwers (2013) pp.59 and 160 with further reading (a comparison is with British and Cyrenaic chariot use).
20. Tyrtaeus and Archaic poetry generally discussed by Schwartz (2013a) pp.115–23.
21. The similarities between the passages of Xenophon and Homer is stressed by Krentz (2013) p.147.

22. On use of *otheo* and related words in Homer, Krentz (2013) pp.146–47.
23. File intervals form the mainstay of much of the hoplite orthodoxy, the assumption being that the use of the hoplite shield requires a very close order, and that this very close order distinguishes the hoplite phalanx from earlier or other contemporary formations. This chain of assumptions fails at each stage – see further in Chapter 3. Pritchett (1971) pp.134–54 collects evidence for file intervals, for which, however, there is none (no concrete evidence with actual figures) before the Hellenistic period. The comparison with New Guineans appears in van Wees (2004). Bardunias and Ray (2016) pp.109–12 make the very interesting observation that a crowd can self-organize into a close-order formation, driven by individual desires such as for protection and security, just as well as it can be forced into such a formation by drill, and such a formation will look much like a rigidly drilled one. I agree with this analysis, though the differences will become apparent when the undrilled formation attempts to change facing or formation or perform any more complex manoeuvre.
24. Van Wees (2004) pp.154–58 describes this type of fighting in Homer, with more examples. The type of fighting described is equivalent to that sometimes called (today) the 'dynamic stand-off', on which see further below.
25. Hanson (1989) p.xxiv for example assigns this more individual style of fighting to 'Egyptians, Hittites, Persians or tribal forces', though he also sees them as not heavily armed or armoured and dependent on missile weapons, and more importantly as not 'seeking quite literally to push the other off the battlefield'. For the 'dynamic stand-off' model, as applied to Roman and Carthaginian combat in the Punic Wars, see Sabin (2000).
26. The idea that Greek (military) history was largely driven by technological changes in weapons and equipment, which underlies much of the orthodox view of the hoplite, has been rightly criticized as 'technological determinism' – see Echeverría (2010). I will address some of these matters below.
27. Dendra panoply – Snodgrass (1999) pp.24–25 and pl.9. Argos panoply – Snodgrass (1999) pp.41–42 and pl.17. Archaic arms and armour – Snodgrass (1999) pp.48–89.
28. For the Greek debt to the Carians for at least some of their equipment, Brouwers (2013) pp.75 and 161; also pp.98–101 for the Greek debt to other Anatolian peoples (Lydians, Phrygians). The historical reality of the Boeotian shield has often been doubted, for example by Snodgrass (1964) and (1999) pp.55 and 97, who doubts the existence of either Boeotian or earlier dipylon types. Greenhalgh (1973) sees the dipylon as real, but the Boeotian as an artistic fiction, a mix of dipylon and hoplite aspis. Here I follow van Wees (2004) pp.50–52 and Boardman (1983). For a modern reconstruction and possible use of the shield, see https://www.youtube.com/watch?v=IeKuy36OG_g.
29. The total unsuitability of the hoplite shield and other equipment for any other form of fighting than a close-order rigid phalanx is a central tenet of of the 'orthodox' position – see Lorimer (1947) and all restatements of the position since, for example Schwartz (2013a); Cartledge (2013) p.77. See also the comments of Snodgrass (2013) pp.91–92 arguing for a form of gradualism in the adoption of the two-handled shield, with earlier lighter and later less-flexible variations. Viggiano (2013) pp.113–14 credits to the introduction of the hoplite shield a 'fundamental change in the social and political structure of the polis', asserting that 'use of the shield makes sense only in the context of a phalanx' (by which he means a Classical-style close-order phalanx). Schwartz (2013b) and Viggiano (2013) attempt to make a case for the impracticality of the hoplite shield, but for example Viggiano's (2013) p.116 suggestion that part of the unsuitability of the hoplite shield for general fighting was because 'the bearer could wield the shield with the left arm only, as opposed to the warrior who could shift a single-grip shield from one hand to the other to relieve its weight' seems so far divorced from reality as to undermine the entire argument. Schwartz (2013b) p.163 repeats the same strange idea. Shields have always been carried on the left arm or hand, and weapons in the right.
30. Schwartz (2013a) pp.53–54 makes the comparison with Danish riot police shields. A search on YouTube for 'hoplite single combat' will reveal a number of examples of those who experiment (with varying degrees of verisimilitude) with the techniques.

31. Mobility and protection e.g. Viggiano and van Wees (2013) p.59; van Wees (2000) pp.130–31. Wheeler (2007) p.199 agrees that 'the notion of hoplites helpless outside the phalanx is a myth', though a myth that has proven remarkably resilient, unfortunately.
32. Viggiano and van Wees (2013) pp.63–70 for other early examples of throwing spears and mixed forces in early vase paintings; that the spears on this vase are throwing spears, Snodgrass (1964) p.138; Snodgrass (1999) p.57. Schwartz (2013a) pp.123–30 on the Chigi vase and other Archaic representations.
33. Statistics for the proportion of archers to hoplites (spearmen), if not their organization, in Sparta can be derived from the numerous lead figurines dedicated to Artemis Orthia – see Wace (1929), showing a decrease in the proportion of archers from the seventh to sixth centuries.
34. The importance of fleet rowers in the development of more democratic constitutions is stressed by Strauss (1996).
35. To Aristotle's version we can add Thucydides' account of early Greek history (Thuc. 1.10–19). Thucydides downplays the importance of land warfare as just 'the usual border contests' with no major expeditions other than the Lelantine War between Chalcis and Eretria (around 700). Note that neither Thucydides nor Aristotle mention the impact of a revolutionary shield design. For Archaic cavalry (and in particular their common role as mounted infantry, *hippobatai*), Brouwers (2013) pp.76–78, suggesting the Argive shield may originally have been a cavalry shield, and pp.108–09 that changes of equipment in the late sixth century reflect the declining role of *hippobatai* and increase of true infantry.
36. Mercenary service in the Archaic period – Raaflaub (2004), Hale (2013). Hale makes the suggestion that mercenary service in the East is perhaps the context for the origin of hoplite and phalanx, rather than Greece itself. Luraghi (2006) gathers the evidence (plundered chariot fittings found in Ionian sanctuaries) for Greek mercenary service in Assyrian armies from the eighth century.
37. It is an open question of course how much reliance should be placed on Aristotle's theory of the progress of constitutions – not much for van Wees (2002) pp.72–77, a lot for Hanson (1995) p.237. The date of the change envisaged by Aristotle is also disputed – as late as the end of the sixth century according to van Wees.
38. Wheeler (2007) pp.205–06 interprets *syntaxis* as 'continuity of files' (that is, presumably, close order) and 'depth', but I think something more must be meant (something encompassed by 'drill'), since depth and (the potential for) close order are common features of any infantry formation.
39. Cartledge (2013) pp.76–77 on the date of Spartan military reforms, about which there is much debate.
40. Viggiano and van Wees (2013) pp.67–68 on the Chigi vase. Snodgrass (1999) p.58 identifies this as a formal phalanx; van Wees (200) pp.136 f. considers the formation more informal. On the flute or pipe player, Krentz (2013) pp.138–39 – the notion (going back to Delbrück and before) that 'the piper is nothing other than the tactical formation' is untenable. Wheeler (2007) p.197 sees Tyrtaeus and the Chigi vase as depicting 'linear tactics before creation of a deep phalanx' but I see little evidence in poem or vase for depth or its absence either way. So far as we can tell, all infantry formations, everywhere, at least before the adoption of firearms, formed deep, so it seems reasonable to assume that Greek ones did also. Brouwers (2013) p.93 similarly finds that the Chigi vase does not depict a phalanx because the men are 'not in any kind of tightly-knit formation'. See also Brouwers and Konijnendijk (2020). I think the formation is clearly quite tightly knit and the problem lies with trying to divide depictions into two groups, phalanx and non-phalanx, which depends on interpreting a phalanx as something unique and distinctive, something clearly different from other types of formation. Matthew (2012) pp.26–28 argues that the Chigi spears are throwing spears (reasonably enough, since the men carry second spears in their left hands); this is noted with approval by Brouwers (2013) p.165 but again, it is a false dichotomy to claim that such men cannot therefore be in 'a phalanx', beyond the fact that such a phalanx will not be identical in every detail to those seen in the later fifth century. Note also that although the men are depicted marching or running in step, this is most likely artistic convention – horses are also usually shown in step.

41. Viggiano and van Wees (2013) pp.63–70 for other examples of apparent close order in early vase paintings, interpreting them as looser formations than the later phalanx. Van Wees (2000) pp.140–42 expands on this argument. For the Amathus bowl and its significance see Hale (2013) pp.182–84; Schwartz (2013a) pp.130–35.
42. Wheeler (2007) p.187 makes the point that 'hopliticization' was not universal. The Edwardian origin of the conception of the Greek phalanx with its supposedly unique fighting style is pointed out by Krentz (2013) pp.143–48. The desire to seek the Classical phalanx in scattered Archaic evidence has been aptly termed 'an attempt to "discover the phalanx" in the sources; to identify a closed formation or a specific kind of heavy-armed warrior in the scattered pieces of literary, iconographic, and archaeological evidence', Echeverría (2012) p.291.
43. Snodgrass (1999) p.59 for antecedents to Greek equipment (which he finds rather lacking); Witowski (2018) for a more positive view. On Persian armour (compared with Herodotus' depiction of it at Plataea), Charles (2012); Konijnendijk (2012). The Persian army and fighting techniques generally, Manning (2021). It may be that a specific body of Persians at Plataea were unarmoured, or that there was a particular reason – the Persians are said to have set off from their camp in pursuit of what they thought was a fleeing enemy, so may not have been fully equipped. See Raaflaub (2013) for the influence of other Near Eastern civilizations on Greece in military matters (finding in the end their influence rather limited). Snodgrass (1964) for more details, particularly possible Assyrian influence on the hoplite shield. Wheeler (2007) p.193 recognizes that 'the phalanx need not be a Greek peculiarity' but is pessimistic about our knowledge of other examples. Raaflaub remarks (p.96 and n.14) on the lack of detailed studies (in English) of Persian or Assyrian armies. To the references he cites we can now add Manning (2021), which includes a review of previous scholarship, along with popular works such as Sekunda (1992) on the Persians and Healy (1991) on the Assyrians. Raaflaub (pp.100–01) finds in depictions of Assyrian armies 'evidence for mass battle but nothing that would suggest dense formation', and compare Snodgrass (2013) p.91 'van Wees did *not* deny the presence of mass combat in the *Iliad*, but that of *massed* combat, a different thing'. This seems to be setting a lot of store by the question of precisely how close together men stood, a question it seems to me very much of secondary importance. The presence of formations seems more important than the details of their drill, and they surely cannot be denied for Assyrians or Persians (see Fagan (2009) for the former). A context for Assyrian influence on Greek equipment is provided by Greek mercenary service in the East, Hale (2013) p.180. See also Bardunias and Ray (2016) pp.30–31 for Assyrian and Urartian parallels to the hoplite shield. See also Brouwers (2013) for further reading on 'much-needed context' concerning earlier Egyptian and Near Eastern infantry formations.
44. In this interpretation I am closer to the views of Raaflaub (1999) than van Wees (2004); but all reject the orthodoxy – just the timing differs. The need to 'explain the phalanx' – as e.g. Wheeler (2007) on p.195 sees as the objective, see also Echeverría (2012) p.291 – presupposes that 'the phalanx' is something unique and different, and in need of explanation.

Chapter 2
1. Lazenby and Whitehead (1996) for the name of the shield. For Etruscan and Roman shields see for example Connolly (2012) pp.94–100.
2. Summary of early Greek arms and armour in Viggiano and van Wees (2013). For descriptions and definitions of the shield and links to further references, Schwarz (2013b) pp.157–61. Snodgrass (1999) pp.53–55 (calling it, however, a *hoplon*).
3. The argument from blazons, Lorimer (1947) and pl.19; Snodgrass (1999) pp.54–55.
4. See for example Viggiano and van Wees (2013) pp.63–70 with illustrations.
5. Snodgrass (2013) pp.88–93 on the chronology of offerings at Olympia and their significance. Brouwers (2013) pp.95–98 – with the observation that for example greaves are rarely found in matched pairs, so we do not know how representative of actual equipment such finds are.
6. Range of sizes – Schwartz (2013b) p.159. Schwartz expends effort and mathematics needlessly on the story of Arrian (*Anab.* 1.19.4) that Milesian soldiers paddled across to an island on their upturned shields, assuming this to mean using them like coracles. They could just

have easily have used them as buoyancy aides – depictions in vase paintings do not suggest Argive shields (if that is what these were), though large and concave, were quite that large and concave.

7. Construction methods – Blyth (1982) pp.9–13; Schwartz (2013b) p.158. De Groote (2016) examines and tests the construction and strength of the shield, with emphasis on its resistance to penetration by spear thrusts (c.f. Plut., *Mor.* 219c); quantified results are given, showing that shields could be penetrated and/or split by spear thrusts but that the domed shape of the shield helped protect against this, and the bronze facing, although too thin to provide much resistance to penetration, did protect against splitting. The domed shape also allowed the central section of the shield to be thinner, and therefore lighter. Interestingly, Schwartz suggests that the grain would run horizontally, offering flexibility 'in case of simultaneous pressure on the sides of the shield', and finds simultaneous pressure on top and bottom 'an unlikely scenario'. In the view of Bardunias and Ray (2016), however, the shield was designed specifically with pressure on top and bottom in mind (as the shield was pressed against the shoulders and thighs of its bearer – which is surely correct, if the shield was ever so pressed, which I doubt, see Chapter 8). See Bardunias and Ray (2016) pp.32–34 for more on the construction of the shield.

8. Woods – Krentz (2013) p.136; citing the reconstruction by Blyth (1982) of an Etruscan example. Schwarz (2013b) p.158 for the same examples. Schwartz (2013a) pp.28–29 on wood and construction.

9. Snodgrass (1964) pp.63–64 is of the view that most shields were not bronze-faced. It seems difficult to settle this question with any degree of certainty, but I tend to favour a bronze facing being common, but not universal. The rim may have been the first part to be so protected, to protect against splitting when struck with edged weapons, Schwartz (2013b) p.159. De Groote (2016) for protection against splitting.

10. Champions of the hoplite orthodoxy such as Hanson (1999) and (1995) assign weights of over 30kg to total hoplite equipment, 'an incredible burden to endure for the ancient infantryman'. But this figure, as shown by Krentz (2010a) pp.45–50, (2010b) pp.190–97 and (2013) pp.135–36, goes back to nineteenth-century (AD) German historians and was little more than a guess, described by contemporaries as 'arbitrary estimates' and 'purely hypothetical'. Krentz estimates the total weight of hoplite equipment, based on surviving examples, as 14–21kg (or 18–22kg for the earlier period), of which the shield accounts for 3.2–6.8kg based on extant shields from Italy. See also Jarva (1995) for similar conclusions. Schwartz (2013b) pp.160–61 quotes 6.75kg for the Bomarzo shield and suggests 7–8kg as the norm. Note similarly that Hanson (2013) p.265 in the section 'Was Hoplite Armor Heavy' references Xen., *Mem.* 3.10.9–14 (amongst other passages) as evidence for 'the weight, discomfort, and clumsiness of their hoplite armour – a ubiquitous theme throughout Greek literature'. This passage, which I quote below, in fact states that badly fitting armour was 'uncomfortable and irksome' but that 'the good fit ... may almost be called an accessory rather than an encumbrance', which is quite the opposite of Hanson's assertion.

11. Krentz (2013) p.136 lists several estimates of shield weight ranging from 3.2kg (the Etruscan example referenced above, without a bronze facing) to 9kg made by some hoplite re-enactors. Average figures seem close to 7kg. For views on the ease of use of spears and shields, a flood of opinions are available by searching 'hoplite spear and shield' (for example) on YouTube (YouTube videos being transient, there is little point linking to particular examples here). A search for 'hoplite single combat' will reveal plenty of examples of vigorous single combats in hoplite kit that would appear to invalidate the theory that such a thing was difficult or impossible. Schwartz (2013b) pp.165–68 argues, from the relatively small stature of ancient Greeks based on skeletal remains, that hoplite equipment would have been more arduous to wield for Greeks than for modern adults; but stature is a poor measure of 'toughness, stamina, strength, and resilience', the factors Schwartz identifies as most important. Bardunias and Ray (2016) pp.78–79 makes this point well. See also Schwartz (2013a) pp.98–101 (with opposite conclusions). Another telling comparison is with the full plate armour ('harness') of the Medieval knight, often supposed by modern scholars and in the popular imagination

to be almost impossibly unwieldy, yet shown by modern practitioners to be anything but – search 'knight armour agility' on YouTube for examples.

12. Schwartz (2013b) pp.161–62 stresses the corporal nature of this punishment and downplays the element of disgrace, which may be fair; but simply carrying a shield cannot have been so very terrible a burden, since after all on the march a hoplite (or his servant) must routinely have carried the shield all day. That it was uncomfortable to stand holding a shield (or that, in another instance, a hoplite struggling up a hill might prefer not to carry his shield, Xen., *Anab.* 3.4.47–49) is not evidence for the shield's 'very real problem of sheer weight' (Schwartz p.162). Few soldiers in the whole of human history have willingly carried their kit, or turned down an offer for someone else to take it (as Eurip., *Heracl.* 720–26). For the 'at ease' position, Schwartz (2013b) p.168 with references; this was the basis of the Athenian general Chabrias' ruse where he repelled a Spartan army by having his men stand at ease, thus showing their confidence (Diod.16.32.5, Polyaen. 2.1.2).

13. Schwartz (2013a) pp.46–49 discusses the *hoplitodromos* (attempting to downplay the relevance it has to the encumbrance of the shield). See also pp.49–53 for the Pyrrhic dance. Modern experiments on running in hoplite gear such as Donlan and Thompson (1976) and (1979) suffer from the usual problems with such experimental efforts, in particular the level of fitness of the test subjects. *Rhipsaspia* – Schwartz (2013a) pp.147–55 but using the offence to argue that 'the shield was hopelessly heavy and clumsy and spectacularly out of place outside the highly specialised fighting environment of the phalanx' (p.154). Hoping to save one's life by throwing away a shield while running away clearly indicates no such thing; while if true, every hoplite on the losing side in a battle would throw away his shield, and the jokes and criticisms (in Aristophanes for example) aimed at those who did so would be meaningless.

14. Wheeler (2007) p.196; Goldsworthy (1996) pp.209–11 for Roman shield weights (with further references).

15. An examination of the Viking shield, with reconstructions and thoughts on use, is online at http://www.hurstwic.org/history/articles/manufacturing/text/viking_shields.htm.

16. The comparison with (Danish) riot police is in Schwartz (2013b) pp.164–65 and (2013a) pp.53–54. Schwartz quotes his police informant to the effect that 'the [riot] shield was deemed too heavy, large and awkward to be wielded freely, and to be put to offensive use' – it is not clear what 'offensive use' means in this context. A shield as large as 1 metre in diameter need not be swung around wildly to protect the whole body of the man behind it – that is presumably the point of making it so large. The comparison with a combat rifle is also made by Blyth (1982) but dismissed as 'altogether useless' by Schwartz (2013a) p.35 on the grounds that with an average weight of 4kg, a rifle is lighter than a shield. I do not agree that this makes the comparison useless. Weights of modern combat equipment are available online: https://en.wikipedia.org/wiki/SA80; https://en.wikipedia.org/wiki/Lee%E2%80%93Enfield; see also the discussion at https://www.thinkdefence.co.uk/overburdened-infantry-soldier/.

17. Strabo 14.2.27 repeats Herodotus' claim of a Carian origin for such handles. There are some interesting second-century AD usages in Lucian: 'Yes, and you saw in the gymnasium a bronze disk like a small buckler [*aspis*], but without handle [*ochanon*] or straps [*telamonas*]', Lucian, *Anacharsis* 27; of a painting of Alexander, 'On the other side of the picture, more Loves playing among Alexander's armour; two are carrying his spear, as porters do a heavy beam; two more grasp the handles [*ochanon*] of the shield, tugging it along with another reclining on it, playing king, I suppose', Lucian, *Herodotus and Aetion* 5. But Pausanias describes a jumping-weight: 'They are half of a circle, not an exact circle but elliptical, and made so that the fingers pass through as they do through the handle of a shield [*ochanon aspidos*]' (5.26.3) (which sounds more like an *antilabe*).

18. For the carrying arrangements of the Macedonian shield, Taylor (2020) esp. pp.61–63.

19. Krentz (2013) pp.139–40 for the rope or cord. Pittman (2007) pp.70–72 suggests hoplites grabbing the cords of their neighbours, so also e.g. Brouwers (2013) p.109. I find this improbable since it would have made it impossible to manoeuvre the shield, though it might have happened in specific circumstances, such as when facing cavalry. Matthew (2012)

pp.42–43 records various other suggestions, none of which are entirely convincing. Depictions showing the *antilabe* separate from this cord would materially affect its practical use, if any.

20. For Egyptian shields, Shannahan (2014); it appears that Egyptians had replaced their earlier tower shields with round shields long before the time of Cunaxa, but their shields were of similar construction (wooden with a covering in the Egyptians' case of hide), and distinguished by Xenophon as *aspides*, as opposed to the wicker *gerrai* of the Persians. Xenophon may be mistaken in identifying these as Egyptians, though their actual nationality does not matter greatly for these purposes. Hale (2013) for the suggestion that hoplite equipment was brought back from the East by mercenaries serving abroad. Krentz (2013) p.148 suggests plausibly that large shields may have been used, at first, by poorer, less well-equipped men, while the bronze-armoured aristocratic *promachoi* may have continued to use lighter shields like the Boeotian (below).

21. See Chapter 1 note 28 on the Boeotian shield; van Wees (2004) pp.50–52. Viggiano and van Wees (2013) p.63 for an example of a painting of Boeotian-shielded warriors fighting an aspis-carrying warrior. Snodgrass (1999) p.55 for the view that it 'can never have existed'.

22. Holding the shield by three points – the *porpax*, the *antilabe* and the rim resting on the shoulder – is suggested by Schwartz (2013b) pp.162–63 as the best way. I prefer to think that it is is one of several ways, based on vase paintings and the experience of re-enactors. Schwartz goes on (p.163) to assert, based on vase paintings, that this was 'the normal grip and defensive stance' (I think there are too many vase paintings depicting other ways of holding the shield to bear this out) and most strangely that re-enactors 'have assured me that this way of handling the shield is not only the logical but indeed the *only possible* way'. Other re-enactors clearly do not agree, nor did ancient painters of vases. See also (contra) Bardunias and Ray (2016) pp.31–32; (pro) Matthew (2012) p.40.

23. For this point see also Viggiano and van Wees (2013) pp.58–59 and drawings, p.60; van Wees (2000) p.128. Viggiano (2013) p.117, however, seems to repeat the assertion that Thucydides means that the man's front was protected by his neighbour's shield, quoting with approval Greenhalgh (1973) p.72, who states that 'the significance of Thucydides' observation is that some lateral, not frontal, protection was obtained from the next man's shield and that it was vital not to allow a gap to develop', with which I agree – but this is contrary to the shared shield theory, not a supporting argument for it. Hanson (2013) p.258 repeats (again) the orthodox view with the usual references; a passage of Plutarch he quotes (in n.7) however (Plut., *Pel.* 1.5, 'the Greek lawgivers punish him who casts away his shield, not him who throws down his sword or spear, thus teaching that his own defence from harm, rather than the infliction of harm upon the enemy, should be every man's first care, and particularly if he governs a city or commands an army') does not support this conclusion at all. On the combat stance see also Bardunias and Ray (2016) pp.16–18 and Matthew (2012) esp. pp.44–59. See also van Wees (2000) pp.126–28 and (2004) pp.168–69, pointing out that such a stance puts the hoplite behind the centre of his shield, not its right side. A glance at any photograph of re-enactors in hoplite equipment will confirm this – the shield easily covers the whole of the body, as viewed from the front – see for example Matthew (2012) plates 18.1, 21.1. See also (for example) Randall (2011) for some practical experiments and findings.

24. On the shield apron or curtain, as intended protection against arrows: Anderson (1970) p.17; Snodgrass (1999) pp.103–04; Bardunias and Ray (2016) p.34.

25. Blazons – Snodgrass (1999) p.55; p.67 for city insignia; van Wees (2004) pp.53–54; Anderson (1970) pp.17–20.

26. For hoplite activities outside the phalanx see Rawlings (2000), and (all too briefly) Chapter 6 of this volume. This point seems to me conclusive – the hoplite simply did not fight exclusively in the phalanx, so the assertion that the shield was only suitable for such fighting is untenable. The point is made by e.g. Krentz (2013) p.138 and cannot be stressed enough.

27. Typology of armour is set out by Jarva (1995), identifying four types (bronze bell, bronze plate, muscled and linen). See Snodgrass (2013) for the chronology of armour use. Schwartz (2013a) pp.66–81, with usual emphasis on awkwardness, difficulty of wearing etc; Bardunias

and Ray (2016) pp.42–46 for a summary, with useful comments on leather and linen armours. Snodgrass (1999) remains the basic account (for all arms and armour).
28. It has been suggested that as few as one in ten hoplites wore metal body armour based on numbers of finds – Jarva (1995) pp.111–13 and 124–80.
29. This armour is 'Type IV' in the scheme of Jarva (1995). Aldrete, Bartell and Aldrete (2013) for detailed discussion. Note that armour of this sort was used also by the Etruscans and is frequently depicted in fourth-century Etruscan tombs, see Gleba (2012).
30. For modern reconstructions of the linen cuirass see Aldrete, Bartell and Aldrete (2013).
31. The construction method by gluing, along with the penetration tests, are detailed in Aldrete, Bartell and Aldrete (2013); it is also possible that the armour was thickly woven from multiple strands, or thickened with other substances, Gleba (2012) p.47.
32. Snodgrass (1999) pp.90–92 on composite armour.
33. There is a tendency to view armour and protection or tactics in strictly functional terms (e.g. Hanson (1991) pp.63–67), and that heavier (or lighter) armour must represent specific tactical requirements (in this case, the presence of bronze body armour being evidence for close-quarters phalanx fighting). However, there are many other factors – social, economic, technical and (not least) to do with custom and fashion – driving the use or abandonment of armour and other military equipment. Note also, relevant to the 'chicken and egg' discussion of Chapter 1, that hoplite armour became lighter at the same time as the phalanx formation became more formal and specialized for hand-to-hand fighting.
34. Helmets are magnificently (and expensively) catalogued and illustrated in Hixenbaugh and Valdman (2019); for a succinct account with a much-followed diagram of typology, Connolly (2012) pp.60–63. Schwartz (2013a) pp.55–66 is also a good summary; Bardunias and Ray (2016) pp.40–42; Snodgrass (1999) pp.51–53; 67–70.
35. Snodgrass (1999) p.51 for construction of the Corinthian helmet. Note that the identification of this helmet as the Corinthian helmet of Herodotus is not certain, but the use of the name for this type of helmet is firmly established in modern usage – we cannot be certain what, if anything, the Greeks called the type. Schwartz (2013a) p.61 reports experiments with the field of view of a Corinthian helmet, typically inconclusive. Certainly the helmet restricted field of view but soldiers in many other eras – particularly the Medieval European mounted knight – have accepted some restriction of vision in return for better perceived protection.
36. See Sekunda (1986) p.6 and pl.A1. See Chapter 5 for other indications of rank.
37. For greaves, Schwartz (2013a) pp.75–77; Bardunias and Ray (2016) pp.46–47; Snodgrass (1999).
38. Spear hold and balance – Bardunias and Ray (2016) pp.14–16; Matthew (2012) pp.1–18 and *passim*. Schwartz (2013a) pp.81–85 for spears, particularly the throwing spear (or javelin).
39. Summary of the grip issue in Krentz (2013) pp.141–42. Anderson (1991) p.31 favours the overarm hold; Hanson (1989) pp.162–64 supports the underarm; Lazenby (1991) pp.92–93 points out that the underarm hold is usually shown in duels, and the overarm in formation or mass battles. Matthew (2012) esp. pp.71–92 argues for a high or couched underarm position, but see also Bardunias and Ray (2016) pp.19–21 and the criticism of Matthew's statistics of artistic depictions in Ray (2014). Matthew (2012) pp.26–28 also argues that depictions of the overarm hold, such as the Chigi vase, are in fact depictions of throwing spears, not thrusting spears; I can see that this is the case on some, but not on all, occasions, and I believe that making such a strict distinction between thrusting and throwing spears is unfounded.
40. Matthew (2012) devotes an entire chapter (pp.60–70) to the perceived difficulties of changing grip. Schwartz (2013a) pp.90–91 on the question, reasonably pointing out that the left hand could be used, as it long had been for carrying a throwing spear. I am perplexed that nobody seems to have found reversing the grip while holding the spear to be simple, although (e.g.) Randall (2011) also found it could be done without difficulty (the details of his technique are not totally clear). See also De Groote (2018). Schwartz identifies *metalepsis* of Pol. 6.25.9 as being a name for this manoeuvre, though Polybius' other uses of the word (2.33.4; 9.20.2; 31.13.3; 38.5.9) show it is just a general word for 'change' or 'switch'.
41. Spear shafts – Matthew (2012) pp.6–14.

42. Spear blades or heads – Matthew (2012) pp.2–4; sauroter – Matthew (2012) pp.4–5.
43. On the mechanics of the *ankyle* see Bardunias and Ray (2016) pp.13–14; Matthew (2012) pp.27–29.
44. Swords are briefly covered by Bardunias and Ray (2016) pp.22–23; Schwartz (2013a) pp.85–86; Snodgrass (1999) *passim*.
45. See Sekunda (2000) pp.58–59 and Pl.F for typical clothing (and other details).

Chapter 3

1. For tactical writers, their sources and their origins, see Rance and Sekunda (2017) and especially the introduction, Rance (2017).
2. For the Macedonian phalanx, which has an extensive literature of its own, see in general Taylor (2020); a suggested reconstruction of the fighting techniques of the Macedonian phalanx is provided in ch.9.
3. For the Spartan army and organization, see Lazenby (2012); van Wees (2004) Appendix 2, pp.243–49; Anderson (1970) Appendix, pp.225–51; Sekunda (1998). Some of the most discussed issues concerning the Spartan army are the extent to which organization changed during the fifth and fourth centuries (on which see below), and the size of the army units listed, in particular whether they were inclusive or exclusive of the *perioikoi* – I do not feel qualified to offer a conclusion on this latter point and instead refer to the works referenced.
4. For these fourth century reforms, see for example van Wees (2004) pp.243–49; very possibly connected to the heavy Spartan losses suffered at the Battle of Leuctra. Lazenby (2012) e.g. pp.54–55 argues for continuity not reform.
5. Lazenby (2012) p.9 for the 'fiftieth'.
6. On the *enomotia* and age classes, Lazenby (2012) especially pp.9–11, 15–16, Anderson (1970) pp.241–44, with further details and discussion.
7. This complex problem is discussed at length in Lazenby (2012) Ch.1 esp. pp.9–11, Anderson (1970) Appendix; van Wees (2004) pp.243–48; Hawkins (2010). Sekunda (2014a) suggests that *perioikoi* were mingled with Spartiates in the fourth century *morai* (seeing these as a fouth-century innovation) to make up for declining Spartiate manpower.
8. For front rank men, Anderson (1970) pp.73, 97, 105; rear rank men, Anderson (1970) pp.95–96, 102, 174–75, and see also Chapter 8 below for their role in combat.
9. For the *Hippeis*, Lazenby (2012) pp.9–16 and *passim*; Anderson (1970) pp.245–49.
10. Sciritae (Skiritai) – Lazenby (2012) pp.51–54, 14–18 and *passim*; Anderson (1970) pp.249–51; Konijnendijk (2018) pp.161–62.
11. For the *neodamodeis*, Lazenby (2012) pp.18–19 (and *passim*), identifying them as the newly enrolled and freed helots.
12. Konijnendijk (2018) pp.184–85 on *ouragoi*. As a more formal rank, or at least common practice, this seems to have developed only in the Macedonian army and the Hellenistic period, but the principle of having experienced men in the rear may be more common.
13. Organization of the Ten Thousand is not covered explicitly by Anderson (1970), though pp.94–110 and the Appendix on the Spartans deal with that of armies generally.
14. The Athenian army is included in Anderson (1970) e.g. pp.94–98; van Wees (2004) pp.99–101; more detail in Crowley (2012) pp.35–39 and see also Sekunda (1992).
15. Crowley (2012) pp.40–69 for Athenian low-level organization – which other than the file, which had some formal existence (see below), would have been provided by pre-existing social 'primary groups' rather than by military units at any level.
16. On the importance of good order – *eutaxia* – see Crowley (2012) pp.49–53. Konijnendijk (2018) pp.54–55 is sceptical of the level of order in all (non-Spartan) hoplite armies and even the existence of formal ranks and files. Certainly we must be wary of ascribing Spartan, or modern, levels of drill to most Greek armies.
17. Anderson (1970) esp. pp.94–105 on tactical training in the *Cyropaedia*. Christesen (2006) suggests this was intended as a model for proposed reforms of Sparta's army.
18. Anderson (1970) p.100 for Xenophon's figures and inconsistency.
19. See also Taylor (2020) Ch.1 for consideration of this passage, and further references.

20. See Anderson (1970) pp.161–64 on the Sacred Band. For other *epilektoi* Hunt (2007) pp.144–45; Konijnendijk (2018) pp.153–62, with particular emphasis on their tactical role. Leuctra is examined by Anderson (1970) pp.192–220 and see also Chapter 7 in this volume.
21. Sekunda (1992) pp.4–6 for Persian units, and Manning (2021). Anderson (1970) pp.98–104 for the evidence of the *Cyropaedia*.
22. Pritchett (1971) pp.134–54 gathers evidence for Classical file intervals, although the only actual hard evidence (with stated figures) comes from the Hellenistic tacticians and applies to the Macedonian phalanx; these figures cannot simply be applied to the earlier phalanx (as Matthew (2012) pp.179–96 attempts to do, adopting the one cubit *synaspismos* of the Macedonian phalanx as the standard hoplite interval). Bardunias and Ray (2016) pp.117–20, Schwartz (2013a) pp.157–67, Anderson (1970) pp.100–01 for further discussion.
23. This pragmatic approach to the size of the cubit is the one I adopted also in Taylor (2020), which contains (p.30) references to further discussion.
24. See Taylor (2020) pp.85–97 for the intervals of the Macedonian phalanx and ch.9 for their fighting style. Matthew (2012) pp.179–82 reaches the opposite conclusion, that the one cubit spacing was used only by Classical hoplites and was not possible for Macedonians. Note that Matthew (2012), table p.180, takes words used in Classical authors as if they refer to the Hellenistic intervals, but I am sure is incorrect in reading this precise technical usage into what (as his examples show) is a highly variable vocabulary, see further below.
25. On doubling, Matthew (2012) pp.171–72 (noting the doubling drills usually described are really Hellenistic). Bardunias and Ray (2016) pp.117–20 (and discussion of degrees of shield overlapping).
26. Van Wees (2004) pp.185–86 for the 'open order' interpretation, with Krentz (1985b) and (1994), and (2013) p.140 (noting also his own earlier misprint of 6m and 3m instead of 6ft and 3ft, which serves as a salutary reminder). See Taylor (2020) pp.422–23 (n.5) for further discussion. In this I agree with Matthew (2012) p.181 that Greek formations were probably more flexible and variable than most scholars allow.
27. Data on depths are collected and examined by Matthew (2012) pp.172–79; Pritchett (1971) pp.137–39. Schwartz (2013a) pp.167–71; Konijendijk (2018) pp.126–38.
28. Hunt (1997) for the helots at Plataea; Cawkwell (1983) p.387 suggests (less plausibly, based on later practice) that Spartans fronted units of allies; Matthew (2012) pp.175–76 and 178–79.
29. Note that Xenophon's language is imprecise – the formation is 'in fours', which could mean four ranks or four files (that is, with subunits in four files) – see Anderson (1970) pp.315–16 n.32.
30. I agree here with Matthew (2012) pp.176–77 that the lack of uniformity of depths is striking, and that attempts such as Pritchett (1971) 1 p.139 to achieve mathematical precision (by adding officers to the files to make up numbers) are implausible. Konijendijk (2018) pp.129–30 on lack of uniformity (eight deep is the modal depth recorded, but still outnumbered 1.5:1 by all other depths).
31. This follows a general trend throughout ancient history whereby weaker or less reliable, but more numerous, forces would often form deeper rather than wider – see Sabin (2009) pp.31–32. Of course this does not apply to every case – the deeply formed Thebans could certainly have considered themselves individually equal to their shallow Spartan enemies.
32. For example Wheeler (2007) pp.192–93 attempts to use depth to define 'the phalanx' (the Classical formation) in opposition to other formations – 'not all linear formations of heavy infantry constitute a phalanx' – using 'the cohesion of the mass' (on which see Chapter 8 in this volume) and depth to distinguish the phalanx, while admitting that 'No hard rule can be established for how much depth distinguishes a linear from a phalanx formation'. I would go further, and say that depth does not constitute a distinction at all (and that in fact there is no real distinction, and 'phalanx' is just a Greek word for a linear formation).
33. See Taylor (2020) ch.3 for discussion of these manoeuvres in the Macedonian phalanx. On hoplite formation changes – Anderson (1970) pp.98–101 (and figs I–IV), pointing out the problems of the need to leave space for these manoeuvres, which means formations could not have been quite so easily altered as Xenophon implies. Bardunias and Ray (2016) pp.114–17; Matthew (2012) pp.171–72.

34. For these drills see also Bardunias and Ray (2016) pp.114–17; Anderson (1970) pp.101–08 and figs I–IV.
35. For the movements at Thymbrara, Anderson (1970) pp.165–91 and fig. VII (p.400). For outflanking manoeuvres generally, Bardunias and Ray (2016) pp.165–68.
36. Columns and lines, and the means of moving between them, are considered by Anderson (1970) pp.94–110; Matthew (2012) pp.171–72; Bardunias and Ray (2016) pp.113–17.
37. See the works referenced in n.36 above.

Chapter 4
1. Van Wees (2004) pp.153–62 for Homeric warfare. Archaic infantry will have been a much more exclusive body than the larger armies we see in the later fifth century, made up of a more exclusive aristocratic class; Brouwers (2013) p.165 and Salmon (1977).
2. As we saw in Chapter 1, the concept of the 'middling farmer' forming the core of the Archaic Greek phalanx has been most forcibly championed by Hanson (1983), (1991) and particularly (1995). Here I follow the analysis of van Wees (2004), (2007) and (2013). See also Pritchard (2010) pp.21–27 for a summary of the debate.
3. See, for all these figures and calculations, van Wees (2007) p.276 and further references; van Wees (2013) pp.229–32. Van Wees sets out his alternative to the 'middling farmer' model at (2013) pp.237–45. The identification of *zeugitai* and the hoplite class is not absolutely certain, and certainly *thetes* could on occasion serve as hoplites (or marines) e.g. Thuc. 6.43.1. See also Crowley (2012) pp.22–26.
4. Lazenby (2012) ch.1 esp. pp.18–27 for these classes in Sparta; van Wees (2004) pp.83–85; Hawkins (2010) (for the fourth century).
5. Hunt (2007) pp.125–27 considers the relative standing of various parts of the armed forces and the ways this reflected political standing in the state; Crowley (2012) pp.70–79 on the Athenian army and tribal system, noting that the tribes themselves did not greatly promote unit cohesion.
6. Dalby (1992) for the Ten Thousand, and for the 'army as *polis*' phenomenon more generally, with other examples, Hornblower (2007) pp.30–34.
7. Van Wees (2007) p.293; (2004) pp.192–95 on *andreia*.
8. The question of Greek views of Persian spear length is considered by Manning (2021) pp.279–82, 296–99. Manning is rightly cautious about spotting sexual subtext, and Herodotus himself of course could swing both ways, saying of Mede attacks at Thermopylae that 'they made it clear to everyone ... that among so many people there were few real men' (Hdt. 7.210.2), while at Plataea 'the Persians were neither less valorous nor weaker' but lacked skill and armour (Hdt. 9.62.3).
9. Hornblower (2007) pp.42–47 for women in Greek warfare, with further discussion of these and other examples; on the fate of captured women in a town or city, Garlan (1975) pp.46–47.
10. Hunt (2007) pp.139–40 on use of slaves and helots in armies. Aristotle, *Politics* 1269a–b makes similar points about the reasons for not encouraging the slaves.
11. For the *neodamodeis*, Lazenby (2012) pp.18–19; 60–61.
12. Plataea – Lazenby (2012) pp.120–39; helots in the army generally, pp.22–26 and *passim*.
13. For servants (sometimes *hypaspistai*, 'shield bearers') carrying shield and supplies, Schwartz (2013b) p.168; as slaves, Hunt (1998) pp.166–68; see also pp.138–40 (slaves).
14. Doubt has been cast (e.g. by Rosenstein (2010)) on whether early Roman and Etruscan forces really used a Greek-style phalanx, but I think this is based on an overly restrictive concept of what constitutes a phalanx. Torelli (2001) for an account of the Etruscans.and Connolly (2012) pp.94–100 for Romans, Etruscans and other Italians.
15. Wheeler (2007) pp.187–88 on non-phalanx Greeks; Macedonians, Taylor (2020) ch.1.
16. For the Thracian peltast the starting point remains Best (1969).
17. See Hunt (2007) pp.137–44 on manpower sources in general.
18. Van Wees (2013) pp.240–45 considers these numbers and their implications for the exclusivity of the hoplite class (Eritrea's population, pp.240–41 and n.72).

19. This is the argument of van Wees (2013) pp.240–45. See also for numbers Bardunias and Ray (2016) pp.66–67.
20. Lazenby (2012) p.23 (and pp.9–10 for contemporary force sizes). As Lazenby notes, *oliganthropia* here specifically means shortage of Spartiates, full citizens, rather than depopulation; see also Cartledge (1979) pp.315–17. Christesen (2006) suggests that Xenophon's *Cyropaedia* was a blueprint for military reforms that could allow Sparta to overcome some of its military deficiencies (particularly shortage of manpower and shortage of decent cavalry) by enrolling the existing hoplites as cavalry and recruiting poorer men to serve as hoplites.
21. The basic reference for Greek mercenary service is still Parke (1933), supplemented by Trundle (2004). For mercenary service in the Archaic period, Parke (1933) pp.3–23, and, arguing it may provide the context for the origin of hoplite and phalanx, Hale (2013); he also argues that this spirit of overseas adventuring provides a better explanation for the origins of the *polis* than does the traditional picture of independent farmers. See also Luraghi (2006). For an alternative take on mercenaries in the fourth century, Rop (2019).
22. Hale (2013) pp.184–85 for the Abu Simbel mercenaries and other examples, and van Wees (2004) pl.26 and p.232. For Greek mercenary service in Egypt in the fourth century, Ropp (2019) ch.4 and 6.
23. Mercenary armies and generals were not like *condottieri*, freelancers without loyalty to their home states; Pritchett (1973) 2 pp.59–116; Kallet (1983). For the fourth century see Rop (2019) ch.2 and throughout. Mercenary service with the permission of the home city was the norm in the Hellenistic period – Taylor (2020) pp.187–98 with further references. Note also that cities would generally put their own commanders over any mercenaries hired, Parke (1933) p.73.
24. Fourth-century mercenary hoplites alongside citizens – Hunt (2007) pp.141–43.
25. For opposing views on this question see Whitehead (1991) and McKechnie (1994).
26. While the attitude Xenophon attributes to Pheraulas has commonly been taken to be typical of Greeks (e.g. Anderson (1970) p.84), we must remember the context – imaginary Persians in Xenophon's didactic story – and the specifics – use of swords (Konijnendijk (2018) pp.59–60). I think there is sufficient evidence elsewhere to support the principle, however. Of course this does not mean that the Greeks were right, and that weapon skill did not matter (as Konijnendijk's examples, pp.62–64, demonstrate).
27. Hunt (2007) pp.132–37 on training (and the lack thereof); Pritchett (1974) 2 pp.208–31. Konijnendijk (2018) pp.39–71. On *hoplomachia*, Wheeler (1982) and (1983); Konijnendijk (2018) pp.66–69.
28. For a view of the *hoplitodromos* and Pyrrhic dance that still emphasizes the weight of the panoply, Schwarz (2013a) pp.46–49.
29. Van Wees (2004) pp.89–95 on training generally; Bardunias and Ray (2016) pp.81–83 on physical training.
30. For the *agoge*, Lazenby (2012) pp.34–35 (with further references); van Wees (2007) pp.290–91. *Syssitia* – Lazenby (2012) pp.17–18.
31. For the antiquity of the *ephebeia*, Vidal Naquet (1986), Siewert (1977); Hunt (2007) p.134. Further discussion in Sekunda (1992) and van Wees (2004) pp.93–95; Crowley (2012) pp.25–26; Konijnendijk (2018) pp.42–43.
32. Hunt (2007) pp.131–32; Konijnendijk (2018) pp.47–49 on military discipline, with further references.
33. For Spartan corporal punishment, though perhaps chiefly of helots (which could include hoplites, since helots were frequently recruited), Hornblower (2000). Whether Spartiates would strike Spartiates is (p.73) less clear.
34. Examples of Persian (or other Eastern) armies operating 'under the lash': Hdt. 7.22.1 (digging Xerxes' canal); Hdt. 7.56.1 (crossing the bridge of boats); Hdt. 7.103.3f. (how Persians might be made to fight against greater numbers); Hdt. 7.223.2–3 (Thermopylae); Xen., *Anab.* 3.4.25 (barbarian missile users).
35. Examples of discipline (or the lack thereof) are collected by Pritchett (1974) 2 pp.232–45. See also van Wees (2004) pp.108–13; Hornblower (2007) pp.34–35; Hunt (2007) pp.130–32.

36. Van Wees (2004) pp.102–04 for the call-up. See also van Wees (2013) p.242 (quoting earlier references) for the argument that recruitment from the *katalogos* meant chosing men from the wealthier (and numerically more restricted) classes, while age-class mobilization included poorer men and therefore produced, in the fourth century, larger armies. Crowley (2012) pp.27–35 for further details and references (and p.34 for Syracusan mobilization).
37. For details of the ways in which Athens funded its wars see Pritchard (2019) (with references also to his earlier papers). In Athens the major cost was of course the fleet, as it was for other Greek cities that sought to emulate or surpass Athens' naval supremacy, which generally required Persian funding.
38. Pay – van Wees (2004) pp.71–74, 236–38. Gabrielsen (2007) pp.256–60 on the financial burden of armed forces (chiefly navies); pp. 265–66 on pay to cavalry.
39. Gabrielsen (2007) pp.256–60 and note 38 above. Both the total costs of armed forces and the total income to be gained from plunder are difficult to quantify, but I believe at any rate that it is unlikely that plunder could fund (rather than partly offset the costs of) a campaign.
40. Plunder – Krentz (2007) pp.170–73, 180–83 (note the importance of sale of slaves, rather than goods); van Wees (2007) pp.282–83 with some figures for sacked cities, but noting that in general the sums seized must have been small. Hanson (1983) downplays the amount of agricultural damage that could be done. Pritchett (1971) pp.93–100 and ch.4 on legal ownership of plunder; p.90 on *laphyropolai*.
41. Van Wees (2007) pp.284–86; stressing on the strategic level that economic motives were probably less important than matters of status, honour and longstanding feuds (between cities).
42. Hale (2013) p.180 for these passages; Hale considers mercenaries, as long-service professionals, the most likely origin of hoplite tactics, though I suspect that mercenaries were hired not so much for their tactical abilities as for their availability and willingness.
43. Crowley (2012) is the classic examination of this question, concentrating on the Athenian hoplite, though many of the points he makes must be more broadly applicable. Van Wees (2004) p.96 for Nestor's recommendation.
44. Crowley (2012) esp. pp.43–46 for the importance of the deme in the Athenian army.
45. Crowley (2012) pp.74–78 on the lack of regimentalism or tribal solidarity in the Athenian army.
46. Crowley (2012) pp.96–100 on the role of religion. Van Wees (2004) p.195 for the ephebic oath, and pp.192–95 for motivation in general; p.98 for the Spartan oath, with pp.243–44 (the oath is that sworn by the Greek army before Plataea, of very doubtful historicity, but this clause seems, from the officers mentioned, to be specifically Spartan). Bardunias and Ray (2016) pp.91–94 have some further interesting thoughts on motivation.
47. On this topic see Ogden (1996), with extensive discussion and references, and criticism of the standard model (set out pp.108–09). Many Ancient Greeks seem to have been somewhat reticent about sexual relationships between adult men, despite the widespread normality of relations between men and adolescents. Similar reticence about something known to be widespread and normal is typical of attitudes in a military setting in other ages; compare the quote, attributed to Winston Churchill, that British naval tradition was 'nothing but rum, sodomy and the lash' (echoing an earlier saying about 'rum, bum and bacca') – although sodomy was technically a capital offence in the Royal Navy, it was assumed to be widespread aboard ship.
48. Crowley (2012) pp.94–95 for the epitaph and further discussion. For those wishing to pursue a Freudian take on this matter, Sagan (1979). Although two examples may not be enough to draw general conclusions, it does raise the possibility that hoplites sometimes kept a score of the number of men they had killed. Even in the modern world, with its more widespread aversion to killing, similar sentiments to those expressed by Xenophon are to be found. Perhaps the closest parallel is in aerial combat, where the enemy combatant, being largely hidden inside his flying machine, is safely dehumanized. From the birth of air warfare in the First World War, the worth of a pilot was reckoned by his number of 'victories' (enemy machines destroyed, and therefore, in that pre-parachute age, approximating to number of

men killed). In the Second World War, the tendency for fighter pilots 'claiming to have killed more than all that were really slain' was common in debriefs after aerial combat, leading to greatly exaggerated claims of enemy aircraft losses.
49. Crowley (2012) pp.92–96 for attitudes to violence and killing. Bardunias and Ray (2016) pp.101–03 also discuss the psychology of violence among hoplites. Clearchus is one of several figures in antiquity who have been retrospectively and posthumously diagnosed as suffering from PTSD. See Rees and Crowley (2015) for counter-arguments – PTSD (as diagnosed today) is connected to the particular features of modern warfare, and the particular disconnect between modern peacetime culture and warfare. That there was mental trauma of some sort in some cases cannot be denied, the 'hysterical blindness' of Epizelus after Marathon (Hdt. 6.117.2) being the most-cited case (a case which Herodotus regarded as remarkable). Bardunias and Ray (p.102) opine that 'Modern studies suggest that only a very few soldiers are able to take an enemy's life without remorse, and there is no reason to think that was any different for hoplites.' I think there are many reasons to think it was different for hoplites, not least the totally different (from modern experience) cultural setting in which war was waged, and the explicit statements to the contrary in ancient sources. If there is, as Bardunias and Ray suggest, 'a cultural and likely genetic imperative' not to kill, presumably applicable across the whole of human history, it has time and again proven remarkably easy to overcome.
50. For the 'hoplite's creed', and manliness and virtue in general, Crowley (2012) pp.65, 86–88, 92–94 with further reading; Roisman (2003); Yoshitake (2010). This aspect is particularly well known from Athens (as the most literate society) and the degree of emphasis was perhaps an Athenian peculiarity, but it seems reasonable to suppose that the basic values were common to all Greeks. Socrates' quote could be taken as evidence of each man having a fixed place in the file (see Chapter 3), though I think a more general principle is meant.

Chapter 5

1. For officers and command in Greek armies, Pritchett (1974) 2 pp.4–132 gathers the evidence. Wheeler (1991); Hamel (1998) specifically on *strategoi* in Athens; Lazenby (1985) pp.5–20 for the Spartan army; Anderson (1970) pp.225–51 for the fourth century.
2. On the Athenian chain of command (particularly the *strategoi*), Hamel (1998); Fornara (1971); Crowley (2012) pp.27–39.
3. Lazenby (2012) pp.6–8 and *passim* for the command structure; pp.61–64 on Amompharetus; Anderson (1970) pp.67–68; van Wees (2004) pp.95–101.
4. Wheeler (1991) pp.151–52 for Paches, and generals' punishments.
5. Lazenby (2012) p.29 for nobility and officers; p.32 for governors.
6. Wheeler (1991) pp.133–35, though pointing out (e.g. n.65) the many problems with Herodotus' account of these matters.
7. Wheeler (1991) pp.138–39 for recognition of generals.
8. Anderson (1970) pp.69–70. There were some generals, the wily Agesilaus among them, capable of deliberately manipulating sacrifices and religion to their advantage.
9. Anderson (1970) p.68, with other examples.
10. Russell (1999) p.117. The reality of the *scytale*, at least in this form, is not universally accepted.
11. For officer training, Wheeler (1991) pp.137–38; Lazenby (2012) p.34 f.
12. That officers did not form a separate officer class – Wheeler (1991) p.140 and n.91; Hunt (2007) p.131, officers, men and badges of rank. Triple plume – Anderson (1970) p.40 thinks it is a badge of rank. Transverse crest – Wheeler (1991) p.141 n.95; Alcibiades *et al.* – Wheeler (1991) p.141 argues that these are badges of rank. For the *bakterion* as badge of rank (officer, or at least Spartiate), Hornblower (2000), and note also the comparison with modern military batons (field marshals' batons, officers' swagger sticks, *bâtons de commandement*), which are symbolic of ability and readiness to administer corporal punishment, but also of the control of drill manoeuvres, which is doubtless significant in the Spartan context.
13. Wheeler (1991) p.142 and n.101 for officers' pay.
14. Pre-battle speeches collected in Pritchett (1994); Wheeler (2007) p.203.

15. The religious aspects both of generalship and of Greek military practice generally are sadly under-represented in this book, the only excuse being that such a massive subject really requires specialist treatment; see Lonis (1979).
16. On maps (and the limitations on campaigning imposed by for example lack of accurate timekeeping), Anderson (1970) pp.67–68.
17. Anderson (1970) pp.67–68 for these examples.
18. For the Homeric warrior code governing such behaviour, Wheeler (1991) p.122; Latacz (1977) pp.153–59.
19. On hoplite generalship, see in general Wheeler (1991) and (2007) (the title of the latter section 'The emergence of generalship, 479–362 BC' shows that Wheeler does see a development in tactics across the Classical period, though he notes (p.215) that 'Greek tactics between 479 and 362 saw changes of degree and scale, not kind', and before these dates Greek tactics are unknown to us). Achilles and Odysseus – Wheeler (1991) p.137. For Greek conceptions of tactics see Konijnendijk (2018), and in particular his comments on the nature of Epaminondas' tactical innovations, pp.34–37; there is no agreement on what these innovations were, or even if they are real, and no clear conception of what 'good' tactics might have been. I am extremely wary of the notion of the brilliant tactician (in the modern sense), as it seems impossible to judge good tactics from bad other than by the outcome, victory or defeat, which in practice relied on many factors other than the tactical tricks of the general. Epaminondas was highly regarded in antiquity for his generalship, but this does not appear to be because of his tactical tricks in battle, on which the sources are confused (Diodorus and Plutarch) or silent (Xenophon). See also Chapters 6 and 7 of this volume.
20. For discipline, Pritchett (1974) 2 pp.232–45, and above in Chapter 4; Wheeler (1991) pp.145–50 for tactical command.
21. Wheeler (1991) p.129.
22. Hanson (1989) ch.9 pp.107–16 in favour of idea that hoplite generals led by example, from the very front, but became more battle managers in the fourth century (as also Wheeler); Wheeler (1991) pp.131–35 for the tactical positioning of officers.
23. Hanson (1989) pp.113–14 stresses the number of generals killed in action as evidence of their front-rank role; but Wheeler (1991) pp.146–50 points out that many of these were not in the pitched battle, formed phalanx stage of the fighting, but the confusion of rout and pursuit, or indeed in other circumstances than pitched battle (as the examples of Cleon and Brasidas show). Strangely, Hanson includes among his list of officer casualties Clearchus, of the Ten Thousand, who was captured and killed by the Persians when he went to negotiate with them, hardly an example of a battlefield casualty or evidence for fighting at the front.
24. Mounted generals – Wheeler (1991) pp.141–42.
25. Wheeler (1991) pp.149–50 considers aspects of this matter, also casting doubt on the *sphagia* (goat sacrifice) really being performed between the two armies; it does appear, as he points out, that there must have been much variability in the position and role of the general, and that personal leadership (and so leader casualties) was often most evident in the period after the breaking of one or another phalanx. For the mechanics of battlefield command see Konijnendikjk (2018) pp.138–53, with table p.149 of revised orders issued in battle indicating that sending orders in the thick of battle was quite possible.
26. Heralds – Anderson (1970) p.79 seeing their use as an exception to the general rule (of passing orders or announcing them directly).
27. Taking this as evidence of relatively egalitarian relations between officers and men, Hunt (2007) p.131; Pritchett (1974) 2 pp.243–45.
28. For Cleon's order, Anderson (1970) pp.80–81 and Anderson (1965).
29. For use of the *salpinx* Krentz (1991), with examples of these other uses and further references.
30. Anderson (1970) pp.82–83 on signals.
31. Anderson (1970) pp.71–83 on the passing of orders, the passing of the watchword and in general the method of passing orders (particularly as illustrated in the *Cyropaedia*). However, as Hunt (2007) p.130 n.97 notes, the Spartan army, on which Anderson's chapter is mostly based, is definitely not typical of Greek armies generally.

Chapter 6

1. As an example of this trend see Lagos and Karyanos (2019), stressing the importance of light infantry at Marathon (completely ignored by Herodotus). Plataea is another battle where light infantry were present in huge, but apparently strangely ineffectual, numbers; see below. A tendency to downplay the role of light infantry and cavalry and stress the role of hoplites can certainly be found in ancient sources and in modern historians; but most modern historians will also at least reference the roles played by forces other than the hoplites. The comments of Konijnendijk (2018) pp.95–106, while a useful corrective, perhaps overstate the extent of the (modern) problem. See below for details on particular arms.
2. See Wheeler (2007) pp.188–92 on 'agonal' warfare and the tension between trickery and open battle. Echeverría (2011) makes a useful distinction between 'tactics' and 'cultural tactics', the latter including all those features that could bring advantage while still fitting within the cultural requirements and expectations of warfare. See further below.
3. Specialist light infantry and (especially) peltasts are a large topic in themselves and their treatment here is necessarily cursory. An introduction in Hunt (2007) pp.119–24. For peltasts see Best (1969). Parke (1933) covers peltasts, particularly, as mercenaries.
4. Van Wees (2004) pp.62–65 on light infantry, including the costs of bows and javelins.
5. Van Wees (2004) pp.62–64 on use of rowers; Anderson (1970) p.114 suggesting it is an innovation.
6. For great detail on the Battle of Lechaeum see Sekunda and Burliga (2014), and specifically for the events of the battle, Konecny (2014).
7. Anderson (1970) pp.122–23 for this episode, noting also that Brasidas had earlier employed similar tactics ('But wherever they charged they found the young men ready to dash out against them, while Brasidas with his picked company sustained their onset', Thuc. 4.127.2).
8. That defeats of hoplites by light infantry only appearing from the late fifth century may be an accident of the survival of evidence is pointed out by Hunt (2007) pp.108–09; van Wees (2000); Anderson (1970) pp.111–64; against the orthodox view of a decline of hoplites, exemplified by Hanson (1995) pp.321–49.
9. Like light infantry, the subject of cavalry is too vast to address in detail here. Detailed studies include Bugh (1988); Spence (1993); Worley (1994); Sidnell (2007).
10. Van Wees (2004) pp.65–68 for the functions of cavalry, with p.67 n.18 for further references.
11. For Marathon see Krentz (2010a), seeing this battle as the start (rather than the endpoint of a centuries-long process) of the development of the ordered phalanx, along with discussion of the usual controversies such as location of the Persian cavalry. Also Billows (2010); Lazenby (1993). Wheeler (2010) for the place of Marathon in military history.
12. For Plataea, Pritchett (1971) 1 pp.101–21 for the (probable) topography; Cartledge (2013) (with further references); Lazenby (1993).
13. There are some doubts about the historicity of this episode, Anderson (1970) p.118 n.27; nevertheless, the tactical point Xenophon is making still stands.
14. That hoplites were capable of and required to fight in many circumstances other than pitched battle is obvious, yet seems overlooked by those who wish to designate the hoplite as uniquely heavily burdened and the phalanx as uniquely rigid and unwieldy. Typical is for example Schwarz (2013a), who writes as if hoplites only ever fought in pitched battle, or Matthew (2012), ostensibly about Greek hoplites at war, but concentrating entirely on the pitched battle. For discussion of the many other types of fighting, see Rawlings (2000).
15. Van Wees (2004) pp.106 and 280 n.20 for transport. For theoretical calculations of carrying and baggage capacity of an ancient (Macedonian) army, Engels (1978), though some of his conclusions must be treated with caution.
16. Van Wees (2004) p.106 on water. Attempts have been made to estimate the size of Xerxes' Persian army based on its water needs, though I think there are too many imponderables to reach any firm conclusions (other than that it must have been a lot smaller than Herodotus claims).
17. The modern 'Spartathlon' is a 246km race between the two cities, in memory of the achievement of the Athenian message runner Pheidippides, who managed the distance in

a day and a half. The modern record is twenty hours and twenty-five minutes. Obviously an equipped army in any sort of formation would move more slowly; even so, three days is a remarkable achievement.
18. Krentz (2007) pp.159–62 for these and other figures.
19. On camps, Krentz (2007) pp.165–66; Anderson (1970) pp.43–66.
20. Krentz (2007) p.164 and Anderson (1970) pp.61–62.
21. Hoplite activity outside the phalanx (and by extension, outside pitched battle) is well covered by Rawlings (2000). See also Krentz (2007) pp.167–73, and further below on the pitched battle. Ober (1991) on the role of terrain in warfare. That hoplites mostly fought *on* the plains, incidentally, is not evidence that they fought *for* the plains – plains simply provide a convenient natural location to deploy a large army, the objective being to defeat the enemy army, not to win possession of the plain.
22. For sieges see the list of sieges and outcomes in Krentz (2007) p.180; Ober (1991) pp.180–88; Hanson (1983) proposed that the agricultural damage an army could do would be limited but I believe his case is greatly overstated; as we have already seen, cutting down trees is a common feature of many campaigns (see also the Amathus bowl), and the objective was likely not to permanently destroy all agricultural activity, but just to damage this year's crops.
23. Frontier defences in Krentz (2007) pp.167–68; Ober (1985) specifically for the defences of Attica, which were the most sophisticated and complete.The primary purpose of such defences, as with most frontier defences, would be to delay and deter raids rather than to stop full-scale invasions.
24. On passes, Hanson (1983) pp.88–102; Krentz (2007) pp.167–68; Pritchett (1982). For Thermopylae, a battle subject to as much mythologizing today as it was in antiquity, Cartledge (2006) (note that Billows (2010) claims modestly only that Marathon changed Western civilization; the title of Cartledge's work has Thermopylae changing the world); Lazenby (1993).
25. For sieges and siegecraft in general see Strauss (2007) pp.237–47.
26. See Krentz (2007) pp.176–80 with statistics and table p.180; note also (p.168) the statistics of forty-seven battles during the Peloponnesian War, but 101 'poliorcetic incidents' (sieges and assaults).
27. Hale (2013) pp.181–82, Luraghi (2006), and Brouwers (2013) cover this 'Viking' period of Greek overseas adventure.
28. For an overview of naval warfare across this period, Strauss (2007) pp.223–36.
29. Krentz (2007) pp.147–50 on the importance of shipborne and amphibious operations, with table (p.149) of the numbers of men carried by ship – of which the largest figures are the 5,000-plus men of the Athenian expeditions to Syracuse. Naval campaigns – van Wees (2004) pp.221–26; naval transport and crews – van Wees (2004) pp.209–14. For marine equipment we should note that the fourth-century reforms of Iphicrates (see Chapter 9) may be related to marines (but even if so, this goes only to prove that regularly equipped hoplites had served as marines previously).
30. See the statistics gathered in Krentz (2007) p.149. The average number of men per ship is 'more than thirty', the highest around eighty (this includes both triremes and transports).
31. For all aspects of ship design, especially triremes, see Morrison and Williams (1968) and Morrison, Coates and Rankov (2000).
32. On agricultural devastation, van Wees (2004) pp.121–26. Hanson (1983); Thorne (2001). The recent trend (following Hanson) has been to downplay the damage that could be inflicted by agricultural devastation.
33. For *epiteichismos*, Krentz (2007) pp.177–78.
34. Echeverría (2011) pp.52–53 for the importance of campaigns as a demonstration of strength, often in the full expectation that the enemy would be unwilling to come out and fight (e.g. Thuc. 2.11.3; 5.7.3; 6.11.4).
35. Agonal war is a central part of the 'hoplite orthodoxy' – see for specifics Ober (1999). For criticisms of the concept see e.g. Krentz (2000) and (2002), Dayton (2006), Konijnendijk (2018). Current thought is, I believe, strongly against the idea, though the influence of e.g.

Hanson (1999) remains pervasive. However, while a strict conception of agonal war must be abandoned, a set of customs, norms and common practices still served to mould the nature of Greek warfare, as it does warfare in all cultures; see below. On stratagems and trickery in Greek warfare, Wheeler (1988a) pp.25–49; Sheldon (2011); Krentz (2000), with a catalogue of known 'deceptions' (pp.183–99). Those as close to the 'old days' as the orator Demosthenes (mid-fourth century) could refer to Spartans being 'so old-fashioned, or rather such good citizens, that they never used money to buy an advantage from anyone, but their fighting was of the fair and open kind' (Dem. 9.48), but this in the context of making a contrast with Philip of Macedon's ability to wage something closer to total war due to his siege capabilities, and Demosthenes' rhetorical purpose may outweigh strict historical accuracy.

36. For such surprise attacks, van Wees (2004) pp.132–33; Bardunias and Ray (2016) pp.61–63.
37. War aims – Hanson (2000); Bardunias and Ray (2016) pp.65–65; van Wees (2007) pp.281–87 for the importance of competition (and of wealth as an expression of such competition). Border disputes between communities over land could be said to be part of the normal human condition, and not necessarily closely related to the actual economic or agricultural value of the land in question. Pritchard (2019) puts Athens' warmaking (greatly expanded during the fifth century) in the context of public finance.
38. The ubiquity of war is not necessarily established by Plato's quote, since that was in the specific context of Cretan institutions. However, other evidence is plentiful – see Hanson (2013) pp.261–62, and I do not think we need to doubt that wars between Greek cities were frequent.
39. The idea that pitched battle was the major – even the only important – form of warfare for the Classical Greeks is not uncontroversial; it forms the backbone of the 'orthodox' view of Greek warfare, see for example Anderson (1970) pp.1–2, on which see the criticisms of Konijnendijk (2018) pp.18–19. In terms of relative frequency, non-pitched-battle combat greatly exceeded that in pitched battle (Echeverría (2011) pp.48–49; Krentz (2007)). While recognizing the validity of such criticism, I still believe, on balance, that we are right to see pitched battles as especially important. Other forms of warfare (low-intensity warfare, sieges and assaults, and naval warfare) all, however, also had a vital role to play, and if they receive only cursory treatment in this book it is more because the subject matter (the hoplite phalanx) dictates this, than because this is a statement on the nature of Greek warfare.
40. Hanson (1989) is the main proponent of the 'Western Way of War' theory, an idea that has proven popular, though it has (rightly) generated some academic opposition. The nature of battle in other cultures is beyond the scope of this work; see for example, with criticism of Hanson's thesis, Lynn (2004) esp. ch.1 and 2.
41. Manning (2021), Sekunda (1992), Head (1992) for Persian armies and warfare.
42. Van Wees (2004) pp.133–34 on the 'Battle of Champions' (perhaps a unique occurrence) and in general that such restrictions, if they are real at all, were special, localized arrangements and not evidence of widespread practices. Under 'Protocols of battle', however (p.134), is included (and dismissed) the offer and acceptance of pitched battle, on which see Chapter 7. Similarly Konijnendijk (2018) p.77.

Chapter 7
1. For the mechanics of battle in general see Wheeler (2007) pp.202–13 on 'The mechanics of hoplite battle', and Hanson (1989), with caution, and Chapter 8 of this volume.
2. 'On the Possibility of Reconstructing Marathon and Other Ancient Battles' is Whatley (1964). The 'inverted pyramid' is identified by Sabin (2009) p.xi, while pp.6–11 examine Whatley's techniques – Sabin answers the question as to the possibility of reconstructing such battles in the affirmative, and offers a new and innovative technique (a wargame model) for doing so. See Konijnendijk (2018) pp.24–38 for comments on the numerous (and largely unsuccessful) attempts to 'reconstruct' the tactics of Leuctra. See also Whitby (2007). For lists of battles, Schwartz (2013a) 'Appendix: battle inventory'; table in Krentz (2007) p.169 (compare also the list of sieges, p.180). Rees (2016) provides an accessible account of a number

of battles and sieges. Needless to say there are no clear rules on what counts as a battle, and how much information about a battle is needed to include it in such a list.
3. For the topographic approach to ancient battles see particularly (in English) Pritchett (1969); see also Butera and Sears (2019) for a very practical guide to actually visiting the battlefields in question.
4. Wellington's comment is contained in a letter to John Croker (8 August 1815), quoted in *The Waterloo Letters* (1891) ed. H.T. Sibome.
5. For some general observations on numbers see Hanson (2007) pp.7–8 – noting the tradition going back to Delbrück (*History of the Art of War Within the Framework of Political History: Antiquity*, English translation 1974) of extreme scepticism of numbers in ancient sources. Krentz (2007) p.170 suggests that Greeks would not fight if outnumbered more than 3:2 and cites the case of Cleandridas (Polyaen. 2.10.4), who had to disguise the size of his army to encourage his enemies to accept battle.
6. Wheeler (2007) p.188 for examples of the constant (and by no means uniquely Greek) tension between 'brawn and brains'. On Athena and Ares, Deacy (2000) – the distinction between the two is not hard and fast in practice, needless to say. On the 'missile ban', Wheeler (2007) p.191 and (1987). See also comments on agonal warfare in Chapter 6 and note 35.
7. Hall (2007) in general, and p.95 for this incident. For more detail on conceptions of the morality of such matters see Dover (1974).
8. Wheeler (2007) pp.202–03 for the offer of battle, but it is not necessary to equate this with 'the agonal clash of rival phalanxes' nor to see it as reflecting an earlier formal battle 'by appointment'. See also Echeverría (2011) pp.49–50; Konijnendijk (2018) pp.72–86. Battles throughout history have frequently taken place by mutual consent and with fairly formal-seeming offer, deployment and commencement – this is after all part of the definition of a pitched battle.
9. Scouting and intelligence – Pritchett (1971) 1 pp.127–33; Russell (1999) pp.10–19.
10. Echeverría (2011) pp.54–58 for practical considerations of deployment; Konijndijk (2018) ch.4 considers the matter in detail.
11. See Konijnendijk (2018) pp.116–26 for a sceptical view of the importance of the right wing. Certainly there are exceptions, and perhaps this is another rule more honoured in the breach than the observance; nevertheless I think the principle of positions in the line having different status is clear enough.
12. See Chapter 3 above for more on depth, with Matthew (2012) pp.172–79; Schwartz (2013a) pp.167–71; Bardunias and Ray (2016) pp.120–21.
13. For Marathon see Krentz (2010a), though he is noncommittal on this particular question.
14. As Wheeler (2007) pp.203–04 observes, the only occasion where we clearly hear of pre-battle light infantry skirmishing is at Syracuse (Thuc. 6.69.2); whether it took place more often and simply went unrecorded is unclear, though Thucydides' language on this occasion suggests he was describing something very familiar. On light infantry in battle, Wheeler (2007) pp.220–21.
15. For cavalry in battle, Wheeler (2007) pp.221–22, and see also Chapter 6 note 9.
16. Krentz (2013) p.141 for the paean, and in greater detail Furley and Bremer (2001). Wheeler (2007) p.204 for sacrifice, paean and trumpet; Jameson (1991) and Parker (2000) for the sacrifice.
17. Krentz (2010a) pp.143–52 (and 219–20) offers a defence of the running charge (and I agree that it was a physical possibility); he also offers a reason for it, to begin the battle before the Persian cavalry could intervene, though why they did not intervene during the ensuing 'long' fight is less clear. I still believe there are few enough examples of any heavy infantry forces running into action over such a distance to make the story doubtful, though Cunaxa might offer a parallel (although there was no fighting).
18. For 'good order' (*eutaxia*), and its opposite, disorder (*ataxia*), and its central importance in explaining fifth- and fourth-century hoplite success or failure, see Crowley (2012) pp.49–53, and further below.

19. On the phenomenon of the victorious right wing and responses to it. Wheeler (2007) pp.216–17; Sabin (2009) pp.104–05 for the 'revolving door' metaphor. Echeverría (2011) pp.65–66 on tactical manoeuvres (seeing all but *kyklosis* as fourth-century innovations), and pp.68–70 on the importance of the right wing.
20. For outflanking and encircling manoeuvres, Bardunias and Ray (2016) pp.165–68; for this example, Anderson (1970) pp.144–46.
21. Leuctra is (along with Marathon) the most extreme example of the 'inverted pyramid'. See discussion in Schwartz (2013a) pp.169–71, and Konijnendijk (2019) pp.24–38. Given the variant sources, the tendency to believe none of them (Xenophon has as many detractors as Diodorus and Plutarch, though his sources and personal knowledge are surely better) and the lack of clarity on the nature of the deployment and manoeuvres, it does seem hopeless attempting any detailed reconstruction. I am inclined to agree that any tactical innovations were quite minor (simply attacking with the left wing rather than the right seemed innovative enough to contemporaries, simple though the change seems to us).
22. For Leuctra see Anderson (1970) pp.192–220; Lazenby (2012) pp.176–88; Pritchett (1965) 1; Devine (1983); Wheeler (2007) p.217 and n.132 for further references to the modern literature and problems with the sources. See also the comments on the problems of modern (and 'Prussian') scholarship (and further bibliography) in Konijnendijk (2018) pp.24–38. The oblique advance incidentally does not require the Theban right to withdraw, merely for them not to advance at the same rate as the Theban left (or at all) – see Konijnendijk (2018) p.29 for the varied understandings of what was, I suspect, a very simple phenomenon. Posting the best forces on the left was itself no great innovation (see Thuc. 3.108 for an earlier Spartan example) – the difference here is the intent, to directly confront the best enemy forces.
23. Second Mantinea is discussed by Bardunias and Ray (2016) pp.73–78; Anderson (1970) pp.221–24. Pritchett (1969) pp.37–72 for the topography. Devine (1983) for the 'wedge' interpretation, against which Buckler (1985). The *paragoge epi keras*, *'paragoge* to the wing', may be a description of a formal drill manoeuvre or may be meant in a less formal sense, Konijnendijk (2018) pp.52–54; certainly Xenophon uses these words in formal (tactical) and less formal ways, though in this case I see no reason he should not have meant the tactical manoeuvre, which I believe the Thebans would have been quite capable of at this point.
24. Anderson (1970) pp.178–81 for the reserves at Thymbrara; Wheeler (2007) p.219 for reserves generally, though he includes flank marching and ambushing forces in his definition of reserves.
25. For Persian armies see Chapter 6 n.41. Raaflaub (2013) pp.98–100 provides a brief discussion. For the individual battles, among many other works, see Krentz (2010a) for Marathon; Matthew and Trundle (2013) for Thermopylae; Konijnendijk (2012) for analysis of Plataea and the relative strengths and weaknesses of Greeks and Persians; and further references below. Manning (2021), Sekunda (1992), Head (1992) for the Persian army.
26. On *sparabarai*, Head (1992) pp.22–27; Sekunda (1992) pp.18–19; Manning (2021) pp.295, 303–35. By the fourth century, these infantry were possibly replaced by men with smaller shields (*takabara*), Manning (2021) p.310, following Sekunda. Manning pp.303–07 proposes a model of Persian combat techniques (by analogy with eighteenth-century AD infantry, due to the importance of firepower).
27. Persian numbers – see Sabin (2009) pp.96–97 (Plataea), and pp.17–27 for an alternative methodology to the usual efforts based on source criticism or logistic considerations. Persian numbers are discussed in the works on the Persian army referenced above.
28. Feigned retreat is discussed by Schwartz (2013a) pp.136–39 and see also Bardunias and Ray (2016) pp.168–72. This manoeuvre has generally caused some perplexity among modern historians and fits well with none of the usual models of hoplite warfare.
29. For examples of use of the whip to encourage Persian military units see Chapter 4 n.34
30. On the length of Persian spears see also Chapter 4 n.8. Manning (2021) pp.279–82, 296–99 discusses this issue, and points out that while the military importance of the Persian short spears is commonly accepted by modern writers, other facts recorded by Herodotus,

such as (Hdt. 3.12) that Persians skulls were thin because of the felt hat (*tiara*) they wore, while Egyptian skulls were thick through going bareheaded, do not much feature in modern analyses.
31. This would explain later Persian efforts to improve the equipment of their infantry by providing small shields to the other ranks – see Sekunda (1992) pp.18–19. For the possible different effective ranges of Greek and Persian spears, Matthew (2012) pp.88–91.
32. Anderson (1970) pp.192–220 on Leuctra and pp.221–24 on Mantinea. The role of cavalry at these battles is emphasized by, for example, Sidnell (2007) pp.62–74. See also Chapter 6 n.9 for general works on cavalry.
33. The quotes are from Ober (1991) p.173 (with further reading as of that time in n.1); the 'geographical paradox' continues to puzzle more recent authors, such as Wheeler (2007) p.202. The willingness of armies to fight on terrain suitable to both sides was due to common interests and values, so should not be seen as too surprising. For terrain in general, Pritchett (1985) 4 pp.76–85; Konijnendijk (2018) pp.72–94; Bardunias and Ray (2016) pp.180–88.
34. Fieldworks and barrier features (walls) are discussed in Bardunias and Ray (2016) pp.51–59.
35. For the importance of terrain in Greek battles, Ray (2009) pp.298–97 and Ray (2012) pp.213–21, finding that in 21 per cent of battles listed, terrain played a significant role (and a decisive role in 50 per cent of these cases).
36. Hanson (1989) p.185 was influential in establishing the concept of 'a certain regularity to Greek battle: charge, collision, hand-to-hand combat, push, and eventual rout' – these are sometimes identified as distinct 'phases' (though Hanson warns against identifying them as '"segments" of distinct action'), and some such features must naturally have occurred in all battles, Greek or otherwise, as there is nothing distinctively Greek about any of these aside from the supposed 'push', on which see Chapter 8. However Hanson goes on: 'This sequence of events is borne out by ancient observers who developed a vocabulary to describe what they saw or heard: the charge (*ephodos* or *epidrome*), the clash of spears (*doratismos*), the hand-to-hand struggle (*en chersi*), the push (*othismos*) and the collapse (*trope*).' The problem is that several of these words are used very rarely if at all of hoplite battles or by our main Classical authors – *doratismos* for example occurs only in Plutarch (twice), and *othismos* just once (in a relevant context), in Thucydides. These neat categories – clash of spears, push and so on – are a modern conception, and the Greek words applied to them – *doratismos*, *othismos* – were not so applied by the Greeks themselves. What is needed is an analysis of the combat vocabulary of Herodotus, Thucydides and Xenophon, equivalent to that of Livy (and Polybius and Caesar) in Koon (2010); this need is only partly met by Pritchett (1991) 4, 'The Pitched Battle'.
37. Konijnendijk (2018) pp.178–88 discusses reasons for victory and defeat, with emphasis on the role of fear. As he aptly puts it (p.179), 'The Greeks appear to have blamed such sudden collapses of the battle line on the gods, but in practice it seems to have been more commonly caused by the Spartans.' See also Echeverría (2011) pp.70–73.
38. See the casualty statistics gathered in Krentz (1985a). That the Spartan preference for not engaging in long pursuits (Thuc. 5.73.4) has somehow been elevated by modern authors into a Greek preference for not pursuing or killing their fellow hoplites, has been pointed out often in recent years. See Konijnendijk (2018) pp.188–206 for numerous examples of pursuits.
39. Konijnendijk (2018) pp.188–205 emphasizes the importance of killing, which can hardly be doubted, but was still surely a means not an end.
40. For the trophy (*tropaion*), Krentz (2002); Pritchett (1974) 1 pp.246–75. The treatment of battle dead is covered by Pritchett (1985) 4 pp.94–259.
41. For details of the retrieval of battle dead see Vaughn (1991).
42. For burial and commemoration practices in Athens, Clairmont (1983); Low (2010).
43. Wounded – Krentz (2007) pp.183–85; Bardunias and Ray (2016) pp.83–86; Gabriel and Metz (1991) p.87 bravely attempt to quantify the numbers and proportion of wounded, though the task is difficult. Matthew (2012) pp.103–12 on the nature of the wounds likely to be suffered. Krentz (1985a) and Wheeler (2007) pp.212–13 give estimates for the survival rates of the wounded. For transport of sick and wounded, Sternberg (1999).

Chapter 8

1. Keegan (1976) for *The Face of Battle*. Hanson (1989) for the first major and explicit application of this approach to Greek (hoplite) battle. Wheeler (2007) p.187 offers criticism of the approach (quoting also an earlier piece, and previewing Wheeler (2011)), but at least some of his criticisms are unfounded ('post-Second World War theories of unit cohesion' are not really central to the approach, though see Crowley (2012) pp.5–21 for an example of their application to Athenian hoplites).
2. For a summary of the history and origins of the scrum theory see Krentz (2013). A relatively recent statement of the opposition to the literal scrum view is Goldsworthy (1997), which also contains references to earlier discussion (the literature on which is extensive). Krentz (2013) pp.143–48 provides a very useful summary of the argument, and spells out its origins in early twentieth-century comparisons to rugby. Other (frequently repetitive) discussions of the phenomenon include (in the literal scrum camp): Holladay (1982); Anderson (1984); Pritchett (1985) 4; Hanson (1989); Lazenby (1991); Luginbill (1994); Schwartz (2013a); Bardunias and Ray (2016); against the literal scrum: Fraser (1942); Krentz (1985b) and (1994); Goldsworthy (1997); van Wees (2000) and (2004); with a foot in both camps: Matthew (2012). The use of the Greek word *othismos* to describe this model of combat is unfortunate – see comments in note 6 below. We might note in passing that the English word 'scrum' is itself derived, via 'scrimmage', from 'skirmish' – so the English word has a military origin.
3. The 'heretical' or 'revisionist' view in this case is best exemplified by Goldsworthy (1997), Krentz (2013) (and several earlier papers) and van Wees (2004), though it has its origins further back, with Fraser (1942).
4. On the Greek (specifically Polybian) use of metaphors of weight see Koon (2010) pp.64–68, contrasting with the Latin usage of nouns and verbs of movement (*impetus*, *concurrere* and related words). See also Lendon (1999); also Koon (2010) pp.32–33 on Polybian language and pp.35–36 on the way it was interpreted by Livy, himself writing in the Roman tradition. I think that the continued dominance of metaphors of pushing and weight in modern English 'combat rhetoric' fatally undermines the argument that such language must derive from some underlying literal physical reality. Strictly speaking, 'weight' is itself a metaphor of course, since weight is the force acting on a mass toward the centre of another mass (that is, on Earth, downwards), but using 'weight' metaphorically (or metonymically) in military and non-military contexts in English (and in Greek) is wholly familiar. Force in a real scrum (in rugby) is generated by leaning forward and pushing with the legs; the mass of a rugby forward is important only in terms of inertia and momentum.
5. The main quote is from Fields (2008). This phenomenon is not confined to popular histories – in Hanson (1989) discussion of 'the push' (Hanson being a hoplite scrum advocate) is indexed under 'push, the (othismos aspidon)'. The quote from the opponent of the orthodoxy is from van Wees (2004) p.184.
6. Greek combat vocabulary is collected by Pritchett (1991) 4 'The Pitched Battle' – see pp.65–66 for *othismos* (taking examples of the verb and noun interchangeably, as he does also for other vocabulary). Hanson (2013) p.263 offers that 'It is hard to accept the repeated references to the *ôthismos* (the "push") or its more frequent verbal forms (*ôtheô*) are merely figurative', and adds (p.272 n.21), 'I do not know why some see much significance in an ancient author's choice of either the verb (*ôtheô*) or the synonymous abstract noun (*ôthismos*).' Referring to the 'repeated references' to *ôthismos* obscures the fact that the word appears only three times in relevant contexts, see below; while whether the verb and noun are really synonymous is surely one of the points at issue. Compare Hanson (2013) p.266, 'it is remarkable that both the poet Tyrtaios and historian Xenophon, composing three centuries apart, alike speak of some sort of *ôthismos*'. Tyrtaios never uses the word *ôthismos*, and Xenophon uses the word just once, but never in the context of battle. This adoption of a Greek word into English to mean something it is not certain it meant in Greek is, I feel, unfortunate.
7. That there are only three uses of the word *othismos* in a relevant context in our main authors would probably come as a surprise to many. Fraser (1942) noted that the scrum theory of combat was based on 'but three literary references', but of these (Thuc. 1.6.70; Thuc. 4.96.2;

Polyaen. 2.3.4), only one uses *othismos*, and there are many more references to 'pushing' in some form using other words (as discussed below) that he does not consider. Pritchett (1985) p.66 n.200 called Fraser's a 'strange article … [which] claims there are only three literary references to pushing'. However, Fraser was perhaps accidentally correct, if we consider just use of the word *othismos*. Pritchett (1985) p.29 states that 'The *othismos* is as common in Homer as it is in later hoplite warfare although the noun is not used' (so in fact, the word '*othismos*' does not occur in Homer at all). As to whether pushing appears in Homer, Krentz (2013) p.146 deals with Pritchett's arguments on this point, and see below. Schwartz (2013a) p.185 opines that 'The examples of *othismos* meaning bodily push are too many and too unambiguous to be safely ignored or explained away' (which as usual is just begging the question). He then goes on to state that 'of the 41 battles in the Inventory, 12 contain *explicit* references to *othismos*, making up for 29.27% of the battle narratives', and lists the battles in question. In fact, only two of the twelve battles he lists contain the word *othismos*, and only one of those is a battle between Greek hoplites. I would expect an 'explicit' reference to *othismos* to actually use the word *othismos*. Schwartz, like Hanson (see note 6), clearly takes the verb *otheo* and the noun *othismos* to be completely interchangeable in usage and meaning, whereas I think it clear (see below) that *otheo* has a number of meanings, including 'force', 'drive off' or 'defeat' (as well as 'push'), while *othismos* is used for 'melee' or 'struggle' (as well as 'pushing') – quite distinct meanings and usages.

8. Bardunias and Ray (2016) p.137 make the point that a scrum requires pressure from both sides, or else there is no pressure and no pushing.

9. The passages are: Cassius Dio 41.42.5 – crowds boarding ships; 47.44.41 – 'much *othismos*' (and much *xiphismos*, 'swordplay') between Roman legionaries at the Battle of Philippi (42); note also the Xenophontic phrase 'they wounded and were wounded, slew and were slain' at 47.45.1; 58.5 – crowds around a door; 62b.15.6 – '*othismos*, fighting and uproar' of crowds around brothels; 74.1.4 – crowds around the Senate. Josephus: *Antiquitates Judaicae* 19.86 – crowds around the palace; *De bello Judaico* 2.327 – crowding about gates. Dionysius of Halicarnassus: 7.35.5 – riotous scenes between plebeians and patricians; 9.48.2 – riots in Rome with '*othismos* of bodies'; 11.38.4 – riots in Rome. Appian: *Mithridatic Wars* 10.71 – crowding at a gate.

10. For these passage see Taylor (2020) pp.309–19, with further discussion. Note that the Macedonian phalanx described by the Hellenistic tacticians is a fundamentally different formation from the Greek hoplite phalanx because of the adoption of the two-handed sarissa or pike.

11. The passages in Procopius are: 1.7.27 (Persian forces occupy a siege ramp at the siege at Amida); 2.25.20 (Persians and Heruls); 2.27.16 (Persians and Romans (Byzantines)); 3.19.22 (Vandals); 4.3.10 (Romans); 5.18.13 (Romans defending Belisarius); 6.27.10, 11 (Romans, barbarians); 7.5.11 (ditto); 7.22.5 (Antae); 8.11.44 (Romans on ladders); 8.11.54 (a riot); 8.23.32 (naval battle); 8.29.18 (Gothic cavalry repulsed by Romans); 8.32.17 (Gothic cavalry).

12. Note also that there is some doubt about the depth Xenophon intends for Cyrus' Persians. Xenophon (*Cyrop.* 6.3.21) specifies that they be drawn up 'having the *lochos* in twos' (*eis duo echontas ton lochon*) – ordinarily, and by comparison with Xen., *Hell.* 6.4.12, this would mean 'with each *lochos* in two files', that is, twelve ranks deep. However, Xenophon is not entirely consistent in his usage – see Anderson (1970) p.174 and n.32 (p.315). I think Xenophon's rhetorical point is stronger if he means 'two ranks', but a case could be made either way. It does not, I think, materially affect the point under discussion here.

13. For use of *otheo* in Homer see Krentz (2013) pp.146–47, who also raises the similarity between Homer and Xenophon referred to below.

14. For shield clashing of this sort, van Wees (2004) pp.189–90 and (2000) pp.131–32. For the literal shoving view of shield use see for example Hanson (1991). A comparison is often made with Roman shield use at Zama, on which see further below. Matthew (2012) p.207 for other examples of 'clashing shields', though he deals oddly with these passages, taking for example the *symbalontes tas aspidas* of Xen., *Hell.* 4.3.19 to mean the phalanxes 'were pressed "shield

against shield'" and concluding that it supports the literal scrum model. I do not see how he reaches this conclusion.

15. For some reason, Polyaenus' story about Epaminondas is widely quoted as evidence in favour of the scrum theory while the Iphicrates story, despite 'explicitly' mentioning *othismos* (by the criteria of Schwartz (2013a) p.185 – the verb *otheo* is used), is not. Given that the same story is told of both commanders, I suspect it is just a common anecdote that could be attached to any general (see the very similar story told of Alexander, Polyaenus 4.3.8). But even assuming the stories have some basis in fact, I do not see either how taking 'one more step' could decide a battle whatever form the fighting took, nor that the *otheo* of Iphicrates must be taken literally here any more than in other cases – the comparison of the two versions suggests that *otheo* carries a similar meaning to 'they won' in the Epaminondas version. Schwartz (2013a) pp.189–90 for use of this passage (the Epaminondas version), though his explanation – 'inspiring his men to break the deadlock and drive back the Spartans by throwing in their last reserves' – does not accord with his purpose of supporting a literal push at all (presumably he means 'last reserves of strength').

16. See the list of meanings for *otheo* given in Liddell-Scott-Jones' *Greek-English Lexicon*: 'thrust, push, rush, throw, thrust, stuff, force, force back, thrust out, banish, hurry, press forward, crowd, throng, jostle'. There is a poetic use of *otheo* which provides an interesting case study of the uses of translation. Aristophanes' *Wasps* includes a description of the chorus of wasps, representing the Athenian hoplites at Marathon, defeating the Persians using the verb *otheo*, which Hanson (1989) p.172 translates as: 'we *pushed* them with the gods until evening' (Hanson's italics). Hanson uses this line as a further argument for the literal scrum. But this translation is different from other renderings of this line – for example, O'Neill's 1938 translation is: 'However, by the help of the gods, we drove off the foe towards evening', while Barrett in the Penguin Classics (freer, verse) translation offers: 'But when the shades of evening fell we had them on the run.' As we can see from L-S-J, 'drove off' or 'had them on the run' are reasonable free translations within the range of meanings for *otheo* in contemporary authors and Homer; Hanson, by giving (and emphasizing) a completely literal translation (at least he does not resort to 'shoved' on this occasion), has again begged the question. Wheeler (2007) p.211, strangely, adduces this particular example as 'significant evidence against the view that pushing is only a metaphor'. His other piece of significant evidence is that Herodotus (8.78) could refer to an *othismos* of words – which is a metaphor! 'Metaphors contrive figurative usage from real practice', he remarks, but it is not clear why this makes the hoplite usage less likely to also be metaphorical – why must the hoplite usage be the literal one? It seems more likely that expressions for arguments ('an *othismos* of words') and expressions for combat ('much *othismos*' in battle) are both metaphors derived from the underlying literal meaning of the word, just as for the English 'push (back)'. That *othismos* has literal pushing among its range of meanings is not in dispute.

17. The quote is from Hanson (1989) p.175. See also Koon (2010) p.56 and ch.7 on Zama, Polybius and Livy.

18. The 1949 Livy translation is from Frank Gardener Moore, *Livy. Books XXVIII–XXX With An English Translation*; that of 1850 is Cyrus Edmonds, *Livy. History of Rome by Titus Livius, books twenty-seven to thirty-six*.

19. Examples of the use of translation in support of a particular interpretation are common. For instance, Lazenby (1991) references a number of the examples I have given above and translates *otheo* as 'shove' in every case. As Koon (2010) p.39 sagely observes, 'a translator necessarily has to put the original language into target language using the conceptual tools at his disposal. The predominant model of combat will influence a translation, but we should limit how much a translation influences the model of combat.'

20. We might also note that the seventeenth-century AD tactical manuals that followed the Hellenistic tradition also give the rear ranks an exhortatory, not shoving, role. For example Edmonds, *Remembrancer of the City of London* p.326 (quoted in Anderson (1970) p.96), 'it must be provided that the bringer up or last rank ... be little inferior [to the first rank] ... that

they may both know when to reprehend their former Ranks, and urge them forward, if they see them declining or yielding upon false occasions'.

21. For a detailed examination of these passages and their implications for the Macedonian phalanx, and a possible model of how this 'pushing' worked in practice in the way the Macedonians fought, and how this differed from the Classical phalanx, see Taylor (2020) Ch. 8 (pp.298–344).

22. *Diakopto*, cutting through, is a Xenophontic usage not found in Thucydides, who uses it just for cutting through the wooden bar of a gate (twice); 2.4.4 and 4.111.2. Polybius uses the word frequently but rarely in a military context. The concept of breaking through the line, however, is common; see below.

23. See Taylor (2020) Ch. 8 for consideration of pike fighting in later periods and the way that formations could be 'pushed' forward while individuals still maintained separation. Modern reconstructions involving a strictly literal interpretation of the common Early Modern expression 'push of pike' can be found by searching on YouTube for 'push of pike' – these bear a close resemblance to the attempted reconstructions of *'othismos'* but are, I am certain, just as far divorced from reality. Indeed, I do not believe that anyone who has witnessed a modern re-enactors' 'push of pike' could believe for one moment that this is how seventeenth-century infantry actually fought. For a more plausible version see the 2006 Spanish movie *Alatriste*, with its recreation of the Battle of Rocroi (although the heroes still rush out to perform solitary 'deeds of valour').

24. This alternative model – 'crowd crush othismos' – is proposed in Bardunias and Ray (2016) pp.132–38.

25. Bardunias and Ray (2016) p.133 point out that the three-quarter stance of Schwartz (2013a) p.193 is incompatible with mass pushing. This point seems not well understood by proponents of the scrum – for example, Hanson (2013) p.263 refers to 'being pushed often into an enemy line, while keeping the shield chest high to protect both oneself and the man on the left, does not preclude individual battle skill in stabbing the enemy, keeping one's balance, and avoiding incoming blows'. In a crush such as would develop at the fronts of two blocks eight men deep pushing, there would be no freedom of movement at all, for man or shield, and the only stabbing possible would have been at the immobile faces of the men several ranks back. Similarly, Schwarz (2013a) p.91 discusses the hoplite's method of fighting, 'cowering cautiously behind ... his shield he would attempt to get a quick jab in against his immediate enemy ... any attack entailed at least a partial and temporary exposure of the hoplite as he twisted his upper right body sideways and forwards'. Yet Schwartz favours the scrum model and his discussion of weapon use (pp.192–94) does not accord with this description (a difficulty he gets around by suggesting that the scrum may have been temporary or intermittent). For the stance used in combat see Bardunias and Ray (2016) pp.16–18 and Matthew (2012) esp. pp.44–59.

26. Bardunias and Ray (2016) p.130 argue against the notion of the high-speed collision, which is proposed by for example Hanson (1989) pp.152–59. I find it difficult to reconcile the various elements of Hanson's model – a charge at the run during which the rear ranks are pushing the men in front in the back, a collision, but one in which some men pass into gaps between enemy files and so enter into the enemy formation, while still being pushed in the back by their own men, a period of spear fighting (*'doratismos'*), yet increasing pressure 'since they were striving to force back the entire enemy mass which was itself trying to press forward' (p.172). I think that the physical realities of this model have not been well elucidated or understood.

27. Bardunias and Ray (2016) pp.34–35 for the 'form follows function' argument, and pp.132–38 for the function envisaged (providing a breathing space for the hoplite). Note also that for example Schwartz (2013a) pp.192–93 uses a similar argument, but for a rather different function, thus demonstrating that the form cannot be mapped one-to-one to a single function. So also De Groote (2016) finds (p.210) that 'The shallow dome provided the hoplite shield with a structure that was considerably more efficient at dispersing impact energy evenly

around the shield, giving it a greater ability to withstand strikes from thrusting spears, as compared with a similar flat, round wooden shield.' So perhaps form did follow function, but opinions vary on what the function was.

28. Cawkwell (quoted in Krentz (1985b) p.50) raised this objection most colourfully, calling the scrum 'wildest folly' since 'the front ranks would have been better able to use their teeth than their weapons when a broad shield was jammed against the back with the weight of seven men'. Bardunias and Ray (2016) p.137 consider that spears and swords (especially the short Spartan *encheiridion*) could still be used in such a press, and that 'the strike of a spear from the rear ranks would require very little range of motion to find targets', which I can well imagine. It is difficult to devise practical tests of whether such a form of combat was practical, but my suspicion is that with every man fixed immobile in place with only his right arm free, casualties would be very high, and that every man killed or wounded (by a blow to the head or face, presumably) would then be left hanging, pinioned in the scrum, with unknown but surely deleterious effects on the scrum's stability.

29. Bardunias and Ray (2016) pp.131–32 suggest combat naturally closed up as spears became broken. The 'shields smashed to pieces' of Xenophon is one of the pieces of evidence proposed for the scrum (on the grounds that only the considerable forces generated by a mass shove could so smash a shield), though Bardunias and Ray report of their experiments with replica shields that 'not a single aspis was deformed' in their simulated crush. The breaking of shields then seems to be an occasional accident, and we do not know the circumstances that could cause a shield to break (whether for example shield-clashing in combat, or splitting after penetration by spears, would be sufficient).

30. See Goldsworthy (1996) pp.228–35 for discussion of this question as it applies to Roman armies. Classical scholars will frequently defer on this point to Keegan's comment in *The Face of Battle* that cavalry are simply unable to charge into contact with infantry, though Keegan does not provide any support for the assertion. Most discussion of the issue is outside our period; the works of Nosworthy, *Battle Tactics of Napoleon and His Enemies* and *The Anatomy of Victory: Battle Tactics, 1689–1763* are particularly useful for the seventeenth and eighteenth centuries. One of the best, and few really clear, descriptions of a cavalry charge is also one of the latest, and is to be found in Churchill's account (in *The River War*) of the charge of the 21st Lancers at Omdurman (1898), in which he participated – here the cavalry charged into and passed right through a deep but apparently quite loosely formed formation of infantry which they came across unexpectedly. The game theory version of the 'game of chicken' is well described in Colman, *Game Theory and its Applications in the Social and Biological Sciences*.

31. For the charge and possibility of a collision see e.g. Wheeler (2007) p.209. Wheeler is, like many, apparently in two minds on this matter. On p.209 he states that 'Two phalanxes charging at each other could not smash together in a horrendous crash', but at p.206 that 'Eighteenth-century debates over the relative virtues of column and line formations produced a theory that eight deep produced the maximum effect of shock when the attacking line collided with the defenders'' (and goes on to apply this to hoplites). I think it certain that eighteenth-century infantry never literally collided, as Wheeler notes also, p.210. Eighteenth-century comparisons do allow us to see that depth was important even where collision, pushing and weight are certainly not involved. As noted above, Hanson (1989) pp.152–59 favours a collision, 'in the case of the Greeks – and perhaps among the Greeks alone'. This is based on 'a fair reading of the ancient accounts' (a 'fair reading' that does not seem at all fair, nor in this case even literal), though only one ancient source is actually adduced to support the idea, the 'clashing of shields' of Xen., *Hell.* 4.3.19 which is not evidence for this at all. See the comments of Krentz (2013) pp.142–43. The idea is restated by (for example) Viggiano (2013) p.118 and n.41, using American Football as an analogy (which makes a change from rugby). The problem with all these sports analogies is the lack of weapons and lethal force in sport. If the quarterback carried a spear instead of a ball, and threatened to stab in the face anyone who got too close, I think there would be far fewer sacks in American Football. Whether the charge into full contact is a necessary component of the scrum model or an optional extra is unclear from reading the various interpretations.

32. See Goldwswothy (1996) pp.201–06 for the charge to contact in Roman armies.
33. On Chabrias' ploy, Anderson (1963); Buckler (1972). The interpretation of Matthew (2012) pp.217–19 (that Chabrias' men adopted a strong close-order formation) misses the point of the story.
34. Bardunias and Ray (2016) p.130 point out that it is easy for a running formation to come to a stop in a short space, having performed experiments with re-enactors to demonstrate the point.
35. For the spear-fighting stage of combat, Bardunias and Ray (2016) pp.130–32. Matthew (2012) devotes considerable effort to quantifying spear reach, pp.71–92.
36. For the 'dynamic stand-off' model of Roman combat see Sabin (2000), discussed also in Koon (2010) pp.21–22. The low-intensity nature of fighting in this model is deduced both from the psychology of close-quarters combat in other better-attested periods, and by the need to find a form of combat that could be continuous across the perhaps considerable length of time taken by a Roman battle.
37. See Chapter 2 and note 38 for the hold of the spear, and Matthew (2012) pp.15–18 (and ff.). I am unconvinced by Matthew's arguments in favour of the 'couched' (high overhand) hold.
38. Matthew (2012) pp.211–17 suggests that the close-order formation (which he sees, with commendable precision, as being 45–50cm per man) combined with the underarm (couched) spear hold would prevent an enemy colliding with the front rank, had they wished to. I do not see why it would not be possible to parry aside the proferred spear points, and am more inclined to think that men would not collide because they did not wish to. On the depth of Cyrus' Persians see note 12 above.
39. See Chapter 7 note 36 for the modern use of *doratismos* to describe a phase of fighting.
40. For Roman close-quarters infantry combat see Goldsworthy (1996) pp.191–227, esp. pp.217–18 for the use of the shield as a weapon (for individual punching).
41. Bardunias and Ray (2016) pp.131–32 describe this tendency to close up, though I do not agree that this would result in a scrum or crowd crush.
42. See Goldsworthy (1996) pp.176–83 for depth in Roman armies, tentatively suggesting three to eight deep; Pompey's army at Pharsalus formed ten deep.
43. The quote is from van Wees (2004) p.189. Compare, among several examples, 'One cannot help but feel that morale, as well as fighting, would be more effectively boosted by comrades actually taking a part in the grisly work, rather than merely standing idly around', Schwartz (2013a) p.195. Sarcastic comments on the lack of contribution of the rear ranks are not an entirely new phenomenon – compare the sixteenth-century words of John Smythe in *Instructions, Observations and orders Mylitarie*, quoted in Taylor (2020) p.321, 'all the rest of the ranks of both the squadrons must by such an unskilfull kind of fighting stand still and look on and cry aime, until the first ranke of each squadron hath fought their bellies full'. Smythe's solution was that the whole formation engage in a mass advance, equivalent to that used by the Macedonian phalanx and proposed as a possible model for the hoplite advance above, but without engaging in a scrum (see discussion in Taylor (2020) pp.323–29).
44. Krentz (1985a) for the casualty statistics.
45. Wheeler (2007) pp.210–11 on the Hellenistic comparison. He notes that 'Shoving the enemy or pushing comrades forward with the shield does not come into question' for the Macedonian phalanx. I think the important point is that, as Wheeler observes, the Hellenistic phalanx cannot have pushed (in the traditional rugby scrum way) because of its pikes, yet it is clearly described as pushing in some sense ('bearing forward') – which means that there is something wrong with the orthodox understanding of the form this pushing took. Wheeler (2007) pp.206–07 dismisses du Picq's (1987) p.169 assertion that rear ranks would not and could not push on the ranks in front (du Picq argues that they would in practice push them over rather than forwards), 'if the principle of shock (a physical collision of attackers with defenders) is conceded'. But the whole point is that it is not conceded. 'Shock' (along with 'push', 'hand-to-hand' and others) is another deep-seated military metaphor in English (and other modern European languages), just as *othein/othismos* was in Greek, and whether it represents an underlying literal physical reality, in any period, is, to say the least, disputed. Du Picq at

least rejected the notion: '*le choc, l'impulsion physique de la masse était un mot. – On savait qu'en penser*' (1987 p.109 = p.62 of the 1880 French edition – the 1987 translation slightly obscures the meaning).
46. Goldsworthy (1997) stresses the morale and psychological benefit of deep formations. Van Wees (2004) p.191 offers that 'The "weight" of a formation was not the physical but the psychological pressure which it brought to bear', but qualifies this with 'the deeper the enemy's formation, the larger the number of soldiers ready to take the place of casualties' – which is unlikely to be a true explanation as we have seen, and fails to follow through on the psychological pressure argument. Schwartz (2013a) expresses some admiration for Goldsworthy's ideas but then opines that the ranks beyond the first were therefore 'nothing more than a road block' and were 'merely standing idly around', which suggests he has not understood the concept at all. Konijnendijk (2018) pp.133–38 also favours the psychological (morale) advantage as well as resistance to breakthrough.
47. The ability of columns to cross broken terrain is stressed by Goldsworthy (1997) pp.6–14. The objections of Schwartz (2013a) pp.196–97 seem to miss the point, concentrating on the tendency of the phalanx to lose order due to differential rates of advance, rather than actual obstacles – though on this point Goldsworthy (1996) p.178 quotes the seventeenth-century AD writer Raimondo Monteuccoli noting that a 'hedgerow', a single line, will develop gaps as the braver men proceed resolutely and others hang back: 'Great breaks occur in it, which miraculously encourages the enemy.'
48. This is broadly the explanation for depth given by du Picq (1987); he points out the difference between the use of deep formations by the Greeks, and of multiple separate lines by the Romans, the latter being a superior way of achieving a similar objective, superior because in the Roman formation the rear units were far enough away from the front line fighting to be insulated from the fear and danger of the losing side, so they were not swept away in the defeat of the front as rear ranks in a Greek (or similar) phalanx would be.
49. That formations rout from the rear is a point made by for example du Picq in his much-quoted analysis of ancient (and more recent) combat (du Picq (1987) e.g. pp.79, 89, 114, 116, 169, 171); Hanson (1989) pp.189–90 agrees; Wheeler (2007) p.211 appears to disagree, it is unclear why, presumably because there is no explicit statement to this effect in ancient sources, but it is I believe implicit in the desire to have brave, experienced men in the rear rank. I also take it as a true observation, if only from practicality – the men in front cannot run away until the men behind make way for them to do so.
50. For panic in ancient armies in general see Wheeler (1988b).
51. The quotes are from Roger Boyle, Earl of Orrery, *A treatise of the art of war dedicated to the King's Most Excellent Majesty* (1677).
52. Pritchett (1985) 4 pp.46–51 gathers references to duration, but unfortunately without distinguishing between types of combat (skirmishes, pitched battles, pursuits) or when the measurements started or ended. Schwartz (2013a) pp.201–25 provides a good discussion of the topic, which I broadly follow here.
53. See Schwartz (2013a) pp.217–25 for problems with sources, particularly Diodorus, and with stock expressions of duration.
54. Matthew (2012) pp.119–29 describes an experiment with re-enactors to determine endurance and accuracy (of strikes with spears); the test period of fifteen minutes was reached easily by some subjects, others stopping due to wrist fatigue or boredom, but the only physical activity required was striking repeatedly with a spear. Matthew admits that the test has many limitations, and that (p.125) 'the test results cannot be regarded as an accurate reflection of the endurance levels of combatants in the fifth and fourth centuries BC'; durations for enduring such spear thrusting (particularly the overhead hold) were as low as 'only a few minutes'. Even so, Matthew proposes (p.128) 'a limit of endurance [for fighting] of between thirty and sixty minutes'. I do not find these results particularly illuminating.
55. For the dynamic stand-off and the concept of the battlefield clock see Sabin (2000). To illustrate the difficulty of multiple hour-long combats, Sabin observes that 'even if we assume that just 5 per cent of the troops were in the front rank, and that they struck their adversaries

only every five seconds, and that less than 1 per cent of these attacks caused death or mortal injury, each army would suffer 5 per cent fatalities every ten minutes.' To give a similar example, we might imagine two eight-rank-deep lines fighting each other, and causing, over the course of the combat, 50 per cent casualties each (including killed and wounded – much higher casualties than are ever attested, but taking a worst case to illustrate the point). If the battle lasted, say, one hour twenty minutes, then each file leader (whoever is file leader at the time) is scoring an effective hit (killing or wounding) just once every twenty minutes. If the combat goes on longer, then hits are naturally even less frequent – in a two-hour melee, there would be one hit every thirty minutes, and so on. With historically lower levels of casualties (closer to 5 per cent), effective hits would be even more infrequent. Fighting with such exceptionally low levels of lethality or effectiveness bears no relation to the usual depictions of fighting on film or in the simulated fights of re-enactors, nor to such descriptions of hand-to-hand fighting as we have from other eras.
56. It is becoming traditional to quote Hornblower (1996) p.306, 'Only an unusually arrogant scholar could claim to know exactly what kind of thing went on in a hoplite battle', which has not of course prevented the publication of numerous books and articles claiming, with suitable caveats, to know exactly this. Etruscan and early Roman warfare is a relatively neglected field (in comparison with the Classical hoplite) in need of deeper exploration – starting points would be Cornell (1995); Connolly (2012) pp.91–100; Torelli (2001) 'Warfare', pp.558–65.

Chapter 9
1. See Taylor (2020) for an account of the Macedonian phalanx, with further bibliography.
2. The battle is discussed in Butera and Sears (2019) pp.157–79; Hammond (1938) for the topography; Sears and Willeke (2016) argue that Alexander did charge at the head of cavalry.
3. For Iphicrates' reforms see most recently Konijnendijk (2014), summarizing earlier arguments of Anderson (1970) pp.129–31, who places the reforms in the context of Egypt, and Best (1969), who doubts the reforms ever really happened, among others. Other views, notably Sekunda (2014b), are available in the same volume (Sekunda and Burliga (2014)). See also Ueda-Sarson (2002a), and (2002b) for the other types of infantry. Opinion remains, it is fair to say, divided.
4. Taylor (2020) pp.18–21 for discussion of the Iphicratean peltast and the Macedonian phalangite. Other interpretations are discussed by Sekunda (2014b); Matthew (2015) pp.11–16.
5. Alexander's battles have a large bibliography of their own, with accounts in any of the (numerous) biographies of Alexander. A useful collection of references and reading are Pietrykowski (2009) and Sabin (2009) pp.125–43, with further details in Devine (1985), (1986) and (1987).
6. See Taylor (2020) ch. 5 for all these points.
7. For *thureophoroi* and *peltophoroi* see Sekunda (2007) pp.339–43; Ueda-Sarson (2002b); Taylor (2020) pp.184–87.
8. Taylor (2020) pp.56–57 and *passim* for these peltasts. Although modern, and ancient literary, attention tends to focus on the great Macedonian-based armies of the kingdoms, the reality of warfare in Greece continued to be minor wars between *poleis*, fought by city armies raised much as they long had been – see Ma (2000).
9. The words used by Polybius in the Sellasia passage are *piezo* for the pushing of the Spartans' courage, and *exotheo* for the pushing of the Macedonians' weight. Taylor (2020) pp.254–55 for the battle and its tactical implications, and pp.62–63 for the carrying arrangements of the shield.
10. Because the adoption of *thureophoros* equipment is not well understood, some modern scholars have interpreted Philopoemen's reform as the first adoption of Classical hoplite equipment by the Achaeans, and rejected it on those grounds – see Snodgrass (2013) p.92.
11. Snodgrass (2013) p.93 comments on this gradual decline of the hoplite, and notes that this may well shed light on the gradual nature of the hoplite's earlier rise.

Abbreviations and Translations

Aelian = *The Tactics of Aelian*, C. Matthew (Barnsley: Pen and Sword)
Aristot., *Const. Ath.* = Aristotle, *Athenian Constitution*, H. Rackham (ed.) (Cambridge, MA:, Harvard University Press; London: William Heinemann Ltd)
Arr., *Anab.* = Arrian, *Anabasis of Alexander*, P.A. Brunt (Cambridge, MA: Harvard University Press; London: William Heinemann Ltd)
Arrian, *Tactics* = author's translation after J.G. DeVoto (Chicago: Ares Publishers)
Asclep. = Asclepiodotus, *Tactics*, C.H. Oldfather and W.A. Oldfather (Cambridge, MA: Harvard University Press; London: William Heinemann Ltd)
Diod. = Diodorus Siculus, C.H. Oldfather (Cambridge, MA: Harvard University Press; London: William Heinemann Ltd)
Euripides, *Heracles* = E.P. Coleridge (ed.) (New York: Random House)
Euripides, *Phoenissae* = E.P. Coleridge (ed.) (New York: Random House)
Hdt. = Herodotus, *The Histories*, A.D. Godley (Cambridge, MA: Harvard University Press)
Homer, *Iliad* = Samuel Butler (London: Longmans, Green and Co.; New York and Bombay: A.T. Murray; Cambridge, MA: Harvard University Press; London: William Heinemann Ltd)
Paus. = Pausanias, *Description of Greece*, W.H.S. Jones and H.A. Ormerod (Cambridge, MA: Harvard University Press; London: William Heinemann Ltd)
Plato, *Laws* = R.G. Bury (Cambridge, MA: Harvard University Press; London: William Heinemann Ltd)
Plut., *De Herod* = Plutarch, *De Herodoti malignitate* (Plutarch's Morals), ed. W.W. Goodwin (Boston: Little, Brown, and Company)
Plut., *Mor.* = Plutarch, *Moralia*, ed. W.W. Goodwin (Boston: Little, Brown, and Company.
Plut., *Pelop.* = Plutarch, *Life of Pelopidas* (*Plutarch's Lives*), Bernadotte Perrin (Cambridge, MA: Harvard University Press; London: William Heinemann Ltd)
Pol. = Polybius, *Histories*, E.S. Shuckburgh (London, New York: Macmillan); and W.R. Paton (Cambridge, MA: Harvard University Press)
Polyaen. = Polyaenus, *Stratagems*, attalus.org after R. Shepherd
Strabo, *Geography* = ed. H.L. Jones (Cambridge, MA: Harvard University Press; London: William Heinemann Ltd)
Thuc. = Thucydides, *The Peloponnesian War* (London: J.M. Dent; New York: E.P. Dutton)
Xen., *Ages.* = Xenophon, *Agesilaus*, E.C. Marchant, G.W. Bowersock (Cambridge, MA: Harvard University Press; London: William Heinemann Ltd)
Xen., *Anab.* = Xenophon, *Anabasis*, Carleton L. Brownson (Cambridge, MA: Harvard University Press; London: William Heinemann Ltd)
Xen., *Const. Lac.* = Xenophon, *Constitution of the Lacedaimonians*, E.C. Marchant, G.W. Bowersock (Cambridge, MA: Harvard University Press; London: William Heinemann Ltd)
Xen., *Cyrop.* = Xenophon, *Cyropaedia*, Walter Miller (Cambridge, MA: Harvard University Press; London: William Heinemann Ltd)
Xen., *Hell.* = Xenophon, *Hellenica*, Carleton L. Brownson (Cambridge, MA: Harvard University Press; London: William Heinemann Ltd)
Xen., *Hipp* = Xenophon, *On the Cavalry Commander*, E.C. Marchant, G.W. Bowersock (Cambridge, MA: Harvard University Press; London: William Heinemann Ltd)
Xen., *Horse* = Xenophon, *Horsemanship*, E.C. Marchant, G.W. Bowersock (Cambridge, MA: Harvard University Press; London: William Heinemann Ltd)
Xen., *Mem.* = Xenophon, *Memorabilia*, E.C. Marchant (Cambridge, MA: Harvard University Press; London: William Heinemann Ltd)

Bibliography

The following bibliography makes no claims to being comprehensive; in particular the reader will note the almost total absence of works in languages other than English. The intent is only to provide references where needed for points raised in the text, and to act as a starting point for further reading. I have tried to refer where possible to recent and generally available works rather than obscure articles. Sabin, van Wees and Whitby, *The Cambridge History of Greek and Roman Warfare* (2007), is particularly useful in this respect, both for its content and also for its much more extensive bibliography which will allow readers to delve deeper into the topics covered in this book.

ADCOCK, F.E. (1957), *The Greek and Macedonian Art of War*
ALDRETE, G.S., BARTELL, S. & ALDRETE, A. (2013), *Reconstructing Ancient Linen Body Armor: Unraveling the Linothorax Mystery*
ANDERSON, J.K. (1963), 'The statue of Chabrias', *American Journal of Archaeology* 67 4, pp.411–13
– (1965), 'Cleon's orders at Amphipolis', *Journal of Hellenic Studies* 85, pp.1–4
– (1970), *Military Theory and Practice in the Age of Xenophon*
– (1984), 'Hoplites and heresies: a note', *Journal of Hellenic Studies* 104, p.152
– (1991), 'Hoplite Weapons and Offensive Arms', in Hanson (ed.) (1991), pp.15–37
ANDREWES, A. (1956), *The Greek Tyrants*
BARDUNIAS, P.M. & RAY, F.E. (2016), *Hoplites at War: A Comprehensive Analysis of Heavy Infantry Combat in the Greek World 750–100 BCE*
BEST, J.G.P. (1969), *Thracian Peltasts and their Influence on Greek Warfare*
BILLOWS, R.A. (2010), *Marathon: The Battle that Changed Western Civilization*
BLYTH, P.H. (1982), 'The Structure of a Hoplite Shield in the Museo Gregoriano Etrusco', *Bolletino dei Musei E'Gallerie Pontifice* 3, pp.5–21
BOARDMAN, J. (1983), 'Symbol and Story in Greek Geometric Art', in W.G. Moon (ed.), *Ancient Greek Art and Iconography*, pp.15–36.
BROUWERS, J. (2013), *Henchmen of Ares: Warriors and Warfare in Early Greece*
– (2014), 'Phalanx and fallacies: ways forward in the study of ancient Greek warfare', *Ancient World Magazine*, online at https://www.ancientworldmagazine.com/articles/phalanx-fallacies-ways-forward-study-ancient-greek-warfare/
BROUWERS, J. & KONIJNENDIJK, R. (2020), 'The Chigi Vase: Ceci n'est pas une phalange', *Ancient Warfare Magazine*, online at https://www.ancientworldmagazine.com/articles/chigi-vase/
BUCKLER, J. (1972), 'A second look at the monument of Chabrias', *Hesperia* 41 4, pp.466–74
– (1985), 'Epameinondas and the *embolon*', *Phoenix* 39, pp.134–43
BUGH, G.R. (1988), *The Horsemen of Athens*
BUTERA, C.J. & SEARS, M.A. (2019), *Battles and Battlefields of Ancient Greece: A Guide to their History, Topography and Archaeology*
CARTLEDGE, P. (1979), *Sparta and Lakonia*
– (2006), *Thermopylae: The Battle that Changed the World*
– (2013a), 'Hoplitai/Politai: Refighting Ancient Battles', in Kagan and Viggiano (eds) (2013), pp.74–84
– (2013b), *After Thermopylae: the Oath of Plataea and the End of the Greco-Persian Wars*

CAWKWELL, G. (1983), 'The decline of Sparta', *Classical Quarterly* 33.2, pp.385–400
– (1989), 'Orthodoxy and Hoplites', *Classical Quarterly* 39, pp.375–89
CHARLES, M. (2012), 'Herodotus, Body Armour and Achaemenid Infantry', *Historia: Zeitschrift Für Alte Geschichte* 61(3), pp.257–69
CHRISTESEN, P. (2006), 'Xenophon's "Cyropaedia" and Military Reform in Sparta', *Journal of Hellenic Studies* 126, pp.47–65
CLAIRMONT, C.W. (1983), *Patrios Nomos: Public Burial in Athens during the Fifth and Fourth Centuries BC*
CONNOLLY, P. (2012), *Greece and Rome at War* (3rd ed.)
CORNELL, T.J. (1995), *The Beginnings of Rome: Italy and Rome from the Bronze Age to the Punic Wars, c. 1000 – 264 BC*
CROWLEY, J. (2012), *The Psychology of the Athenian Hoplite: The Culture of Combat in Classical Athens*
DALBY, A. (1992), 'Greeks abroad; social organization and food among the Ten Thousand', *Journal of Hellenic Studies* 112, pp.16–30
DAYTON, J.C. (2006), *The Athletes of War: An Evaluation of the Agonistic Elements in Greek Warfare*
DEACY, S. (2000), 'Athena and Ares: War, violence and warlike deities', in van Wees (ed.) (2000), pp.285–98
DE GROOTE, K.R. (2016), '"Twas when my shield turned traitor!" Establishing the combat effectiveness of the Greek hoplite shield', *Oxford Journal Of Archaeology* 35 (2), pp.197–212
– (2018), '"All your strength is in your spears" – How hoplites wielded their *dory*', *Ancient Warfare* 12 1, pp.34–40
DEVINE, A.M. (1983), 'Embolon: a study in tactical terminology', *Phoenix* 37, pp.201–17
– (1985), 'Grand tactics at the battle of Issus', *Ancient World* 12, pp.39–59
– (1986), 'The battle of Gaugamela: a tactical and source-critical study', *Ancient World* 13, pp.87–115
– (1988), 'A pawn-sacrifice at the battle of the Granicus: the origins of a favourite stratagem of Alexander the Great', *Ancient World* 18, pp.13–20
DONLAN, W. & THOMPSON, J. (1976), 'The charge at Marathon: Herodotus 6.112', *Classical Journal* 71, pp.339–43
– (1979) 'The charge at Marathon again', *Classical World* 72, pp.419–20
DOVER, K.J. (1974), *Greek Popular Morality in the Time of Plato and Aristotle*
DU PICQ, A. (1987), *Battle Studies*, in *Roots of Strategy: Book 2*
ECHEVERRÍA, F. (2010), 'Weapons, technological determinism, and Ancient warfare', in Fagan and Trundle (eds) (2009), pp.21–56
– (2011), '*Taktikè technè*: the neglected element in Classical hoplite battles', *Ancient Society* 41, pp.45–82
– (2012), 'Hoplite and phalanx in Archaic and Classical Greece: a reassessment', *Classical Philology* 107 4, pp.291–318
ENGELS, D.W. (1978), *Alexander the Great and the Logistics of the Macedonian Army*
FAGAN, G. (2009), '"I Fell upon Him like a Furious Arrow": Toward a Reconstruction of the Assyrian Tactical System', in Fagan and Trundle (eds) (2009), pp.81–100
FAGAN, G. & TRUNDLE, M. (eds) (2009), *New Perspectives on Ancient Warfare*
FIELDS, N. (2008), *Syracuse 415–413 BC: Destruction Of The Athenian Imperial Fleet*
FORNARA, C.W. (1971), *The Athenian Board of Generals from 501–404 BC*
FORSDYKE, S. (2006), 'Land, Labor and Economy in Solonian Athens: Breaking the Impasse between Archaeology and History', in J.H. Blok and A.P.M.H .Lardinois (eds), *Solon of Athens: New Historical and Philological Approaches*
FOXHALL, L. (1997), 'A view from the top: evaluating the Solonian property classes', in L.G. Mitchell and P.J.Rhodes (eds), *The Development of the Polis in Archaic Greece*, pp.113–35
– (2013), 'Can We See the Hoplite Revolution on the Ground? Archaeological Landscapes, Material Culture, and Social Status in Early Greece', in Kagan and Viggiano (eds) (2013), pp.194–221

FRASER, A.D. (1942), 'The Myth of the Phalanx Scrimmage', *Classical World* 36, pp.15–16
FURLEY, W.D. & BREMER, J.M. (2001), *Greek Hymns: Selected Cult Songs from the Archaic to the Hellenistic Period*
GABRIEL, R.A. & METZ, K.S. (1991), *From Sumer to Rome: The Military Capabilities of Ancient Armies*
GABRIELSEN, V. (2007), 'Warfare and the state', in Sabin, van Wees and Whitby (eds) (2007), pp.248–72
GARLAN, Y. (1975), *War in the Ancient World: a Social History*
GLEBA, M. (2012), 'Linen-clad Etruscan Warriors', in M.L. Mosch (ed.), *Wearing the Cloak: Dressing the Soldier in Roman Times*, pp.45–55
GOLDSWORTHY, A.K. (1996), *The Roman Army at War 100 BC–AD 200*
– (1997), 'The Othismos, Myths and Heresies: The Nature of Hoplite Battle', *War In History* 4, 1, pp.1–26
GREENHALGH, P.A.L. (1973), *Early Greek Warfare: Horsemen and Chariots in the Homeric and Archaic Ages*
GROTE, G. (1846), *A History of Greece*
GRUNDY, G.B. (1911), *Thucydides and the History of his Age*
HALE, J.R. (2013), 'Not Patriots, Not Farmers, Not Amateurs: Greek Soldiers of Fortune and the Origins of Hoplite Warfare', in Kagan and Viggiano (eds) (2013), pp.176–93
HALL, J.M. (2007), 'International Relations', in Sabin, van Wees and Whitby (eds) (2007), pp.85–107
HAMEL, D. (1998), *Athenian Generals: Military Authority in the Classical Period*
HAMMOND, N.G.L. (1938), 'The Two Battles of Chaeronea', *Klio* 31 pp. 187-218
HANSON, V.D. (1983, new edition 1998), *Warfare and Agriculture in Classical Greece*
– (1989, new edition 2000), *The Western Way of War: Infantry Battle in Classical Greece*
– (1991), 'Hoplite Technology in Phalanx Battle', in Hanson (ed.) (1991), pp.63–84
– (ed.) (1991), *Hoplites: The Classical Greek Battle Experience*
– (1995), *The Other Greeks: The Family Farm and the Agrarian Roots of Western Civilization*
– (2000), 'Hoplite battle as Ancient Greek warfare: when, where and why?', in van Wees (ed.) (2000), pp.201–32
– (2007), 'The Modern historiography of Ancient warfare', in Sabin, van Wees and Whitby (eds) (2007), pp.3–21
– (2013), 'The Hoplite Narrative', in Kagan and Viggiano (eds) (2013), pp.256–75
HAWKINS, C. (2010), 'Spartans and Perioikoi: The organization and ideology of the Lakedaimonian army in the Fourth Century BCE', *Greek, Roman and Byzantine Studies* 51, pp.401–34
HEAD, D. (1992), *The Achaemenid Persian Army*
HEALY, M. (1991), *The Ancient Assyrians*
HIXENBAUGH, R. & VALDMAN, A. (illus.) (2019), *Ancient Greek Helmets: A Complete Guide and Catalog*
HOLLADAY, A.J. (1982), 'Hoplites and heresies', *Journal of Hellenic Studies* 102, pp.94–103
HORNBLOWER, S. (1991), (1996), (2008), *A Commentary on Thucydides*, vols 1, 2, 3
– (2000), 'Sticks, stones and Spartans: the sociology of Spartan violence', in van Wees (ed.) (2000), pp.57–82
– (2007), 'Warfare in literature; the paradox of war', in Sabin, van Wees and Whitby (eds) (2007), pp.22–53
HUNT, P. (1997), 'Helots at the Battle of Plataea', *Historia* 46, pp.129–44
– (1998), *Slaves, Warfare and Ideology in the Greek Historians*
– (2007), 'Military Forces', in Sabin, van Wees and Whitby (eds) (2007), pp.108–46
JAMESON, M.H. (1991), 'Sacrifice before battle', in Hanson (ed.) (1991)
JARVA, E. (1995), *Archaialogia on Archaic Greek Body Armour*
KAGAN, D. & VIGGIANO, G.F. (eds) (2013), *Men of Bronze: Hoplite Warfare in Ancient Greece*
KAGAN, D. & VIGGIANO, G.F. (2013), 'The Hoplite Debate', in Kagan and Viggiano (eds) (2013), pp.1–56

KALLET, L. (1983), 'Iphicrates, Timotheos and Athens, 371–360 BC', *Greek, Roman and Byzantine Studies* 24, pp.239–52

KEEGAN, J. (1976), *The Face of Battle: A Study of Agincourt, Waterloo and the Somme*

KONECNY, A, (2014), 'The Battle of Lechaeum, Early Summer, 390 BC', in Sekunda and Burliga (eds) (2014), Ch.1

KONIJNENDIJK, R. (2012), '"Neither the less valorous nor the weaker": Persian military might and the Battle of Plataia', *Historia Zeitschrift für Alte Geschichte* 61.1, pp.1–17

– (2014), 'Iphikrates the Innovator and the Historiography of Lechaeum', in Sekunda and Burliga (eds) (2014), Ch.4

– (2018) *Classical Greek Tactics: A Cultural History*

KOON, S. (2010), *Infantry Combat in Livy's Battle Narratives*

KRENTZ, P. (1985a), 'Casualties in hoplite battles', *Greek, Roman and Byzantine Studies* 26, pp.13–21

– (1985b), 'The Nature of Hoplite Battle', *Classical Antiquity* 4, pp.50–61

– (1991), 'The *salpinx* in Greek battle', in Hanson (ed.) (1991), pp.110–20

– (1994), 'Continuing the *Othismos* on *Othismos*', *Ancient History Bulletin* 8, pp.45–49

– (2000), 'Deception in Archaic and Classical Greek Warfare', in Van Wees (ed.) (2000), pp.167–200

– (2002), 'Fighting by the Rules: The Invention of the Hoplite *Agôn*', *Hesperia* 71, pp.23–39

– (2007), 'War', in Sabin, van Wees and Whitby (eds) (2007), pp.147–85

– (2010a), *The Battle of Marathon*

– (2010b), 'A Cup by Douris and the Battle of Marathon', in Fagan and Trundle (eds), (2009), pp.188–90

– (2013) 'Hoplite Hell: How Hoplites Fought', in Kagan and Viggiano (eds) (2013), pp.134–56

LAGOS, K & KARYANOS, F. (2019), *Who Really Won the Battle of Marathon?: A bold reappraisal of one of history's most famous battles*

LATACZ, J. (1977), *Kampfparanese, Kampfdarstellung, und Kampfwirklichkeit in der Ilias, bei Kallinos und Tyrtaios*

LAZENBY, J.F. (1985, reprinted 2012), *The Spartan Army*

– (1991), 'The Killing Zone', in Hanson (ed.) (1991), pp.87–109

– (1993), *The Defence of Greece, 490–479 BC*

LAZENBY, J.F. & WHITEHEAD, D. (1996), 'The Myth of the Hoplite's Hoplon', *Classical Quarterly* 46, pp.27–33

LENDON, J.E. (1999), 'The rhetoric of combat: Greek military theory and Roman culture in Julius Caesar's battle descriptions', *Classical Antiquity* 18.1, pp.273–329

– (2005), *Soldiers and Ghosts: A History of Battle in Classical Antiquity*

LONIS, R. (1979), *Guerre et Religion en Grèce á l'Époque Classique*

LORIMER, H.L. (1947), 'The Hoplite Phalanx with Special Reference to the Poems of Archilochus and Tyrtaeus', *The Annual of the British School at Athens*, pp.76–138

LOW, P. (2010), 'Commemoration of the war dead in classical Athens: remembering defeat and victory', in Pritchard (ed.) (2010), pp.341–58

LUGINBILL, R.D. (1994), '*Othismos*: the importance of the mass-shove in hoplite warfare', *Phoenix* 48, pp.51–61

LURAGHI, N. (2006), 'Traders, Pirates, Warriors: The Proto-History of Greek Mercenary Soldiers in the Eastern Mediterranean', *Phoenix* 60.1, pp.21–47

LYNN, J.A. (2004), *Battle: A History of Combat and Culture from Ancient Greece to Modern America*

MA, J. (2000), 'Fighting *poleis* of the Hellenistic World', in van Wees (ed.) (2000), pp.337–76

MANNING, S. (2021), *Armed Force in the Teispid-Achaemenid Empire: Past Approaches, Future Prospects*

MATTHEW, C. (2012), *A Storm of Spears: Understanding the Greek Hoplite at War*

– (2015), *An Invincible Beast: understanding the Hellenistic pike-phalanx at war*

MATTHEW, C. & TRUNDLE, M. (2013), *Beyond the Gates of Fire: New Perspectives on the Battle of Thermopylae*

MCKECHNIE, P. (1994), 'Greek Mercenary Troops and Their Equipment', *Historia: Zeitschrift für Alte Geschichte* 43.3, pp.297–305
MORRISON, J.S., COATES, J.F. & RANKOV, N.B. (2000), *The Athenian Trireme: The History and Reconstruction of an Ancient Greek Warship* (2nd ed.)
MORRISON, J.S. & WILLIAMS, R.T. (1968), *Greek Oared Ships 900–322 BC*
NAGY, G. (1997), 'The Shield of Achilles: Ends of the *Iliad* and beginnings of the polis', in S. Langdon (ed.), *New Light on a Dark Age: Exploring the Culture of Geometric Greece*
NILSSON, M.P. (1929), 'Die Hoplitentaktik und das Staatswesen', *Klio* 22, pp.240–49
OBER, J. (1985), *Fortress Attica: Defence of the Athenian Land Frontier, 404–322 BC*
– (1991), 'Hoplites and obstacles', in Hanson (ed.) (1991), pp.173–96
– (1999), 'The rules of war in Classical Greece', in Ober (ed.), *The Athenian Revolution: Essays on Ancient Greek Democracy and Political Theory*, pp.53–71
OGDEN, D. (1996), 'Homosexuality and warfare in Ancient Greece', in A.B. Lloyd (ed.), *Battle in Antiquity*, pp.107–68
PARKE, H.W. (1933, reprinted 1981), *Greek Mercenary Soldiers: From the Earliest Times to the Battle of Issus*
PARKER, R. (2000), 'Sacrifice and battle', in van Wees (ed.) (2000), pp.299–314
PIETRYKOWSKI, J. (2009), *Great Battles of the Hellenistic World*
PITTMAN, A. (2007), '"With Your Shield or On It": Combat Applications of the Greek Hoplite Spear and Shield', in B. Molloy (ed.), *The Cutting Edge: Studies in Ancient and Medieval Combat*, pp.64–76
PRITCHARD, D.M. (ed.) (2010), *War, Democracy and Culture in Classical Athens*
– (2010), 'The symbiosis between democracy and war: the case of ancient Athens', in Pritchard (ed.) (2010), pp.1–62
– (2019) *Athenian Democracy at War*
PRITCHETT, W.K (1971), (1974), (1979), (1985), (1991), *The Greek State at War*, vols 1, 2, 3, 4, 5
– (1969), *Studies in Ancient Greek Topography, Part II – Battlefields*
– (1982), *Studies in Ancient Greek Topography, Part IV – Passes*
– (1994), 'The general's exhortation in Greek warfare', in Pritchett, *Essays in Greek History*, pp.27–109
RAAFLAUB, K.A. (1997), 'Soldiers, citizens and the evolution of the early Greek polis', in L.G. Mitchell and P.J. Rhodes (eds), *The Development of the Polis in Archaic Greece*, pp.26–31
– (1999) 'Archaic and Classical Greece', in K.A. Raaflaub and N. Rosenstein (eds), *War and Society in the Ancient and Medieval Worlds*
– (2004), 'Archaic Greek Aristocrats as Carriers of Cultural Interaction', in R. Rollinger and C. Ulf (eds), *Commerce and Monetary Systems in the Ancient World: Means of Transmission and Cultural Interaction*, pp.197–217
– (2013), 'Early Greek Infantry Fighting in a Mediterranean Context', in Kagan and Viggiano (eds) (2013), pp.95–111
RANCE, P. (2017) 'Introduction', in Rance and Sekunda (eds.) (2017) pp. 9–64
RANCE, P. & SEKUNDA, N.V. (eds) (2017), *Greek Taktika: Ancient Military Writing and its Heritage*
RANDALL, K. (2011), 'Hoplite phalanx mechanics: investigation of footwork, spacing and shield coverage', https://www.academia.edu/922527
RAWLINGS, L. (2000), 'Alternative Agonies: Hoplite Martial and Combat Experiences beyond the Phalanx', in H. van Wees (ed.), *War and Violence in Ancient Greece*
– (2007), *The Ancient Greeks at War*
RAY, F.E. (2009), *Land Battles in 5th Century BC Greece: A History and Analysis of 173 Engagements*
– (2012) *Greek and Macedonian Land Battles of the 4th Century BC: A History and Analysis of 187 Engagements*
– (2014), 'Some observations regarding the analysis of artistic data in Christopher Matthew's flawed analysis of the mechanics of hoplite combat', online at hollow-lakedaimon.blogspot.com
REES, O. (2016), *Great Battles of the Classical Greek World*

REES, O. & CROWLEY, J. (2015), 'Was there mental trauma in ancient warfare? PTSD in Ancient Greece', *Ancient Warfare* 9:4
ROP, J. (2019), *Greek Military Service in the Ancient Near East, 401–330 BCE*
ROISMAN, J. (2003), 'The rhetoric of courage in the Athenian orators', in R. Rosen and I. Sluiter (eds), *Andreia: Studies in Manliness and Courage in Classical Antiquity*, pp.127–43
ROSENSTEIN, N (2010), 'Phalanges in Rome?', in Fagan and Trundle (eds) (2009), pp.289–304
RUSSELL, F.S. (1999), *Information Gathering in Classical Greece*
SABIN, P. (2000), 'The Face of Roman Battle', *Journal of Roman Studies* 90, pp.1–17
– (2009), *Lost Battles: Reconstructing the Great Clashes of the Ancient World*
SABIN, P., VAN WEES, H. & WHITBY, M. (eds) (2007), *The Cambridge History of Greek and Roman Warfare*
SAGAN, E. (1979), *The Lust to Annihilate: A Psychoanalytic Study of Violence in Ancient Greek Culture*
SALMON, J.B. (1977), 'Political hoplites?', *Journal of Hellenic Studies* 97, pp.87–122
SCHWARTZ, A. (2013a), *Reinstating the Hoplite: Arms, Armour and Phalanx Fighting in Archaic and Classical Greece*
– (2013b), 'Large Weapons, Small Greeks: The Practical Limitations of Hoplite Weapons and Equipment', in Kagan and Viggiano (eds) (2013), pp.157–75
SEARS, M.A. & WILLEKES, C. (2016), 'Alexander's cavalry charge at Chaeronea, 338 BCE', *Journal of Military History* 80, pp.117–35
SEKUNDA, N.V. (1986), *The Ancient Greeks*
– (1992), *The Persian Army 560–300 BC*
– (1998), *The Spartan Army*
– (1992), 'Athenian demography and military strength 338–322 BC', *Annual of the British School at Athens* 87, pp.311–55
– (2000), *Greek Hoplite 480–323 BC: Weapons, Armour, Tactics*
– (2007), 'Military forces: Land forces', in Sabin, van Wees and Whitby (eds) (2007), pp.325–57
– (2014a), 'The Composition of the Lacedaemonian *Mora* at Lechaeum', in Sekunda and Burliga (eds) (2014), Ch.2
– (2014b), 'The Chronology of the Iphicratean Peltast Reform', in Sekunda and Burliga (eds) (2014), Ch.7
SEKUNDA, N.V. & BURLIGA, B. (eds) (2014), *Iphicrates, Peltasts and Lechaeum*
SHANNAHAN, J. (2014), 'Two Notes on the Battle of Cunaxa', *Ancient History Bulletin* 28 1–2 pp.61–81
SHELDON, R.M. (2011), *Ambush: Surprise Attack in Ancient Greek Warfare*
SIDNELL, P. (2007), *Warhorse: Cavalry in Ancient Warfare*
SIEWERT, P. (1977), 'The ephebic oath in fifth-century Athens', *Journal of Hellenic Studies* 97, pp.102–11
SNODGRASS, A.M. (1964), *Early Greek Armour and Weapons*
– (1965), 'The Hoplite Reform and History', *Journal of Hellenic Studies* 85, pp.110–22
– (1974), 'A Historical Homeric Society?', *Journal of Hellenic Studies* 94, pp.114–25
– (1999), *Arms and Armour of the Greeks* (new ed.)
– (2013), 'Setting the Frame Chronologically', in Kagan and Viggiano (eds) (2013), pp.85–94
SPENCE, I.G. (1993), *The Cavalry of Classical Greece: A Social and Military History*
STERNBERG, R.H. (1999), 'The transport of sick and wounded soldiers in Classical Greece', *Phoenix* 53, pp.191–205
STRAUSS, B.S. (1996), 'The Athenian Trireme, School of Democracy', in J. Ober and C.W. Hedrick (eds), *Dēmokratia: A Conversation on Democracies, Ancient and Modern*, pp.313–25
– (2007), 'Naval battles and sieges', in Sabin, van Wees and Whitby (eds) (2007), pp.223–47
TAYLOR, R. (2020), *The Macedonian Phalanx: Equipment, organization and tactics from Philip and Alexander to the Roman conquest*
THORNE, J. (2001), 'Warfare and agriculture: the economic impact of devastation in Classical Greece', *GRBS* 42, pp.225–53

TORELLI, M. (ed.) (2001), *The Etruscans*
TRUNDLE, M. (2004), *Greek Mercenaries: From the Late Archaic Period to Alexander*
UEDA-SARSON, L. (2002a), 'The Reforms of Iphikrates', *Slingshot* 222, pp.30–36, online at http://lukeuedasarson.com/Iphikrates1.html
– (2002b), 'Infantry of the Successors', *Slingshot* 223, pp.23–28, online at http://lukeuedasarson.com/Iphikrates2.html
VAN WEES, H. (1994), 'The Homeric Way of War: The Iliad and the Hoplite Phalanx', *Greece & Rome* 41.2
– (2000), 'The Development of the Hoplite Phalanx: Iconography and Reality in the Seventh Century', in van Wees (ed.) (2000), pp.125–66
– (ed.) (2000), *War and Violence in Ancient Greece*
– (2002), 'Tyrants, oligarchs and citizen militias', in A. Chaniotis and P. Ducrey (eds), *Army and Power in the Ancient World*, pp.61–82
– (2004), *Greek Warfare: Myths and Realities*
– (2007), 'War and Society', in Sabin, van Wees and Whitby (eds) (2007), pp.273–99
– (2013), 'Farmers and Hoplites: Models of Historical Development', in Kagan and Viggiano (eds) (2013), pp.222–55
VAUGHN, P. (1991), 'The identification and retrieval of the hoplite battle-dead', in Hanson (ed.) (1991), pp.38–67
VIDAL NAQUET, P. (1986), *The Black Hunter: Forms of Thought and Forms of Society in the Greek World*
VIGGIANO, G.F. (2013), 'The Hoplite Revolution and the Rise of the Polis', in Kagan and Viggiano (eds) (2013), pp.112–33
VIGGIANO, G.F. & VAN WEES, H. (2013), 'The Arms, Armour and Iconography of Early Greek Hoplite Warfare', in Kagan and Viggiano (eds) (2013), pp.57–73
WACE, A.J.B (1929), 'Lead figurines', in R.M. Dawkins (ed.), *The Sanctuary of Artemis Orthia at Sparta*
WHATLEY, N. (1964), 'On the possibility of reconstructing Marathon and other ancient battles', *Journal of Hellenic Studies* 84, pp.119–39
WHEELER, E. (1982), '*Hoplomachia* and Greek dances in arms', *Greek, Roman and Byzantine Studies* 22, pp.223–33
– (1983), 'The *hoplomachoi* and Vegetius' Spartan drillmasters', *Chiron* 13, pp.1–20
– (1987), 'Ephorus and the prohibition of missiles', *Transactions of the American Philological Association* 117, pp.157–82
– (1988a), *Stratagem and the Vocabulary of Military Trickery*
– (1988b), 'Πολλὰ κενὰ τοῦ πολέμου: The History of a Greek Proverb,' *Greek, Roman and Byzantine Studies* 29, pp.153–84
– (1991), 'The general as hoplite', in Hanson (ed.) (1991), pp.121–70
– (2007), 'Battle: Land battle', in Sabin, van Wees and Whitby (eds) (2007), pp.186–224
– (2010), 'Present but absent: Marathon in the tradition of Western military thought', in K. Buraselis and E. Koulakiotis (eds), *Marathon the Day After: Symposium Proceedings, Delphi 2–4 July 2010*, pp.241–67
– (2011), 'Greece: Mad Hatters and March Hares', in L. Brice and J. Roberts (eds), *Recent Directions in the Military History of the Ancient World*, pp.53–104
WHITBY, M. (2007), 'Reconstructing ancient warfare', in Sabin, van Wees and Whitby (eds) (2007), pp.54–81
WHITEHEAD, D. (1991), 'Who Equipped Mercenary Troops in Classical Greece?', *Historia: Zeitschrift Für Alte Geschichte* 40, pp.105–13
WITOWSKI, J. (2018), 'Several remarks about the Near-Eastern contribution to early Archaic Greek warfare', *Studies In Ancient Art And Civilization* 22, pp.2–22
WORLEY, L.A. (1994), *Hippeis: The Cavalry of Ancient Greece*
YOSHITAKE, S. (2010), '*Aretē* and the achievements of the war dead: the logic of praise in the Athenian funeral oration', in Pritchard (ed.) (2010), pp.359–77

Index

Abydus (battle of), 381
Achilles, 21, 90, 158, 167, 183, 237–8, 238, 297–8, 300, 407
Aegitium (battle of), 261–2, 364, 375–6
Age classes, 106, 112, 116, 161, 174, 181, 198, 200, 261, 264, 313, 473
Agesilaus, 96, 108, 136, 143, 163, 171, 203–204, 218, 228–9, 240–1, 270–1, 277, 286, 294–6, 300, 304, 318, 340, 344, 354, 377–8, 385, 397, 402, 407, 410, 430, 432
Agis, 94, 104, 109, 196, 217–18, 235–6, 240, 242, 246, 285, 324, 342, 369–70
Agoge, 188, 224
Agon, agonal warfare, 238, 254–5, 297, 315
Alcibiades, 195, 198, 227, 381
Alope (battle of), 309
Amathus bowl, 43
Ambush, 219, 280, 296–300, 305, 320, 323–6, 351, 379
Amphipolis (battle of), 4, 130, 214, 218, 239–40, 249, 326, 351, 371, 378, 380, 384
Anastrophe, 136, 143–5, 150
Ankyle, 31, 93
Antilabe, 12, 26–7, 63–6
Archers, 22, 30–3, 49, 60, 163, 174, 179, 186, 244, 258, 260, 262, 268, 289, 291–2, 332, 337, 355, 360
Archidamus, 129, 241, 378, 402, 426, 430, 457
Argives, 42, 109, 124, 168, 199–200, 235–6, 246, 282, 289, 298, 304, 329, 339, 342–4, 347, 369–70, 377–8, 382, 397, 401, 407, 412, 430, 458
Argive shield, 26–7, 29, 31, 34, 56, 67–8, 71–2
Argos panoply, 13, 25, 73, 86, 89
Armour, 73–85
  'Tube and yoke', 77, 79, 84
  see also Linothorax, Spolas, Thorax
Artisans, 10, 13, 162, 179
Assyrians, 15, 33, 46, 64, 78, 279, 297, 403
Athenians, 4–5 (and *passim*)
  numbers, 161, 174
  officers, 216–22
  organization, 116–20
  recruitment, 160–1, 164–5, 197–200
  service, 201–203
  see also Ephebes
Auteretai (self-rowing), 290–1

Baggage, 108, 115, 246, 262, 272, 274, 276–8, 333, 344, 352
Bell cuirass, 73–4, 85
Blazon (on shield), 56, 59, 68, 71–2, 95
Boeotians, 124, 136, 188, 236, 313–14, 317–18, 328–9, 332–3, 335, 349, 366, 372, 381, 386, 397, 401, 458, 465, 473
  see also Thebans
Boeotian shield, 29, 44, 67–8, 71
Brasidas, 111, 170–1, 176, 196, 239, 276, 292–3, 303–304, 321, 326, 328, 340–1, 351, 378, 380, 384
Burial of dead, 188, 388–9, 390

Camps, 278–80
Carians, 26, 39, 63–4, 66, 71, 85, 89, 165–6, 172, 177, 478
Carthaginians, 128, 351, 373, 397, 415–16, 437, 476
Cavalry,
  fighting style, 361, 436
  versus hoplites, 265–71, 362–8
Chabrias, 283, 432
Charging, 427–33
Chariots, 10, 16–17, 24, 175, 186, 269, 353–4, 373, 403, 421
Chigi Vase, 11–12, 31–3, 42–3, 73–4, 89, 173
Citizens, 156–65
Clearchus, 135, 213, 229, 245
Cleombrotus, 20, 54, 103, 200, 241, 283, 314, 325, 340, 348, 426
Clothing, 71, 94–7, 166–7, 179, 251
Conscription *see* Recruitment
Corinth (battle of), 279, 374
Corinthians, 175, 235, 261, 267, 344, 347, 363, 365–6, 371, 373, 376, 385, 387, 401, 457–8
Coronea (battle of), 18, 30, 96, 144, 241, 318, 329, 340, 344, 347, 354, 377–8, 385, 395, 397, 402, 404, 406–407, 410–11, 426, 430, 433, 438, 440
Courage (*andreia*), 166, 213, 304
Crests (on helmets), 17–18, 26, 54, 64, 89, 118, 227
Crete, Cretans, 159, 179, 183, 289
Crimisus (battle of), 373, 397, 437
Cromnus (battle of), 129

Cubit 127–32
Cunaxa (battle of), 4, 66–7, 78, 112, 140, 179, 185, 250–2, 273, 276, 278, 306, 327, 336, 338–9, 353, 362, 382, 421, 430, 454, 457
Customs *see* Burial of dead; *Nomoi*; Rules; Trophies
Cynoscephalae (battle of), 368
Cyrus, 8, 99, 121, 141, 146–8, 153, 182, 189, 204, 216, 226, 231–3, 245–7, 250, 271, 403–405, 415
Cyrus the Younger, 97, 112–13, 125, 135, 138, 178–80, 248, 250–1, 272, 276, 278, 327, 336, 338, 362, 454

Dark Ages, 45, 73
Delium (battle of), 3, 96, 123–4, 134, 137, 139, 214, 236, 302, 313–15, 317, 330, 332, 336, 348, 366, 372–3, 380–1, 394–6, 401, 438, 446
Delphi, 221, 222, 344, 472
Demosthenes (general), 236, 258–9, 261–2, 277, 298–9, 323–4, 374–6, 379
Demosthenes (orator), 179–81, 192–3, 201
Dendra panoply, 24, 73
Depth, 9, 133–44, 441–54
Dionysius, 179–80
Discipline, 39–44, 192–7
  see also Drill; Manoeuvres; Training
Doratismos, 437
Doric (dialect), 299
Drill, 126–55
  see also Depth; Formations; Intervals; Manoeuvres; Training
Duration (of battles), 454–9

Egyptians, 17, 33, 46, 66, 77–8, 85, 140, 159, 177–8, 206, 425
  at Thymbrara, 403–406, 438, 446
Encumbrance, 59–62, 80–2, 90, 93
Enomotarchos, 102, 112, 114, 140, 151, 216, 245
Enomotia, 100–12, 136, 150, 215–16, 245–6
Epaminondas, 4, 196, 250, 253, 325, 330, 347–8, 350, 367–8, 386, 410, 421, 452–3
Ephebes, *ephebeia*, 184, 190–1, 198, 210, 264
Ephesus (battle of), 203, 385, 387–8, 390
Ephors, 103, 110, 200, 217–20, 224
Epibatai see Marines
Epilektoi, 123–4, 130, 187, 208, 466
Epiteichismos, 295–6
Eteocles and Polyneices, 27, 408, 433–4
  see also Single combat
Etruscans, 55, 57, 87, 173, 306, 461

'Face of Battle', 312, 391–2
Farmers, 10–14, 36, 158–9, 173, 179–80, 295, 463
Feigned retreats, 358, 361, 436, 466, 474

File closers *see Ouragoi*
File leaders, 109, 115, 119, 124, 139, 141, 144–5, 148, 150, 216, 460
Formations, 114, 129–30, 140–3, 148–55, 236, 276, 340, 346–7, 352, 368, 420–1, 445–6, 452

Garrisons, 161, 174, 179–80, 313, 469
Generalship, 118–20, 188–90, 225–7, 230, 237, 297, 300
Greaves, 35, 54, 64, 84, 89–90, 97, 475
Gymnetes, 5, 7, 34, 361

Haliartus (battle of), 219, 236, 322, 370–1, 387
Helmets, 13, 66, 86–9
  see also Crests
Helots, 3, 34, 106, 111–12, 137, 156, 161–2, 169–72, 175–6, 188, 235, 255–6, 279, 285, 301, 474
Heralds, 108, 140, 244–5, 298, 316–7, 319–21, 384–7
Hippeis, 109–12, 124, 160, 240, 266
Homeric warfare, 10–12, 15–24
Homoioi, 35, 104, 156, 222
Homosexuality, 210–12
Hoplites, *passim*
  definition, 5–7, 478–480
  'Hoplite debate', 9–15, 35, 39, 41, 44, 59, 66, 174, 393
  'Hoplite revolution', 10, 15, 37–8, 48–9
Hoplitodromos, 61, 185–6
Hoplomachia, hoplomachoi, 150, 184–6, 191, 225
'Hoplon', 6–7, 54

Idomene (battle of), 299, 324
Illyrians 174, 276, 413
  helmet, 86–7
Intervals (of files), 20–1, 46, 49, 126–33, 141–2, 422–3
Ionia, Ionians, 2–3, 26, 33, 39, 85, 177, 288, 291, 340, 357, 377–8
  dialect, 299
Iphicrates, 6, 29, 79, 178, 195, 221, 253, 260–1, 263, 292, 410, 467–9

Javelins, 4, 5, 31–2, 78, 92, 171, 244, 257–9, 261–3, 265, 268, 292, 353, 370, 409, 439, 477

Lacedaemonians *see* Spartans
Laodocium (battle of), 344, 385
Lechaeum (battle of), 260–1, 263, 269, 296, 364, 372
Leonidas, 110, 219, 302, 358–9, 395
Leuctra (battle of), 20, 102, 106, 109, 123–4, 136, 138–9, 177, 188, 199, 241, 250, 252,

266, 283, 285, 302, 314, 325, 330, 333–5, 340, 347–9, 367, 371, 386, 397, 402, 405, 410, 421, 426, 452
Light infantry *see* Peltasts; *Psiloi*
Linen, 54, 59, 71, 73, 76–80, 84–5, 355
Linothorax, 77
Lochagos, 101–24, 134, 190, 216–17, 330
Lochos, 100–19, 141–2, 151, 190, 246
Lydians, 33, 39, 140, 146, 148, 272, 352
Lyncestis (battle of), 321, 334, 371

Macedonian phalanx, 8, 16–18, 21, 45, 79, 99–100, 105, 127–9, 132–3, 277, 291, 353, 409–23, 443–8, 463–77
Macedonians, 99–100, 123, 128, 173, 208, 222, 254–5, 271, 276, 302, 369, 390, 464–72
Manoeuvres, 140–8, 252, 341–54
Manpower, 174–7, 313–5, 356
Mantinea, first battle of, 3, 42–3, 65, 69, 101, 103–105, 109–11, 124, 132–4, 137, 171, 196, 203, 217, 233, 240, 242, 246, 313, 324, 329–30, 339, 341–3, 345, 365, 369, 371, 375, 377, 382, 391, 412, 458
Mantinea, second battle of, 4, 136, 143, 304, 325, 330, 350, 367–8, 386
Maps, 234
Marathon (battle of), 3, 61, 117, 134, 137, 140, 167, 186, 220–1, 239, 267, 278, 310–11, 328–9, 331, 337, 339, 341, 357, 388
Marines, 72, 196, 288–93
Markets, 202, 271
Medes, 33, 167, 221, 278, 328, 331, 337, 357–8, 403
  see also Persians
Mercenaries, 2, 33, 37, 43, 85, 177–82, 187–8, 193, 200–202, 206–207, 230, 264, 289, 373, 432, 437, 469–71
  see also 'Ten Thousand', the
Messenian War, 39–40
Metoikoi (metics), 161, 174
Miletus (battle of), 344, 378
Miltiades, 117, 220–1, 298, 331, 401
Morai, 101–12, 171, 200, 208, 222–3, 240, 294, 314
Munychia (battle of), 135, 331, 370, 386
Muscled cuirass, 75
Mycale (battle of), 354, 361–2, 373
Mycenaeans, 1–2, 16–17, 24, 74
Mytilene (battle of), 385

Nemea (battle of), 4, 116–17, 136–8, 234, 246, 302, 324, 329–30, 335, 345–8, 375, 383, 458
Neodamodeis, 111–12, 161, 170–1, 176, 329
Nicias, 165, 199, 219, 275, 277, 322

Nisaea (battle of), 384
Nomoi, 317–19, 324, 328, 335
Numbers *see* Manpower

Ochanon, 26, 63–5
Odysseus, 30–1, 157, 192, 237–8, 297–8, 300, 315
Offering battle, 282, 293, 320–3, 371
Officers, 215–24
Oliganthropia, 176–7, 301
  see also Manpower
Olpae (battle of), 323, 379, 454
Olympia, 56, 71, 74, 200
Olynthus (battle of), 383
Othismos, otheo (meaning), 391–427
  among cavalry, 398–400, 405, 418–19
  othismos aspidon, 394–5, 399, 438
  'othismos debate', 394–5
  see also Pushing (in combat)
Ouragoi (file closers), 109, 115, 119, 138, 141, 351, 381, 415, 417–19, 431, 448, 459–60
Outflanking *see* Manoeuvres

Paean, 164, 234, 244, 324, 336–9, 375, 377
Panic, 252, 323–4, 337, 341, 364, 366–7, 378–82, 420, 450–1
Panormus (battle of), 386
Paragoge, 140–3, 151–3, 190, 245, 350
Passes (defence of), 11, 282–4, 371–2
Pay, 163, 178, 182, 188, 193, 201–202, 205, 228, 271–2
  see also Mercenaries
Pelopidas, 124, 211, 241, 253, 340, 347–8, 354, 367, 382, 466
Pelta, pelte, 4, 29, 174, 256, 467–9
Peltasts, 4–6, 34, 61, 83, 113, 129, 174, 178, 203, 244, 248, 256–64, 269–70, 283, 296, 313–14, 333, 353, 371–2, 383, 385, 401, 404
  Hellenistic peltasts, 84, 468, 473
  Iphicratean reform, 6–7, 467–8
Pericles, 3, 167–8, 176, 202, 209, 213, 294, 388
Perioikoi, 106, 111–12, 161, 163, 176, 313, 474
Perizoma, 76
Persians:
  equipment, 46, 167, 355
  style of fighting, 41, 361, 396, 436
  versus hoplites, 354–62
  see also Marathon; Mycale; Plataea; Thermopylae
Phalanges, 8, 17, 19, 21–2, 30, 32, 45, 49, 237
Phalanx, *passim*
  definition, 7–9, 478–80
  usage, 8
  see also Depth; Drill; Intervals; Formations; Manoeuvres

Philip II of Macedon, 4, 18, 38, 80, 98, 123, 128, 173, 180, 271, 274, 287, 390, 463, 468–9
Philopoemen, 475–6
Phyle (battle of), 299–300
Plataea (battle of), 3, 34–5, 40–1, 48, 50–1, 124–5, 130, 137, 139–40, 166, 171, 175, 202–203, 217, 222, 242–3, 255–6, 267–9, 274, 279, 319, 328, 331–2, 334–6, 338, 357, 360–2, 364, 373, 382, 396, 435–6
Plataea, siege of, 285–6
Plunder, 85, 194, 204–205, 244, 263, 266, 270–3, 293–6, 301, 386, 454
Poleis, 2, 11, 13, 209, 302
Polemarchoi, 101–104, 108–109, 112, 215–17, 220, 222, 239–43, 245–6, 261, 328, 385, 402
Porpax, 12, 26–7, 63–6
Potidaea (battle of), 214, 249–50, 321, 344, 363
Promachoi, 22, 24, 28, 45, 74, 139, 158, 334, 437, 444
Prizes, 202–205, 390
Psamtik (Psammetichus), 85, 177
Psiloi, 5, 7, 34, 61, 171, 173, 256–9, 308, 313–15, 363, 470
Pteruges, 75–6
Pursuit, 186, 212–13, 263, 269, 304, 330, 339, 344, 347, 350, 363–5, 370, 378, 380–4, 442, 451, 454–9, 461, 471
Pushing (in combat), 48, 59–60, 377, 391–462
 by Romans, 415–17
 in Macedonian phalanx, 417–20
 use of shield for, 46, 64–7, 69
 see also Othismos, otheo (meaning); Scrum (in combat)
Pygela (battle of), 385
Pylos (battle of), 4, 57, 170, 292–3, 374, 401, 455
Pyrrhic dance, 185–6, 433–4

Rations, 201–202, 217, 228, 271–5
Recruitment, 197–200
Religion, 42–3, 188, 210, 222–3, 233–4, 251, 277, 317, 319, 334–7
Reserves, 253, 315, 350–3, 442, 459, 461
Rhipsaspia, 61, 166, 194–5
Romans, 5, 31, 55, 61, 93, 173, 208, 278, 351, 410–14, 431–2, 435, 442–3, 445, 457, 476–7
 infantry at Zama, 415–17, 438
 shields, 61, 70, 72
Rules, 11–12, 204, 297, 303, 305, 308, 315–20, 480
 see also Agon; Nomoi; Customs
Running, 61, 186, 264, 337–41, 397, 422–3, 429–33
 see also Hoplitodromos; Rhipsaspia

Sacred Band, 123–4, 200, 208, 210–12, 348–9, 367, 444, 465–6
Sacrifices, 108, 136, 143, 204, 217, 222–3, 232–4, 242, 270, 295, 319, 324, 329, 334–7, 346, 360, 390
 see also Religion
Sarissa, 128, 463, 466, 469, 474–6
Sauroter, 92
Sciritae, 101, 104, 111–12, 240, 242, 329, 342, 378, 412
Scouting, 111, 235, 246, 266, 318, 324–5, 327, 367, 375
Scrum (in combat), 11, 377, 392–427
 see also Othismos, otheo (meaning); Pushing (in combat)
Scytale, 224
Siegecraft, 3, 279, 284–7, 464
Shields, 24–30, 54–73 (and *passim*)
 apron (on shield), 70
 see also Antilabe; Argive shield; Blazon; Boeotian shield; 'Hoplon'; *Ochanon*; *Pelta*; *Porpax*; *Rhipsaspia*; *Telamon*
Shield clashing, 69, 407–408, 419–20, 431–3, 438, 462
 see also Othismos aspidon
Signals, signalling, 247–50, 325
 see also Trumpets
Single combat, 17, 19–20, 27–9, 31, 42–3, 68, 237, 408, 426, 434–5, 437–8
Skirmishing, 93, 172, 259–60, 265, 267, 332–5, 362, 457, 473
Slaves, slavery, 159, 169–71, 178, 192–3, 197, 204, 227, 273, 288, 295, 305, 389
 see also Helots
Socrates:
 on armour, 80–1
 on courage, 268–9
 on generalship, 118–19, 188–9, 217, 225–6, 230–1
 on rear ranks, 406
 on training, 184–5, 225
Solon, 160, 175
Solygeia (battle of), 266–7, 321, 336, 366, 371, 376–7, 387
Sparabara, 355–6
Sparta, Spartans, 3–4 (and *passim*)
 numbers, 174–7
 organization, 100–12
 officers, 215–22
 recruitment, 161–3
 see also Agoge; Ephors; Helots; *Perioikoi*; Spartiates
Spartiates, 100, 106, 161–2, 171, 176–7, 188, 218–22, 224
 see also Spartans
Spartolus (battle of), 262, 334, 364, 379

Spears, 90–3, 467 (and *passim*)
  grip, 91, 436–7
  spear fighting, 377–8, 433, 437–9
  throwing, 31–2, 42, 45, 49, 93, 257
Speeches, 228, 232–4, 245, 303, 326, 332, 370, 380
Sphacteria (battle of), 88, 217, 247, 258–60, 299, 364, 375–6, 401, 455
Spolas, 83–4, 96
Stratagems, 237–8, 315–16, 323–6
Strategoi, 113, 115, 117–19, 197–8, 210, 216–22, 228, 239, 248, 272, 328, 338
  see also Generalship
Stratus (battle of), 323, 379
Subunits (of phalanx), 100–26
Swords, 54, 93–4, 425–6, 437–8, 467
Synaspismos, 18, 128–9, 132, 422, 475
Syracuse, siege of, 4, 120, 135, 165, 172, 199, 218, 251, 273, 275–7, 279–80, 286, 289, 310, 322, 331, 333, 335, 340, 351–2, 366, 373–4, 389, 401, 451, 457

Tacticians, 16, 21, 98–100, 127–31, 141–4, 189, 225, 247, 409, 418–19, 422, 442, 449, 475
Taxiarchoi, 117–19, 122, 135, 137, 141, 153, 189, 193, 197, 203, 210, 216–17, 227–8, 232, 239, 246
'Tearless Battle', 378–9, 430
Tegyra (battle of), 241, 354, 367, 382
'Ten Thousand', the, 83–4, 97, 112–14, 116, 135, 150–1, 164–5, 179, 194, 205, 213–14, 216, 220, 223, 228–30, 248, 252, 262–3, 272–4, 279, 336, 338, 362, 389, 420–1
Telamon, 12, 65–6
Terrain, 280, 311, 320, 368–76
Thebans, 3–4, 123, 279, 283, 286, 301–302, 315, 321, 385–6, 412, 458
  battles and tactics, 130, 136–8, 241, 253, 325, 334, 340, 344–8, 354, 367, 370–1, 380, 382, 394–5, 397, 401–402, 405, 430, 465–6
  deep formations, 123, 134, 136, 139, 151, 252, 330, 444, 452–3
  see also Boeotians; Sacred Band
Thermopylae (battle of), 3, 46, 110, 203, 210, 219, 234–5, 255, 283–4, 318, 324–5, 357–60, 375, 436, 466
Thessalians, 173, 328
Thetes, 160, 289–91
Thorax, 73, 80–4
  see also Armour; *Linothorax*; *Spolas*
Thracians, 31, 174, 257, 272, 308, 320
  helmet, 88
Thureophoroi, 5, 468, 472–3
Thymbrara (battle of), 146, 352, 403–405, 438, 446
Timoleon, 128, 373, 397, 437
Training, 182–91
Trophies, 134, 292, 296, 310, 315–17, 321, 341, 344, 363–4, 371, 380, 384–6, 390, 452
Trumpets (*salpinx*), 87, 128, 244–5, 247–9, 335, 338, 437
Tyrtaeus, 14, 18, 24, 32, 50, 139, 210

'Unshielded side', 68–70, 132–3, 342

Virtue (*arete*), 213, 303

Watchwords, 248, 250–1
Weight (of equipment) *see* Encumbrance
Weight (in combat), 421–3
'Western Way of War', 12, 305–306, 480
Women, 96, 118, 157, 165–70, 176, 293, 360, 439
Wounded, 223, 241, 261, 389, 426
Wrestling, 186, 397–8, 433–5, 440

Zeugitai, 160